Calculus for the Managerial, Life, and Social Sciences

THE PRINDLE, WEBER & SCHMIDT SERIES IN DEVELOPMENTAL, BUSINESS, AND TECHNICAL MATHEMATICS

Baley and Holstege, *Algebra: A First Course, Third Edition*
Cass and O'Connor, *Beginning Algebra*
Cass and O'Connor, *Beginning Algebra with Fundamentals*
Cass and O'Connor, *Fundamentals with Elements of Algebra, Second Edition*
Ewen and Nelson, *Elementary Technical Mathematics, Fifth Edition*
Geltner and Peterson, *Geometry for College Students, Second Edition*
Johnston, Willis, and Hughes, *Developmental Mathematics, Third Edition*
Johnston, Willis, and Lazaris, *Essential Algebra, Sixth Edition*
Johnston, Willis, and Lazaris, *Essential Arithmetic, Sixth Edition*
Johnston, Willis, and Lazaris, *Intermediate Algebra, Fifth Edition*
Jordan and Palow, *Integrated Arithmetic and Algebra*
Kennedy and Green, *Prealgebra for College Students*
Lee, *Self-Paced Business Mathematics, Fourth Edition*
McCready, *Business Mathematics, Sixth Edition*
McKeague, *Basic Mathematics, Third Edition*
McKeague, *Introductory Mathematics*
McKeague, *Prealgebra, Second Edition*
Mangan, *Arithmetic for Self-Study, Second Edition*
Perez, Weltman, Byrum, and Polito, *Beginning Algebra: A Worktext*
Perez, Weltman, Byrum, and Polito, *Intermediate Algebra: A Worktext*
Pierce and Tebeaux, *Operational Mathematics for Business, Second Edition*
Proga, *Arithmetic and Algebra, Third Edition*
Proga, *Basic Mathematics, Third Edition*
Rogers, Haney, and Laird, *Fundamentals of Business Mathematics*
Szymanski, *Algebra Facts*
Szymanski, *Math Facts: Survival Guide to Basic Mathematics*
Wood, Capell, and Hall, *Developmental Mathematics, Fourth Edition*

THE PRINDLE, WEBER & SCHMIDT SERIES IN PRECALCULUS, LIBERAL ARTS MATHEMATICS, AND TEACHER TRAINING

Barnett, *Analytic Trigonometry with Applications, Fifth Edition*
Bean, Sharp, and Sharp, *Precalculus*
Boye, Kavanaugh, and Williams, *Elementary Algebra*
Boye, Kavanaugh, and Williams, *Intermediate Algebra*
Davis, Moran, and Murphy, *Precalculus in Context: Functioning in the Real World*
Drooyan and Franklin, *Intermediate Algebra, Seventh Edition*
Franklin and Drooyan, *Modeling, Functions and Graphs*
Fraser, *Elementary Algebra*
Fraser, *Intermediate Algebra: An Early Functions Approach*
Gantner and Gantner, *Trigonometry*
Gobran, *Beginning Algebra, Fifth Edition*
Grady, Drooyan, Beckenbach, *College Algebra, Eighth Edition*
Hall, *Beginning Algebra*
Hall, *Intermediate Algebra*
Hall, *Algebra for College Students, Second Edition*
Hall, *College Algebra with Applications, Third Edition*
Holder, *A Primer for Calculus, Sixth Edition*

Huff and Peterson, *College Algebra Activities for the TI-81 Graphics Calculator*
Johnson and Mowry, *Mathematics: A Practical Odyssey*
Kaufmann, *Elementary Algebra for College Students, Fourth Edition*
Kaufmann, *Intermediate Algebra for College Students, Fourth Edition*
Kaufmann, *Elementary and Intermediate Algebra: A Combined Approach*
Kaufmann, *Algebra for College Students, Fourth Edition*
Kaufmann, *Algebra with Trigonometry for College Students, Third Edition*
Kaufmann, *College Algebra, Third Edition*
Kaufmann, *Trigonometry, Second Edition*
Kaufmann, *College Algebra and Trigonometry, Third Edition*
Kaufmann, *Precalculus, Second Edition*
Lavoie, *Discovering Mathematics*
McCown and Sequeira, *Patterns in Mathematics: From Counting to Chaos*
Rice and Strange, *Plane Trigonometry, Sixth Edition*
Riddle, *Analytic Geometry, Fifth Edition*
Ruud and Shell, *Prelude to Calculus, Second Edition*
Sgroi and Sgroi, *Mathematics for Elementary School Teachers*
Swokowski and Cole, *Fundamentals of College Algebra, Eighth Edition*
Swokowski and Cole, *Fundamentals of Algebra and Trigonometry, Eighth Edition*
Swokowski and Cole, *Fundamentals of Trigonometry, Eighth Edition*
Swokowski and Cole, *Algebra and Trigonometry with Analytic Geometry, Eighth Edition*
Swokowski and Cole, *Precalculus: Functions and Graphs, Seventh Edition*
Weltman and Perez, *Beginning Algebra, Second Edition*
Weltman and Perez, *Intermediate Algebra, Third Edition*

THE PRINDLE, WEBER & SCHMIDT SERIES IN CALCULUS AND UPPER-DIVISION MATHEMATICS

Althoen and Bumcrot, *Introduction to Discrete Mathematics*
Andrilli and Hecker, *Elementary Linear Algebra*
Burden and Faires, *Numerical Analysis, Fifth Edition*
Crooke and Ratcliffe, *A Guidebook to Calculus with Mathematica*
Cullen, *An Introduction to Numerical Linear Algebra*
Cullen, *Linear Algebra and Differential Equations, Second Edition*
Denton and Nasby, *Finite Mathematics, Preliminary Edition*
Dick and Patton, *Calculus*
Dick and Patton, *Single Variable Calculus*
Dick and Patton, *Technology in Calculus: A Sourcebook of Activities*
Edgar, *A First Course in Number Theory*
Eves, *In Mathematical Circles*
Eves, *Mathematical Circles Revisited*
Eves, *Mathematical Circles Squared*
Eves, *Return to Mathematical Circles*
Faires and Burden, *Numerical Methods*
Finizio and Ladas, *Introduction to Differential Equations*
Finizio and Ladas, *Ordinary Differential Equations with Modern Applications, Third Edition*
Fletcher, Hoyle, and Patty, *Foundations of Discrete Mathematics*
Fletcher and Patty, *Foundations of Higher Mathematics, Second Edition*

Gilbert and Gilbert, *Elements of Modern Algebra, Third Edition*
Gordon, *Calculus and the Computer*
Hartfiel and Hobbs, *Elementary Linear Algebra*
Hartig, *Guidebook to Linear Algebra for Theorist*
Hill, Ellis, and Lodi, *Calculus Illustrated*
Hillman and Alexanderson, *Abstract Algebra: A First Undergraduate Course, Fifth Edition*
Humi and Miller, *Boundary-Value Problems and Partial Differential Equations*
Laufer, *Discrete Mathematics and Applied Modern Algebra*
Leinbach, *Calculus Laboratories Using Derive*
Maron and Lopez, *Numerical Analysis, Third Edition*
Miech, *Calculus with Mathcad*
Mizrahi and Sullivan, *Calculus with Analytic Geometry, Third Edition*
Molluzzo and Buckley, *A First Course in Discrete Mathematics*
Nicholson, *Elementary Linear Algebra with Applications, Third Edition*
Nicholson, *Introduction to Abstract Algebra*
O'Neil, *Advanced Engineering Mathematics, Third Edition*
Pence, *Calculus Activities for Graphic Calculators*
Pence, *Calculus Activities for the TI-81 Graphic Calculator, Second Edition*
Plybon, *An Introduction to Applied Numerical Analysis*
Powers, *Elementary Differential Equations, Brief Edition*
Powers, *Elementary Differential Equations with Boundary-Value Problems*
Prescience Corporation, *The Student Edition of Theorist*
Riddle, *Calculus and Analytic Geometry, Fourth Edition*
Schelin and Bange, *Mathematical Analysis for Business and Economics, Second Edition*
Sentilles, *Applying Calculus in Economics and Life Science*
Swokowski, Olinick, and Pence, *Calculus, Sixth Edition*
Swokowski, Olinick, and Pence, *Calculus of a Single Variable, Second Edition*

Swokowski, *Calculus, Fifth Edition (Late Trigonometry Version)*
Swokowski, *Elements of Calculus with Analytic Geometry: High School Edition*
Tan, *Applied Finite Mathematics, Fourth Edition*
Tan, *Calculus for the Managerial, Life, and Social Sciences, Third Edition*
Tan, *Applied Calculus, Third Edition*
Tan, *College Mathematics, Third Edition*
Trim, *Applied Partial Differential Equations*
Venit and Bishop, *Elementary Linear Algebra, Third Edition*
Venit and Bishop, *Elementary Linear Algebra, Alternate Second Edition*
Wattenberg, *Calculus in a Real and Complex World*
Wiggins, *Problem Solver for Finite Mathematics and Calculus*
Zill, *Calculus, Third Edition*
Zill, *A First Course in Differential Equations, Fifth Edition*
Zill and Cullen, *Differential Equations with Boundary-Value Problems, Third Edition*
Zill and Cullen, *Advanced Engineering Mathematics*

THE PRINDLE, WEBER & SCHMIDT SERIES IN ADVANCED MATHEMATICS

Ehrlich, *Fundamental Concepts of Abstract Algebra*
Judson, *Abstract Algebra: Theory and Applications*
Kirkwood, *An Introduction to Analysis*
Patty, *Foundations of Topology*
Ruckle, *Modern Analysis: Measure Theory and Functional Analysis with Applications*
Sieradski, *An Introduction to Topology and Homotopy*
Steinberger, *Algebra*
Strayer, *Elementary Number Theory*
Troutman, *Boundary-Value Problems of Applied Mathematics*

To Pat, Bill, and Michael

Calculus for the Managerial, Life, and Social Sciences

THIRD EDITION

S. T. TAN

Stonehill College

PWS Publishing Company

BOSTON

PWS PUBLISHING COMPANY
20 Park Plaza, Boston, MA 02116-4324

Copyright © 1994 by PWS Publishing Company
Copyright © 1990 by PWS-KENT Publishing Company
Copyright © 1987 by PWS Publishers

All rights reserved. No part of this book may be reproduced, stored in a retrieval system, or transcribed in any form or by any means—electronic, mechanical, photocopying, recording, or otherwise—without the prior written permission of PWS Publishing Company.

PWS Publishing Company is a division of Wadsworth, Inc.

 I⟨T⟩P™

International Thomson Publishing
The trademark ITP is used under license.

Library of Congress Cataloging-in-Publication Data

Tan, Soo Tang.
    Calculus for the managerial, life, and social sciences / Soo Tang
Tan.—3rd ed.
       p. cm.
    Brief ed. of: Applied calculus. 3rd ed. 1993.
    Includes index.
    ISBN 0-534-93566-4
    1. Calculus.   I. Tan, Soo Tang.   Applied calculus.   II. Title.
QA303.T142   1993   93-8203
515—dc20   CIP

Photo Credits

Chapter 1—Rhoda F. Duren (*p. 2*); *Chapter 2*—Arthur d'Arazien/The Image Bank (*p. 48*); *Chapter 3*—© Dann Coffey 1990/The Image Bank (*p. 128*); *Chapter 4*—© Eric Meola/The Image Bank (*p. 202*); *Chapter 5*—Gregory Heisler/The Image Bank (*p. 282*); *Chapter 6*—Harald Sund/ The Image Bank (*p. 338*); *Chapter 7*—Alain Keler/Sygma (*p. 424*); *Chapter 8*—© 1983 Morton Beebe/The Image Bank (*p. 478*).

Sponsoring Editor: Steve Quigley
Production Coordinator: Robine Andrau
Assistant Editor: Kelle Flannery
Editorial Assistant: John Ward
Production: Del Mar Associates
Interior/Cover Designer: Julia Gecha
Interior Illustrators: Kristi Mendola and Deborah Ivanoff
Cover Art: George Snyder, "Once Their Home," 1990, acrylic on canvas. Used with permission of the artist.
Marketing Manager: Marianne C. P. Rutter
Manufacturing Coordinator: Ruth Graham
Cover Printer: Henry N. Sawyer Company
Typesetter: Jonathan Peck Typographers
Printer and Binder: R. R. Donnelley & Sons/Willard

Printed and bound in the United States of America
94 95 96 97 98 — 10 9 8 7 6 5 4 3 2 1

CONTENTS

*Sections marked with an asterisk are not prerequisites for later material.

Contents

PREFACE

Calculus for the Managerial, Life, and Social Sciences, Third Edition is a brief edition of the author's *Applied Calculus, Third Edition.* This book is suitable for use in a one-semester or two-quarter introductory calculus course for students in the managerial, life, and social sciences.

As with the previous editions, the objective of *Calculus for the Managerial, Life, and Social Sciences, Third Edition,* is twofold: (1) to write a textbook that is readable by students and (2) to make the book a useful teaching tool for instructors. We hope that with the present edition we have come one step closer to realizing our goal. The third edition of this text incorporates many suggestions by users of both earlier editions.

Features

The following list includes some of the many important features of the book:

- **Coverage of Topics** This book contains more than enough material for the usual applied calculus course. Optional sections have been marked with an asterisk in the table of contents, thereby allowing the instructor to be flexible in choosing the topics most suitable for his or her course.

- **Approach** The problem-solving approach is stressed throughout the book. Numerous examples and solved problems are used to amplify each new concept or result in order to facilitate students' comprehension of new material. Figures are used extensively to help students visualize concepts and ideas.

- **Level of Presentation** Our approach is intuitive and we state the results informally. However, we have taken special care to ensure that this approach does not compromise the mathematical content and accuracy. Proofs of certain results are given, but they may be omitted if desired.

- **Applications** The text is application oriented. Many interesting, relevant, and up-to-date applications are drawn from the fields of business, economics, social and behavioral sciences, life sciences, physical sciences, and other fields of general interest. Some of these applications have their source in newspapers, weekly periodicals, and other magazines. Applications are found in the illustrative examples in the main body of the text as well as in the exercise sets. In fact, one goal of the text is to include at least one real-life application in each section (whenever feasible).

- **Exercises** Each section of the text is accompanied by an extensive set of exercises containing an ample set of problems of a routine, computational nature that will help students master new techniques. The routine problems are followed by an extensive set of application-oriented problems that test students' mastery of the topics. Self-check exercises are also included at the end of each section. These exercises give the students a chance to test themselves on their understanding of the material.

■ **Portfolios**   These interviews are designed to convey to the student the real-world experiences of professionals who have a background in mathematics and use it in their professions.

Changes in the Third Edition

■ Chapters 1 and 2 have been streamlined; the precalculus review has been reorganized and condensed. Logarithms are now reviewed right before the material on logarithmic functions in Section 5.2. The sections on limits and continuity have been reorganized and condensed, allowing for an earlier introduction to the derivative.

■ Section 2.6 has been rewritten and now gives a more concise introduction to the derivative along with a step-by-step procedure for finding the derivative.

■ Computer/graphics calculator exercises have been added. Textual examples that make use of graphing utilities are marked with a computer/graphics calculator margin icon, and end-of-section exercises meant to be solved with the use of a computer or graphics calculator are identified by a small boxed "C" in the margin.

■ More introductory real-life illustrations that illustrate the concepts being developed have been included at the beginning of sections. (See Sections 4.1 through 4.6.)

■ In many sections, easier examples have been added to illustrate a new concept or the use of a new technique when it is first introduced. More explanatory side remarks have also been included to help the students. Many new problems have been added, including practice problems at the beginning of the exercise sets, to help students gain confidence and computational skill before they move on to the more difficult problems and applications.

■ Optimization problems that require formulation are now covered in a separate section. New problem-solving guidelines are given with accompanying step-by-step worked examples.

■ Topics on integration have been reorganized. The definite integral is now defined using Riemann sums.

■ Chapter overviews with photographs relating to the applications to be covered have been added. The review exercises for each chapter have also been expanded.

■ Exercises have been paired (evens and odds) where appropriate.

■ Common errors and pitfalls have been highlighted by a caution symbol.

■ Applications have been labeled within each exercise set.

Supplements

■ A *Student's Solutions Manual,* available to both students and instructors, includes the solutions to odd-numbered exercises.

■ An *Instructor's Complete Solutions Manual* includes solutions to all exercises.

■ A *Test Bank with Chapter Tests,* free to adopters of the book, contains sample tests for each chapter.

- *EXPTest* and *ExamBuilder* With these computerized testing systems for IBM PCs and Macintosh, respectively, instructors can select or modify questions from prepared test banks or add their own test items to create any number of tests. These tests can be viewed on screen and printed with typeset-quality mathematical symbols, notation, graphs, and diagrams.

- A *Graphing Calculator Supplement,* written by Robert E. Seaver of Lorain County Community College, available to both students and instructors, further develops selected examples and exercises from the text and includes additional problems for reinforcement. It is specifically geared for use with the TI-line of programmable graphic calculators.

- *Derive* and *Theorist Notebooks,* packaged with the text at additional nominal cost to students, provide more than 1000 book-specific and manipulatable electronic problems to be used in conjunction with popularly used and distributed computer algebra systems.

Acknowledgments

I wish to thank Carol Hay, Nancy Nickerson, and Linda J. Murphy, all from Merrimack College, for their excellent job error checking exercise answers for this edition. I extend my personal appreciation to each of the following reviewers, whose many suggestions have helped make a much improved book:

Reviewers of the previous editions:

Daniel Anderson
University of Iowa

Charles Clever
South Dakota State University

William Coppage
Wright State University

Lyle Dixon
Kansas State University

Bruce Edwards
University of Florida at Gainesville

Charles S. Frady
Georgia State University

Howard Frisinger
Colorado State University

Larry Gerstein
University of California at Santa Barbara

Kedrick Hartfield
Mercer University

Lowell Leake
University of Cincinnati

Maurice Monahan
South Dakota State University

Karla Neal
Louisiana State University

Lloyd Olson
North Dakota State University

Richard Porter
Northeastern University

Priscilla Putnam-Haindl
Jersey City State College

Thomas N. Roe
South Dakota State University

P. L. Sperry
University of South Carolina

Reviewers of the third edition:

Jim Bruening
Southeast Missouri State University

Connie Carruthers
Scottsdale Community College

William J. Carsrud
Gonzaga University

Charles C. Clever
South Dakota State University

Margaret Crider
Tomball College

Michael Lambe
Grossmont College

Richard Nadel
Florida International University

Karla V. Neal
Louisiana State University

Nathan P. Ritchey
Youngstown State University

Ivan Schukei
University of South Carolina—Beaufort

Nora Schukei
University of South Carolina—Beaufort

Gary D. Shaffer
Allegany Community College

Shai Simonson
Stonehill College

W. Allen Smith
Georgia State University

I wish to thank Reverend Robert J. Kruse, vice-president of Stonehill College, for his enthusiastic support of this project. My thanks also go to the editorial and production staffs of PWS Publishing Company, Steve Quigley, Kelle Flannery, and Robine Andrau, for their thoughtful contributions and patient assistance and cooperation during the development and production of this book. Finally, I wish to thank Nancy Sjoberg of Del Mar Associates and Andrea Olshevsky for doing an excellent job ensuring the accuracy and readability of this third edition.

S. T. Tan
Stonehill College

APPLICATIONS

*In **Calculus for the Managerial, Life, and Social Sciences** we attempt to solve a wide variety of problems arising from many diverse fields of study. A small sample of the types of practical problems we will consider follows:*

POPULATION GROWTH A study prepared for a Sunbelt town's Chamber of Commerce projected that the town's population in the next three years would grow according to the rule

$$P(x) = 50,000 + 30x^{3/2} + 20x$$

where $P(x)$ denotes the population x months from now. How fast will the population be increasing nine months from now? Sixteen months from now?

AIR POLLUTION According to the South Coast Air Quality Management District, the level of nitrogen dioxide, a brown gas that impairs breathing, present in the atmosphere on a certain May day in downtown Los Angeles is approximated by

$$A(t) = 0.03t^3(t - 7)^4 + 60.2 \quad (0 \le t \le 7)$$

where $A(t)$ is measured in pollutant standard index and t is measured in hours, with $t = 0$ corresponding to 7 A.M. How fast is the level of nitrogen dioxide increasing at 11 A.M.?

OPTIMAL DRIVING SPEED A truck gets $400/x$ miles per gallon when driven at a constant speed of x miles per hour (between 50 and 70 miles per hour). If the price of fuel is $1 per gallon and the driver is paid $8 an hour, at what speed between 50 and 70 miles per hour is it most economical to drive?

LEARNING CURVES The Eastman Optical Company produces a 35-mm single-lens reflex camera. Eastman's training department determined that after completing the basic training program, a new, previously inexperienced employee would be able to assemble

$$Q(t) = 50 - 30e^{-0.5t}$$

model F cameras per day, t months after the employee began work on the assembly line. How many model F cameras can a new employee assemble per day after basic training? How many model F cameras can the average experienced employee assemble per day?

SUBWAY FARES A city's Metropolitan Transit Authority (MTA) operates a subway line for commuters from a certain suburb to the downtown metropolitan area. Currently, an average of 6000 passengers a day take the trains, paying a fare of $1.50 per ride. The Board of the MTA, contemplating raising the fare to $1.75 per ride in order to generate a larger revenue, engaged the services of a consulting firm. The firm's study revealed that for each 25-cent increase in fare, the ridership would be reduced by an average of 1000 passengers a day. The consulting firm recommended that MTA stick to the current fare of $1.50 per ride, which already yields a maximum revenue. Show that the consultants' recommendations were correct.

ADVERTISING EXPENDITURES The Ross-Simons Company has a monthly advertising budget of $60,000. Their marketing department estimates that if they spend x dollars on newspaper advertising and y dollars on television advertising, then the monthly sales will be given by

$$z = f(x, y) = 90x^{1/4}y^{3/4}$$

dollars. Determine how much money Ross-Simons should spend on newspaper ads and on television ads per month to maximize its monthly sales.

Applications Photo Credits

Population growth—© Zigy Kaluzny/Tony Stone Worldwide; *Air Pollution*—© Vince Streano/Tony Stone Worldwide; *Optimal Driving Speed*—© Fred George/Tony Stone Worldwide; *Learning Curves*—© Steve Niedorf/The Image Bank; *Subway Fares*—© Jon Riley/Tony Stone Worldwide; *Advertising Expenditures*—© Walter Bibikow 1990/The Image Bank

Calculus for the Managerial, Life, and Social Sciences

What sales figure can be predicted for next year? In Example 10, page 38, you will see how the manager of a local sporting goods store used sales figures from the previous years to predict the sales level for next year.

Preliminaries

T he first two sections of this chapter contain a brief review of algebra. We then introduce the Cartesian coordinate system, which allows us to represent points in the plane in terms of ordered pairs of real numbers. This in turn enables us to compute the distance between two points algebraically. This chapter also covers straight lines. The slope of a straight line plays an important role in the study of calculus.

1.1   Precalculus Review I

In Sections 1.1 and 1.2, we review some of the basic concepts and techniques of algebra that are essential in the study of calculus. The material in this review will help you work through the examples and exercises in this book. You can read through this material now and do the exercises in those areas where you feel a little "rusty," or you can review the material on an as-needed basis as you study the text. We begin our review with a discussion of real numbers.

The Real Number Line

The real number system is made up of the set of real numbers together with the usual operations of addition, subtraction, multiplication, and division.

Real numbers may be represented geometrically by points on a line. Such a line is called the **real number,** or **coordinate, line** and can be constructed as follows. Arbitrarily select a point on a straight line to represent the number zero. This point is called the **origin.** If the line is horizontal, then a point at a convenient distance to the right of the origin is chosen to represent the number one. This determines the scale for the number line. Each positive real number lies at an appropriate distance to the right of the origin, and each negative real number lies at an appropriate distance to the left of the origin, as shown in Figure 1.1.

FIGURE 1.1

The real number line

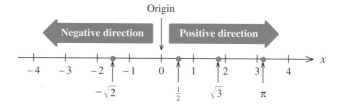

In this manner, a *one-to-one correspondence* is set up between the set of all real numbers and the set of points on the number line. That is, exactly one point on the line is associated with each real number. Conversely, exactly one real number is associated with each point on the line. The real number that is associated with a point on the real number line is called the **coordinate** of that point.

Intervals

Throughout this book, we will often restrict our attention to certain subsets of the set of real numbers. For example, if x denotes the number of cars rolling off a plant assembly line each day, then x must be nonnegative, that is, $x \geq 0$. Taking this example one step further, suppose management decides that the daily production must not exceed 200 cars. Then x must satisfy the inequality $0 \leq x \leq 200$.

More generally, we will be interested in the following subsets of real numbers: open intervals, closed intervals, and half-open intervals. The set of all real numbers that lie *strictly* between two fixed numbers a and b is called an **open interval** (a, b). It consists of all real numbers x that satisfy the inequalities $a < x < b$, and it is called "open" because neither of its endpoints is included in the interval. A **closed interval** contains *both* of its endpoints.

Thus the set of all real numbers x that satisfy the inequalities $a \leq x \leq b$ is the closed interval $[a, b]$. Notice that square brackets are used to indicate that the endpoints are included in this interval. **Half-open intervals** contain only *one* of their endpoints. Thus, the interval $[a, b)$ is the set of all real numbers x that satisfy $a \leq x < b$, whereas the interval $(a, b]$ is described by the inequalities $a < x \leq b$. Examples of these **finite intervals** are illustrated in Table 1.1.

TABLE 1.1
Finite Intervals

Interval	Graph	Example
Open (a, b)		$(-2, 1)$
Closed $[a, b]$		$[-1, 2]$
Half-open $(a, b]$		$\left(\frac{1}{2}, 3\right]$
Half-open $[a, b)$		$\left[-\frac{1}{2}, 3\right)$

In addition to finite intervals, we will encounter **infinite intervals.** Examples of infinite intervals are the half-lines (a, ∞), $[a, \infty)$, $(-\infty, a)$, and $(-\infty, a]$ defined by the set of all real numbers that satisfy $x > a$, $x \geq a$, $x < a$, and $x \leq a$, respectively. The symbol ∞, called *infinity*, is not a real number. It is used here only for notational purposes in conjunction with the definition of infinite intervals. The notation $(-\infty, \infty)$ is used for the set of all real numbers x, since, by definition, the inequalities $-\infty < x < \infty$ hold for any real number x. Infinite intervals are illustrated in Table 1.2.

TABLE 1.2
Infinite Intervals

Interval	Graph	Example
(a, ∞)		$(2, \infty)$
$[a, \infty)$		$[-1, \infty)$
$(-\infty, a)$		$(-\infty, 1)$
$(-\infty, a]$		$\left(-\infty, -\frac{1}{2}\right]$

Properties of Inequalities

In practical applications, intervals are often found by solving one or more inequalities involving a variable. In such situations, the following properties may be used to advantage.

PROPERTIES OF INEQUALITIES

If a, b, and c are any real numbers, then

Example

Property 1 If $a < b$ and $b < c$, then $a < c$.

$2 < 3$ and $3 < 8$, so $2 < 8$

Property 2 If $a < b$, then $a + c < b + c$.

$-5 < -3$, so $-5 + 2 < -3 + 2$; that is, $-3 < -1$

Property 3 If $a < b$ and $c > 0$, then $ac < bc$.

$-5 < -3$, and since $2 > 0$, we have $(2)(-5) < (2)(-3)$; that is, $-10 < -6$

Property 4 If $a < b$ and $c < 0$, then $ac > bc$.

$-2 < 4$, and since $-3 < 0$, we have $(-2)(-3) > (4)(-3)$ that is, $6 > -12$

Similar properties hold if each inequality sign, $<$, between a and b is replaced by $\geq$, $>$, or $\leq$.

EXAMPLE 1   Find the set of real numbers that satisfy $-1 \leq 2x - 5 < 7$.

Solution   Add 5 to each member of the given double inequality, obtaining

$$4 \leq 2x < 12.$$

Next, multiply each member of the resulting double inequality by $\frac{1}{2}$, yielding

$$2 \leq x < 6.$$

Thus the solution is the set of all values of x lying in the interval $[2, 6)$.

EXAMPLE 2   The management of Corbyco, a giant conglomerate, has estimated that x thousand dollars is needed to purchase

$$100{,}000(-1 + \sqrt{1 + 0.001x})$$

shares of common stock of the Starr Communications Company. Determine how much money Corbyco needs in order to purchase at least 100,000 shares of Starr's stock.

Solution   The amount of cash Corbyco needs to purchase at least 100,000 shares is found by solving the inequality

$$100{,}000(-1 + \sqrt{1 + 0.001x}) \geq 100{,}000.$$

Proceeding, we find

$$-1 + \sqrt{1 + 0.001x} \geq 1$$
$$\sqrt{1 + 0.001x} \geq 2$$
$$1 + 0.001x \geq 4 \qquad \text{(Square both sides.)}$$
$$0.001x \geq 3$$
$$x \geq 3000,$$

so Corbyco needs at least $3,000,000. ■

Absolute Value

ABSOLUTE VALUE

The **absolute value** of a number a is denoted by $|a|$ and is defined by

$$|a| = \begin{cases} a & \text{if } a \geq 0 \\ -a & \text{if } a < 0. \end{cases}$$

FIGURE 1.2
The absolute value of a number a

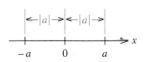

Since $-a$ is a positive number when a is negative, it follows that the absolute value of a number is always nonnegative. For example, $|5| = 5$ and $|-5| = -(-5) = 5$. Geometrically, $|a|$ is the distance between the origin and the point on the number line that represents the number a (Figure 1.2).

ABSOLUTE VALUE PROPERTIES

If a and b are any real numbers, then

Example

Property 5 $|-a| = |a|$ $|-3| = -(-3) = 3 = |3|$

Property 6 $|ab| = |a||b|$ $|(2)(-3)| = |-6| = 6$
$$= |2||-3|$$

Property 7 $\left|\dfrac{a}{b}\right| = \dfrac{|a|}{|b|}, (b \neq 0)$ $\left|\dfrac{(-3)}{(-4)}\right| = \left|\dfrac{3}{4}\right| = \dfrac{3}{4} = \dfrac{|-3|}{|-4|}$

Property 8 $|a + b| \leq |a| + |b|$ $|8 + (-5)| = |3| = 3$
$$\leq |8| + |-5|$$
$$= 13$$

Property 8 is called the **triangle inequality.**

EXAMPLE 3   Evaluate each of the following expressions:

a. $|\pi - 5| + 3$ **b.** $|\sqrt{3} - 2| + |2 - \sqrt{3}|$

Solution

a. Since $\pi - 5 < 0$, we see that $|\pi - 5| = -(\pi - 5)$. Therefore,

$$|\pi - 5| + 3 = -(\pi - 5) + 3 = 8 - \pi.$$

b. Since $\sqrt{3} - 2 < 0$, we see that $|\sqrt{3} - 2| = -(\sqrt{3} - 2)$. Next, observe that $2 - \sqrt{3} > 0$, so $|2 - \sqrt{3}| = 2 - \sqrt{3}$. Therefore,

$$|\sqrt{3} - 2| + |2 - \sqrt{3}| = -(\sqrt{3} - 2) + (2 - \sqrt{3})$$
$$= 4 - 2\sqrt{3} = 2(2 - \sqrt{3}).$$

■

Exponents and Radicals

Recall that if b is any real number and n is a positive integer, then the expression b^n (read "b to the power n") is defined as the number

$$b^n = \underbrace{b \cdot b \cdot b \cdots b}_{n \text{ factors}}.$$

The number b is called the **base,** and the superscript n is called the **power** of the exponential expression b^n. For example,

$$2^5 = 2 \cdot 2 \cdot 2 \cdot 2 \cdot 2 = 32 \quad \text{and} \quad \left(\frac{2}{3}\right)^3 = \left(\frac{2}{3}\right)\left(\frac{2}{3}\right)\left(\frac{2}{3}\right) = \frac{8}{27}.$$

If $b \neq 0$, we define

$$b^0 = 1.$$

For example, $2^0 = 1$ and $(-\pi)^0 = 1$, but the expression 0^0 is undefined.

Next, recall that if n is a positive integer, then the expression $b^{1/n}$ is defined to be the number that, when raised to the nth power, is equal to b. Thus,

$$(b^{1/n})^n = b.$$

Such a number, if it exists, is called the **nth root of b,** also written $\sqrt[n]{b}$.

 Observe that the nth root of a negative number is not defined when n is even. For example, the square root of -2 is not defined, because there is no real number b such that $b^2 = -2$. Also, given a number b, there might be more than one number that satisfies our definition of the nth root. For example, both 3 and -3 squared equal 9, and each is a square root of 9. So to avoid ambiguity, we define $b^{1/n}$ to be the positive nth root of b whenever it exists. Thus, $\sqrt{9} = 9^{1/2} = 3$.

Next, recall that if p/q (p, q, positive integers with $q \neq 0$) is a rational number in lowest terms, then the expression $b^{p/q}$ is defined as the number $(b^{1/q})^p$ or, equivalently, $\sqrt[q]{b^p}$, whenever it exists. Expressions involving negative rational exponents are taken care of by the definition

$$b^{-p/q} = \frac{1}{b^{p/q}}.$$

Examples are

$$2^{3/2} = (2^{1/2})^3 \approx (1.4142)^3 \approx 2.8284$$

and

$$4^{-5/2} = \frac{1}{4^{5/2}} = \frac{1}{(4^{1/2})^5} = \frac{1}{2^5} = \frac{1}{32}.$$

The rules defining the exponential expression a^n where $a > 0$ for all rational values of n are given in Table 1.3.

TABLE 1.3

Definition of a^n ($a > 0$)	Example	Definition of a^n ($a > 0$)	Example
Integer exponent: If n is a positive integer, then $$a^n = a \cdot a \cdot a \cdots a$$ (n factors of a).	$2^5 = 2 \cdot 2 \cdot 2 \cdot 2 \cdot 2$ (5 factors) $= 32$	**Fractional exponent:** **a.** If n is a positive integer, then $$a^{1/n} \text{ or } \sqrt[n]{a}$$ denotes the **nth root** of a.	$16^{1/2} = \sqrt{16}$ $= 4$
Zero exponent: If n is equal to zero, then $$a^0 = 1$$ (0^0 is not defined).	$7^0 = 1$	**b.** If m and n are positive integers, then $$a^{m/n} = \sqrt[n]{a^m} = (\sqrt[n]{a})^m.$$	$8^{2/3} = (\sqrt[3]{8})^2$ $= 4$
Negative exponent: If n is a positive integer, then $$a^{-n} = \frac{1}{a^n} \quad (a \ne 0).$$	$6^{-2} = \frac{1}{6^2}$ $= \frac{1}{36}$	**c.** If m and n are positive integers, then $$a^{-m/n} = \frac{1}{a^{m/n}} \quad (a \ne 0).$$	$9^{-3/2} = \frac{1}{9^{3/2}}$ $= \frac{1}{27}$

The first three definitions are also valid for negative values of a, whereas the fourth definition is valid only for negative values of a when n is odd. Thus,

$$(-8)^{1/3} = \sqrt[3]{-8} = -2 \qquad (n \text{ is odd})$$
$$(-8)^{1/2} \text{ has no real value} \qquad (n \text{ is even})$$

Finally, note that it can be shown that a^n has meaning for *all* real numbers n. For example, using a pocket calculator with a "y^x" key, we see that $2^{\sqrt{2}} \approx 2.665144$.

The five laws of exponents are listed in Table 1.4.

TABLE 1.4
The Laws of Exponents

Law	Example
1. $a^m \cdot a^n = a^{m+n}$	$x^2 \cdot x^3 = x^{2+3} = x^5$
2. $\dfrac{a^m}{a^n} = a^{m-n} \quad (a \ne 0)$	$\dfrac{x^7}{x^4} = x^{7-4} = x^3$
3. $(a^m)^n = a^{m \cdot n}$	$(x^4)^3 = x^{4 \cdot 3} = x^{12}$
4. $(ab)^n = a^n \cdot b^n$	$(2x)^4 = 2^4 \cdot x^4 = 16x^4$
5. $\left(\dfrac{a}{b}\right)^n = \dfrac{a^n}{b^n} \quad (b \ne 0)$	$\left(\dfrac{x}{2}\right)^3 = \dfrac{x^3}{2^3} = \dfrac{x^3}{8}$

These laws are valid for any real numbers a, b, m, and n whenever the quantities are defined.

Remember, $(x^2)^3 \neq x^5$. The correct equation is $(x^2)^3 = x^{2 \cdot 3} = x^6$.

The next several examples illustrate the use of the laws of exponents.

EXAMPLE 4   Simplify the given expressions:

a. $(3x^2)(4x^3)$ **b.** $\dfrac{16^{5/4}}{16^{1/2}}$ **c.** $(6^{2/3})^3$ **d.** $(x^3 y^{-2})^{-2}$ **e.** $\left(\dfrac{y^{3/2}}{x^{1/4}}\right)^{-2}$

Solution

a. $(3x^2)(4x^3) = 12x^{2+3} = 12x^5$ (Law 1)

b. $\dfrac{16^{5/4}}{16^{1/2}} = 16^{5/4 - 1/2} = 16^{3/4} = (\sqrt[4]{16})^3 = 2^3 = 8$ (Law 2)

c. $(6^{2/3})^3 = 6^{2/3 \cdot 3} = 6^{6/3} = 6^2 = 36$ (Law 3)

d. $(x^3 y^{-2})^{-2} = (x^3)^{-2}(y^{-2})^{-2} = x^{(3)(-2)} y^{(-2)(-2)} = x^{-6} y^4 = \dfrac{y^4}{x^6}$ (Law 4)

e. $\left(\dfrac{y^{3/2}}{x^{1/4}}\right)^{-2} = \dfrac{y^{(3/2)(-2)}}{x^{(1/4)(-2)}} = \dfrac{y^{-3}}{x^{-1/2}} = \dfrac{x^{1/2}}{y^3}$ (Law 5)   ■

We can also use the laws of exponents to simplify expressions involving radicals, as illustrated in the next example.

EXAMPLE 5   Simplify the given expression:

a. $\sqrt[4]{16x^4 y^8}$ **b.** $\sqrt{12m^3 n} \cdot \sqrt{3m^5 n}$ **c.** $\dfrac{\sqrt[3]{-27x^6}}{\sqrt[3]{8y^3}}$

Solution

a. $\sqrt[4]{16x^4 y^8} = (16x^4 y^8)^{1/4} = 16^{1/4} \cdot x^{4/4} y^{8/4} = 2xy^2$

b. $\sqrt{12m^3 n} \cdot \sqrt{3m^5 n} = \sqrt{36m^8 n^2} = (36m^8 n^2)^{1/2} = 36^{1/2} \cdot m^{8/2} n^{2/2}$
$\qquad = 6m^4 n$

c. $\dfrac{\sqrt[3]{-27x^6}}{\sqrt[3]{8y^3}} = \dfrac{(-27x^6)^{1/3}}{(8y^3)^{1/3}} = \dfrac{-27^{1/3} x^{6/3}}{8^{1/3} y^{3/3}} = -\dfrac{3x^2}{2y}$ ■

When a radical appears in the numerator or denominator of an algebraic expression, we often try to simplify the expression by eliminating the radical from the numerator or denominator. This process, called **rationalization,** is illustrated in the next two examples.

EXAMPLE 6   Rationalize the denominator of the expression $\dfrac{3x}{2\sqrt{x}}$.

Solution $\dfrac{3x}{2\sqrt{x}} = \dfrac{3x}{2\sqrt{x}} \cdot \dfrac{\sqrt{x}}{\sqrt{x}} = \dfrac{3x\sqrt{x}}{2\sqrt{x^2}} = \dfrac{3x\sqrt{x}}{2x} = \dfrac{3}{2}\sqrt{x}$ ■

EXAMPLE 7   Rationalize the numerator of the expression $\dfrac{3\sqrt{x}}{2x}$.

Solution $\dfrac{3\sqrt{x}}{2x} = \dfrac{3\sqrt{x}}{2x} \cdot \dfrac{\sqrt{x}}{\sqrt{x}} = \dfrac{3\sqrt{x^2}}{2x\sqrt{x}} = \dfrac{3x}{2x\sqrt{x}} = \dfrac{3}{2\sqrt{x}}$ ∎

1.1   EXERCISES

In Exercises 1–4, determine whether the given statement is true or false.

1. $-3 < -20$
2. $-5 \le -5$
3. $\dfrac{2}{3} > \dfrac{5}{6}$
4. $-\dfrac{5}{6} < -\dfrac{11}{12}$

In Exercises 5–10, show the given interval on a number line.

5. $(3, 6)$
6. $(-2, 5]$
7. $[-1, 4)$
8. $\left[-\dfrac{6}{5}, -\dfrac{1}{2}\right]$
9. $(0, \infty)$
10. $(-\infty, 5]$

In Exercises 11–20, find the values of x that satisfy the given inequalities.

11. $2x + 4 < 8$
12. $-6 > 4 + 5x$
13. $-4x \ge 20$
14. $-12 \le -3x$
15. $-6 < x - 2 < 4$
16. $0 \le x + 1 \le 4$
17. $x + 1 > 4$ or $x + 2 < -1$
18. $x + 1 > 2$ or $x - 1 < -2$
19. $x + 3 > 1$ and $x - 2 < 1$
20. $x - 4 \le 1$ and $x + 3 > 2$

In Exercises 21–30, evaluate the given expression.

21. $|-6 + 2|$
22. $4 + |-4|$
23. $\dfrac{|-12 + 4|}{|16 - 12|}$
24. $\left|\dfrac{0.2 - 1.4}{1.6 - 2.4}\right|$
25. $\sqrt{3}|-2| + 3|-\sqrt{3}|$
26. $|-1| + \sqrt{2}|-2|$
27. $|\pi - 1| + 2$
28. $|\pi - 6| - 3$
29. $|\sqrt{2} - 1| + |3 - \sqrt{2}|$
30. $|2\sqrt{3} - 3| - |\sqrt{3} - 4|$

In Exercises 31–36, suppose that a and b are real numbers other than zero and that $a > b$. State whether the given inequality is true or false.

31. $b - a > 0$
32. $\dfrac{a}{b} > 1$
33. $a^2 > b^2$
34. $\dfrac{1}{a} > \dfrac{1}{b}$
35. $a^3 > b^3$
36. $-a < -b$

In Exercises 37–42, determine whether the given statement is true for all real numbers a and b.

37. $|-a| = a$
38. $|b^2| = b^2$
39. $|a - 4| = |4 - a|$
40. $|a + 1| - |a| + 1$
41. $|a + b| = |a| + |b|$
42. $|a - b| = |a| - |b|$

In Exercises 43–58, evaluate the given expression.

43. $27^{2/3}$
44. $8^{-4/3}$
45. $\left(\dfrac{1}{\sqrt{3}}\right)^0$
46. $(7^{1/2})^4$
47. $\left[\left(\dfrac{1}{8}\right)^{1/3}\right]^{-2}$
48. $\left[\left(-\dfrac{1}{3}\right)^2\right]^{-3}$
49. $\left(\dfrac{7^{-5} \cdot 7^2}{7^{-2}}\right)^{-1}$
50. $\left(\dfrac{9}{16}\right)^{-1/2}$
51. $(125^{2/3})^{-1/2}$
52. $\sqrt[3]{2^6}$
53. $\dfrac{\sqrt{32}}{\sqrt{8}}$
54. $\sqrt[3]{\dfrac{-8}{27}}$
55. $\dfrac{16^{5/8}16^{1/2}}{16^{7/8}}$
56. $\left(\dfrac{9^{-3} \cdot 9^5}{9^{-2}}\right)^{-1/2}$
57. $16^{1/4} \cdot (8)^{-1/3}$
58. $\dfrac{6^{2.5} \cdot 6^{-1.9}}{6^{-1.4}}$

In Exercises 59–68, determine whether the given statement is true or false. Give a reason for your choice.

59. $x^4 + 2x^4 = 3x^4$

60. $3^2 \cdot 2^2 = 6^2$

61. $x^3 \cdot 2x^2 = 2x^6$

62. $3^3 + 3 = 3^4$

63. $\dfrac{2^{4x}}{1^{3x}} = 2^{4x-3x}$

64. $(2^2 \cdot 3^2)^2 = 6^4$

65. $\dfrac{1}{4^{-3}} = \dfrac{1}{64}$

66. $\dfrac{4^{3/2}}{2^4} = \dfrac{1}{2}$

67. $(1.2^{1/2})^{-1/2} = 1$

68. $5^{2/3} \cdot (25)^{2/3} = 25$

In Exercises 69–74, rewrite the given expression using positive exponents only.

69. $(xy)^{-2}$

70. $3s^{1/3} \cdot s^{-7/3}$

71. $\dfrac{x^{-1/3}}{x^{1/2}}$

72. $\sqrt{x^{-1}} \cdot \sqrt{9x^{-3}}$

73. $12^0 (s + t)^{-3}$

74. $(x - y)(x^{-1} + y^{-1})$

In Exercises 75–90, simplify the given expression.

75. $\dfrac{x^{7/3}}{x^{-2}}$

76. $(49x^{-2})^{-1/2}$

77. $(x^2 y^{-3})(x^{-5} y^3)$

78. $\dfrac{5x^6 y^3}{2x^2 y^7}$

79. $\dfrac{x^{3/4}}{x^{-1/4}}$

80. $\left(\dfrac{x^3 y^2}{z^2}\right)^2$

81. $\left(\dfrac{x^3}{-27y^{-6}}\right)^{-2/3}$

82. $\left(\dfrac{e^x}{e^{x-2}}\right)^{-1/2}$

83. $\left(\dfrac{x^{-3}}{y^{-2}}\right)^2 \left(\dfrac{y}{x}\right)^4$

84. $\dfrac{(r^n)^4}{r^{5-2n}}$

85. $\sqrt[3]{x^{-2}} \cdot \sqrt{4x^5}$

86. $\sqrt{81x^6 y^{-4}}$

87. $-\sqrt[4]{16x^4 y^8}$

88. $\sqrt[3]{x^{3a+b}}$

89. $\sqrt[6]{64x^8 y^3}$

90. $\sqrt[3]{27r^6} \cdot \sqrt{s^2 t^4}$

In Exercises 91–94, use the fact that $2^{1/2} \approx 1.414$ and $3^{1/2} \approx 1.732$ to evaluate the given expression without using a calculator.

91. $2^{3/2}$

92. $8^{1/2}$

93. $9^{3/4}$

94. $6^{1/2}$

In Exercises 95–98, use the fact that $10^{1/2} \approx 3.162$ and $10^{1/3} \approx 2.154$ to evaluate the given expression without using a calculator.

95. $10^{3/2}$

96. $1000^{3/2}$

97. $10^{2.5}$

98. $(0.0001)^{-1/3}$

In Exercises 99–104, rationalize the denominator of the given expression.

99. $\dfrac{3}{2\sqrt{x}}$

100. $\dfrac{3}{\sqrt{xy}}$

101. $\dfrac{2y}{\sqrt{3y}}$

102. $\dfrac{5x^2}{\sqrt{3x}}$

103. $\dfrac{1}{\sqrt[3]{x}}$

104. $\sqrt{\dfrac{2x}{y}}$

In Exercises 105–110, rationalize the numerator of the given expression.

105. $\dfrac{2\sqrt{x}}{3}$

106. $\dfrac{\sqrt[3]{x}}{24}$

107. $\sqrt{\dfrac{2y}{x}}$

108. $\sqrt[3]{\dfrac{2x}{3y}}$

109. $\dfrac{\sqrt[3]{x^2 z}}{y}$

110. $\dfrac{\sqrt[3]{x^2 y}}{2x}$

111. An advertisement for a certain car states that the EPA fuel economy is 20 mpg city and 27 mpg highway and that the car's fuel-tank capacity is 18.1 gallons. Assuming ideal driving conditions, determine the driving range for the car from the foregoing data.

112. Find the minimum cost C (in dollars), given that
$$5(C - 25) \geq 1.75 + 2.5C.$$

113. Find the maximum profit P (in dollars) given that
$$6(P - 2500) \leq 4(P + 2400).$$

114. The relationship between Celsius (°C) and Fahrenheit (°F) temperatures is given by the formula
$$C = \dfrac{5}{9}(F - 32).$$

 a. If the temperature range for Montreal during the month of January is $-15° < C° < -5°$, find the range in degrees Fahrenheit in Montreal for the same period.

 b. If the temperature range for New York City during the month of June is $63° < F° < 80°$, find the range in degrees Celsius in New York City for the same period.

115. A salesman's monthly commission is 15 percent on all sales over \$12,000. If his goal is to make a commission of at least \$1,000 per month, what minimum monthly sales figures must he attain?

116. The markup on a used car was at least 30 percent of its current wholesale price. If the car was sold for \$2800, what was the maximum wholesale price?

117. A manufacturer of a certain commodity has estimated that her profit in thousands of dollars is given by the expression

$$-6x^2 + 30x - 10,$$

where x (in thousands) is the number of units produced. What production range will enable the manufacturer to realize a profit of at least $14,000 on the commodity?

118. The distribution of income in a certain city can be described by the exponential model $y = (2.8 \cdot 10^{11})(x)^{-1.5}$, where y is the number of families with an income of x or more dollars.

  a. How many families in this city have an income of $20,000 or more?

  b. How many families have an income of $40,000 or more?

  c. How many families have an income of $100,000 or more?

1.2 Precalculus Review II

Operations with Algebraic Expressions

In calculus, we often work with algebraic expressions such as

$$2x^{4/3} - x^{1/3} + 1, \qquad 2x^2 - x - \frac{2}{\sqrt{x}}, \qquad \frac{3xy + 2}{x + 1}, \quad \text{and} \quad 2x^3 + 2x + 1.$$

An algebraic expression of the form ax^n where the coefficient a is a real number and n is a nonnegative integer is called a **monomial,** meaning it consists of one term. For example, $7x^2$ is a monomial. A **polynomial** is a monomial or the sum of two or more monomials. For example,

$$x^2 + 4x + 4, \qquad x^3 + 5, \qquad x^4 + 3x^2 + 3, \quad \text{and} \quad x^2y + xy + y$$

are all polynomials.

Constant terms and terms containing the same variable factor are called **like,** or **similar, terms.** Like terms may be combined by adding or subtracting their numerical coefficients. For example,

$$3x + 7x = 10x \quad \text{and} \quad \frac{1}{2}xy + 3xy = \frac{7}{2}xy.$$

The distributive property of the real number system,

$$ab + ac = a \cdot (b + c),$$

is used to justify this procedure.

To add or subtract two or more algebraic expressions, first remove the parentheses and then combine like terms. The resulting expression is written in order of decreasing degree from left to right.

$\circledR^\dagger$ **EXAMPLE 1**

a. $(2x^4 + 3x^3 + 4x + 6) - (3x^4 + 9x^3 + 3x^2)$

$= 2x^4 + 3x^3 + 4x + 6 - 3x^4 - 9x^3 - 3x^2$ (Remove parentheses.)

$= 2x^4 - 3x^4 + 3x^3 - 9x^3 - 3x^2 + 4x + 6$

$= -x^4 - 6x^3 - 3x^2 + 4x + 6$ (Combine like terms.)

---

†The symbol $\circledR$ indicates that these examples were selected from the calculus portion of the text in order to help you review the algebraic computations you will *actually* be using in calculus.

b. $2t^3 - \{t^2 - [t - (2t - 1)] + 4\}$

$\quad = 2t^3 - \{t^2 - [t - 2t + 1] + 4\}$

$\quad = 2t^3 - \{t^2 - [-t + 1] + 4\}$ (Remove parentheses and combine like terms within brackets.)

$\quad = 2t^3 - \{t^2 + t - 1 + 4\}$ (Remove brackets.)

$\quad = 2t^3 - \{t^2 + t + 3\}$ (Combine like terms within the braces.)

$\quad = 2t^3 - t^2 - t - 3$ (Remove braces.)   ■

An algebraic expression is said to be **simplified** if none of its terms are similar. Observe that when the algebraic expression in Example 1(b) was simplified, the innermost grouping symbols were removed first; that is, the parentheses ( ) were removed first, the brackets [ ] second, and the braces { } third.

When algebraic expressions are multiplied, each term of one algebraic expression is multiplied by each term of the other. The resulting algebraic expression is then simplified.

(R)   **EXAMPLE 2**   Perform the indicated operations:

a. $(x^2 + 1)(3x^2 + 10x + 3)$ **b.** $(e^t + e^{-t})e^t - e^t(e^t - e^{-t})$

Solution

a. $(x^2 + 1)(3x^2 + 10x + 3)$

$\quad = x^2(3x^2 + 10x + 3) + 1(3x^2 + 10x + 3)$

$\quad = 3x^4 + 10x^3 + 3x^2 + 3x^2 + 10x + 3$

$\quad = 3x^4 + 10x^3 + 6x^2 + 10x + 3$

b. $(e^t + e^{-t})e^t - e^t(e^t - e^{-t})$

$\quad = e^{2t} + e^0 - e^{2t} + e^0$

$\quad = e^{2t} - e^{2t} + e^0 + e^0$

$\quad = 1 + 1$ (Recall that $e^0 = 1$.)

$\quad = 2$ ■

Certain product formulas that are frequently used in algebraic computations are given in Table 1.5.

TABLE 1.5

Formula	Example
$(a + b)^2 = a^2 + 2ab + b^2$	$(2x + 3y)^2 = (2x)^2 + 2(2x)(3y) + (3y)^2$ $= 4x^2 + 12xy + 9y^2$
$(a - b)^2 = a^2 - 2ab + b^2$	$(4x - 2y)^2 = (4x)^2 - 2(4x)(2y) + (2y)^2$ $= 16x^2 - 16xy + 4y^2$
$(a + b)(a - b) = a^2 - b^2$	$(2x + y)(2x - y) = (2x)^2 - (y)^2$ $= 4x^2 - y^2$

Factoring

Factoring is the process of expressing an algebraic expression as a product of other algebraic expressions. For example, by applying the distributive property we may write

$$3x^2 - x = x(3x - 1).$$

The first step in factoring an algebraic expression is to check to see whether it contains any common terms. If it does, the greatest common term is then factored out. For example, the common factor of the algebraic expression $2a^2x + 4ax + 6a$ is $2a$, because

$$2a^2x + 4ax + 6a = 2a \cdot ax + 2a \cdot 2x + 2a \cdot 3 = 2a(ax + 2x + 3).$$

Ⓡ   **EXAMPLE 3**   Factor out the greatest common factor in each of the following expressions:

a. $-0.3t^2 + 3t$ **b.** $2x^{3/2} - 3x^{1/2}$ **c.** $2ye^{xy^2} + 2xy^3e^{xy^2}$

d. $4x(x + 1)^{1/2} - 2x^2(\frac{1}{2})(x + 1)^{-1/2}$

Solution

a. $-0.3t^2 + 3t = -0.3t(t - 10)$

b. $2x^{3/2} - 3x^{1/2} = x^{1/2}(2x - 3)$

c. $2ye^{xy^2} + 2xy^3e^{xy^2} = 2ye^{xy^2}(1 + xy^2)$

d. $4x(x + 1)^{1/2} - 2x^2(\frac{1}{2})(x + 1)^{-1/2} = 4x(x + 1)^{1/2} - x^2(x + 1)^{-1/2}$
$$= x(x + 1)^{-1/2}[4(x + 1)^{1/2}(x + 1)^{1/2} - x]$$
$$= x(x + 1)^{-1/2}[4(x + 1) - x]$$
$$= x(x + 1)^{-1/2}(4x + 4 - x) = x(x + 1)^{-1/2}(3x + 4)$$

Here we select $(x + 1)^{-1/2}$ as the common factor because it is "contained" in each algebraic term. In particular, observe that

$$(x + 1)^{-1/2}(x + 1)^{1/2}(x + 1)^{1/2} = (x + 1)^{1/2}.$$ ■

Sometimes an algebraic expression may be factored by regrouping and rearranging its terms and then factoring out a common term. This technique is illustrated in Example 4.

EXAMPLE 4   Factor:

a. $2ax + 2ay + bx + by$ **b.** $3x\sqrt{y} - 4 - 2\sqrt{y} + 6x$

Solution

a. First, factor the common term $2a$ from the first two terms and the common term b from the last two terms. Thus,

$$2ax + 2ay + bx + by = 2a(x + y) + b(x + y).$$

Since $(x + y)$ is common to both terms of the polynomial, we may factor it out. Hence,

$$2a(x + y) + b(x + y) = (x + y)(2a + b).$$

b. $3x\sqrt{y} - 4 - 2\sqrt{y} + 6x = 3x\sqrt{y} - 2\sqrt{y} + 6x - 4$
$$= \sqrt{y}(3x - 2) + 2(3x - 2)$$
$$= (3x - 2)(\sqrt{y} + 2)$$ ■

The first step in factoring a polynomial is to find the common factors. The next step is to express the polynomial as the product of a constant and/or one or more prime polynomials.

Certain product formulas that are useful in factoring binomials and trinomials are listed in Table 1.6.

TABLE 1.6

Formula	Example
Difference of two squares $$x^2 - y^2 = (x + y)(x - y)$$	$$x^2 - 36 = (x + 6)(x - 6)$$ $$8x^2 - 2y^2 = 2(4x^2 - y^2)$$ $$= 2(2x + y)(2x - y)$$ $$9 - a^6 = (3 + a^3)(3 - a^3)$$
Perfect-square trinomial $$x^2 + 2xy + y^2 = (x + y)^2$$ $$x^2 - 2xy + y^2 = (x - y)^2$$	$$x^2 + 8x + 16 = (x + 4)^2$$ $$4x^2 - 4xy + y^2 = (2x - y)^2$$
Sum of two cubes $$x^3 + y^3 = (x + y)(x^2 - xy + y^2)$$	$$z^3 + 27 = z^3 + (3)^3$$ $$= (z + 3)(z^2 - 3z + 9)$$
Difference of two cubes $$x^3 - y^3 = (x - y)(x^2 + xy + y^2)$$	$$8x^3 - y^6 = (2x)^3 - (y^2)^3$$ $$= (2x - y^2)(4x^2 + 2xy^2 + y^4)$$

The factors of the second-degree polynomial with integral coefficients

$$px^2 + qx + r$$

are $(ax + b)(cx + d)$, where $ac = p$, $ad + bc = q$, and $bd = r$. The next example illustrates the procedure used to factor these polynomials.

(R) **EXAMPLE 5**   Factor:

a. $3x^2 + 4x - 4$ **b.** $3x^2 - 6x - 24$

Solution

a. Using trial and error, we find that the correct factorization is

$$3x^2 + 4x - 4 = (3x - 2)(x + 2).$$

b. Since each term has the common factor 3, we have

$$3x^2 - 6x - 24 = 3(x^2 - 2x - 8).$$

Using the trial-and-error method of factorization, we find that

$$x^2 - 2x - 8 = (x - 4)(x + 2).$$

Thus, we have

$$3x^2 - 6x - 24 = 3(x - 4)(x + 2). \qquad \blacksquare$$

Roots of Polynomial Equations

A polynomial equation of degree n in the variable x is an equation of the form

$$a_n x^n + a_{n-1} x^{n-1} + \cdots + a_0 = 0,$$

where n is a nonnegative integer and $a_0, a_1, \ldots, a_n$ are real numbers with $a_n \neq 0$. For example, the equation

$$-2x^5 + 8x^3 - 6x^2 + 3x + 1 = 0$$

is a polynomial equation of degree 5 in x.

The **roots** of a polynomial equation are precisely the values of x that satisfy the given equation.[†] One way of finding the roots of a polynomial equation is to first factor the polynomial and then solve the resulting equation. For example, the polynomial equation

$$x^3 - 3x^2 + 2x = 0$$

may be rewritten in the form

$$x(x^2 - 3x + 2) = 0$$

or $\qquad\qquad x(x - 1)(x - 2) = 0.$

Since the product of two real numbers can be equal to zero if and only if one (or both) of the factors is equal to zero, we have

$$x = 0, \; x - 1 = 0, \text{ or } x - 2 = 0,$$

from which we see that the desired roots are $x = 0$, 1, and 2.

The Quadratic Formula

In general, the problem of finding the roots of a polynomial equation is a difficult one. But the roots of a quadratic equation (a polynomial equation of degree 2) are easily found either by factoring or by using the following quadratic formula.

THE QUADRATIC FORMULA

The solutions of the equation $ax^2 + bx + c = 0$ $(a \neq 0)$ are given by

$$x = \frac{-b \pm \sqrt{b^2 - 4ac}}{2a}.$$

[†]In this book, we are interested only in the *real* roots of an equation.

EXAMPLE 6   Solve each of the following quadratic equations:

a. $2x^2 + 5x - 12 = 0$ **b.** $x^2 = -3x + 8$

Solution

a. The equation is in standard form with $a = 2$, $b = 5$, and $c = -12$. Using the quadratic formula, we find

$$x = \frac{-b \pm \sqrt{b^2 - 4ac}}{2a} = \frac{-5 \pm \sqrt{5^2 - 4(2)(-12)}}{2(2)}$$

$$= \frac{-5 \pm \sqrt{121}}{4} = \frac{-5 \pm 11}{4}$$

$$= -4 \quad \text{or} \quad \frac{3}{2}.$$

This equation can also be solved by factoring. Thus

$$2x^2 + 5x - 12 = (2x - 3)(x + 4) = 0,$$

from which we see that the desired roots are $x = 3/2$ or $x = -4$, as obtained earlier.

b. We first rewrite the given equation in the standard form $x^2 + 3x - 8 = 0$, from which we see that $a = 1$, $b = 3$, and $c = -8$. Using the quadratic formula, we find

$$x = \frac{-b \pm \sqrt{b^2 - 4ac}}{2a} = \frac{-3 \pm \sqrt{3^2 - 4(1)(-8)}}{2(1)}$$

$$= \frac{-3 \pm \sqrt{41}}{2};$$

that is, the solutions are $\dfrac{-3 + \sqrt{41}}{2} \approx 1.7$ and $\dfrac{-3 - \sqrt{41}}{2} \approx -4.7$. In this case, the quadratic formula proves quite handy!   ■

Rational Expressions

Quotients of polynomials are called **rational expressions.** Examples of rational expressions are

$$\frac{6x - 1}{2x + 3}, \qquad \frac{3x^2y^3 - 2xy}{4x}, \quad \text{and} \quad \frac{2}{5ab}.$$

Since rational expressions are quotients in which the variables represent real numbers, the properties of the real numbers apply to rational expressions as well, and operations with rational fractions are performed in the same manner as operations with arithmetic fractions. For example, using the properties of the real number system, we may write

$$\frac{ac}{bc} = \frac{a}{b} \cdot \frac{c}{c} = \frac{a}{b} \cdot 1 = \frac{a}{b},$$

where a, b, and c are any real numbers and b and c are not zero.

Similarly, using the same properties of real numbers, we may write

$$\frac{(x + 2)(x - 3)}{(x - 2)(x - 3)} = \frac{x + 2}{x - 2} \qquad (x \neq 2, 3)$$

after "canceling" the common factors.

 $\dfrac{\cancel{3} + 4x}{\cancel{3}} \neq 1 + 4x$ is an example of incorrect cancellation. Instead we write

$$\frac{3 + 4x}{3} = \frac{3}{3} + \frac{4x}{3} = 1 + \frac{4x}{3}.$$

A rational expression is simplified, or in lowest terms, when the numerator and denominator have no common factors other than 1 and -1.

Ⓡ   **EXAMPLE 7**   Simplify the following expressions:

a. $\dfrac{x^2 + 2x - 3}{x^2 + 4x + 3}$ **b.** $\dfrac{[(t^2 + 4)(2t - 4) - (t^2 - 4t + 4)(2t)]}{(t^2 + 4)^2}$

Solution

a. $\dfrac{x^2 + 2x - 3}{x^2 + 4x + 3} = \dfrac{(x + 3)(x - 1)}{(x + 3)(x + 1)} = \dfrac{x - 1}{x + 1}$

b. $\dfrac{[(t^2 + 4)(2t - 4) - (t^2 - 4t + 4)(2t)]}{(t^2 + 4)^2}$

$= \dfrac{2t^3 - 4t^2 + 8t - 16 - 2t^3 + 8t^2 - 8t}{(t^2 + 4)^2}$ (Carry out the indicated multiplication.)

$= \dfrac{4t^2 - 16}{(t^2 + 4)^2}$ (Combine like terms.)

$= \dfrac{4(t^2 - 4)}{(t^2 + 4)^2}$ (Factor.)   ∎

The operations of multiplication and division are performed with algebraic fractions in the same manner as with arithmetic fractions (see Table 1.7).

TABLE 1.7

Operation	Example
If P, Q, R, and S are polynomials then **Multiplication** $\dfrac{P}{Q} \cdot \dfrac{R}{S} = \dfrac{PR}{QS}$ $(Q, S \neq 0)$	$\dfrac{2x}{y} \cdot \dfrac{(x + 1)}{(y - 1)} = \dfrac{2x(x + 1)}{y(y - 1)} = \dfrac{2x^2 + 2x}{y^2 - y}$
Division $\dfrac{P}{Q} \div \dfrac{R}{S} = \dfrac{P}{Q} \cdot \dfrac{S}{R} = \dfrac{PS}{QR}$ $(Q, R, S \neq 0)$	$\dfrac{x^2 + 3}{y} \div \dfrac{y^2 + 1}{x} = \dfrac{x^2 + 3}{y} \cdot \dfrac{x}{y^2 + 1}$ $= \dfrac{x^3 + 3x}{y^3 + y}$

When rational expressions are multiplied and divided, the resulting expressions should be simplified.

EXAMPLE 8   Perform the indicated operations and simplify:

$$\frac{2x - 8}{x + 2} \cdot \frac{x^2 + 4x + 4}{x^2 - 16}$$

Solution $\dfrac{2x - 8}{x + 2} \cdot \dfrac{x^2 + 4x + 4}{x^2 - 16}$

$$= \frac{2(x - 4)}{x + 2} \cdot \frac{(x + 2)^2}{(x + 4)(x - 4)}$$

$$= \frac{2(x - 4)(x + 2)(x + 2)}{(x + 2)(x + 4)(x - 4)} \qquad \begin{array}{l} \text{[Cancel the common} \\ \text{factors } (x + 2)(x - 4).] \end{array}$$

$$= \frac{2(x + 2)}{x + 4} \qquad \blacksquare$$

For rational expressions, the operations of addition and subtraction are performed by finding a common denominator of the fractions and then adding or subtracting the fractions. Table 1.8 shows the rules for fractions with equal denominators.

TABLE 1.8

Operation	Example
If P, Q, and R are polynomials, then **Addition** $\dfrac{P}{R} + \dfrac{Q}{R} = \dfrac{P + Q}{R}$ $\quad(R \neq 0)$	$\dfrac{2x}{x + 2} + \dfrac{6x}{x + 2} = \dfrac{2x + 6x}{x + 2} = \dfrac{8x}{x + 2}$
Subtraction $\dfrac{P}{R} - \dfrac{Q}{R} = \dfrac{P - Q}{R}$ $\quad(R \neq 0)$	$\dfrac{3y}{y - x} - \dfrac{y}{y - x} = \dfrac{3y - y}{y - x} = \dfrac{2y}{y - x}$

To add or subtract fractions that have different denominators, first find a common denominator, preferably the least common denominator (LCD). Then carry out the indicated operations following the procedure described in Table 1.8.

To find the least common denominator (LCD) of two or more rational expressions,

1. *find the prime factors of each denominator.*

2. *form the product of the different prime factors that occur in the denominators. Each prime factor in this product should be raised to the highest power of that factor appearing in the denominators.*

$$\frac{x}{2 + y} \neq \frac{x}{2} + \frac{x}{y} \; !$$

⑧   **EXAMPLE 9**   Simplify:

a. $\dfrac{2x}{x^2 + 1} + \dfrac{6(3x^2)}{x^3 + 2}$ **b.** $\dfrac{1}{x + h} - \dfrac{1}{x}$

Solution

a. $\dfrac{2x}{x^2 + 1} + \dfrac{6(3x^2)}{x^3 + 2}$

$= \dfrac{2x(x^3 + 2) + 6(3x^2)(x^2 + 1)}{(x^2 + 1)(x^3 + 2)}$ [LCD $= (x^2 + 1)(x^3 + 2)$.]

$= \dfrac{2x^4 + 4x + 18x^4 + 18x^2}{(x^2 + 1)(x^3 + 2)}$

$= \dfrac{20x^4 + 18x^2 + 4x}{(x^2 + 1)(x^3 + 2)}$

$= \dfrac{2x(10x^3 + 9x + 2)}{(x^2 + 1)(x^3 + 2)}$

b. $\dfrac{1}{x + h} - \dfrac{1}{x} = \dfrac{1}{x + h} \cdot \dfrac{x}{x} - \dfrac{1}{x} \cdot \dfrac{x + h}{x + h}$ [LCD $= (x)(x + h)$.]

$= \dfrac{x}{x(x + h)} - \dfrac{x + h}{x(x + h)}$

$= \dfrac{x - x - h}{x(x + h)}$

$= \dfrac{-h}{x(x + h)}$ ∎

Other Algebraic Fractions

The techniques used to simplify rational expressions may also be used to simplify algebraic fractions in which the numerator and denominator are not polynomials, as illustrated in Example 10.

EXAMPLE 10   Simplify:

a. $\dfrac{1 + \dfrac{1}{x + 1}}{x - \dfrac{4}{x}}$ **b.** $\dfrac{x^{-1} + y^{-1}}{x^{-2} - y^{-2}}$

Solution

a. $\dfrac{1 + \dfrac{1}{x + 1}}{x - \dfrac{4}{x}} = \dfrac{1 \cdot \dfrac{x + 1}{x + 1} + \dfrac{1}{x + 1}}{x \cdot \dfrac{x}{x} - \dfrac{4}{x}} = \dfrac{\dfrac{x + 1 + 1}{x + 1}}{\dfrac{x^2 - 4}{x}}$

$= \dfrac{x + 2}{x + 1} \cdot \dfrac{x}{x^2 - 4} = \dfrac{x + 2}{x + 1} \cdot \dfrac{x}{(x + 2)(x - 2)}$

$= \dfrac{x}{(x + 1)(x - 2)}$

b. $\dfrac{x^{-1} + y^{-1}}{x^{-2} - y^{-2}} = \dfrac{\dfrac{1}{x} + \dfrac{1}{y}}{\dfrac{1}{x^2} - \dfrac{1}{y^2}} = \dfrac{\dfrac{y + x}{xy}}{\dfrac{y^2 - x^2}{x^2 y^2}}$ $\left(x^{-n} = \dfrac{1}{x^n} \right)$

$$= \dfrac{y + x}{xy} \cdot \dfrac{x^2 y^2}{y^2 - x^2} = \dfrac{y + x}{xy} \cdot \dfrac{(xy)^2}{(y + x)(y - x)}$$

$$= \dfrac{xy}{y - x} \qquad \blacksquare$$

EXAMPLE 11   Perform the given operations and simplify:

a. $\dfrac{x^2 (2x^2 + 1)^{1/2}}{x - 1} \cdot \dfrac{4x^3 - 6x^2 + x - 2}{x(x - 1)(2x^2 + 1)}$ **b.** $\dfrac{12x^2}{\sqrt{2x^2 + 3}} + 6\sqrt{2x^2 + 3}$

Solution

a. $\dfrac{x^2 (2x^2 + 1)^{1/2}}{x - 1} \cdot \dfrac{4x^3 - 6x^2 + x - 2}{x(x - 1)(2x^2 + 1)}$

$$= \dfrac{x(4x^3 - 6x^2 + x - 2)}{(x - 1)^2 (2x^2 + 1)^{1 - 1/2}}$$

$$= \dfrac{x(4x^3 - 6x^2 + x - 2)}{(x - 1)^2 (2x^2 + 1)^{1/2}}$$

b. $\dfrac{12x^2}{\sqrt{2x^2 + 3}} + 6\sqrt{2x^2 + 3} = \dfrac{12x^2}{(2x^2 + 3)^{1/2}} + 6(2x^2 + 3)^{1/2}$

$$= \dfrac{12x^2 + 6(2x^2 + 3)^{1/2}(2x^2 + 3)^{1/2}}{(2x^2 + 3)^{1/2}}$$

$$= \dfrac{12x^2 + 6(2x^2 + 3)}{(2x^2 + 3)^{1/2}}$$

$$= \dfrac{24x^2 + 18}{(2x^2 + 3)^{1/2}} = \dfrac{6(4x^2 + 3)}{\sqrt{2x^2 + 3}} \qquad \blacksquare$$

Rationalizing Algebraic Fractions

When the denominator of an algebraic fraction contains sums or differences involving radicals, we may **rationalize the denominator**—that is, transform the fraction into an equivalent one with a denominator that does not contain radicals. In doing so, we make use of the fact that

$$(\sqrt{a} + \sqrt{b})(\sqrt{a} - \sqrt{b}) = (\sqrt{a})^2 - (\sqrt{b})^2$$
$$= a - b.$$

This procedure is illustrated in Example 12.

EXAMPLE 12   Rationalize the denominator of $\dfrac{1}{1 + \sqrt{x}}$.

Solution Upon multiplying the numerator and the denominator by $(1 - \sqrt{x})$, we obtain

$$\frac{1}{1 + \sqrt{x}} = \frac{1}{1 + \sqrt{x}} \cdot \frac{1 - \sqrt{x}}{1 - \sqrt{x}}$$

$$= \frac{1 - \sqrt{x}}{1 - (\sqrt{x})^2}$$

$$= \frac{1 - \sqrt{x}}{1 - x}.$$ ∎

In other situations, it may be necessary to rationalize the numerator of an algebraic expression. In calculus, for example, one encounters the following problem.

EXAMPLE 13 Rationalize the numerator: $\dfrac{\sqrt{1 + h} - 1}{h}$

Solution

$$\frac{\sqrt{1 + h} - 1}{h} = \frac{\sqrt{1 + h} - 1}{h} \cdot \frac{\sqrt{1 + h} + 1}{\sqrt{1 + h} + 1}$$

$$= \frac{(\sqrt{1 + h})^2 - (1)^2}{h(\sqrt{1 + h} + 1)}$$

$$= \frac{1 + h - 1}{h(\sqrt{1 + h} + 1)} \qquad ((\sqrt{1 + h})^2 = \sqrt{1 + h} \cdot \sqrt{1 + h}$$
$$= 1 + h)$$

$$= \frac{h}{h(\sqrt{1 + h} + 1)}$$

$$= \frac{1}{\sqrt{1 + h} + 1}$$ ∎

1.2 EXERCISES

In Exercises 1–22, perform the indicated operations and simplify each expression.

1. $(7x^2 - 2x + 5) + (2x^2 + 5x - 4)$

2. $(3x^2 + 5xy + 2y) + (4 - 3xy - 2x^2)$

3. $(5y^2 - 2y + 1) - (y^2 - 3y - 7)$

4. $3(2a - b) - 4(b - 2a)$

5. $x - \{2x - [-x - (1 - x)]\}$

6. $3x^2 - \{x^2 + 1 - x[x - (2x - 1)]\} + 2$

7. $\left(\dfrac{1}{3} - 1 + e\right) - \left(-\dfrac{1}{3} - 1 + e^{-1}\right)$

8. $-\dfrac{3}{4}y - \dfrac{1}{4}x + 100 + \dfrac{1}{2}x + \dfrac{1}{4}y - 120$

9. $3\sqrt{8} + 8 - 2\sqrt{y} + \dfrac{1}{2}\sqrt{x} - \dfrac{3}{4}\sqrt{y}$

10. $\dfrac{8}{9}x^2 + \dfrac{2}{3}x + \dfrac{16}{3}x^2 - \dfrac{16}{3}x - 2x + 2$

11. $(x + 8)(x - 2)$ 12. $(5x + 2)(3x - 4)$

13. $(a + 5)^2$ 14. $(3a - 4b)^2$

15. $(x + 2y)^2$ 16. $(6 - 3x)^2$

17. $(2x + y)(2x - y)$ 18. $(3x + 2)(2 - 3x)$

19. $(x^2 - 1)(2x) - x^2(2x)$

20. $(x^{1/2} + 1)\left(\dfrac{1}{2}x^{-1/2}\right) - (x^{1/2} - 1)\left(\dfrac{1}{2}x^{-1/2}\right)$

21. $2(t + \sqrt{t})^2 - 2t^2$ 22. $2x^2 + (-x + 1)^2$

In Exercises 23–30, factor out the greatest common factor from each expression.

23. $4x^5 - 12x^4 - 6x^3$

24. $4x^2y^2z - 2x^5y^2 + 6x^3y^2z^2$

25. $7a^4 - 42a^2b^2 + 49a^3b$

26. $3x^{2/3} - 2x^{1/3}$ 27. $e^{-x} - xe^{-x}$

28. $2ye^{xy^2} + 2xy^3e^{xy^2}$ 29. $2x^{-5/2} - \dfrac{3}{2}x^{-3/2}$

30. $\dfrac{1}{2}\left(\dfrac{2}{3}u^{3/2} - 2u^{1/2}\right)$

In Exercises 31–44, factor each expression.

31. $6ac + 3bc - 4ad - 2bd$

32. $3x^3 - x^2 + 3x - 1$

33. $4a^2 - b^2$ 34. $12x^2 - 3y^2$

35. $10 - 14x - 12x^2$ 36. $x^2 - 2x - 15$

37. $3x^2 - 6x - 24$ 38. $3x^2 - 4x - 4$

39. $12x^2 - 2x - 30$ 40. $(x + y)^2 - 1$

41. $9x^2 - 16y^2$ 42. $8a^2 - 2ab - 6b^2$

43. $x^6 + 125$ 44. $x^3 - 27$

In Exercises 45–52, perform the indicated operations and simplify the given algebraic expression.

45. $(x^2 + y^2)x - xy(2y)$ 46. $2kr(R - r) - kr^2$

47. $2(x - 1)(2x + 2)^3[4(x - 1) + (2x + 2)]$

48. $5x^2(3x^2 + 1)^4(6x) + (3x^2 + 1)^5(2x)$

49. $4(x - 1)^2(2x + 2)^3(2) + (2x + 2)^4(2)(x - 1)$

50. $(x^2 + 1)(4x^3 - 3x^2 + 2x) - (x^4 - x^3 + x^2)(2x)$

51. $(x^2 + 2)^2[5(x^2 + 2)^2 - 3](2x)$

52. $(x^2 - 4)(x^2 + 4)(2x + 8) - (x^2 + 8x - 4)(4x^3)$

In Exercises 53–58, find the real roots of each of the given equations by factoring.

53. $x^2 + x - 12 = 0$ 54. $3x^2 - x - 4 = 0$

55. $4t^2 + 2t - 2 = 0$ 56. $-6x^2 + x + 12 = 0$

57. $\frac{1}{4}x^2 - x + 1 = 0$ 58. $\frac{1}{2}a^2 + a - 12 = 0$

In Exercises 59–64, use the quadratic formula to solve the given quadratic equation.

59. $4x^2 + 5x - 6 = 0$ 60. $3x^2 - 4x + 1 = 0$

61. $8x^2 - 8x - 3 = 0$ 62. $x^2 - 6x + 6 = 0$

63. $2x^2 + 4x - 3 = 0$ 64. $2x^2 + 7x - 15 = 0$

In Exercises 65–70, simplify the given expression.

65. $\dfrac{x^2 + x - 2}{x^2 - 4}$ 66. $\dfrac{2a^2 - 3ab - 9b^2}{2ab^2 + 3b^3}$

67. $\dfrac{12t^2 + 12t + 3}{4t^2 - 1}$ 68. $\dfrac{x^3 + 2x^2 - 3x}{-2x^2 - x + 3}$

69. $\dfrac{(4x - 1)(3) - (3x + 1)(4)}{(4x - 1)^2}$

70. $\dfrac{(1 + x^2)^2(2) - 2x(2)(1 + x^2)(2x)}{(1 + x^2)^4}$

In Exercises 71–88, perform the indicated operations and simplify each expression.

71. $\dfrac{2a^2 - 2b^2}{b - a} \cdot \dfrac{4a + 4b}{a^2 + 2ab + b^2}$

72. $\dfrac{x^2 - 6x + 9}{x^2 - x - 6} \cdot \dfrac{3x + 6}{2x^2 - 7x + 3}$

73. $\dfrac{3x^2 + 2x - 1}{2x + 6} \div \dfrac{x^2 - 1}{x^2 + 2x - 3}$

74. $\dfrac{3x^2 - 4xy - 4y^2}{x^2y} \div \dfrac{(2y - x)^2}{x^3y}$

75. $\dfrac{58}{3(3t + 2)} + \dfrac{1}{3}$ 76. $\dfrac{a + 1}{3a} + \dfrac{b - 2}{5b}$

77. $\dfrac{2x}{2x - 1} - \dfrac{3x}{2x + 5}$ 78. $\dfrac{-xe^x}{x + 1} + e^x$

79. $\dfrac{4}{x^2 - 9} - \dfrac{5}{x^2 - 6x + 9}$

80. $\dfrac{x}{1 - x} + \dfrac{2x + 3}{x^2 - 1}$

81. $\dfrac{1 + \dfrac{1}{x}}{1 - \dfrac{1}{x}}$ 82. $\dfrac{\dfrac{1}{x} + \dfrac{1}{y}}{1 - \dfrac{1}{xy}}$

83. $\dfrac{4x^2}{2\sqrt{2x^2 + 7}} + \sqrt{2x^2 + 7}$

84. $6(2x + 1)^2\sqrt{x^2 + x} + \dfrac{(2x + 1)^4}{2\sqrt{x^2 + x}}$

85. $\dfrac{2x(x + 1)^{-1/2} - (x + 1)^{1/2}}{x^2}$

86. $\dfrac{(x^2 + 1)^{1/2} - 2x^2(x^2 + 1)^{-1/2}}{1 - x^2}$

87. $\dfrac{(2x + 1)^{1/2} - (x + 2)(2x + 1)^{-1/2}}{2x + 1}$

88. $\dfrac{2(2x - 3)^{1/3} - (x - 1)(2x - 3)^{-2/3}}{(2x - 3)^{2/3}}$

In Exercises 89–94, rationalize the denominator of each expression.

89. $\dfrac{1}{\sqrt{3} - 1}$

90. $\dfrac{1}{\sqrt{x + 5}}$

91. $\dfrac{1}{\sqrt{x} - \sqrt{y}}$

92. $\dfrac{a}{1 - \sqrt{a}}$

93. $\dfrac{\sqrt{a} + \sqrt{b}}{\sqrt{a} - \sqrt{b}}$

94. $\dfrac{2\sqrt{a} + \sqrt{b}}{2\sqrt{a} - \sqrt{b}}$

In Exercises 95–100, rationalize the numerator of each expression.

95. $\dfrac{\sqrt{x}}{3}$

96. $\dfrac{\sqrt[3]{y}}{x}$

97. $\dfrac{1 - \sqrt{3}}{3}$

98. $\dfrac{\sqrt{x} - 1}{x}$

99. $\dfrac{1 + \sqrt{x + 2}}{\sqrt{x + 2}}$

100. $\dfrac{\sqrt{x + 3} - \sqrt{x}}{3}$

1.3   The Cartesian Coordinate System

The Cartesian Coordinate System

In Section 1.1 we saw how a one-to-one correspondence between the set of real numbers and the points on a straight line leads to a coordinate system on a line (a one-dimensional space).

A similar representation for points in a plane (a two-dimensional space) is realized through the **Cartesian coordinate system,** which is constructed as follows: Take two perpendicular lines, one of which is normally chosen to be horizontal. These lines intersect at a point O, called the **origin** (Figure 1.3). The horizontal line is called the **x-axis,** and the vertical line is called the **y-axis.** A number scale is set up along the x-axis with the positive numbers lying to the right of the origin and the negative numbers lying to the left of it. Similarly, a number scale is set up along the y-axis with the positive numbers lying above the origin and the negative numbers lying below it.

The number scales on the two axes need not be the same. Indeed, in many applications different quantities are represented by x and y. For example, x may represent the number of typewriters sold and y the total revenue resulting from the sales. In such cases it is often desirable to choose different number scales to represent the different quantities. Note, however, that the zeros of both number scales coincide at the origin of the two-dimensional coordinate system.

A point in the plane can now be represented uniquely in this coordinate system by an **ordered pair** of numbers; that is, a pair (x, y) where x is the first number and y the second. To see this, let P be any point in the plane (Figure 1.4). Draw perpendiculars from P to the x-axis and y-axis, respectively. Then the number x is precisely the number that corresponds to the point on the x-axis at which the perpendicular through P hits the x-axis. Similarly, y is the number that corresponds to the point on the y-axis at which the perpendicular through P crosses the y-axis.

FIGURE 1.3

The Cartesian coordinate system

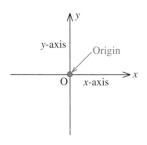

FIGURE 1.4

An ordered pair (x, y)

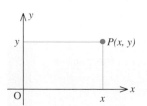

Conversely, given an ordered pair (x, y) with x as the first number and y the second, a point P in the plane is uniquely determined as follows: Locate the point on the x-axis represented by the number x and draw a line through that point parallel to the y-axis. Next, locate the point on the y-axis represented by the number y and draw a line through that point parallel to the x-axis. The point of intersection of these two lines is the point P (Figure 1.4).

In the ordered pair (x, y), x is called the **abscissa,** or **x-coordinate,** y is called the **ordinate,** or **y-coordinate,** and x and y together are referred to as the **coordinates** of the point P.

Letting (a, b) denote the point P with x-coordinate a and y-coordinate b, the points $A = (2, 3)$, $B = (-2, 3)$, $C = (-2, -3)$, $D = (2, -3)$, $E = (3, 2)$, $F = (4, 0)$, and $G = (0, -5)$ are plotted in Figure 1.5. The fact that, in general, $(x, y) \neq (y, x)$ is clearly illustrated by points A and E.

FIGURE 1.5

Several points in the Cartesian plane

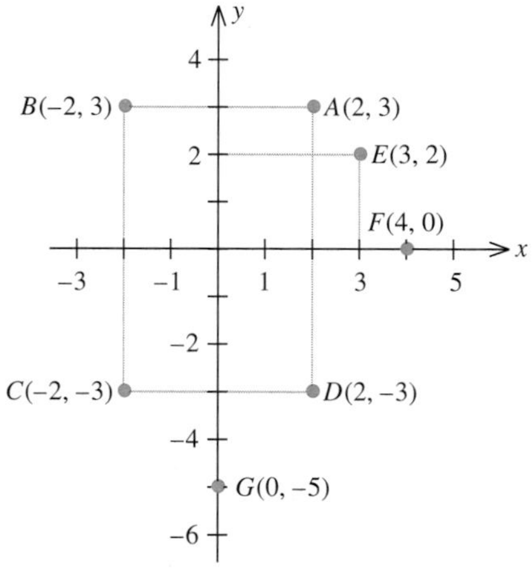

FIGURE 1.6

The four quadrants in the Cartesian plane

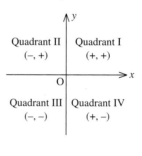

The axes divide the plane into four quadrants. Quadrant I consists of the points (x, y) that satisfy $x > 0$ and $y > 0$; Quadrant II, the points (x, y) where $x < 0$ and $y > 0$; Quadrant III, the points (x, y) where $x < 0$ and $y < 0$; and Quadrant IV, the points (x, y) where $x > 0$ and $y < 0$ (Figure 1.6).

The Distance Formula

One immediate benefit that arises from using the Cartesian coordinate system is that the distance between any two points in the plane may be expressed solely in terms of their coordinates. Suppose, for example, that (x_1, y_1) and (x_2, y_2) are any two points in the plane (Figure 1.7). Then the distance between these two points can be computed using the following formula.

FIGURE 1.7

The distance d between the points (x_1, y_1) and (x_2, y_2)

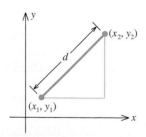

DISTANCE FORMULA	The distance d between two points (x_1, y_1) and (x_2, y_2) in the plane is given by

$$d = \sqrt{(x_2 - x_1)^2 + (y_2 - y_1)^2}. \tag{1}$$

For a proof of this result, see Exercise 35, page 30.

EXAMPLE 1   Find the distance between the points $(-4, 3)$ and $(2, 6)$.

Solution   We have, by the distance formula (1)

$$\begin{aligned} d &= \sqrt{[2 - (-4)]^2 + (6 - 3)^2} \\ &= \sqrt{6^2 + 3^2} \\ &= \sqrt{45} = 3\sqrt{5}. \end{aligned}$$ ∎

Application

EXAMPLE 2   In the following diagram (Figure 1.8) S represents the position of a power relay station located on a straight coastal highway, and M shows the location of a marine biology experimental station on an island. A cable is to be laid connecting the relay station with the experimental station. If the cost of running the cable on land is $2 per running foot and the cost of running the cable under water is $6 per running foot, find the total cost for laying the cable.

FIGURE 1.8

Cable connecting relay station S to experimental station M

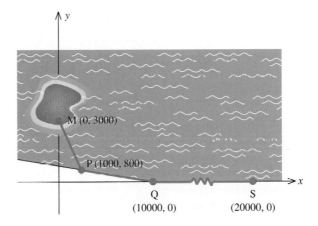

Solution   The length of cable required on land is given by the distance from P to Q plus the distance from Q to S. The distance is

$$\begin{aligned} &\sqrt{(10{,}000 - 1{,}000)^2 + (0 - 800)^2} + \sqrt{(20{,}000 - 10{,}000)^2 + (0 - 0)^2} \\ &= \sqrt{9{,}000^2 + 800^2} + 10{,}000 \\ &= \sqrt{81{,}640{,}000} + 10{,}000 \\ &\approx 19{,}035.49, \end{aligned}$$

or approximately 19,035.49 feet. Next, we see that the length of cable required under water is given by the distance from M to P. This distance is

$$\begin{aligned} \sqrt{(0 - 1{,}000)^2 + (3{,}000 - 800)^2} &= \sqrt{1{,}000^2 + 2{,}200^2} \\ &= \sqrt{5{,}840{,}000} \\ &\approx 2{,}416.61, \end{aligned}$$

or approximately 2,416.61 feet. Therefore, the total cost for laying the cable is

$$2(19{,}035.49) + 6(2{,}416.61) = 52{,}570.64,$$

or approximately $52,571.   ∎

**SELF-CHECK
EXERCISES 1.3**

1. a. Plot the points $A(4, -2)$, $B(2, 3)$, and $C(-3, 1)$.
   b. Find the distance between the points A and B; between B and C; between A and C.
   c. Use the Pythagorean Theorem to show that the triangle with vertices A, B, and C is a right triangle.

c 2. The following figure shows the location of cities A, B, and C. Suppose that a pilot wishes to fly from city A to city C but must make a mandatory stopover in city B. If the single-engine light plane has a range of 650 miles, can she make the trip without refueling in city B?

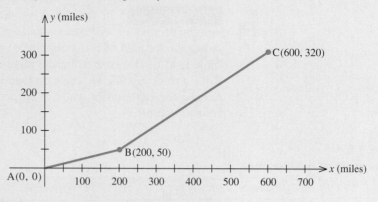

Solutions to Self-Check Exercises 1.3 can be found on page 30.

1.3   EXERCISES

In Exercises 1–6, refer to the following figure and determine the coordinates of the given point and the quadrant in which it is located.

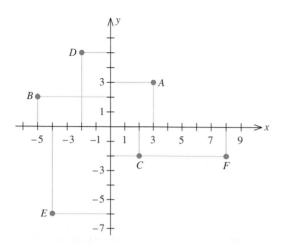

1. A	2. B	3. C
4. D	5. E	6. F

In Exercises 7–12, refer to the following figure.

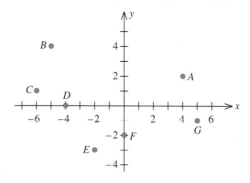

7. Which point has coordinates $(4, 2)$?

8. What are the coordinates of point B?

9. Which points have negative y-coordinates?

10. Which point has a negative x-coordinate and a negative y-coordinate?

11. Which point has an x-coordinate that is equal to zero?

12. Which point has a y-coordinate that is equal to zero?

In Exercises 13–20, sketch a set of coordinate axes and plot the given point.

13. $(-2, 5)$ **14.** $(1, 3)$

15. $(3, -1)$ **16.** $(3, -4)$

17. $(8, -7/2)$ **18.** $(-5/2, 3/2)$

19. $(4.5, -4.5)$ **20.** $(1.2, -3.4)$

In Exercises 21–24, find the distance between the given points.

21. $(1, 3)$ and $(4, 7)$ **22.** $(1, 0)$ and $(4, 4)$

23. $(-1, 3)$ and $(4, 9)$ **24.** $(-2, 1)$ and $(10, 6)$

25. Find the coordinates of the points that are 10 units away from the origin and have a y-coordinate equal to -6.

26. Find the coordinates of the points that are 5 units away from the origin and have an x-coordinate equal to 3.

27. Show that the points $(3, 4), (-3, 7), (-6, 1),$ and $(0, -2)$ form the vertices of a square.

28. Show that the triangle with vertices $(-5, 2), (-2, 5),$ and $(5, -2)$ is a right triangle.

C 29. Distance Traveled   A grand tour of four cities begins at city A and makes successive stops at cities B, C, and D before returning to city A. If the cities are located as shown in the following figure, find the total distance covered on the tour.

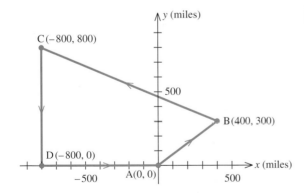

C 30. Delivery Charges   A furniture store offers free set-up and delivery services to all points within a 25-mile radius of its warehouse distribution center. If you live 20 miles east and 14 miles south of the warehouse, will you incur a delivery charge? Justify your answer.

C 31. Travel Time   Towns A, B, C, and D are located as shown in the following figure. Two highways link town A to town D. Route 1 runs from town A to town D via town B, and route 2 runs from town A to town D via

town C. If a saleman wishes to drive from town A to town D and traffic conditions are such that he could expect to average the same speed on either route, which highway should he take in order to arrive in the shortest time?

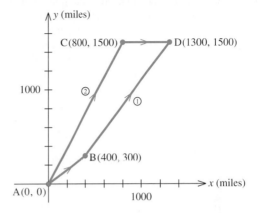

Figure for Exercises 31 and 32

C 32. Minimizing Shipping Costs   Refer to the figure for Exercise 31. Suppose that a fleet of 100 automobiles are to be shipped from an assembly plant in town A to town D. They may be shipped either by freight train along route 1 at a cost of 11 cents per mile per automobile or by truck along route 2 at a cost of $10\frac{1}{2}$ cents per mile per automobile. Which means of transportation minimizes the shipping cost? What is the net savings?

C 33. Consumer Decisions   Mr. Barclay wishes to determine which antenna he should purchase for his home. The TV store has supplied him with the following information:

Range in Miles			
VHF	UHF	Model	Price
30	20	A	$40.00
45	35	B	$50.00
60	40	C	$60.00
75	55	D	$70.00

Barclay wishes to receive channel 17 (VHF), which is located 25 miles east and 35 miles north of his home, and channel 38 (UHF), which is located 20 miles south and 32 miles west of his home. Which model will allow him to receive both channels at the least cost? (Assume that the terrain between Barclay's home and both broadcasting stations is flat.)

C 34. **Calculating the Cost of Laying Cable**   In the following diagram, S represents the position of a power relay station located on a coastal highway, and M shows the location of a marine biology experimental station on an island. A cable is to be laid connecting the relay station with the experimental station. If the cost of running the cable on land is $2 per running foot and the cost of running cable under water is $6 per running foot, find an expression in terms of x that gives the total cost for laying the cable. Use this expression to find the total cost when $x = 900$. When $x = 1000$.

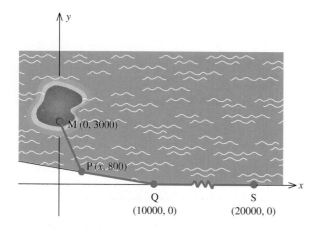

35. Let (x_1, y_1) and (x_2, y_2) be two points lying in the xy-plane. Show that the distance between the two points is given by

$$d = \sqrt{(x_2 - x_1)^2 + (y_2 - y_1)^2}.$$

[*Hint:*   Refer to the following figure and use the Pythagorean Theorem.]

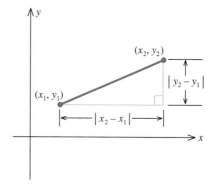

**SOLUTIONS TO
SELF-CHECK
EXERCISES 1.3**

1. **a.** The points are plotted in the following figure:

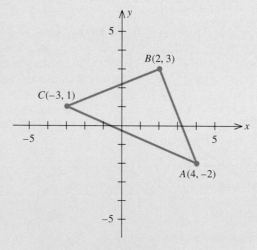

b. The distance between A and B is

$$d(A, B) = \sqrt{(2 - 4)^2 + [3 - (-2)]^2}$$
$$= \sqrt{(-2)^2 + 5^2} = \sqrt{4 + 25} = \sqrt{29}.$$

The distance between B and C is

$$d(B, C) = \sqrt{(-3 - 2)^2 + (1 - 3)^2}$$
$$= \sqrt{(-5)^2 + (-2)^2} = \sqrt{25 + 4} = \sqrt{29}.$$

The distance between A and C is

$$d(A, C) = \sqrt{(-3 - 4)^2 + [1 - (-2)]^2}$$
$$= \sqrt{(-7)^2 + 3^2} = \sqrt{49 + 9} = \sqrt{58}.$$

  c. We will show that

$$[d(A, C)]^2 = [d(A, B)]^2 + [d(B, C)]^2.$$

From (b), we see that $[d(A, B)]^2 = 29$, $[d(B, C)]^2 = 29$, and $[d(A, C)]^2 = 58$, and the desired result follows.

2. The distance between city A and city B is

$$d(A, B) = \sqrt{200^2 + 50^2} \approx 206,$$

or 206 miles. The distance between city B and city C is

$$d(B, C) = \sqrt{[600 - 200]^2 + [320 - 50]^2}$$
$$= \sqrt{400^2 + 270^2} \approx 483,$$

or 483 miles. Therefore, the total distance the pilot would have to cover is 689 miles, so she must refuel in city B.

1.4   Straight Lines

In computing income tax, business firms are allowed by law to depreciate certain assets such as buildings, machines, furniture, automobiles, and so on, over a period of time. Linear depreciation, or the straight-line method, is often used for this purpose. The graph of the straight line shown in Figure 1.9 describes the book value V of a computer that has an initial value of $100,000 and that is being depreciated linearly over five years with a scrap value of $30,000. Note that only the solid portion of the straight line is of interest here.

    The book value of the computer at the end of year t, where t lies between 0 and 5, can be read directly from the graph. But there is one shortcoming in this approach: The result depends on how accurately you draw and read the graph. A better and more accurate method is based on finding an *algebraic* representation of the depreciation line.

    In order to see how a straight line in the xy-plane may be described algebraically, we need to first recall certain properties of straight lines.

FIGURE 1.9
Linear depreciation of an asset

FIGURE 1.10
m is undefined.

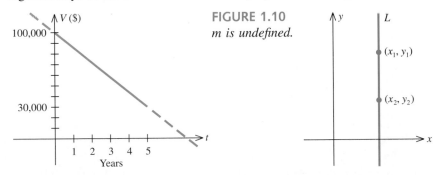

Slope of a Line

Let L denote the unique straight line that passes through the two distinct points (x_1, y_1) and (x_2, y_2). If $x_1 = x_2$, then L is a vertical line and the slope is undefined (Figure 1.10). If $x_1 \neq x_2$, we define the slope of L as follows:

SLOPE OF A NONVERTICAL LINE

If (x_1, y_1) and (x_2, y_2) are any two distinct points on a nonvertical line L, then the slope m of L is given by

$$m = \frac{\Delta y}{\Delta x} = \frac{y_2 - y_1}{x_2 - x_1}. \tag{2}$$

FIGURE 1.11

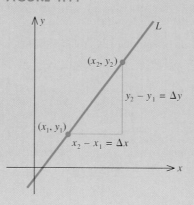

Observe that the slope of a straight line is a constant whenever it is defined. The number $\Delta y = y_2 - y_1$ (Δy is read "delta y") is a measure of the vertical change in y, and $\Delta x = x_2 - x_1$ is a measure of the horizontal change in x as shown in Figure 1.11. From this figure we can see that the slope m of a straight line L is a measure of the *rate of change of y with respect to x.*

FIGURE 1.12

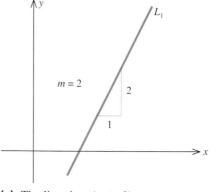

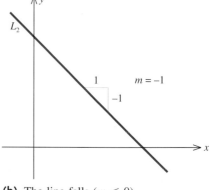

FIGURE 1.13
A family of straight lines

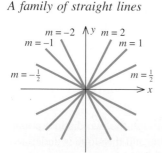

(a) The line rises ($m > 0$)   **(b)** The line falls ($m < 0$)

Figure 1.12a shows a straight line L_1 with slope 2. Observe that L_1 has the property that a unit increase in x results in a 2-unit increase in y. To see this, let $\Delta x = 1$ in (2) so that $m = \Delta y$. Since $m = 2$, we conclude that $\Delta y = 2$. Similarly, Figure 1.12b shows a line L_2 with slope -1. Observe that a straight line with positive slope slants upward from left to right (y increases as x increases), whereas a line with negative slope slants downward from left to right (y decreases as x increases). Finally, Figure 1.13 shows a family of straight lines passing through the origin with indicated slopes.

EXAMPLE 1   Sketch the graph of the straight line that passes through the point $(-2, 5)$ and has slope $-4/3$.

Solution   First plot the point $(-2, 5)$ [Figure 1.14]. Next, recall that a slope of $-4/3$ indicates that an increase of 1 unit in the x-direction produces a *decrease* of $4/3$ units in the y-direction, or equivalently, a 3-unit increase in the x-direction produces a $3(4/3)$, or 4-unit, decrease in the y-direction. Using this information, we plot the point $(1, 1)$ and draw the line through the two points.

FIGURE 1.14

L has slope $-4/3$ and passes through $(-2, 5)$.

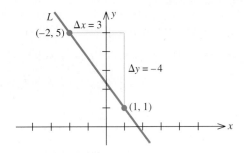

EXAMPLE 2   Find the slope m of the line that passes through the points $(-1, 1)$ and $(5, 3)$.

Solution   Choose (x_1, y_1) to be the point $(-1, 1)$ and (x_2, y_2) to be the point $(5, 3)$. Then, with $x_1 = -1$, $y_1 = 1$, $x_2 = 5$, and $y_2 = 3$, we find

$$m = \frac{y_2 - y_1}{x_2 - x_1} = \frac{3 - 1}{5 - (-1)} = \frac{1}{3} \qquad [\text{Using (2)}]$$

FIGURE 1.15

L passes through $(5, 3)$ and $(-1, 1)$

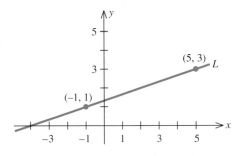

FIGURE 1.16

The slope of the horizontal line L is 0.

(Figure 1.15). Try to verify that the result obtained would have been the same had we chosen the point $(-1, 1)$ to be (x_2, y_2) and the point $(5, 3)$ to be (x_1, y_1).

EXAMPLE 3   Find the slope of the line that passes through the points $(-2, 5)$ and $(3, 5)$.

Solution   The slope of the required line is given by

$$m = \frac{5 - 5}{3 - (-2)} = \frac{0}{5} = 0$$

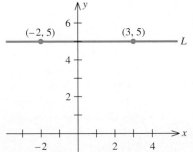

(Figure 1.16).

REMARK   In general, the slope of a horizontal line is zero.

We can use the slope of a straight line to determine whether a line is parallel to another line.

PARALLEL LINES	Two distinct lines are **parallel** if and only if their slopes are equal or their slopes are undefined.

FIGURE 1.17

L_1 and L_2 have the same slope and hence are parallel.

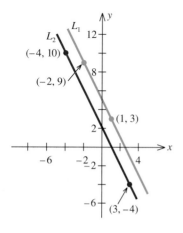

EXAMPLE 4   Let L_1 be a line that passes through the points $(-2, 9)$ and $(1, 3)$, and let L_2 be the line that passes through the points $(-4, 10)$ and $(3, -4)$. Determine whether L_1 and L_2 are parallel.

Solution   The slope m_1 of L_1 is given by

$$m_1 = \frac{3 - 9}{1 - (-2)} = -2.$$

The slope m_2 of L_2 is given by

$$m_2 = \frac{-4 - 10}{3 - (-4)} = -2.$$

Since $m_1 = m_2$, the lines L_1 and L_2 are in fact parallel (Figure 1.17). ■

Equations of Lines

We will now show that every straight line lying in the xy-plane may be represented by an equation involving the variables x and y. One immediate benefit of this is that problems involving straight lines may be solved algebraically.

Let L be a straight line parallel to the y-axis (perpendicular to the x-axis) [Figure 1.18]. Then L crosses the x-axis at some point $(a, 0)$ with the x-coordinate given by $x = a$, where a is some real number. Any other point on L has the form $(a, \bar{y})$, where $\bar{y}$ is an appropriate number. Therefore, the vertical line L is described by the sole condition

$$x = a,$$

and this is accordingly the equation of L. For example, the equation $x = -2$ represents a vertical line 2 units to the left of the y-axis, and the equation $x = 3$ represents a vertical line 3 units to the right of the y-axis (Figure 1.19).

FIGURE 1.18

The vertical line $x = a$

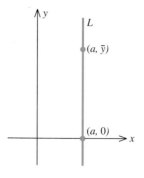

FIGURE 1.19

The vertical lines $x = -2$ and $x = 3$

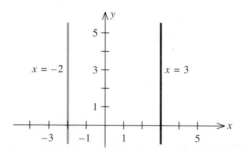

Next, suppose that L is a nonvertical line so that it has a well-defined slope m. Suppose (x_1, y_1) is a fixed point lying on L and (x, y) is a variable point on L distinct from (x_1, y_1) [Figure 1.20].

FIGURE 1.20
L passes through (x_1, y_1) and has slope m.

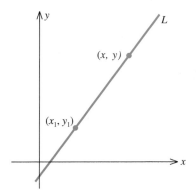

Using (2) with the point $(x_2, y_2) = (x, y)$, we find that the slope of L is given by

$$m = \frac{y - y_1}{x - x_1}.$$

Upon multiplying both sides of the equation by $x - x_1$, we obtain (3).

POINT-SLOPE FORM   An equation of the line that has slope m and passes through the point (x_1, y_1) is given by

$$y - y_1 = m(x - x_1). \qquad (3)$$

Equation (3) is called the **point-slope form of the equation of a line,** since it utilizes a given point (x_1, y_1) on a line and the slope m of the line.

FIGURE 1.21
L passes through (1, 3) and has slope 2.

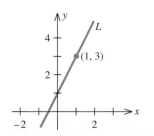

EXAMPLE 5   Find an equation of the line that passes through the point (1, 3) and has slope 2.

Solution   Using the point-slope form of the equation of a line with the point (1, 3) and $m = 2$, we obtain

$$y - 3 = 2(x - 1), \qquad [(y - y_1) = m(x - x_1)]$$

which, when simplified, becomes

$$2x - y + 1 = 0$$

(Figure 1.21).

EXAMPLE 6 Find an equation of the line that passes through the points $(-3, 2)$ and $(4, -1)$.

Solution The slope of the line is given by

$$m = \frac{-1 - 2}{4 - (-3)} = -\frac{3}{7}.$$

Using the point-slope form of the equation of a line with the point $(4, -1)$ and the slope $m = -3/7$, we have

$$y + 1 = -\frac{3}{7}(x - 4) \qquad [(y - y_1) = m(x - x_1)]$$

$$7y + 7 = -3x + 12$$

or $\qquad 3x + 7y - 5 = 0$

(Figure 1.22).

FIGURE 1.22
L passes through $(-3, 2)$ and $(4, -1)$.

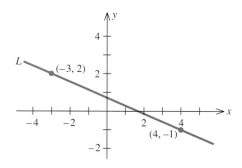

We can use the slope of a straight line to determine whether a line is perpendicular to another line.

PERPENDICULAR LINES

If L_1 and L_2 are two distinct nonvertical lines that have slopes m_1 and m_2, respectively, then L_1 is **perpendicular** to L_2 (written $L_1 \perp L_2$) if and only if

$$m_1 = -\frac{1}{m_2}.$$

If the line L_1 is vertical (so that its slope is undefined), then L_1 is perpendicular to another line, L_2, if and only if L_2 is horizontal (so that its slope is zero). For a proof of these results, see Exercise 69, page 43.

EXAMPLE 7 Find an equation of the line that passes through the point $(3, 1)$ and is perpendicular to the line of Example 5.

Solution Since the slope of the line in Example 5 is 2, the slope of the required line is given by $m = -1/2$, the negative reciprocal of 2. Using the point-slope form of the equation of a line, we obtain

$$y - 1 = -\frac{1}{2}(x - 3) \qquad [(y - y_1) = m(x - x_1)]$$

$$2y - 2 = -x + 3$$

or $\qquad x + 2y - 5 = 0$

(Figure 1.23).

FIGURE 1.23

L_2 is perpendicular to L_1 and passes through (3, 1).

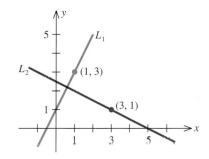

FIGURE 1.24

The line L has x-intercept a and y-intercept b.

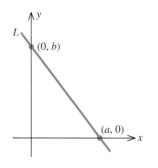

A straight line L that is neither horizontal nor vertical cuts the x-axis and the y-axis at, say, points $(a, 0)$ and $(0, b)$, respectively (Figure 1.24). The numbers a and b are called the **x-intercept** and **y-intercept,** respectively, of L.

Now, let L be a line with slope m and y-intercept b. Using (3), the point-slope form of the equation of a line, with the point $(0, b)$ and slope m, we have

$$y - b = m(x - 0)$$

or $\qquad y = mx + b.$

SLOPE-INTERCEPT FORM

The equation of the line that has slope m and intersects the y-axis at the point $(0, b)$ is given by

$$y = mx + b. \tag{4}$$

EXAMPLE 8   Find an equation of the line that has slope 3 and y-intercept -4.

Solution   Using (4) with $m = 3$ and $b = -4$, we obtain the required equation

$$y = 3x - 4.$$

EXAMPLE 9   Determine the slope and y-intercept of the line whose equation is $3x - 4y = 8$.

Solution   Rewrite the given equation in the slope-intercept form and obtain

$$y = \frac{3}{4}x - 2.$$

Comparing this result with (4), we find $m = 3/4$ and $b = -2$, and we conclude that the slope and y-intercept of the given line are $3/4$ and -2, respectively. ∎

Applications

EXAMPLE 10   The sales manager of a local sporting goods store plotted sales versus time for the last five years and found the points to lie approximately along a straight line (Figure 1.25). By using the points corresponding to the first and fifth years, find an equation of the trend line. What sales figure can be predicted for the sixth year?

FIGURE 1.25

Sales of a sporting goods store

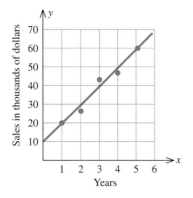

Solution   Using (2) with the points $(1, 20)$ and $(5, 60)$, we find that the slope of the required line is given by

$$m = \frac{60 - 20}{5 - 1} = 10.$$

Next, using the point-slope form of the equation of a line with the point $(1, 20)$ and $m = 10$, we obtain

$$y - 20 = 10(x - 1) \qquad [(y - y_1) = m(x - x_1)]$$

or

$$y = 10x + 10,$$

as the required equation.

The sales figure for the sixth year is obtained by letting $x = 6$ in the last equation, giving

$$y = 70,$$

or $70,000. ∎

EXAMPLE 11   Suppose that an art object purchased for $50,000 is expected to appreciate in value at a constant rate of $5,000 per year for the next five years. Use (4) to write an equation predicting the value of the art object in the next several years. What will its value be three years from the date of purchase?

Solution   Let x denote the time (in years) that has elapsed since the date the object was purchased and let y denote the object's value (in dollars). Then $y = 50,000$ when $x = 0$. Furthermore, the slope of the required equation is given by $m = 5,000$, since each unit increase in x (one year) implies an increase of 5,000 units (dollars) in y. Using (4) with $m = 5,000$ and $b = 50,000$, we obtain

$$y = 5,000x + 50,000. \qquad (y = mx + b)$$

Three years from the date of purchase, the value of the object will be given by

$$y = 5,000(3) + 50,000,$$

or $65,000.   ■

General Equation of a Line

We have considered several forms of the equation of a straight line in the plane. These different forms of the equation are equivalent to each other. In fact, each is a special case of the following equation.

GENERAL FORM OF A LINEAR EQUATION	The equation
	$$Ax + By + C = 0, \tag{5}$$
	where A, B, and C are constants and A and B are not both zero, is called the general form of a linear equation in the variables x and y.

We will now state (without proof) an important result concerning the algebraic representation of straight lines in the plane.

THEOREM 1	An equation of a straight line is a linear equation; conversely, every linear equation represents a straight line.

FIGURE 1.26
To sketch $3x - 4y - 12 = 0$, first find the x-intercept, 4, and the y-intercept, −3.

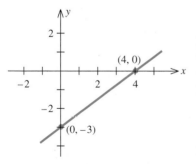

This result justifies the use of the adjective *linear* in describing Equation (5).

EXAMPLE 12   Sketch the straight line represented by the equation

$$3x - 4y - 12 = 0.$$

Solution   Since every straight line is uniquely determined by two distinct points, we need find only two such points through which the line passes in order to sketch it. For convenience let us compute the x- and y-intercepts. Setting $y = 0$, we find $x = 4$; thus the x-intercept is 4. Setting $x = 0$ gives $y = -3$, and the y-intercept is -3. A sketch of the line appears in Figure 1.26.

■

Following is a summary of the common forms of the equations of straight lines discussed in this section.

EQUATIONS OF STRAIGHT LINES	Vertical line:	$x = a$
	Horizontal line:	$y = b$
	Point-slope form:	$y - y_1 = m(x - x_1)$
	Slope-intercept form:	$y = mx + b$
	General form:	$Ax + By + C = 0$

**SELF-CHECK
EXERCISES 1.4**

1. Determine the number a so that the line passing through the points $(a, 2)$ and $(3, 6)$ is parallel to a line with slope 4.

2. Find an equation of the line that passes through the point $(3, -1)$ and is perpendicular to a line with slope $-\frac{1}{2}$.

3. Does the point $(3, -3)$ lie on the line with equation $2x - 3y - 12 = 0$? Sketch the graph of the line.

4. The percentage of people over 65 who have high school diplomas is summarized in the following table:

Year (x)	1960	1965	1970	1975	1980	1985	1990
Percent with diplomas (y)	20	25	30	36	42	47	52

   a. Plot the percentage of people over 65 who have high school diplomas (y) versus the year (x).
   b. Draw the straight line L through the points (1960, 20) and (1990, 52).
   c. Find an equation of the line L.
   d. If the trend continues, estimate the percentage of people over 65 who will have high school diplomas by the year 2000.

Solutions to Self-Check Exercises 1.4 can be found on page 43.

1.4   EXERCISES

In Exercises 1–4, find the slope of the line shown in each figure.

1.

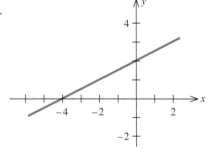

3.

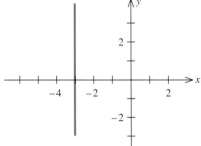

2.

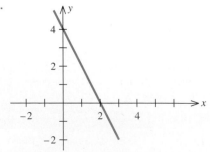

4.

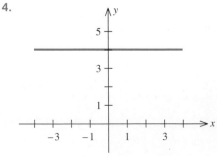

In Exercises 5–10, find the slope of the line that passes through the given pair of points.

5. (4, 3) and (5, 8)   6. (4, 5) and (3, 8)

7. (−2, 3) and (4, 8)   8. (−2, −2) and (4, −4)

9. (a, b) and (c, d)

10. (−a + 1, b − 1) and (a + 1, −b)

11. Given the equation $y = 4x - 3$, answer the following questions.
   a. If x increases by 1 unit, what is the corresponding change in y?
   b. If x decreases by 2 units, what is the corresponding change in y?

12. Given the equation $2x + 3y = 4$, answer the following questions.
   a. Is the slope of the line described by this equation positive or negative?
   b. As x increases in value, does y increase or decrease?
   c. If x decreases by 2 units, what is the corresponding change in y?

In Exercises 13 and 14, determine whether the lines through the given pairs of points are parallel.

13. $A(1, -2)$, $B(-3, -10)$ and $C(1, 5)$, $D(-1, 1)$

14. $A(2, 3)$, $B(2, -2)$ and $C(-2, 4)$, $D(-2, 5)$

In Exercises 15 and 16, determine whether the lines through the given pairs of points are perpendicular.

15. $A(-2, 5)$, $B(4, 2)$ and $C(-1, -2)$, $D(3, 6)$

16. $A(2, 0)$, $B(1, -2)$ and $C(4, 2)$, $D(-8, 4)$

17. If the line passing through the points (1, a) and (4, −2) is parallel to the line passing through the points (2, 8) and (−7, a + 4), what is the value of a?

18. If the line passing through the points (a, 1) and (5, 8) is parallel to the line passing through the points (4, 9) and (a + 2, 1), what is the value of a?

19. Find an equation of the horizontal line that passes through (−4, −3).

20. Find an equation of the vertical line that passes through (0, 5).

In Exercises 21–24, find an equation of the line that passes through the given point and has the indicated slope m.

21. (3, −4); $m = 2$ 22. (2, 4); $m = -1$

23. (−3, 2); $m = 0$ 24. (1, 2); $m = -1/2$

In Exercises 25–28, find an equation of the line that passes through the given points.

25. (2, 4) and (3, 7)   26. (2, 1) and (2, 5)

27. (1, 2) and (−3, −2)   28. (−1, −2) and (3, −4)

In Exercises 29–32, find an equation of the line that has slope m and y-intercept b.

29. $m = 3$; $b = 4$ 30. $m = -2$; $b = -1$

31. $m = 0$; $b = 5$ 32. $m = -1/2$; $b = 3/4$

In Exercises 33–38, write the given equation in the slope-intercept form and then find the slope and y-intercept of the corresponding line.

33. $x - 2y = 0$ 34. $y - 2 = 0$

35. $2x - 3y - 9 = 0$ 36. $3x - 4y + 8 = 0$

37. $2x + 4y = 14$ 38. $5x + 8y - 24 = 0$

39. Find an equation of the line that passes through the point (−2, 2) and is parallel to the line $2x - 4y - 8 = 0$.

40. Find an equation of the line that passes through the point (2, 4) and is perpendicular to the line $3x + 4y - 22 = 0$.

In Exercises 41–46, find an equation of the line that satisfies the given condition.

41. The line parallel to the x-axis and 6 units below it.

42. The line passing through the origin and parallel to the line joining the points (2, 4) and (4, 7).

43. The line passing through the point (a, b) with slope equal to zero.

44. The line passing through (−3, 4) and parallel to the x-axis.

45. The line passing through (−5, −4) and parallel to the line joining (−3, 2) and (6, 8).

46. The line passing through (a, b) with undefined slope.

47. Given that the point $P(-3, 5)$ lies on the line $kx + 3y + 9 = 0$, find k.

48. Given that the point $P(2, -3)$ lies on the line $-2x + ky + 10 = 0$, find k.

In Exercises 49–54, sketch the straight line defined by the given linear equation by finding the x- and y-intercepts. [Hint: See Example 12, page 39.]

49. $3x - 2y + 6 = 0$ 50. $2x - 5y + 10 = 0$

51. $x + 2y - 4 = 0$ 52. $2x + 3y - 15 = 0$

42   CHAPTER 1   Preliminaries

53. $y + 5 = 0$ **54.** $-2x - 8y + 24 = 0$

55. Show that an equation of a line through the points $(a, 0)$ and $(0, b)$ with $a \neq 0$ and $b \neq 0$ can be written in the form

$$\frac{x}{a} + \frac{y}{b} = 1.$$

(Recall that the numbers a and b are the x- and y-intercepts, respectively, of the line. This form of an equation of a line is called the **intercept form**.)

In Exercises 56–59, use the results of Exercise 55 to find an equation of a line with the given x- *and* y-*intercepts.*

56. x-intercept 3; y-intercept 4

57. x-intercept -2; y-intercept -4

58. x-intercept $-\frac{1}{2}$; y-intercept $\frac{3}{4}$

59. x-intercept 4; y-intercept $-\frac{1}{2}$

In Exercises 60 and 61, determine whether the given points lie on a straight line.

60. $A(-1, 7)$, $B(2, -2)$, and $C(5, -9)$

61. $A(-2, 1)$, $B(1, 7)$, and $C(4, 13)$

62. Social Security Contributions   For wages less than the maximum taxable wage base, social security contributions by employees are 7.65 percent of the employee's wages.
 a. Find an equation that expresses the relationship between the wages earned (x) and the social security taxes paid (y) by an employee who earns less than the maximum taxable wage base.
 b. For each additional dollar that an employee earns, by how much is his or her social security contribution increased? (Assume that the employee's wages are less than the maximum taxable wage base.)
 c. What social security contributions will an employee who earns $35,000 (which is less than the maximum taxable wage base) be required to make?

63. College Admissions   Using data compiled by the Admissions Office at Faber University, college admissions officers estimate that 55 percent of the students who are offered admission to the freshman class at the university will actually enroll.
 a. Find an equation that expresses the relationship between the number of students who actually enroll (y) and the number of students who are offered admission to the university (x).
 b. If the desired freshman class size for the upcoming academic year is 1100 students, how many students should be admitted?

64. Weight of Whales   The equation $W = 3.51L - 192$, expressing the relationship between the length L (in feet) and the expected weight W (in British tons) of adult blue whales, was adopted in the late 1960s by the International Whaling Commission.
 a. What is the expected weight of an 80-foot blue whale?
 b. Sketch the straight line that represents the equation.

65. Cost of a Commodity   A manufacturer obtained the following data relating the cost y (in dollars) to the number of units (x) of a commodity produced:

Number of units produced (x)	0	20	40	60	80	100
Cost in dollars (y)	200	208	222	230	242	250

 a. Plot the cost (y) versus the quantity produced (x).
 b. Draw the straight line through the points (0, 200) and (100, 250).
 c. Derive an equation of the straight line of (b).
 d. Taking this equation to be an approximation of the relationship between the cost and the level of production, estimate the cost of producing 54 units of the commodity.

66. Ideal Heights and Weights for Women   The Venus Health Club for Women provides its members with the following table, which gives the average desirable weight for women of a certain height:

Height x (in inches)	60	63	66	69	72
Weight y (in pounds)	108	118	129	140	152

 a. Plot the weight (y) versus the height (x).
 b. Draw a straight line L through the points corresponding to heights of 5 feet and 6 feet.
 c. Derive an equation of the line L.
 d. Using the equation of (c), estimate the average desirable weight for a woman who is 5 feet 5 inches tall.

67. Sales Growth   Metro Department Store's annual sales (in millions of dollars) during the past five years were:

Annual sales (x)	5.8	6.2	7.2	8.4	9.0
Year (y)	1	2	3	4	5

a. Plot the annual sales (y) versus the year (x).
b. Draw a straight line L through the points corresponding to the first and fifth years.
c. Derive an equation of the line L.
d. Using the equation found in (c), estimate Metro's annual sales four years from now ($x = 9$).

68. Show that two distinct lines with equations $a_1 x + b_1 y + c_1 = 0$ and $a_2 x + b_2 y + c_2 = 0$, respectively, are parallel if and only if $a_1 b_2 - b_1 a_2 = 0$. [*Hint:* Write each equation in the slope-intercept form and compare.]

69. Prove that if a line L_1 with slope m_1 is perpendicular to a line L_2 with slope m_2, then $m_1 m_2 = -1$. [*Hint:* Refer to the following figure. Show that $m_1 = b$ and $m_2 = c$. Next, apply the Pythagorean Theorem to triangles OAC, OCB, and OBA to show that $1 = -bc$.]

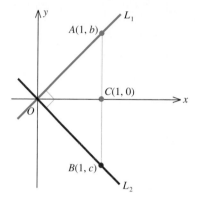

SOLUTIONS TO SELF-CHECK EXERCISES 1.4

1. The slope of the line that passes through the points $(a, 2)$ and $(3, 6)$ is

$$m = \frac{6 - 2}{3 - a}$$

$$= \frac{4}{3 - a}.$$

Since this line is parallel to a line with slope 4, m must be equal to 4; that is,

$$\frac{4}{3 - a} = 4,$$

or, upon multiplying both sides of the equation by $3 - a$,

$$4 = 4(3 - a)$$
$$4 = 12 - 4a$$
$$4a = 8$$

and $a = 2$.

2. Since the required line L is perpendicular to a line with slope $-\frac{1}{2}$, the slope of L is

$$-\frac{1}{-\frac{1}{2}} = 2.$$

Next, using the point-slope form of the equation of a line, we have

$$y - (-1) = 2(x - 3)$$
$$y + 1 = 2x - 6$$

or $y = 2x - 7$.

3. Substituting $x = 3$ and $y = -3$ into the left-hand side of the given equation, we find

$$2(3) - 3(-3) - 12 = 3,$$

which is not equal to zero (the right-hand side). Therefore, $(3, -3)$ does not lie on the line with equation $2x - 3y - 12 = 0$.

Setting $x = 0$, we find $y = -4$, the y-intercept. Next, setting $y = 0$ gives $x = 6$, the x-intercept. We now draw the line passing through the points $(0, -4)$ and $(6, 0)$ as shown.

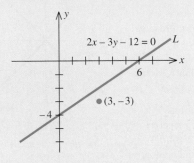

4. **a.** and **b.** See the accompanying figure.

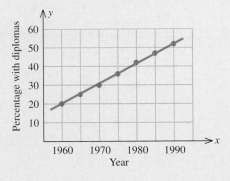

c. The slope of L is

$$m = \frac{52 - 20}{1990 - 1960} = \frac{32}{30} = \frac{16}{15}.$$

Using the point-slope form of the equation of a line with the point $(1960, 20)$, we find

$$y - 20 = \frac{16}{15}(x - 1960) = \frac{16}{15}x - \frac{(16)(1960)}{15}$$

or
$$y = \frac{16}{15}x - \frac{6272}{3} + 20$$

$$= \frac{16}{15}x - \frac{6212}{3}.$$

d. To estimate the percentage of people over 65 who will have high school diplomas by the year 2000, let $x = 2000$ in the equation obtained in (c). Thus, the required estimate is

$$y = \frac{16}{15}(2000) - \frac{6212}{3} \approx 62.67,$$

or approximately 63 percent.

CHAPTER ONE
SUMMARY OF
PRINCIPAL
FORMULAS AND
TERMS

Formulas

1. Quadratic formula $x = \dfrac{-b \pm \sqrt{b^2 - 4ac}}{2a}$

2. Distance between two points $d = \sqrt{(x_2 - x_1)^2 + (y_2 - y_1)^2}$

3. Slope of a line $m = \dfrac{y_2 - y_1}{x_2 - x_1}$

4. Equation of a vertical line $x = a$

5. Equation of a horizontal line $y = b$

6. Point-slope form of the equation of a line $y - y_1 = m(x - x_1)$

7. Slope-intercept form of the equation of a line $y = mx + b$

8. General equation of a line $Ax + By + C = 0$

Terms

Real number (coordinate) line
Open interval
Closed interval
Half-open interval
Finite interval
Infinite interval
Absolute value

Polynomial
Roots of a polynomial equation
Cartesian coordinate system
Ordered pair
Parallel lines
Perpendicular lines

CHAPTER 1 REVIEW EXERCISES

In Exercises 1–4, find the values of x *that satisfy each of the given inequalities.*

1. $-x + 3 \le 2x + 9$

2. $-2 \le 3x + 1 \le 7$

3. $x - 3 > 2$ or $x + 3 < -1$

4. $2x^2 > 50$

In Exercises 5–8, evaluate the given expression.

5. $|-5 + 7| + |-2|$

6. $\left|\dfrac{5 - 12}{-4 - 3}\right|$

7. $|2\pi - 6| - \pi$

8. $|\sqrt{3} - 4| + |4 - 2\sqrt{3}|$

In Exercises 9–14, evaluate the given expression.

9. $\left(\dfrac{9}{4}\right)^{3/2}$

10. $\dfrac{5^6}{5^4}$

11. $(3 \cdot 4)^{-2}$

12. $(-8)^{5/3}$

13. $\dfrac{(3 \cdot 2^{-3})(4 \cdot 3^5)}{2 \cdot 9^3}$

14. $\dfrac{3\sqrt[3]{54}}{\sqrt[3]{18}}$

In Exercises 15–19, simplify the given expression.

15. $\dfrac{4(x^2 + y)^3}{x^2 + y}$

16. $\dfrac{a^6 b^{-5}}{(a^3 b^{-2})^{-3}}$

17. $\dfrac{\sqrt[4]{16x^5 yz}}{\sqrt[4]{81 xyz^5}}$

18. $(2x^3)(-3x^{-2})\left(\dfrac{1}{6}x^{-1/2}\right)$

19. $\left(\dfrac{3xy^2}{4x^3 y}\right)^{-2}\left(\dfrac{3xy^3}{2x^2}\right)^3$

In Exercises 20–23, factor the given expression.

20. $-2\pi^2 r^3 + 100\pi r^2$

21. $2v^3 w + 2vw^3 + 2u^2 vw$

22. $16 - x^2$

23. $12t^3 - 6t^2 - 18t$

In Exercises 24–27, solve the given equation by factoring.

24. $8x^2 + 2x - 3 = 0$ 25. $-6x^2 - 10x + 4 = 0$

26. $-x^3 - 2x^2 + 3x = 0$ 27. $2x^4 + x^2 = 1$

In Exercises 28 and 29, use the quadratic formula to solve the given quadratic equation.

28. $x^2 - 2x - 5 = 0$ 29. $2x^2 + 8x + 7 = 0$

In Exercises 30–33, perform the indicated operations and simplify the given expression.

30. $\dfrac{(t + 6)(60) - (60t + 180)}{(t + 6)^2}$

31. $\dfrac{6x}{2(3x^2 + 2)} + \dfrac{1}{4(x + 2)}$

32. $\dfrac{2}{3}\left(\dfrac{4x}{2x^2 - 1}\right) + 3\left(\dfrac{3}{3x - 1}\right)$

33. $\dfrac{-2x}{\sqrt{x + 1}} + 4\sqrt{x + 1}$

34. Rationalize the numerator.

$$\frac{\sqrt{x} - 1}{x - 1}$$

35. Rationalize the denominator.

$$\frac{\sqrt{x} - 1}{2\sqrt{x}}$$

In Exercises 36 and 37, find the distance between the two given points.

36. $(-2, -3)$ and $(1, -7)$

37. $\left(\dfrac{1}{2}, \sqrt{3}\right)$ and $\left(-\dfrac{1}{2}, 2\sqrt{3}\right)$

In Exercises 38–43, find an equation of the line L that passes through the point $(-2, 4)$ and satisfies the given condition.

38. L is a vertical line.

39. L is a horizontal line.

40. L passes through the point $\left(3, \dfrac{7}{2}\right)$.

41. The x-intercept of L is 3.

42. L is parallel to the line $5x - 2y = 6$.

43. L is perpendicular to the line $4x + 3y = 6$.

44. Find an equation of the straight line that passes through the point $(2, 3)$ and is parallel to the line with equation $3x + 4y - 8 = 0$.

45. Find an equation of the straight line that passes through the point $(-1, 3)$ and is parallel to the line passing through the points $(-3, 4)$ and $(2, 1)$.

46. Find an equation of the line that passes through the point $(-2, -4)$ and is perpendicular to the line with equation $2x - 3y - 24 = 0$.

47. Sketch the graph of the equation $3x - 4y = 24$.

48. Sketch the graph of the line that passes through the point $(3, 2)$ and has slope $-2/3$.

49. Find the minimum cost C (in dollars) given that

$$2(1.5C + 80) \le 2(2.5C - 20).$$

50. Find the maximum revenue R (in dollars) given that

$$12(2R - 320) \le 4(3R + 240).$$

How does the change in the demand for a certain make of tires affect the unit price of the tires? The management of the Titan Tire Company has determined the demand function that relates the unit price of its Super Titan tires to the quantity demanded. In Example 7, page 117, you will see how this function can be used to compute the rate of change of the unit price of the Super Titan tires with respect to the quantity demanded.

Functions, Limits, and the Derivative

I n this chapter we define a *function*, a special relationship between two variables. The concept of a function enables us to describe many relationships that exist in applications. We will also begin the study of differential calculus. Historically, differential calculus was developed in response to the problem of finding the tangent line to an arbitrary curve. But it quickly became apparent that solving this problem provided mathematicians with a method of solving many practical problems involving the rate of change of one quantity with respect to another. The basic tool used in differential calculus is the *derivative* of a function. The concept of the derivative is based, in turn, on a more fundamental notion—that of the *limit* of a function.

Functions and Their Graphs

Functions

A manufacturer would like to know how his company's profit is related to its production level; a biologist would like to know how the size of the population of a certain culture of bacteria will change over time; a psychologist would like to know the relationship between the learning time of an individual and the length of a vocabulary list; and a chemist would like to know how the initial speed of a chemical reaction is related to the amount of substrate used. In each instance we are concerned with the same question: How does one quantity depend upon another? The relationship between two quantities is conveniently described in mathematics by using the concept of a function.

FUNCTION

A **function** is a rule that assigns to each element in a set A one and only one element in a set B.

The set A is called the **domain** of the function. It is customary to denote a function by a letter of the alphabet, such as the letter f. If x is an element in the domain of a function f, then the element in B that f associates with x is written $f(x)$ (read "f of x") and is called the **value of f at x.**

We can think of a function f as a machine. The domain is the set of inputs (raw material) for the machine, the rule describes how the input is to be processed, and the value(s) of the function are the outputs of the machine (Figure 2.1).

We can also think of a function f as a mapping in which an element x in the domain of f is mapped onto a unique element $f(x)$ in B (Figure 2.2).

FIGURE 2.1
A function machine

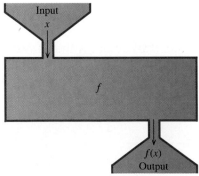

FIGURE 2.2
The function f viewed as a mapping

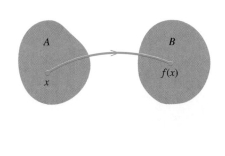

REMARKS

1. It is important to understand that the output $f(x)$ associated with an input x is unique. To appreciate the importance of this uniqueness property, consider a rule that associates with each item x in a department store its selling price y. Then, each x must correspond to *one and only one y*. Notice, however, that different x's may be associated with the same y. In the context of the present example, this says that different items may have the same price.

2. Although the sets A and B that appear in the definition of a function may be quite arbitrary, in this book they will denote sets of real numbers.   □

An example of a function may be taken from the familiar relationship between the area of a circle and its radius. Letting x and y denote the radius and area of a circle, respectively, we have, from elementary geometry,

$$y = \pi x^2. \tag{1}$$

Equation (1) defines y as a function of x, since for each admissible value of x (that is, for each nonnegative number representing the radius of a certain circle) there corresponds precisely one number $y = \pi x^2$ that gives the area of the circle. The rule defining this "area function" may be written as

$$f(x) = \pi x^2. \tag{2}$$

To compute the area of a circle of radius 5 inches, we simply replace x in (2) with the number 5. Thus, the area of the circle is

$$f(5) = \pi 5^2 = 25\pi,$$

or 25π square inches.

In general, to evaluate a function at a specific value of x, we replace x with that value, as illustrated in Examples 1 and 2.

EXAMPLE 1   Let the function f be defined by the rule $f(x) = 2x^2$ $x + 1$. Compute

a. $f(1)$;   **b.** $f(-2)$;   **c.** $f(a)$;   and   **d.** $f(a + h)$.

Solution
a. $f(1) = 2(1)^2 - (1) + 1 = 2 - 1 + 1 = 2$
b. $f(-2) = 2(-2)^2 - (-2) + 1 = 8 + 2 + 1 = 11$
c. $f(a) = 2(a)^2 - (a) + 1 = 2a^2 - a + 1$
d. $f(a + h) = 2(a + h)^2 - (a + h) + 1 = 2a^2 + 4ah + 2h^2 - a - h + 1$

■

EXAMPLE 2   The Thermo-Master Company manufactures an indoor-outdoor thermometer at its Mexico subsidiary. Management estimates that the profit (in dollars) realizable by Thermo-Master in the manufacture and sale of x thermometers per week is

$$P(x) = -0.001x^2 + 8x - 5000.$$

Find Thermo-Master's weekly profit if its level of production is (a) 1000 thermometers per week and (b) 2000 thermometers per week.

Solution
a. The weekly profit realizable by Thermo-Master when the level of production is 1000 units per week is found by evaluating the profit function P at $x = 1000$. Thus,

$$P(1000) = -0.001(1000)^2 + 8(1000) - 5000 = 2000,$$

or $2000.

b. When the level of production is 2000 units per week, the weekly profit is given by

$$P(2000) = -0.001(2000)^2 + 8(2000) - 5000 - 7000,$$

or $7000. ∎

Determining the Domain of a Function

Suppose we are given the function $y = f(x)$.† Then the variable x is called the **independent variable**. The variable y, whose value depends on x, is called the **dependent variable**.

In determining the domain of a function, we need to find what restrictions, if any, are to be placed on the independent variable x. In many practical applications, the domain of a function is dictated by the nature of the problem, as illustrated in Example 3.

EXAMPLE 3   An open box is to be made from a rectangular piece of cardboard 16 inches long and 10 inches wide by cutting away identical squares (x by x inches) from each corner and folding up the resulting flaps (Figure 2.3). Find an expression that gives the volume V of the box as a function of x. What is the domain of the function?

FIGURE 2.3

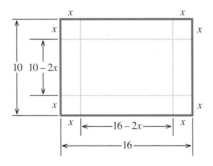

(a) The box is constructed by cutting x-by-x-inch squares from each corner.

(b) The dimensions of the resulting box are $(10 - 2x)$ by $(16 - 2x)$ by x inches.

Solution   The dimensions of the box are $(16 - 2x)$ inches long, $(10 - 2x)$ inches wide, and x inches high, so its volume (in cubic inches) is given by

$$V = f(x) = (16 - 2x)(10 - 2x)x \qquad \text{(Length · width · height)}$$
$$= (160 - 52x + 4x^2)x$$
$$= 4x^3 - 52x^2 + 160x.$$

Since the length of each side of the box must be greater than or equal to zero, we see that

$$16 - 2x \geq 0, \qquad 10 - 2x \geq 0, \quad \text{and} \quad x \geq 0$$

simultaneously; that is,

$$x \leq 8, \qquad x \leq 5, \quad \text{and} \quad x \geq 0.$$

---

†It is customary to refer to a function f as $f(x)$ or by the equation $y = f(x)$ defining it.

All three inequalities are satisfied simultaneously provided that $0 \leq x \leq 5$. Thus, the domain of the function f is the interval $[0, 5]$.   ■

In general, if a function is defined by a rule relating x to $f(x)$ without specific mention of its domain, it is understood that the domain will consist of all values of x for which $f(x)$ is a real number. In this connection, you should keep in mind that (1) division by zero is not permitted and (2) the square root of a negative number is not defined.

EXAMPLE 4   Find the domain of each of the functions defined by the following equations:

a. $f(x) = \sqrt{x - 1}$ **b.** $f(x) = \dfrac{1}{x^2 - 4}$ **c.** $f(x) = x^2 + 3$

Solution
a. Since the square root of a negative number is undefined, it is necessary that $x - 1 \geq 0$. The inequality is satisfied by the set of real numbers $x \geq 1$. Thus, the domain of f is the interval $[1, \infty)$.
b. The only restriction on x is that $x^2 - 4$ be different from zero, since division by zero is not allowed. But, $(x^2 - 4) = (x + 2)(x - 2) = 0$ if $x = -2$ or $x = 2$. Thus, the domain of f in this case consists of the intervals $(-\infty, -2)$, $(-2, 2)$, and $(2, \infty)$.
c. Here, any real number satisfies the equation, so the domain of f is the set of all real numbers.   ■

Graphs of Functions

If f is a function with domain A, then corresponding to each real number x in A there is precisely one real number $f(x)$. We can also express this fact by using **ordered pairs** of real numbers. Write each number x in A as the first member of an ordered pair and each number $f(x)$ corresponding to x as the second member of the ordered pair. This gives exactly one ordered pair $(x, f(x))$ for each x in A. This observation leads to an alternative definition of a function f:

A function f with domain A is the set of all
ordered pairs $(x, f(x))$ where x belongs to A.

Observe that the condition that there be one and only one number $f(x)$ corresponding to each number x in A translates into the requirement that *no two ordered pairs have the same first number*.
Since ordered pairs of real numbers correspond to points in the plane, we have found a way to exhibit a function graphically.

GRAPH OF A
FUNCTION OF
ONE VARIABLE

The **graph of a function** f is the set of all points (x, y) in the xy-plane such that x is in the domain of f and $y = f(x)$.

Much information about the graph of a function can be gained by plotting a few points on its graph. Later on we will develop more systematic and sophisticated techniques for graphing functions.

EXAMPLE 5   Sketch the graph of the function defined by the equation $y = x^2 + 1$.

FIGURE 2.4

The graph of $y = x^2 + 1$ is a parabola.

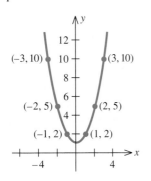

Solution   The domain of the function is the set of all real numbers. By assigning several values to the variable x and computing the corresponding values for y, we obtain the following solutions to the equation $y = x^2 + 1$:

x	−3	−2	−1	0	1	2	3
y	10	5	2	1	2	5	10

By plotting these points and then connecting them with a smooth curve, we obtain the graph of $y = f(x)$, which is a parabola (Figure 2.4).   ∎

Some functions are defined in a piecewise fashion, as Examples 6 and 7 show.

EXAMPLE 6   The Madison Finance Company plans to open two branch offices two years from now in two separate locations: an industrial complex and a newly developed commercial center in the city. As a result of these expansion plans, Madison's total deposits during the next five years are expected to grow in accordance with the rule

$$f(x) = \begin{cases} \sqrt{2x} + 20 & 0 \le x \le 2 \\ \frac{1}{2}x^2 + 20 & 2 < x \le 5, \end{cases}$$

where $y = f(x)$ gives the total amount of money (in millions of dollars) on deposit with Madison in year x ($x = 0$ corresponds to the present). Sketch the graph of the function f.

Solution   The function f is defined in a piecewise fashion on the interval $[0, 5]$. In the subdomain $[0, 2]$, the rule for f is given by $f(x) = \sqrt{2x} + 20$. The values of $f(x)$ corresponding to $x = 0$, 1, and 2 may be tabulated as follows:

x	0	1	2
f(x)	20	21.4	22

Next, in the subdomain $(2, 5]$, the rule for f is given by $f(x) = \frac{1}{2}x^2 + 20$. The values of $f(x)$ corresponding to $x = 3$, 4, and 5 are shown in the following table.

FIGURE 2.5

We obtain the graph of the function $y = f(x)$ by graphing $y = \sqrt{2x} + 20$ over $[0, 2]$ and $y = \frac{1}{2}x^2 + 20$ over $(2, 5]$.

x	3	4	5
f(x)	24.5	28	32.5

Using the values of $f(x)$ in this table, we sketch the graph of the function f as shown in Figure 2.5. ■

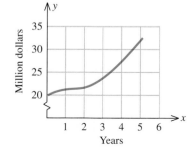

Years

EXAMPLE 7   Sketch the graph of the function f defined by

$$f(x) = \begin{cases} -x & \text{if } x < 0 \\ \sqrt{x} & \text{if } x \geq 0. \end{cases}$$

Solution   The function f is defined in a piecewise fashion on the set of all real numbers. In the subdomain $(-\infty, 0)$, the rule for f is given by $f(x) = -x$. The equation $y = -x$ is a linear equation in the slope-intercept form (with slope -1 and intercept 0). Therefore the graph of f corresponding to the subdomain $(-\infty, 0)$ is the half-line shown in Figure 2.6. Next, in the subdomain $[0, \infty)$, the rule for f is given by $f(x) = \sqrt{x}$. The values of $f(x)$ corresponding to $x = 0, 1, 2, 3, 4, 9,$ and 16 are shown in the following table.

x	0	1	2	3	4	9	16
f(x)	0	1	$\sqrt{2}$	$\sqrt{3}$	2	3	4

Using these values, we sketch the graph of the function f as shown in Figure 2.6.

FIGURE 2.6

The graph of $y = f(x)$ is obtained by graphing $y = -x$ over $(-\infty, 0)$ and $y = \sqrt{x}$ over $[0, \infty)$.

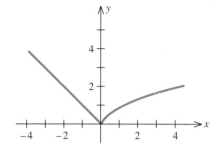

■

The Vertical-Line Test

Although it is true that every function f of a variable x has a graph in the xy-plane, it is important to realize that not every curve in the xy-plane is the graph of a function. For example, consider the curve depicted in Figure 2.7. This is the graph of the equation $x = y^2$. In general, the *graph of an equation* is the set of all ordered pairs (x, y) that satisfy the given equation. Observe that the points $(9, -3)$ and $(9, 3)$ both lie on the curve. This implies that the number $x = 9$ is associated with *two* numbers: $y = -3$ and $y = 3$. But this clearly

FIGURE 2.7

Since a vertical line passes through the curve at more than one point, we deduce that it is not the graph of a function.

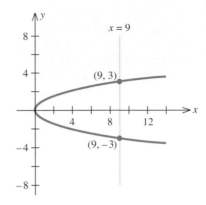

violates the uniqueness property of a function. Thus we conclude that the curve under consideration cannot be the graph of a function.

This example suggests the following test for determining when a curve is the graph of a function.

VERTICAL-LINE TEST   A curve in the xy-plane is the graph of a function $y = f(x)$ if and only if each vertical line intersects it in at most one point.

EXAMPLE 8   Determine which of the curves shown in Figure 2.8 are the graphs of functions of x.

FIGURE 2.8

The vertical-line test can be used to determine which of these curves are graphs of functions.

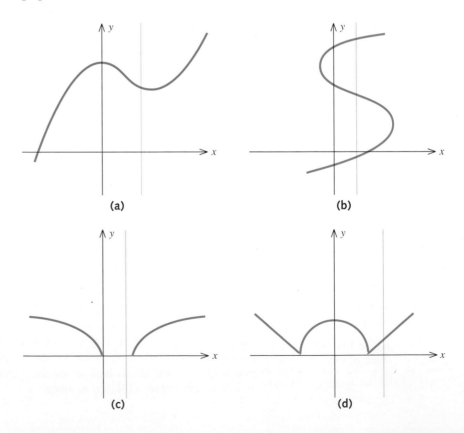

Solution   The curves depicted in Figures 2.8a, c, and d are graphs of functions, because each curve satisfies the requirement that each vertical line intersects the curve in at most one point. Note that the vertical line shown in Figure 2.8c does not intersect the graph, because the point on the *x*-axis through which this line passes does not lie in the domain of the function. The curve depicted in Figure 2.8b is *not* the graph of a function, because the vertical line shown there intersects the graph at three points.   ■

**SELF-CHECK
EXERCISES 2.1**

1. Let *f* be the function defined by

$$f(x) = \frac{\sqrt{x + 1}}{x}.$$

   a. Find the domain of *f*.
   b. Compute $f(3)$.
   c. Compute $f(a + h)$.

c 2. Statistics obtained by Amoco Corporation show that more and more motorists are pumping their own gas. The following function gives self-serve sales as a percentage of all U.S. gas sales:

$$f(t) = \begin{cases} 6t + 17 & 0 \le t \le 6 \\ 15.98(t - 6)^{1/4} + 53 & 6 < t \le 20. \end{cases}$$

   Here *t* is measured in years, with $t = 0$ corresponding to the beginning of 1974.
   a. Sketch the graph of the function *f*.
   b. What percentage of all gas sales at the beginning of 1978 were self-serve? At the beginning of 1994?

3. Let $f(x) = \sqrt{2x + 1} + 2$. Determine whether the point (4, 6) lies on the graph of *f*.

Solutions to Self-Check Exercises 2.1 can be found on page 60.

2.1   EXERCISES

1. Let *f* be the function defined by $f(x) = 5x + 6$. Find $f(3), f(-3), f(a), f(-a)$, and $f(a + 3)$.

2. Let *f* be the function defined by $f(x) = 4x - 3$. Find $f(4), f(\frac{1}{4}), f(0), f(a)$, and $f(a + 1)$.

3. Let *g* be the function defined by $g(x) = 3x^2 - 6x - 3$. Find $g(0), g(-1), g(a), g(-a)$, and $g(x + 1)$.

4. Let *h* be the function defined by $h(x) = x^3 - x^2 + x + 1$. Find $h(-5), h(0), h(a)$, and $h(-a)$.

5. Let *s* be the function defined by $s(t) = 2t/(t^2 - 1)$. Find $s(4), s(0), s(a), s(2 + a)$, and $s(t + 1)$.

6. Let *g* be the function defined by $g(u) = (3u - 2)^{3/2}$. Find $g(1), g(6), g(\frac{11}{3})$, and $g(u + 1)$.

7. Let *f* be the function defined by $f(t) = 2t^2/\sqrt{t - 1}$. Find $f(2), f(a), f(x + 1)$, and $f(x - 1)$.

8. Let *f* be the function defined by $f(x) = 2 + 2\sqrt{5 - x}$. Find $f(-4), f(1), f(\frac{11}{4})$, and $f(x + 5)$.

9. Let *f* be the function defined by

$$f(x) = \begin{cases} x^2 + 1 & \text{if } x \le 0 \\ \sqrt{x} & \text{if } x > 0. \end{cases}$$

   Find $f(-2), f(0)$, and $f(1)$.

10. Let g be the function defined by

$$g(x) = \begin{cases} -\dfrac{1}{2}x + 1 & \text{if } x < 2 \\ \sqrt{x - 2} & \text{if } x \geq 2. \end{cases}$$

Find $g(-2)$, $g(0)$, $g(2)$, and $g(4)$.

11. Let f be the function defined by

$$f(x) = \begin{cases} -\dfrac{1}{2}x^2 + 3 & \text{if } x < 1 \\ 2x^2 + 1 & \text{if } x \geq 1. \end{cases}$$

Find $f(-1)$, $f(0)$, $f(1)$, and $f(2)$.

12. Let f be the function defined by

$$f(x) = \begin{cases} 2 + \sqrt{1 - x} & \text{if } x \leq 1 \\ \dfrac{1}{1 - x} & \text{if } x > 1. \end{cases}$$

Find $f(0)$, $f(1)$, and $f(2)$.

In Exercises 13–16, determine whether the given point lies on the graph of the function.

13. $(2, \sqrt{3})$; $g(x) = \sqrt{x^2 - 1}$

14. $(3, 3)$; $f(x) = \dfrac{x + 1}{\sqrt{x^2 + 7}} + 2$

15. $(-2, -3)$; $f(t) = \dfrac{|t - 1|}{t + 1}$

16. $\left(-3, -\dfrac{1}{13}\right)$; $h(t) = \dfrac{|t + 1|}{t^3 + 1}$

In Exercises 17–30, find the domain of the given function.

17. $f(x) = x^2 + 3$

18. $f(x) = 7 - x^2$

19. $f(x) = \dfrac{3x + 1}{x^2}$

20. $g(x) = \dfrac{2x + 1}{x - 1}$

21. $f(x) = \sqrt{x^2 + 1}$

22. $f(x) = \sqrt{x - 5}$

23. $f(x) = \sqrt{5 - x}$

24. $g(x) = \sqrt{2x^2 + 3}$

25. $f(x) = \dfrac{x}{x^2 - 1}$

26. $f(x) = \dfrac{1}{x^2 + x - 2}$

27. $f(x) = (x + 3)^{3/2}$

28. $g(x) = 2(x - 1)^{5/2}$

29. $f(x) = \dfrac{\sqrt{1 - x}}{x^2 - 4}$

30. $f(x) = \dfrac{\sqrt{x - 1}}{(x + 2)(x - 3)}$

31. Let f be a function defined by the rule $f(x) = x^2 - x - 6$.
  a. Find the domain of f.
  b. Compute $f(x)$ for $x = -3, -2, -1, 0, \frac{1}{2}, 1, 2, 3$.
  c. Use the results obtained in (a) and (b) to sketch the graph of f.

32. Let f be a function defined by the rule $f(x) = 2x^2 + x - 3$.
  a. Find the domain of f.
  b. Compute $f(x)$ for $x = -3, -2, -1, -\frac{1}{2}, 0, 1, 2, 3$.
  c. Use the results obtained in (a) and (b) to sketch the graph of f.

In Exercises 33–44, sketch the graph of the function f with the given rule.

33. $f(x) = 2x^2 + 1$

34. $f(x) = 9 - x^2$

35. $f(x) = 2 + \sqrt{x}$

36. $g(x) = 4 - \sqrt{x}$

37. $f(x) = \sqrt{1 - x}$

38. $f(x) = \sqrt{x - 1}$

39. $f(x) = |x| - 1$

40. $f(x) = |x| + 1$

41. $f(x) = \begin{cases} x & \text{if } x < 0 \\ 2x + 1 & \text{if } x \geq 0 \end{cases}$

42. $f(x) = \begin{cases} 4 - x & \text{if } x < 2 \\ 2x - 2 & \text{if } x \geq 2 \end{cases}$

43. $f(x) = \begin{cases} -x + 1 & \text{if } x \leq 1 \\ x^2 - 1 & \text{if } x > 1 \end{cases}$

44. $f(x) = \begin{cases} -x - 1 & \text{if } x < -1 \\ 0 & \text{if } -1 \leq x \leq 1 \\ x + 1 & \text{if } x > 1 \end{cases}$

In Exercises 45–52, use the vertical-line test to determine whether the given graph represents y as a function of x.

45.

46.

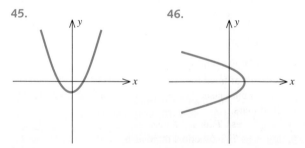

47.

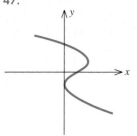

48.

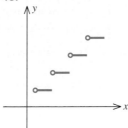

49.

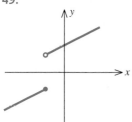

50.

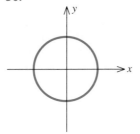

51.

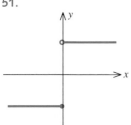

52.

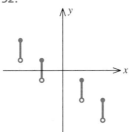

53. The circumference of a circle is given by $C(r) = 2\pi r$, where r is the radius of the circle. What is the circumference of a circle with a 5-inch radius?

54. The volume of a sphere of radius r is given by $V(r) = \frac{4}{3}\pi r^3$. Compute $V(2.1)$ and $V(2)$. What does the quantity $V(2.1) - V(2)$ measure?

55. **Consumption Function** The consumption function in a certain economy is given by the equation

$$C(y) = 0.75y + 6,$$

where $C(y)$ is the personal consumption expenditure, y is the disposable personal income, and both $C(y)$ and y are measured in billions of dollars. Find $C(0)$, $C(50)$, and $C(100)$.

56. **Sales Taxes** In a certain state, the sales tax T on the amount of taxable goods is 6 percent of the value of the goods purchased (x), where both T and x are measured in dollars.
   a. Express T as a function of x.
   b. Find $T(200)$ and $T(5.65)$.

57. **Surface Area of a Single-Celled Organism** The surface area S of a single-celled organism may be found by multiplying 4π times the square of the radius r of the cell. Express S as a function of r.

58. **Friend's Rule** Friend's rule, a method for calculating pediatric drug dosages, is based on a child's age. If a denotes the adult dosage (in mg) and if t is the age of the child (in years), then the child's dosage is given by

$$D(t) = \frac{2}{25}ta.$$

If the adult dose of a substance is 500 mg, how much should a four-year-old child receive?

59. **COLAs** Social security recipients receive an automatic cost-of-living adjustment (COLA) once each year. Their monthly benefit is increased by the amount that consumer prices increased during the preceding year. Suppose that consumer prices increased by 5.3 percent during the preceding year.
   a. Express the adjusted monthly benefit of a social security recipient as a function of his or her current monthly benefit.
   b. If Mr. Harrington's monthly social security benefit is now $620, what will his adjusted monthly benefit be?

60. **Cost of Renting a Truck** The Ace Truck Leasing Company leases a certain size truck at $30 a day and 15 cents a mile, whereas the Acme Truck Leasing Company leases the same size truck at $25 a day and 20 cents a mile.
   a. Find the daily cost of leasing from each company as a function of the number of miles driven.
   b. Sketch the graphs of the two functions on the same set of axes.
   c. Which company should a customer rent a truck from for one day if she plans to drive at most 70 miles and wishes to minimize her cost?

61. **Linear Depreciation** A new machine was purchased by the National Textile Company for $120,000. For income tax purposes, the machine is depreciated linearly over ten years; that is, the book value of the machine decreases at a constant rate, so that at the end of ten years the book value is zero.
   a. Express the book value of the machine (V) as a function of the age, in years, of the machine (n).
   b. Sketch the graph of the function in (a).
   c. Find the book value of the machine at the end of the sixth year.
   d. Find the rate at which the machine is being depreciated each year.

62. **Linear Depreciation**   Refer to Exercise 61. An office building worth $1,000,000 when completed in 1980 was depreciated linearly over 50 years. What will the book value of the building be in 1999? In 2003? (Assume that the book value of the building will be zero at the end of the 50th year.)

63. **Boyle's Law**   As a consequence of Boyle's law, the pressure P of a fixed sample of gas held at a constant temperature is related to the volume V of the gas by the rule

$$P = f(V) = \frac{k}{V},$$

where k is a constant. What is the domain of the function f? Sketch the graph of the function f.

64. **Poiseuille's Law**   According to a law discovered by the nineteenth-century physician Poiseuille, the velocity (in centimeters per second) of blood r centimeters from the central axis of an artery is given by

$$v(r) = k(R^2 - r^2),$$

where k is a constant and R is the radius of the artery. Suppose that for a certain artery, $k = 1000$ and $R = 0.2$ so that $v(r) = 1000(0.04 - r^2)$.
a. What is the domain of the function $v(r)$?
b. Compute $v(0)$, $v(0.1)$, and $v(0.2)$ and interpret your results.

65. **Population Growth**   A study prepared for a certain Sunbelt town's Chamber of Commerce projected that the population of the town in the next three years will grow according to the rule

$$P(x) = 50,000 + 30x^{3/2} + 20x,$$

where $P(x)$ denotes the population x months from now. By how much will the population increase during the next 9 months? During the next 16 months?

66. **Worker Efficiency**   An efficiency study conducted for the Elektra Electronics Company showed that the number of "Space Commander" walkie-talkies assembled by the average worker t hours after starting work at 8:00 A.M. is given by

$$N(t) = -t^3 + 6t^2 + 15t \qquad (0 \le t \le 4).$$

How many walkie-talkies can an average worker be expected to assemble between 8:00 and 9:00 A.M.? Between 9:00 and 10:00 A.M.?

67. **Learning Curves**   The Emory Secretarial School finds from experience that the average student taking Advanced Typing will progress according to the rule

$$N(t) = \frac{60t + 180}{t + 6},$$

where $N(t)$ measures the number of words per minute the student can type after t weeks in the course. How fast can the average student be expected to type after two weeks in the course? After four weeks in the course?

68. **Politics**   Political scientists have discovered the following empirical rule, known as the "cube rule," which gives the relationship between the proportion of seats in the House of Representatives won by Democratic candidates $s(x)$ and the proportion of popular votes x received by the Democratic presidential candidate:

$$s(x) = \frac{x^3}{x^3 + (1 - x)^3} \qquad (0 \le x \le 1).$$

Compute $s(0.6)$ and interpret your result.

69. **Home Shopping Industry**   According to industry sources, revenue from the home shopping industry for the years since its inception may be approximated by the function

$$R(t) = \begin{cases} -0.03t^3 + 0.25t^2 - 0.12t & 0 \le t \le 3 \\ 0.57t - 0.63 & 3 < t \le 11, \end{cases}$$

where $R(t)$ measures the revenue in billions of dollars and t is measured in years, with $t = 0$ corresponding to the beginning of 1984. What was the revenue at the beginning of 1985? At the beginning of 1993?

70. **Postal Regulations**   The postage for first-class mail is 29 cents for the first ounce or fraction thereof, and 23 cents for each additional ounce or fraction thereof. Any parcel not exceeding 12 ounces may be sent by first-class mail. Letting x denote the weight of a parcel in ounces, and $f(x)$ the postage in cents, complete the following description of the "postage function" f:

$$f(x) = \begin{cases} 29 & \text{if } 0 < x \le 1 \\ 52 & \text{if } 1 < x \le 2 \\ \dots & \\ ? & \text{if } 11 < x \le 12. \end{cases}$$

a. What is the domain of f?
b. Sketch the graph of f.

SOLUTIONS TO
SELF-CHECK
EXERCISES 2.1

1. **a.** The expression under the radical sign must be nonnegative, so $x + 1 \ge 0$ or $x \ge -1$. Also, $x \ne 0$, because division by zero is not permitted. Therefore, the domain of f is $[-1, 0) \cup (0, \infty)$.

b. $f(3) = \dfrac{\sqrt{3+1}}{3} = \dfrac{\sqrt{4}}{3} = \dfrac{2}{3}$

c. $f(a+h) = \dfrac{\sqrt{(a+h)+1}}{a+h} = \dfrac{\sqrt{a+h+1}}{a+h}$

2. a. For t in the subdomain $[0, 6]$, the rule for f is given by $f(t) = 6t + 17$. The equation $y = 6t + 17$ is a linear equation, so that portion of the graph of f is the line segment joining the points $(0, 17)$ and $(6, 53)$. Next, in the subdomain $(6, 20]$, the rule for f is given by $f(t) = 15.98(t - 6)^{1/4} + 53$. Using a calculator, we construct the following table of values of $f(t)$ for selected values of t.

t	6	8	10	12	14	16	18	20
f(t)	53	72	75.6	78	79.9	81.4	82.7	83.9

We have included $t = 6$ in the table, although it does not lie in the subdomain of the function under consideration, in order to help us obtain a better sketch of that portion of the graph of f in the subdomain $(6, 20]$. The graph of f is as follows.

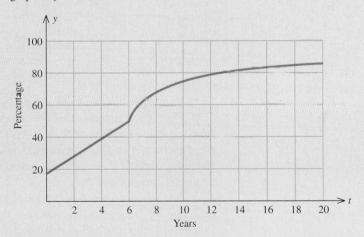

b. The percentage of all self-serve gas sales at the beginning of 1978 is found by evaluating f at $t = 4$. Since this point lies in the interval $[0, 6]$, we use the rule $f(t) = 6t + 17$ and find

$$f(4) = 6(4) + 17,$$

giving 41 percent as the required figure. The percentage of all self-serve gas sales at the beginning of 1994 is given by

$$f(20) = 15.98(20 - 6)^{1/4} + 53,$$

or approximately 83.9 percent.

3. A point (x, y) lies on the graph of the function f if and only if the coordinates satisfy the equation $y = f(x)$. Now

$$f(4) = \sqrt{2(4) + 1} + 2 = \sqrt{9} + 2 = 5 \neq 6,$$

and we conclude that the given point does not lie on the graph of f.

The Algebra of Functions

The Sum, Difference, Product, and Quotient of Functions

Let $S(t)$ and $R(t)$ denote, respectively, the federal government's spending and revenue at any time t, measured in billions of dollars. The graphs of these functions for the period between 1981 and 1991 are shown in Figure 2.9.

FIGURE 2.9
$S(t) - R(t)$ gives the federal budget deficit at any time t.

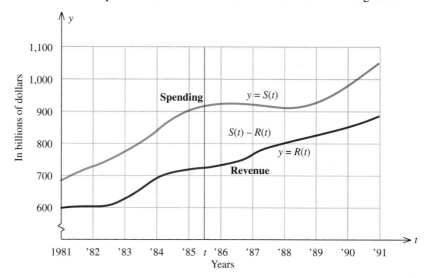

The budget deficit at any time t is given by $S(t) - R(t)$ billion dollars. This observation suggests that we can define a function D whose value at any time t is given by $D(t) = S(t) - R(t)$. The function D, the *difference* of the two functions S and R, is written $D = S - R$, and may be called the "deficit function," since it gives the budget deficit at any time t. It has the same domain as the functions S and R.

Most functions are built up from other, generally simpler, functions. For example, we may view the function $f(x) = 2x + 4$ as the sum of the two functions $g(x) = 2x$ and $h(x) = 4$. The function $g(x) = 2x$ may, in turn, be viewed as the product of the functions $p(x) = 2$ and $q(x) = x$.

In general, given the functions f and g, we define the **sum, difference, product,** and **quotient** functions as $f + g$, $f - g$, fg, and f/g, respectively, with values defined by

$$(f + g)(x) = f(x) + g(x) \qquad \text{(Sum)}$$

$$(f - g)(x) = f(x) - g(x) \qquad \text{(Difference)}$$

$$(fg)(x) = f(x)g(x) \qquad \text{(Product)}$$

$$\left(\frac{f}{g}\right)(x) = \frac{f(x)}{g(x)} \qquad (g(x) \neq 0). \qquad \text{(Quotient)}$$

Thus, the rules for the sum, difference, product, and quotient of two functions f and g are obtained by simply adding, subtracting, multiplying, and dividing the respective expressions for $f(x)$ and $g(x)$.

EXAMPLE 1   Let f and g be functions defined by the rules $f(x) = \frac{1}{2}x^2$ and $g(x) = 2x + 1$, respectively. Find the rules for the sum s, the difference d, the product p, and the quotient q of the functions f and g.

Solution   The required rules are given by

$$s(x) = (f + g)(x) = f(x) + g(x) = \frac{1}{2}x^2 + 2x + 1$$

$$d(x) = (f - g)(x) = f(x) - g(x) = \frac{1}{2}x^2 - (2x + 1) = \frac{1}{2}x^2 - 2x - 1$$

$$p(x) = (fg)(x) = f(x)g(x) = \frac{1}{2}x^2(2x + 1)$$

and $$q(x) = \left(\frac{f}{g}\right)(x) = \frac{f(x)}{g(x)} = \frac{\frac{1}{2}x^2}{2x + 1} = \frac{x^2}{2(2x + 1)} \qquad \left(x \neq -\frac{1}{2}\right). \qquad \blacksquare$$

Applications

The mathematical formulation of a problem arising from a practical situation often leads to an expression that involves the combination of functions. Consider, for example, the costs incurred in operating a business. Costs that remain more or less constant regardless of the firm's level of activity are called **fixed costs**. Examples of fixed costs are rental fees and executive salaries. On the other hand, costs that vary with production or sales are called **variable costs**. Examples of variable costs are wages and costs of raw materials. The **total cost** of operating a business is thus given by the *sum* of the variable costs and the fixed costs, as illustrated in the next example.

EXAMPLE 2   Suppose that Puritron, a manufacturer of water filters, has a monthly fixed cost of $10,000 and a variable cost of

$$-0.0001x^2 + 10x \qquad (0 \leq x \leq 40{,}000)$$

dollars, where x denotes the number of filters manufactured per month. Find a function C that gives the total cost incurred by Puritron in the manufacture of x filters.

Solution   Puritron's monthly fixed cost is always $10,000, regardless of the level of production, and it is described by the constant function $F(x) = 10{,}000$. Next, the variable cost is described by the function $V(x) = -0.0001x^2 + 10x$. Since the total cost incurred by Puritron at any level of production is the sum of the variable cost and the fixed cost, we see that the required total cost function is given by

$$C(x) = V(x) + F(x)$$
$$= -0.0001x^2 + 10x + 10{,}000 \qquad (0 \leq x \leq 40{,}000). \qquad \blacksquare$$

Next, the **total profit** realized by a firm in operating a business is the *difference* between the total revenue realized and the total cost incurred; that is, $P(x) = R(x) - C(x)$.

EXAMPLE 3   Refer to Example 2. Suppose that the total revenue realized by Puritron from the sale of x water filters is given by the total revenue function

$$R(x) = -0.0005x^2 + 20x \qquad (0 \le x \le 40,000).$$

a. Find the total profit function, that is, the function that describes the total profit Puritron realizes in manufacturing and selling x water filters per month.
b. What is the profit when the level of production is 10,000 filters per month.

Solution
a. The total profit realized by Puritron in manufacturing and selling x water filters per month is the difference between the total revenue realized and the total cost incurred. Thus, the required total profit function is given by

$$\begin{aligned} P(x) &= R(x) - C(x) \\ &= (-0.0005x^2 + 20x) - (-0.0001x^2 + 10x + 10,000) \\ &= -0.0004x^2 + 10x - 10,000. \end{aligned}$$

b. The profit realized by Puritron when the level of production is 10,000 filters per month is

$$P(10,000) = -0.0004(10,000)^2 + 10(10,000) - 10,000 = 50,000,$$

or $50,000 per month. ∎

Composition of Functions

Another way to build up a function from other functions is through a process known as the *composition of functions.* Consider, for example, the function h, whose rule is given by $h(x) = \sqrt{x^2 - 1}$. Let f and g be functions defined by the rules $f(x) = x^2 - 1$ and $g(x) = \sqrt{x}$. Evaluating the function g at the point $f(x)$ [remember that for each real number x in the domain of f, $f(x)$ is simply a real number], we find that

$$g(f(x)) = \sqrt{f(x)} = \sqrt{x^2 - 1},$$

which is just the rule defining the function h!

In general, the **composition of a function g with a function f** is defined as the function h whose value is found by evaluating the function g at the point $f(x)$; that is,

$$h(x) = g(f(x)).$$

The function h is called a **composite function** and is commonly denoted by $h = g \circ f$ (read "g circle f" or "g composed with f"). The interpretation of the function h as a machine is illustrated in Figure 2.10, and its interpretation as a mapping is shown in Figure 2.11.

EXAMPLE 4   Let $f(x) = x^2 - 1$ and $g(x) = \sqrt{x} + 1$. Compute

a. the rule for the composite function $g \circ f$, and
b. the rule for the composite function $f \circ g$.

FIGURE 2.10

The composite function
$h = g \circ f$ *is found by*
evaluating f at x and then
g *at* $f(x)$, *in that order.*

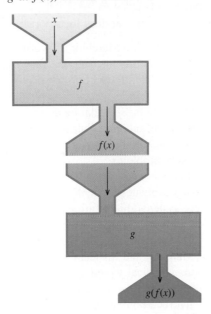

FIGURE 2.11

The function $h = g \circ f$ *viewed*
as a mapping

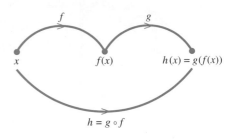

Solution

a. To find the rule for the composite function $g \circ f$, evaluate the function g at $f(x)$. Therefore,

$$(g \circ f)(x) = g(f(x)) = \sqrt{f(x)} + 1 = \sqrt{x^2 - 1} + 1.$$

b. To find the rule for the composite function $f \circ g$, evaluate the function f at $g(x)$. Thus,

$$(f \circ g)(x) = f(g(x)) = (g(x))^2 - 1 = (\sqrt{x} + 1)^2 - 1$$
$$= x + 2\sqrt{x} + 1 - 1 = x + 2\sqrt{x}. \qquad \blacksquare$$

Example 4 reminds us that, in general, $g \circ f$ is different from $f \circ g$, and care must be taken regarding the order when computing a composite function.

EXAMPLE 5   An environmental impact study conducted for Oxnard's City Environmental Management Department indicates that, under existing environmental protection laws, the level of carbon monoxide present in the air due to pollution from automobile exhaust will be $0.01x^{2/3}$ parts per million when the number of motor vehicles is x thousand. A separate study conducted by a state government agency estimates that t years from now the number of motor vehicles in Oxnard will be $0.2t^2 + 4t + 64$ thousand.

a. Find an expression for the concentration of carbon monoxide in the air due to automobile exhaust t years from now.

b. What will the level of concentration be five years from now?

Michael Marchlik

Title: Project Manager
Institution: Ebasco Services Incorporated

Ebasco is a diversified engineering and construction company. In addition to designing and building electric generating facilities, the company provides clients with safety and risk assessment studies. Their studies include analyses of potential hazards, recommendations for safe work practices, and evaluations of responses to emergencies in chemical plants, fuel-storage depots, nuclear facilities, and hazardous waste sites. As a project manager, Marchlik deals directly with clients to help them understand and implement Ebasco's recommendations to ensure a safer environment.

Calculus wasn't Mike Marchlik's favorite college subject. In fact, it was not until he started his first job that the "lights went on" and he realized how using calculus allowed him to solve real problems in his everyday work.

Marchlik emphasizes that he doesn't do "number crunching" himself. "I don't work out integrals, but in my work I use computer models that do that." The important issue is not computation but how the answers relate to client problems.

Marchlik's clients typically process highly toxic or explosive materials. Ebasco evaluates a client site, such as a chemical plant, to determine how safety systems might fail and what the probable consequences might be.

To avoid a major disaster like the one that occurred in Bhopal, India, in 1984, Marchlik and his team might be asked to determine how quickly a poisonous chemical would spread if a leak occurred. Or they might help avert a disaster like the one that rocked the Houston area in 1989. In that incident, hydrocarbon vapor exploded at a Phillips 66 chemical plant, shaking office buildings in Houston, 12 miles away. Several people were killed and hundreds were injured. Property damage totaled $1.39 billion.

In assessing risks for a fuel-storage depot, Ebasco considers variables such as weather conditions, including probable wind speed, the flow properties of a gas, and possible ignition sources. With today's more powerful computers, the models can involve a system of very complex equations to project likely scenarios.

Mathematical models vary, however. One model might forecast how much gas will flow out of a hole and how quickly it will disperse. Another model might project where the gas will go depending on local factors such as temperature and wind speed. Choosing the right model is essential. A model based on flat terrain when the client's storage depot is set among hills is going to produce the wrong answer.

Marchlik and his team run several models together to come up with their projections. Each model uses "equations that have to be integrated to come up with solutions." The bottom line? Marchlik stresses that "calculus is at the very heart" of Ebasco's risk-assessment work.

Solution

a. The level of carbon monoxide present in the air due to pollution from automobile exhaust is described by the function $g(x) = 0.01x^{2/3}$, where x is the number (in thousands) of motor vehicles. But the number of motor vehicles x (in thousands) t years from now may be estimated by the rule $f(t) = 0.2t^2 + 4t + 64$. Therefore, the concentration of carbon monoxide due to automobile exhaust t years from now is given by

$$C(t) = g(f(t)) = 0.01(0.2t^2 + 4t + 64)^{2/3}$$

parts per million.

b. The level of concentration five years from now will be

$$C(5) = 0.01[0.2(5)^2 + 4(5) + 64]^{2/3}$$
$$= (0.01)89^{2/3} \approx 0.20,$$

or 0.20 parts per million.   ∎

SELF-CHECK EXERCISES 2.2

1. Let f and g be functions defined by the rules $f(x) = \sqrt{x} + 1$ and $g(x) = \dfrac{x}{1 + x}$, respectively. Find the rules for

   a. the sum s, the difference d, the product p, and the quotient q of f and g.
   b. the composite functions $f \circ g$ and $g \circ f$.

2. Health-care spending per person by the private sector comprises payments by individuals, corporations, and their insurance companies, and is approximated by the function

   $$f(t) = 2.5t^2 + 31.3t + 406 \qquad (0 \le t \le 20),$$

   where $f(t)$ is measured in dollars and t is measured in years, with $t = 0$ corresponding to the beginning of 1975. The corresponding government spending, comprising expenditures for medicaid, medicare, and other federal, state, and local government public health care, is

   $$g(t) = 1.4t^2 + 29.6t + 251 \qquad (0 \le t \le 20),$$

   where t has the same meaning as before.

   a. Find a function that gives the difference between private and government health-care spending per person at any time t.
   b. What was the difference between private and government expenditures per person at the beginning of 1985? At the beginning of 1993?

Solutions to Self-Check Exercises 2.2 can be found on page 69.

2.2   EXERCISES

In Exercises 1–8, let $f(x) = x^3 + 5$, $g(x) = x^2 - 2$, and $h(x) = 2x + 4$. Find the rule for each function.

1. $f + g$

2. $f - g$

3. fg

4. gf

5. $\dfrac{f}{g}$

6. $\dfrac{f - g}{h}$

7. $\dfrac{fg}{h}$

8. fgh

In Exercises 9–18, let $f(x) = x - 1$, $g(x) = \sqrt{x + 1}$, and $h(x) = 2x^3 - 1$. Find the rule for each function.

9. $f + g$ 10. $g - f$ 11. fg

12. gf 13. $\dfrac{g}{h}$ 14. $\dfrac{h}{g}$

15. $\dfrac{fg}{h}$ 16. $\dfrac{fh}{g}$ 17. $\dfrac{f - h}{g}$

18. $\dfrac{gh}{g - f}$

In Exercises 19–24, find the rules for $f + g$, $f - g$, fg, and $\dfrac{f}{g}$.

19. $f(x) = x^2 + 5$; $g(x) = \sqrt{x} - 2$

20. $f(x) = \sqrt{x - 1}$; $g(x) = x^3 + 1$

21. $f(x) = \sqrt{x + 3}$; $g(x) = \dfrac{1}{x - 1}$

22. $f(x) = \dfrac{1}{x^2 + 1}$; $g(x) = \dfrac{1}{x^2 - 1}$

23. $f(x) = \dfrac{x + 1}{x - 1}$; $g(x) = \dfrac{x + 2}{x - 2}$

24. $f(x) = x^2 + 1$; $g(x) = \sqrt{x + 1}$

In Exercises 25–30, find the rules for the composite functions $f \circ g$ and $g \circ f$.

25. $f(x) = x^2 + x + 1$; $g(x) = x^2$

26. $f(x) = 3x^2 + 2x + 1$; $g(x) = x + 3$

27. $f(x) = \sqrt{x} + 1$; $g(x) = x^2 - 1$

28. $f(x) = 2\sqrt{x} + 3$; $g(x) = x^2 + 1$

29. $f(x) = \dfrac{x}{x^2 + 1}$; $g(x) = \dfrac{1}{x}$

30. $f(x) = \sqrt{x + 1}$; $g(x) = \dfrac{1}{x - 1}$

In Exercises 31–34, evaluate $h(2)$ where $h = g \circ f$.

31. $f(x) = x^2 + x + 1$; $g(x) = x^2$

32. $f(x) = \sqrt[3]{x^2 - 1}$; $g(x) = 3x^3 + 1$

33. $f(x) = \dfrac{1}{2x + 1}$; $g(x) = \sqrt{x}$

34. $f(x) = \dfrac{1}{x - 1}$; $g(x) = x^2 + 1$

In Exercises 35–42, find functions f and g such that $h = g \circ f$. (Note: The answer is not unique.)

35. $h(x) = (2x^3 + x^2 + 1)^5$

36. $h(x) = (3x^2 - 4)^{-3}$

37. $h(x) = \sqrt{x^2 - 1}$ 38. $h(x) = (2x - 3)^{3/2}$

39. $h(x) = \dfrac{1}{x^2 - 1}$ 40. $h(x) = \dfrac{1}{\sqrt{x^2 - 4}}$

41. $h(x) = \dfrac{1}{(3x^2 + 2)^{3/2}}$

42. $h(x) = \dfrac{1}{\sqrt{2x + 1}} + \sqrt{2x + 1}$

In Exercises 43–46, find $f(a + h) - f(a)$ for the given function. Simplify your answer.

43. $f(x) = 3x + 4$ 44. $f(x) = -\tfrac{1}{2}x + 3$

45. $f(x) = 4 - x^2$ 46. $f(x) = x^2 - 2x + 1$

47. If $f(x) = x^2 + 1$, find and simplify

$$\frac{f(a + h) - f(a)}{h} \qquad (h \neq 0).$$

48. If $f(x) = 1/x$, find and simplify

$$\frac{f(a + h) - f(a)}{h} \qquad (h \neq 0).$$

49. **Manufacturing Costs**   TMI, Inc., a manufacturer of blank audiocassette tapes, has a monthly fixed cost of \$12,100 and a variable cost of 60 cents per tape. Find a function C that gives the total cost incurred by TMI in the manufacture of x tapes per month.

50. **Cost of Producing Word Processors**   Apollo, Inc., manufactures its word processors at a variable cost of

$$V(x) = 0.000003x^3 - 0.03x^2 + 200x$$

dollars, where x denotes the number of units manufactured per month. The monthly fixed cost attributable to the division that produces these word processors is \$100,000. Find a function C that gives the total cost incurred by the manufacture of x word processors. What is the total cost incurred in producing 2000 units per month?

C 51. **Cost of Producing Word Processors**   Refer to Exercise 50. Suppose the total revenue realized by Apollo, Inc., from the sale of x word processors is given by the total revenue function

$$R(x) = -0.1x^2 + 500x \qquad (0 \leq x \leq 5000),$$

where x is measured in dollars.

a. Find the total profit function.
b. What is the profit when 1500 units are produced and sold each month?

[C] 52. **Overcrowding of Prisons** The 1980s saw a trend toward old-fashioned punitive deterrence as opposed to the more liberal penal policies and community-based corrections popular in the 1960s and early 1970s. As a result, prisons became more crowded and the gap between the number of people in prison and the prison capacity widened. Based on figures from the U.S. Department of Justice, the number of prisoners (in thousands) in federal and state prisons is approximated by the function

$$N(t) = 3.5t^2 + 26.7t + 436.2 \qquad (0 \le t \le 10),$$

where t is measured in years and $t = 0$ corresponds to 1983. The number of inmates for which prisons were designed is given by

$$C(t) = 24.3t + 365 \qquad (0 \le t \le 10),$$

where $C(t)$ is measured in thousands and t has the same meaning as before.
a. Find an expression that shows the gap between the number of prisoners and the number of inmates for which the prisons were designed at any time t.
b. Find the gap at the beginning of 1983. At the beginning of 1986.

53. **Effect of Mortgage Rates on Housing Starts** A study prepared for the National Association of Realtors estimated that the number of housing starts per year over the next five years will be

$$N(r) = \frac{7}{1 + 0.02r^2}$$

million units, where r (percent) is the mortgage rate. Suppose that the mortgage rate over the next t months is

$$r(t) = \frac{10t + 150}{t + 10} \qquad (0 \le t \le 24)$$

percent per year.

a. Find an expression for the number of housing starts per year as a function of t, t months from now.
b. Using the result from (a), determine the number of housing starts at present; 12 months from now; 18 months from now.

[C] 54. **Hotel Occupancy Rate** The occupancy rate of the all-suite Wonderland Hotel, located near an amusement park, is given by the function

$$r(t) = \frac{10}{81}t^3 - \frac{10}{3}t^2 + \frac{200}{9}t + 55 \qquad (0 \le t \le 11),$$

where t is measured in months and $t = 0$ corresponds to the beginning of January. Management has estimated that the monthly revenue (in thousands of dollars) is approximated by the function

$$R(r) = -\frac{3}{5000}r^3 + \frac{9}{50}r^2 \qquad (0 \le r \le 100),$$

where r is the occupancy rate.
a. What is the hotel's occupancy rate at the beginning of January? At the beginning of June?
b. What is the hotel's monthly revenue at the beginning of January? At the beginning of June? [*Hint:* Compute $R(r(0))$ and $R(r(5))$.]

[C] 55. **Housing Starts and Construction Jobs** The president of a major housing construction firm reports that the number of construction jobs (in millions) created is given by

$$N(x) = 1.42x,$$

where x denotes the number of housing starts. Suppose that the number of housing starts in the next t months is expected to be

$$x(t) = \frac{7(t + 10)^2}{(t + 10)^2 + 2(t + 15)^2}$$

million units per year. Find an expression for the number of jobs created per month in the next t months. How many jobs will have been created 6 months from now? 12 months from now?

SOLUTIONS TO SELF-CHECK EXERCISES 2.2

1. a. $s(x) = f(x) + g(x) = \sqrt{x} + 1 + \dfrac{x}{1 + x}$

$d(x) = f(x) - g(x) = \sqrt{x} + 1 - \dfrac{x}{1 + x}$

$p(x) = f(x)g(x) = (\sqrt{x} + 1) \cdot \dfrac{x}{1 + x} = \dfrac{x(\sqrt{x} + 1)}{1 + x}$

$q(x) = \dfrac{f(x)}{g(x)} = \dfrac{\sqrt{x} + 1}{\dfrac{x}{1 + x}} = \dfrac{(\sqrt{x} + 1)(1 + x)}{x}$

b. $(f \circ g)(x) = f(g(x)) = \sqrt{\dfrac{x}{1+x} + 1}$

$(g \circ f)(x) = g(f(x)) = \dfrac{\sqrt{x}+1}{1+(\sqrt{x}+1)} = \dfrac{\sqrt{x}+1}{\sqrt{x}+2}$

2. a. The difference between private and government health-care spending per person at any time t is given by the function d with the rule

$$d(t) = f(t) - g(t) = (2.5t^2 + 31.3t + 406)$$
$$- (1.4t^2 + 29.6t + 251)$$
$$= 1.1t^2 + 1.7t + 155.$$

b. The difference between private and government expenditures per person at the beginning of 1985 is given by

$$d(10) = 1.1(10)^2 + 1.7(10) + 155,$$

or $282 per person.

The difference between private and government expenditures per person at the beginning of 1993 is given by

$$d(18) = 1.1(18)^2 + 1.7(18) + 155,$$

or $542 per person.

2.3   Functions and Mathematical Models

Mathematical Models

Before we can apply mathematics to solving real-world problems, we must be able to formulate those problems in the language of mathematics. This process is referred to as **mathematical modeling.** A *mathematical model* may describe precisely the problem under consideration, or, more likely than not, it may provide only an acceptable approximation of the problem. For example, the accumulated amount A at the end of t years when a sum of P dollars is deposited in a fixed bank account, earning interest at the rate of r percent per year compounded m times a year, is given *exactly* by the formula (model)

$$A = P\left(1 + \frac{r}{m}\right)^{mt}.$$

On the other hand, the size of a cancer tumor may be *approximated* by the volume of a sphere,

$$V = \frac{4}{3}\pi r^3,$$

where r is the radius of the tumor in centimeters.

The many techniques used in constructing mathematical models of practical problems range from theoretical consideration of the problem on the one extreme to an interpretation of data associated with the problem on the other. The model for the accumulated amount of a fixed bank account mentioned earlier may be derived theoretically (see Chapter 5). Later we will see how linear equations (models) can be constructed from a given set of data points.

In calculus, we are concerned primarily with how one (dependent) variable depends on one or more (independent) variables. Consequently, most of our mathematical models will involve functions of one or more variables.[†] Once a function has been constructed to describe a specific real-world problem, a host of questions pertaining to the problem may be answered by analyzing the function (mathematical model). For example, if we have a function that gives the population of a certain culture of bacteria at any time t, then we can determine how fast the population is increasing or decreasing at any time t, and so on. Conversely, if we have a model that gives the rate of change of the cost of producing a certain item as a function of the level of production, and if we know the fixed cost incurred in producing this item, then we can find the total cost incurred in producing a certain number of those items.

Before going on, let us look at an actual mathematical model used for projecting the growth of the number of people enrolled in health maintenance organizations (HMOs). The model is derived from the data using the *least squares technique*. (We will not, however, pursue the actual construction of the model.)

EXAMPLE 1   The number of people enrolled in health maintenance organizations (HMOs) from 1980 to 1985 is given in the following table.

Year	1980	1981	1982	1983	1984	1985
Number of people (in millions)	9.7	10.8	11.9	12.5	17	21.1

A mathematical model approximating the number of people, $N(t)$, enrolled in HMOs during this period is

$$N(t) = 0.48t^2 - 0.18t + 10.02,$$

where t is measured in years and $t = 0$ corresponds to 1980.

a. Sketch the graph of the function N to see how the model compares with the actual data.
b. Assuming that this trend continues, use the model to predict how many people will be enrolled in HMOs at the beginning of 1995.

Solution
a. The graph of the function N is shown in Figure 2.12.
b. The number of people enrolled in HMOs at the beginning of 1995 is given by

$$N(15) = 0.48(15)^2 - 0.18(15) + 10.02$$
$$= 115.32,$$

or approximately 115.32 million people.

[†]Functions of more than one variable will be studied later.

FIGURE 2.12
The graph of $y = N(t)$ approximates the number of people enrolled in HMOs from 1980 to 1985.

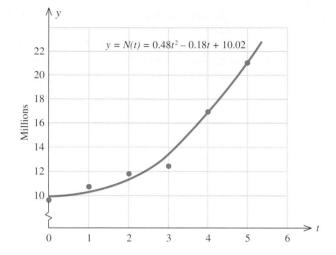

$y = N(t) = 0.48t^2 - 0.18t + 10.02$

We will discuss several mathematical models from the field of economics later on in this section, but first we will review some important functions that are the bases for many mathematical models.

Polynomial Functions

We begin by recalling a special class of functions, *polynomial functions.*

A POLYNOMIAL FUNCTION

A **polynomial function** of degree n is a function of the form

$$f(x) = a_0 x^n + a_1 x^{n-1} + \cdots + a_{n-1} x + a_n \qquad (a_0 \neq 0),$$

where $a_0, a_1, \ldots, a_n$ are constants and n is a nonnegative integer.

For example, the functions

$$f(x) = 4x^5 - 3x^4 + x^2 - x + 8$$

and

$$g(x) = 0.001x^3 - 2x^2 + 20x + 400$$

are polynomial functions of degrees 5 and 3, respectively. Observe that a polynomial function is defined everywhere so that it has domain $(-\infty, \infty)$.

A polynomial function of degree 1 ($n = 1$)

$$f(x) = a_0 x + a_1 \qquad (a_0 \neq 0)$$

is the equation of a straight line in the slope-intercept form with slope $m = a_0$ and y-intercept $b = a_1$ (see Section 1.4). For this reason, a polynomial function of degree 1 is called a **linear function.**

A polynomial function of degree 2 is referred to as a **quadratic function.** A polynomial function of degree 3 is called a **cubic function,** and so on. The mathematical model in Example 1 involves a quadratic function.

Rational and Power Functions

Another important class of functions is rational functions. A **rational function** is simply the quotient of two polynomials. Examples of rational functions are

$$F(x) = \frac{3x^3 + x^2 - x + 1}{x - 2}$$

and

$$G(x) = \frac{x^2 + 1}{x^2 - 1}.$$

In general, a rational function has the form

$$R(x) = \frac{f(x)}{g(x)},$$

where $f(x)$ and $g(x)$ are polynomial functions. Since division by zero is not allowed, we conclude that the domain of a rational function is the set of all real numbers except the zeros of g, that is, the roots of the equation $g(x) = 0$. Thus, the domain of the function F is the set of all numbers except $x = 2$, whereas the domain of the function G is the set of all numbers except those that satisfy $x^2 - 1 = 0$ or $x = \pm 1$.

Functions of the form

$$f(x) = x^r,$$

where r is any real number, are called **power functions.** We encountered examples of power functions earlier in our work. For example, the functions

$$f(x) = \sqrt{x} = x^{1/2} \quad \text{and} \quad g(x) = \frac{1}{x^2} = x^{-2}$$

are power functions.

Many of the functions we will encounter later will involve combinations of the functions introduced here. For example, the following functions may be viewed as suitable combinations of such functions:

$$f(x) = \sqrt{\frac{1 - x^2}{1 + x^2}},$$

$$g(x) = \sqrt{x^2 - 3x + 4},$$

and

$$h(x) = (1 + 2x)^{1/2} + \frac{1}{(x^2 + 2)^{3/2}}.$$

As with polynomials of degree 3 or greater, analyzing the properties of these functions is facilitated by using the tools of calculus, to be developed later on.

 EXAMPLE 2   A study of driving costs based on 1992 model compact (six-cylinder) cars found that the average cost (car payments, gas, insurance, upkeep, and depreciation), measured in cents per mile, is approximated by the function

$$C(x) = \frac{2095}{x^{2.2}} + 20.08,$$

where x (in thousands) denotes the number of miles the car is driven in one year. Using this model, estimate the average cost of driving a compact car 10,000 miles a year; 20,000 miles a year.

Solution   The average cost of driving a compact car 10,000 miles a year is given by

$$C(10) = \frac{2095}{10^{2.2}} + 20.08$$

$$\approx 33.3,$$

or approximately 33 cents per mile. The average cost of driving 20,000 miles a year is given by

$$C(20) = \frac{2095}{20^{2.2}} + 20.08$$

$$\approx 23.0,$$

or approximately 23 cents per mile.   ■

Some Economic Models

In the remainder of this section we will look at some economic models.

In a free market economy, consumer demand for a particular commodity depends on the commodity's unit price. A **demand equation** expresses the relationship between the unit price and the quantity demanded. The graph of the demand equation is called a **demand curve.** In general, the quantity demanded of a commodity decreases as the commodity's unit price increases, and vice versa. Accordingly, a **demand function** defined by $p = f(x)$, where p measures the unit price and x measures the number of units of the commodity in question, is generally characterized as a decreasing function of x; that is, $p = f(x)$ decreases as x increases. Since both x and p assume only nonnegative values, the demand curve is that part of the graph of $f(x)$ that lies in the first quadrant (Figure 2.13).

In a competitive market, a relationship also exists between the unit price of a commodity and the commodity's availability in the market. In general, an increase in the commodity's unit price induces the producer to increase the supply of the commodity. Conversely, a decrease in the unit price generally leads to a drop in the supply. The equation that expresses the relation between the unit price and the quantity supplied is called a **supply equation,** and its graph is called a **supply curve.** A **supply function** defined by $p = f(x)$ is generally characterized as an increasing function of x; that is, $p = f(x)$ increases as x increases. Since both x and p assume only nonnegative values, the supply curve is that part of the graph of $f(x)$ that lies in the first quadrant (Figure 2.14).

Under pure competition, the price of a commodity will eventually settle at a level dictated by the following condition: the supply of the commodity be equal to the demand for it. If the price is too high, the consumer will not buy, and if the price is too low, the supplier will not produce. **Market equilibrium** prevails when the quantity produced is equal to the quantity demanded. The quantity produced at market equilibrium is called the **equilibrium quantity,** and the corresponding price is called the **equilibrium price.**

FIGURE 2.13
A demand curve

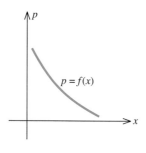

FIGURE 2.14
A supply curve

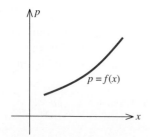

FIGURE 2.15
Market equilibrium corresponds to (x_0, p_0), the point at which the supply and demand curves intersect.

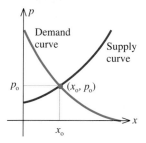

Market equilibrium corresponds to the point at which the demand curve and the supply curve intersect. In Figure 2.15, x_0 represents the equilibrium quantity and p_0 the equilibrium price. The point (x_0, p_0) lies on the supply curve and therefore satisfies the supply equation. At the same time, it also lies on the demand curve and therefore satisfies the demand equation. Thus, to find the point (x_0, p_0), and hence the equilibrium quantity and price, we solve the demand and supply equations simultaneously for x and p. For meaningful solutions, x and p must both be positive.

EXAMPLE 3 The demand function for a certain brand of video cassette is given by

$$p = d(x) = -0.01x^2 - 0.2x + 8,$$

and the corresponding supply function is given by

$$p = s(x) = 0.01x^2 + 0.1x + 3,$$

where p is expressed in dollars and x is measured in units of a thousand. Find the equilibrium quantity and price.

Solution We solve the following system of equations:

$$p = -0.01x^2 - 0.2x + 8$$

and

$$p = 0.01x^2 + 0.1x + 3.$$

Substituting the first equation into the second yields

$$-0.01x^2 - 0.2x + 8 = 0.01x^2 + 0.1x + 3,$$

which is equivalent to

$$0.02x^2 + 0.3x - 5 = 0$$
$$2x^2 + 30x - 500 = 0$$
$$x^2 + 15x - 250 = 0$$

or

$$(x + 25)(x - 10) = 0.$$

FIGURE 2.16
The supply curve and the demand curve intersect at the point (10, 5).

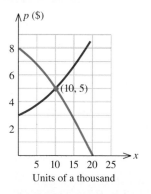

Units of a thousand

Thus, $x = -25$ or $x = 10$. Since x must be nonnegative, the root $x = -25$ is rejected. Therefore, the equilibrium quantity is 10,000 video cassettes. The equilibrium price is given by

$$p = 0.01(10)^2 + 0.1(10) + 3 = 5,$$

or $5 per video cassette (Figure 2.16). ∎

SELF-CHECK EXERCISES 2.3

1. Thomas Young has suggested the following rule for calculating the dosage of medicine for children from ages 1 to 12 years. If a denotes the adult dosage (in mg), and t is the age of the child (in years), then the child's dosage is given by

$$D(t) = \frac{at}{t + 12}.$$

If the adult dose of a substance is 500 mg, how much should a four-year-old child receive?

2. The demand function for Mrs. Baker's cookies is given by

$$d(x) = -\frac{2}{15}x + 4,$$

where $d(x)$ is the wholesale price in dollars per pound and x is the quantity demanded each week, measured in thousands of pounds. The supply function for the cookies is given by

$$s(x) = \frac{1}{75}x^2 + \frac{1}{10}x + \frac{3}{2},$$

where $s(x)$ is the wholesale price in dollars per pound and x is the quantity, in thousands of pounds, that will be made available in the market per week by the supplier.

a. Sketch the graphs of the functions d and s.
b. Find the equilibrium quantity and price.

Solutions to Self-Check Exercises 2.3 can be found on page 78.

2.3 EXERCISES

In Exercises 1–6, determine whether the given function is a polynomial function, a rational function, or some other function. State the degree of each polynomial function.

1. $f(x) = 3x^6 - 2x^2 + 1$

2. $f(x) = \dfrac{x^2 - 9}{x - 3}$

3. $G(x) = 2(x^2 - 3)^3$

4. $H(x) = 2x^{-3} + 5x^{-2} + 6$

5. $f(t) = 2t^2 + 3\sqrt{t}$

6. $f(r) = \dfrac{6r}{(r^3 - 8)}$

7. Disposable Income Economists define the *disposable annual income* for an individual by the equation $D = (1 - r)T$, where T is the individual's total income and r is the net rate at which he or she is taxed. What is the disposable income for an individual whose income is $40,000 and whose net tax rate is 28 percent?

8. Drug Dosages A method sometimes used by pediatricians to calculate the dosage of medicine for children is based on the child's surface area. If a denotes the adult dosage (in mg), and S is the surface area of the child (in square meters), then the child's dosage is given by

$$D(S) = \frac{Sa}{1.7}.$$

If the adult dose of a substance is 500 mg, how much should a child whose surface area is 0.4 square meters receive?

9. Cowling's Rule Cowling's Rule is a method for calculating pediatric drug dosages. If a denotes the adult dosage (in mg) and t is the age of the child (in years), then the child's dosage is given by

$$D(t) = \left(\frac{t + 1}{24}\right)a.$$

If the adult dose of a substance is 500 mg, how much should a four-year-old child receive?

10. Worker Efficiency An efficiency study showed that the average worker at Delphi Electronics assembled cordless telephones at the rate of

$$f(t) = -\frac{3}{2}t^2 + 6t + 10 \qquad (0 \le t \le 4)$$

phones per hour, t hours after starting work during the morning shift. At what rate does the average worker assemble telephones 2 hours after starting work?

11. Revenue Functions The revenue (in dollars) realized by Apollo, Inc., from the sale of its electronic correcting typewriters is given by

$$R(x) = -0.1x^2 + 500x,$$

where x denotes the number of units manufactured per month. What is Apollo's revenue when 1000 units are produced?

c **12. Effect of Advertising on Sales** The quarterly profit of Cunningham Realty depends on the amount of money

x spent on advertising per quarter according to the rule

$$P(x) = -\frac{1}{8}x^2 + 7x + 30 \qquad (0 \le x \le 50),$$

where $P(x)$ and x are measured in thousands of dollars. What is Cunningham's profit when its quarterly advertising budget is $28,000?

13. **Reaction of a Frog to a Drug** Experiments conducted by A. J. Clark suggest that the response $R(x)$ of a frog's heart muscle to the injection of x units of acetylcholine (as a percentage of the maximum possible effect of the drug) may be approximated by the rational function

$$R(x) = \frac{100x}{b + x} \qquad (x \ge 0),$$

where b is a positive constant that depends on the particular frog.
a. If a concentration of 40 units of acetylcholine produces a response of 50 percent for a certain frog, find the "response function" for this frog.
b. Using the model found in (a), find the response of the frog's heart muscle when 60 units of acetylcholine are administered.

14. **Linear Depreciation** In computing income tax, business firms are allowed by law to depreciate certain assets such as buildings, machines, furniture, automobiles, and so on, over a period of time. The linear depreciation, or straight-line method, is often used for this purpose. Suppose that an asset has an initial value of C and is to be depreciated linearly over n years with a scrap value of S. Show that the book value of the asset at any time t ($0 \le t \le n$) is given by the linear function

$$V(t) = C - \frac{(C - S)}{n}t.$$

[*Hint:* Find an equation of the straight line that passes through the points $(0, C)$ and (n, S). Then rewrite the equation in the slope-intercept form.]

15. **Linear Depreciation** Using the linear depreciation model of Exercise 14, find the book value of a printing machine at the end of the second year if its initial value is $100,000 and it is depreciated linearly over five years with a scrap value of $30,000.

C 16. **Price of Ivory** According to the World Wildlife Fund, a group in the forefront of the fight against illegal ivory trade, the price of ivory (in dollars per kilo) compiled from a variety of legal and black market sources is approximated by the function

$$f(t) = \begin{cases} 8.37t + 7.44 & 0 \le t \le 8 \\ 2.84t + 51.68 & 8 < t \le 30, \end{cases}$$

where t is measured in years and $t = 0$ corresponds to the beginning of 1970.
a. Sketch the graph of the function f.
b. What was the price of ivory at the beginning of 1970? At the beginning of 1990?

17. **Price of Automobile Parts** For years, automobile manufacturers had a monopoly on the replacement-parts market, particularly for sheet metal parts such as fenders, doors, and hoods, the parts most often damaged in a crash. Beginning in the late 1970s, however, competition appeared on the scene. In a report conducted by an insurance company to study the effects of the competition, the price of an OEM (original equipment manufacturer) fender for a particular 1983 model car was found to be

$$f(t) = \frac{110}{\frac{1}{2}t + 1} \qquad (0 \le t \le 2),$$

where $f(t)$ is measured in dollars and t is in years. Over the same period of time, the price of a non-OEM fender for the car was found to be

$$g(t) = 26\left(\frac{1}{4}t^2 - 1\right)^2 + 52 \qquad (0 \le t \le 2),$$

where $g(t)$ is also measured in dollars. Find a function $h(t)$ that gives the difference in price between an OEM fender and a non-OEM fender. Compute $h(0)$, $h(1)$, and $h(2)$. What does the result of your computation seem to say about the price gap between OEM and non-OEM fenders over the two years?

For the demand equations in Exercises 18–21, where x represents the quantity demanded in units of a thousand and p is the unit price in dollars, (a) sketch the demand curve and (b) determine the quantity demanded when the unit price is set at $p.

18. $p = -x^2 + 36; \quad p = 11$

19. $p = -x^2 + 16; \quad p = 7$

20. $p = \sqrt{9 - x^2}; \quad p = 2$

21. $p = \sqrt{18 - x^2}; \quad p = 3$

22. **Demand for Smoke Alarms** The demand function for the Sentinel smoke alarm is given by

$$p = \frac{30}{0.02x^2 + 1} \qquad (0 \le x \le 10),$$

where x (measured in units of a thousand) is the quantity demanded per week and p is the unit price in dollars. Sketch the graph of the demand function. What is the unit price that corresponds to a quantity demanded of 10,000 units?

C **23. Demand for Commodities** Assume that the demand function for a certain commodity has the form

$$p = \sqrt{-ax^2 + b} \qquad (a \geq 0, b \geq 0),$$

where x is the quantity demanded, measured in units of a thousand, and p is the unit price in dollars. Suppose that the quantity demanded is 6000 ($x = 6$) when the unit price is \$8 and 8000 ($x = 8$) when the unit price is \$6. Determine the demand equation. What is the quantity demanded when the unit price is set at \$7.50?

For the supply equations in Exercises 24–27, where x is the quantity supplied in units of a thousand and p is the unit price in dollars, (a) sketch the supply curve and (b) determine the price at which the supplier will make 2000 units of the commodity available in the market.

24. $p = 2x^2 + 18$

25. $p = x^2 + 16x + 40$

26. $p = x^3 + x + 10$

27. $p = x^3 + 2x + 3$

28. Supply Functions The supply function for the Luminar desk lamp is given by

$$p = 0.1x^2 + 0.5x + 15,$$

where x is the quantity supplied (in thousands) and p is the unit price in dollars. Sketch the graph of the supply function. What unit price will induce the supplier to make 5000 lamps available in the marketplace?

29. Supply Functions Suppliers of transistor radios will market 10,000 units when the unit price is \$20 and 62,500 units when the unit price is \$35. Determine the supply function if it is known to have the form

$$p = a\sqrt{x} + b \qquad (a \geq 0, b \geq 0),$$

where x is the quantity supplied and p is the unit price in dollars. Sketch the graph of the supply function. What unit price will induce the supplier to make 40,000 transistor radios available in the marketplace?

For each pair of supply and demand equations in Exercises 30–33, where x represents the quantity demanded in units of a thousand and p the unit price in dollars, find the equilibrium quantity and the equilibrium price.

30. $p = -2x^2 + 80$ and $p = 15x + 30$

31. $p = -x^2 - 2x + 100$ and $p = 8x + 25$

32. $11p + 3x - 66 = 0$ and $2p^2 + p - x = 10$

33. $p = 60 - 2x^2$ and $p = x^2 + 9x + 30$

34. Market Equilibrium The weekly demand and supply functions for Sportsman 5 × 7 tents are given by

$$p = -0.1x^2 - x + 40$$

and

$$p = 0.1x^2 + 2x + 20,$$

respectively, where p is measured in dollars and x is measured in units of a hundred. Find the equilibrium quantity and price.

35. Market Equilibrium The management of the Titan Tire Company has determined that the weekly demand and supply functions for their Super Titan tires are given by

$$p = 144 - x^2$$

and

$$p = 48 + \frac{1}{2}x^2,$$

respectively, where p is measured in dollars and x is measured in units of a thousand. Find the equilibrium quantity and price.

SOLUTIONS TO SELF-CHECK EXERCISES 2.3

1. Since the adult dose of the substance is 500 mg, $a = 500$; thus the rule in this case is

$$D(t) = \frac{500t}{t + 12}.$$

A four-year-old should receive

$$D(4) = \frac{500(4)}{4 + 12},$$

or 125 mg of the substance.

2. a. The graphs of the functions d and s are shown in the following figure.

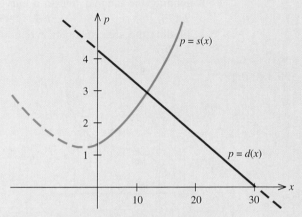

b. Solve the following system of equations:

$$p = -\frac{2}{15}x + 4$$

$$p = \frac{1}{75}x^2 + \frac{1}{10}x + \frac{3}{2}$$

Substituting the first equation into the second yields

$$\frac{1}{75}x^2 + \frac{1}{10}x + \frac{3}{2} = -\frac{2}{15}x + 4,$$

$$\frac{1}{75}x^2 + \left(\frac{1}{10} + \frac{2}{15}\right)x - \frac{5}{2} = 0$$

or $$\frac{1}{75}x^2 + \frac{7}{30}x - \frac{5}{2} = 0.$$

Multiplying both sides of the last equation by 150, we have

$$2x^2 + 35x - 375 = 0$$

or $$(2x - 15)(x + 25) = 0.$$

Thus, $x = -25$ or $x = \frac{15}{2} = 7.5$. Since x must be nonnegative, we take $x = 7.5$, and the equilibrium quantity is 7500 pounds. The equilibrium price is given by

$$p = -\frac{2}{15}\left(\frac{15}{2}\right) + 4,$$

or $3 per pound.

2.4

Limits

Introduction to Calculus

Historically, the development of calculus by Isaac Newton (1642–1727) and Gottfried Wilhelm Leibniz (1646–1716) resulted from the investigation of the following problems:

1. Finding the tangent line to a curve at a given point on the curve (Figure 2.17a).
2. Finding the area of a planar region bounded by an arbitrary curve (Figure 2.17b).

FIGURE 2.17

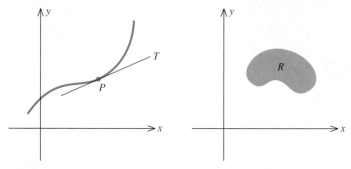

(a) What is the slope of the tangent line T at point P?

(b) What is the area of region R?

The tangent-line problem might appear to be unrelated to any practical applications of mathematics, but as you will see later on, the problem of finding the *rate of change* of one quantity with respect to another is mathematically equivalent to the geometric problem of finding the slope of the *tangent line* to a curve at a given point on the curve. It is precisely the discovery of the relationship between these two problems that spurred the development of calculus in the seventeenth century and made it such an indispensable tool for solving practical problems. The following are a few examples of such problems.

■ Finding the velocity of an object
■ Finding the rate of change of a bacteria population with respect to time
■ Finding the rate of change of a company's profit with respect to time
■ Finding the rate of change of a travel agency's revenue with respect to the agency's expenditure for advertising

The study of the tangent-line problem led to the creation of *differential calculus,* which relies on the concept of the *derivative* of a function. The study of the area problem led to the creation of *integral calculus,* which relies on the concept of the *antiderivative,* or *integral,* of a function. (The derivative of a function and the integral of a function are intimately related, as you will see in Section 6.4.) Both the derivative of a function and the integral of a function are defined in terms of a more fundamental concept—the limit—our next topic.

A Real-Life Example

Suppose that a car traveling along a straight road covers a distance of s feet in t seconds, where s and t are related by the function

$$s = f(t) = 2t^2 \qquad (0 \le t \le 10). \qquad (3)$$

Then the position of the car at time $t = 0, 1, 2, 3, \ldots, 10$, measured from its initial position, is

$$f(0) = 0, \quad f(1) = 2, \quad f(2) = 8, \quad f(3) = 18, \ldots, \quad f(10) = 200$$

feet (Figure 2.18).

FIGURE 2.18

Position of a car at time t

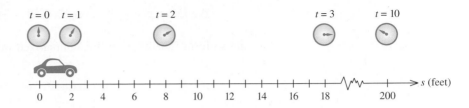

Suppose we want to know the velocity of the car when $t = 2$. This is just the speed of the car that would be indicated on the speedometer at that precise instant. Offhand, calculating this quantity using only Equation (3) appears to be an impossible task, but consider what quantities we *can* compute using this relationship. Obviously, we can compute the position of the car at any time t, as we did earlier for some selected values of t. Using these values we can then compute the *average velocity* of the car over an interval of time. For example, the average velocity of the car over the interval [2, 4] is given by

$$\frac{\text{distance covered}}{\text{time elapsed}} = \frac{f(4) - f(2)}{4 - 2}$$

$$= \frac{2(4^2) - 2(2^2)}{2}$$

$$= \frac{32 - 8}{2} = 12,$$

or 12 ft/sec. This is not quite the velocity of the car at $t = 2$, but it is an approximation. Can we do better? Intuitively, the smaller the time interval we pick (with endpoint $t = 2$),[†] the better the average velocity over that time interval will approximate the actual velocity of the car at $t = 2$.

To formulate this process mathematically, let $t > 2$. Then the distance covered by the car over the time interval [2, t] is $(f(t) - f(2))$ feet, and the time elapsed is $(t - 2)$ seconds. Therefore, the average velocity of the car over the time interval [2, t] is given by

$$\frac{f(t) - f(2)}{t - 2} = \frac{2t^2 - 2(2)^2}{t - 2}$$

$$= \frac{2(t^2 - 4)}{t - 2}. \tag{4}$$

Equation (4) may be used to generate a sequence of average velocities of the car over some appropriate intervals. By choosing the values of t closer and closer to $t = 2$, we obtain a sequence that approximates with increasing accuracy the *instantaneous velocity* of the car at $t = 2$.

Before going on, let us consider a sample calculation. Taking the sequence $t = 3, 2.5, 2.1, 2.01, 2.001$, and 2.0001, which evidently approaches $t = 2$, we obtain (with the help of a calculator)

the average velocity over [2, 3] is $\dfrac{2[(3)^2 - 4]}{3 - 2} = 10$, or 10 ft/sec,

---

[†]Actually, any interval containing $t = 2$ will suffice.

the average velocity over [2, 2.5] is $\dfrac{2[(2.5)^2 - 4]}{2.5 - 2} = 9$, or 9 ft/sec,

and so forth. These results are summarized in Table 2.1.

TABLE 2.1

t approaches 2 from the right

t	3	2.5	2.1	2.01	2.001	2.0001
Average velocity over [2, t]	10	9	8.2	8.02	8.002	8.0002

average velocity approaches 8 from the right

From Table 2.1 we see that the average velocities of the car over the time intervals [2, 3], [2, 2.5], ..., [2, 2.0001] seem to approach the number 8. These computations suggest that the instantaneous velocity of the car at $t = 2$ is 8 ft/sec.

This example illustrates the mechanics of the limit process. Let us summarize what we have done. Given the function

$$g(t) = \frac{2(t^2 - 4)}{t - 2},$$

which is just the function in (4), we were required to determine the value that the function g approached as t approached the fixed number 2. We took the sequence of values of t approaching $t = 2$ from the right-hand side and saw that $g(t)$ approached the number 8 as t approached $t = 2$. Similarly, if we take the sequence $t = 1, 1.5, 1.9, 1.99, 1.999,$ and 1.9999, which approaches $t = 2$ from the left, we obtain the results given in Table 2.2.

TABLE 2.2

t approaches 2 from the left

t	1	1.5	1.9	1.99	1.999	1.9999
g(t)	6	7	7.8	7.98	7.998	7.9998

$g(t)$ approaches 8 from the left

Observe that $g(t)$ approaches the number 8 as t approaches $t = 2$, this time from the left-hand side. In this situation, we say that the *limit of the function*

$$g(t) = \frac{2(t^2 - 4)}{t - 2}$$

as t approaches 2 is 8, written

$$\lim_{t \to 2} \frac{2(t^2 - 4)}{t - 2} = 8.$$

The graph of the function g shown in Figure 2.19 confirms this observation.

FIGURE 2.19
As t approaches $t = 2$ from either direction, $g(t)$ approaches $y = 8$.

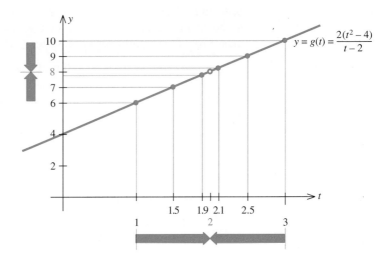

$$y = g(t) = \frac{2(t^2 - 4)}{t - 2}$$

Observe that the point $t = 2$ is not in the domain of the function g (for this reason, the point $(2, 8)$ is missing from the graph of g). This, however, is inconsequential because the value, if any, of $g(t)$ at $t = 2$ plays no role in computing the limit.

This example leads to the following informal definition.

LIMIT OF A FUNCTION

The function f has the limit L as x approaches a, written

$$\lim_{x \to a} f(x) = L,$$

if the values $f(x)$ can be made as close to the number L as we please by taking x sufficiently close to (but not equal to) a.

Evaluating the Limit of a Function

Let us now consider some examples involving the computation of limits.

FIGURE 2.20
$f(x)$ is close to 8 whenever x is close to 2.

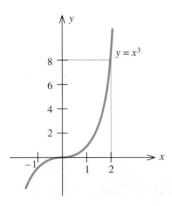

EXAMPLE 1 Let $f(x) = x^3$ and evaluate $\lim_{x \to 2} f(x)$.

Solution The graph of f is shown in Figure 2.20. You can see that $f(x)$ can be made as close to the number 8 as we please by taking x sufficiently close to 2. Therefore,

$$\lim_{x \to 2} x^3 = 8.$$

EXAMPLE 2 Let

$$g(x) = \begin{cases} x + 2 & \text{if } x \neq 1 \\ 1 & \text{if } x = 1. \end{cases}$$

Evaluate $\lim_{x \to 1} g(x)$.

FIGURE 2.21
$$\lim_{x \to 1} g(x) = 3$$

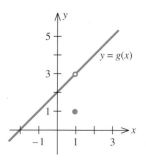

Solution   The domain of g is the set of all real numbers. From the graph of g shown in Figure 2.21, we see that $g(x)$ can be made as close to 3 as we please by taking x sufficiently close to 1. Therefore

$$\lim_{x \to 1} g(x) = 3.$$

Observe that $g(1) = 1$, which is not equal to the limit of the function g as x approaches 1. (Once again, the value of $g(x)$ at $x = 1$ has no bearing on the existence or value of the limit of g as x approaches 1.)   ∎

EXAMPLE 3   Evaluate the limit of the following functions as x approaches the indicated point.

a. $f(x) = \begin{cases} -1 & \text{if } x < 0 \\ 1 & \text{if } x \geq 0 \end{cases}; \quad x = 0$ **b.** $g(x) = \dfrac{1}{x^2}; \quad x = 0$

Solution   The graphs of the functions f and g are shown in Figure 2.22.

FIGURE 2.22

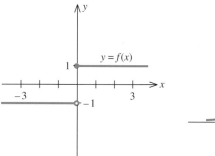

(a) $\lim_{x \to 0} f(x)$ does not exist   **(b)** $\lim_{x \to 0} g(x)$ does not exist

a. Referring to Figure 2.22a, we see that no matter how close x is to $x = 0$, $f(x)$ takes on the values 1 or -1 depending on whether x is positive or negative. Thus, there is no *single* real number L that $f(x)$ approaches as x approaches zero. We conclude that the limit of $f(x)$ does *not* exist as x approaches zero.
b. Referring to Figure 2.22b, we see that as x approaches $x = 0$ (from either side), $g(x)$ increases without bound and thus does not approach any specific real number. We conclude, accordingly, that the limit of $g(x)$ does not exist as x approaches zero.   ∎

Until now, we have relied on knowing the actual values of a function or the graph of a function near $x = a$ to help us evaluate the limit of the function $f(x)$ as x approaches a. The following properties of limits, which we list without proof, enable us to evaluate limits of functions algebraically.

THEOREM 1
PROPERTIES OF
LIMITS

Suppose that

$$\lim_{x \to a} f(x) = L \quad \text{and} \quad \lim_{x \to a} g(x) = M.$$

Then

i. $\displaystyle\lim_{x\to a}[f(x)]^r = [\lim_{x\to a} f(x)]^r = L^r$ $(r$, a real number$)$

ii. $\displaystyle\lim_{x\to a} cf(x) = c\lim_{x\to a} f(x) = cL$ $(c$, a real number$)$

iii. $\displaystyle\lim_{x\to a}[f(x) \pm g(x)] = \lim_{x\to a} f(x) \pm \lim_{x\to a} g(x) = L \pm M$

iv. $\displaystyle\lim_{x\to a}[f(x)g(x)] = [\lim_{x\to a} f(x)][\lim_{x\to a} g(x)] = LM$

v. $\displaystyle\lim_{x\to a}\frac{f(x)}{g(x)} = \frac{\displaystyle\lim_{x\to a} f(x)}{\displaystyle\lim_{x\to a} g(x)} = \frac{L}{M}$ (Provided $M \neq 0$)

EXAMPLE 4   Use Theorem 1 to evaluate the following limits.

a. $\displaystyle\lim_{x\to 2} x^3$ **b.** $\displaystyle\lim_{x\to 4} 5x^{3/2}$ **c.** $\displaystyle\lim_{x\to 1} (5x^4 - 2)$

d. $\displaystyle\lim_{x\to 3} 2x^3 \sqrt{x^2 + 7}$ **e.** $\displaystyle\lim_{x\to 2} \frac{2x^2 + 1}{x + 1}$

Solution

a. $\displaystyle\lim_{x\to 2} x^3 = [\lim_{x\to 2} x]^3$ [Property (i)]

$\qquad\qquad = 2^3 = 8$ $(\displaystyle\lim_{x\to 2} x = 2)$

b. $\displaystyle\lim_{x\to 4} 5x^{3/2} = 5[\lim_{x\to 4} x^{3/2}]$ [Property (ii)]

$\qquad\qquad = 5(4)^{3/2} = 40$

c. $\displaystyle\lim_{x\to 1}(5x^4 - 2) = \lim_{x\to 1} 5x^4 - \lim_{x\to 1} 2$ [Property (iii)]

To evaluate $\displaystyle\lim_{x\to 1} 2$, observe that the constant function $g(x) = 2$ has value 2 for all values of x. Therefore, $g(x)$ must approach the limit 2 as x approaches $x = 1$ (or any other point for that matter!). Therefore,

$$\lim_{x\to 1}(5x^4 - 2) = 5(1)^4 - 2$$
$$= 3.$$

d. $\displaystyle\lim_{x\to 3} 2x^3 \sqrt{x^2 + 7} = 2\lim_{x\to 3} x^3 \sqrt{x^2 + 7}$ [Property (ii)]

$\qquad\qquad = 2\lim_{x\to 3} x^3 \lim_{x\to 3} \sqrt{x^2 + 7}$ [Property (iv)]

$\qquad\qquad = 2(3)^3 \sqrt{3^2 + 7}$ [Property (i)]

$\qquad\qquad = 2(27)\sqrt{16} = 216$

e. $\displaystyle\lim_{x\to 2} \frac{2x^2 + 1}{x + 1} = \frac{\displaystyle\lim_{x\to 2}(2x^2 + 1)}{\displaystyle\lim_{x\to 2}(x + 1)}$ [Property (v)]

$\qquad\qquad = \frac{2(2)^2 + 1}{2 + 1} = \frac{9}{3} = 3$ ■

Indeterminate Forms

Let us emphasize once again that Property (v) of limits is valid only when the limit of the function that appears in the denominator is not equal to zero at the point in question.

If the numerator has a limit different from zero and the denominator has a limit equal to zero, then the limit of the quotient does not exist at the point in question. This is the case with the function $g(x) = 1/x^2$ in Example 3b. Here, as x approaches zero, the numerator approaches 1 but the denominator approaches zero, so that the quotient becomes arbitrarily large. Thus, as observed earlier, the limit does not exist.

Next, consider

$$\lim_{x \to 2} \frac{2(x^2 - 4)}{x - 2},$$

which we evaluated earlier by looking at the values of the function for x near $x = 2$. If we attempt to evaluate this expression by applying Property (v) of limits, we see that both the numerator and denominator of the function

$$\frac{2(x^2 - 4)}{x - 2}$$

approach zero as x approaches 2; that is, we obtain an expression of the form $0/0$. In this event, we say that the limit of the quotient $f(x)/g(x)$ as x approaches 2 has the **indeterminate form 0/0.**

We will need to evaluate limits of this type when we discuss the derivative of a function, a fundamental concept in the study of calculus. As the name suggests, the meaningless expression $0/0$ does not provide us with a solution to our problem. One strategy that can be used to solve this type of problem follows.

STRATEGY FOR EVALUATING INDETERMINATE FORMS	**1.** Replace the given function with an appropriate one that takes on the same values as the original function everywhere except at $x = a$. **2.** Evaluate the limit of this function as x approaches a.

Examples 5 and 6 illustrate this strategy.

EXAMPLE 5   Evaluate

$$\lim_{x \to 2} \frac{2(x^2 - 4)}{x - 2}.$$

Solution   Since both the numerator and the denominator of this expression approach zero as x approaches 2, we have the indeterminate form $0/0$. We rewrite

$$\frac{2(x^2 - 4)}{x - 2} = \frac{2(x - 2)(x + 2)}{(x - 2)},$$

which, upon canceling the common factors, is equivalent to $2(x + 2)$. Next, we replace $2(x^2 - 4)/(x - 2)$ with $2(x + 2)$ and take the limit as x approaches 2, obtaining

$$\lim_{x \to 2} \frac{2(x^2 - 4)}{x - 2} = \lim_{x \to 2} 2(x + 2) = 8.$$ ■

REMARK   Notice that this is the same limit that we evaluated earlier when we discussed the instantaneous velocity of a car at a specified time.   □

EXAMPLE 6   Evaluate

$$\lim_{h \to 0} \frac{\sqrt{1 + h} - 1}{h}.$$

Solution   Letting h approach zero, we obtain the indeterminate form $0/0$. Next, we rationalize the numerator of the quotient (see page 23) by multiplying both the numerator and the denominator by the expression $(\sqrt{1 + h} + 1)$, obtaining

$$\frac{\sqrt{1 + h} - 1}{h} = \frac{(\sqrt{1 + h} - 1)(\sqrt{1 + h} + 1)}{h(\sqrt{1 + h} + 1)}$$

$$= \frac{1 + h - 1}{h(\sqrt{1 + h} + 1)} \qquad [(\sqrt{a} - \sqrt{b})(\sqrt{a} + \sqrt{b}) = a - b]$$

$$= \frac{h}{h(\sqrt{1 + h} + 1)}$$

$$= \frac{1}{\sqrt{1 + h} + 1}.$$

Therefore,

$$\lim_{h \to 0} \frac{\sqrt{1 + h} - 1}{h} = \lim_{h \to 0} \frac{1}{\sqrt{1 + h} + 1}$$

$$= \frac{1}{\sqrt{1} + 1} = \frac{1}{2}.$$ ■

Limits at Infinity

In certain applications we need to know the behavior of f as x increases without bound (gets larger and larger) or decreases without bound (gets smaller and smaller). For example, suppose that we are given the function

$$f(x) = \frac{2x^2}{1 + x^2}$$

and we want to determine what happens to $f(x)$ as x gets larger and larger. Picking the sequence of numbers 1, 2, 5, 10, 100, and 1000 and computing the corresponding values of $f(x)$, we obtain the following table of values:

x	1	2	5	10	100	1000
$f(x)$	1	1.6	1.92	1.98	1.9998	1.999998

FIGURE 2.23

The graph of $y = \dfrac{2x^2}{1 + x^2}$ has a horizontal asymptote at $y = 2$.

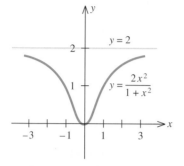

From the table, we see that as x gets larger and larger, $f(x)$ gets closer and closer to 2. The graph of the function f shown in Figure 2.23 confirms this observation. We call the line $y = 2$ a **horizontal asymptote.**[†]

In this situation we say that the limit of the function

$$f(x) = \frac{2x^2}{1 + x^2}$$

as x increases without bound is 2, written

$$\lim_{x \to \infty} \frac{2x^2}{1 + x^2} = 2.$$

In the general case, the following definition is applicable:

**LIMIT OF
A FUNCTION
AT INFINITY**

The function f has the limit L as x increases without bound (or, as x approaches infinity), written

$$\lim_{x \to \infty} f(x) = L,$$

if $f(x)$ can be made arbitrarily close to L by taking x large enough.

Similarly, the function f has the limit M as x decreases without bound (or, as x approaches negative infinity), written

$$\lim_{x \to -\infty} f(x) = M,$$

if $f(x)$ can be made arbitrarily close to M by taking x to be negative and sufficiently large in absolute value.

EXAMPLE 7   Let f and g be the functions

$$f(x) = \begin{cases} -1 & \text{if } x < 0 \\ 1 & \text{if } x \geq 0 \end{cases} \quad \text{and} \quad g(x) = \frac{1}{x^2}.$$

Evaluate

a. $\lim\limits_{x \to \infty} f(x)$ and $\lim\limits_{x \to -\infty} f(x)$ **b.** $\lim\limits_{x \to \infty} g(x)$ and $\lim\limits_{x \to -\infty} g(x)$

Solution   The graphs of $f(x)$ and $g(x)$ are shown in Figure 2.24. Referring to the graphs of the respective functions, we see that

a. $\lim\limits_{x \to \infty} f(x) = 1$ and $\lim\limits_{x \to -\infty} f(x) = -1$

b. $\lim\limits_{x \to \infty} \dfrac{1}{x^2} = 0$ and $\lim\limits_{x \to -\infty} \dfrac{1}{x^2} = 0.$

[†]We will discuss asymptotes in greater detail in Section 4.4.

FIGURE 2.24

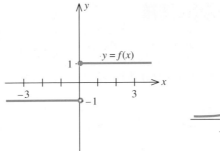

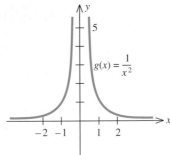

(a) $\lim_{x \to \infty} f(x) = 1$ and

$\lim_{x \to -\infty} f(x) = -1$

(b) $\lim_{x \to \infty} g(x) = 0$ and

$\lim_{x \to -\infty} g(x) = 0$ ∎

All of the properties of limits listed in Theorem 1 are valid when a is replaced by ∞ or $-\infty$. In addition, we have the following property for the limit at infinity.

THEOREM 2

$$\lim_{x \to \infty} \frac{1}{x^n} = 0 \quad \text{for all } n > 0$$

and $\quad \lim_{x \to -\infty} \frac{1}{x^n} = 0 \quad \text{for all } n > 0$, provided that $\frac{1}{x^n}$ is defined.

In evaluating the limit at infinity of a rational function, the following technique is often used: *Divide the numerator and denominator of the expression by x^n, where n is the highest power present in the denominator of the expression.*

EXAMPLE 8   Evaluate

$$\lim_{x \to \infty} \frac{x^2 - x + 3}{2x^3 + 1}.$$

Solution   Since the limits of both the numerator and the denominator do not exist as x approaches infinity, the property pertaining to the limit of a quotient [Property (v)] is not applicable. Let us divide the numerator and denominator of the rational expression by x^3, obtaining

$$\lim_{x \to \infty} \frac{x^2 - x + 3}{2x^3 + 1} = \lim_{x \to \infty} \frac{\dfrac{1}{x} - \dfrac{1}{x^2} + \dfrac{3}{x^3}}{2 + \dfrac{1}{x^3}}$$

$$= \frac{0}{2} \quad \text{(Using Theorem 2)}$$

$$= 0. \qquad ∎$$

EXAMPLE 9   Let

$$f(x) = \frac{3x^2 + 8x - 4}{2x^2 + 4x - 5}.$$

Compute $\lim_{x \to \infty} f(x)$ if it exists.

Solution   Again, we see that Property (v) is not applicable. Dividing the numerator and the denominator by x^2, we obtain

$$\lim_{x \to \infty} \frac{3x^2 + 8x - 4}{2x^2 + 4x - 5} = \lim_{x \to \infty} \frac{3 + \dfrac{8}{x} - \dfrac{4}{x^2}}{2 + \dfrac{4}{x} - \dfrac{5}{x^2}}$$

$$= \frac{\lim_{x \to \infty} 3 + 8 \lim_{x \to \infty} \dfrac{1}{x} - 4 \lim_{x \to \infty} \dfrac{1}{x^2}}{\lim_{x \to \infty} 2 + 4 \lim_{x \to \infty} \dfrac{1}{x} - 5 \lim_{x \to \infty} \dfrac{1}{x^2}}$$

$$= \frac{3 + 0 - 0}{2 + 0 - 0}$$

$$= \frac{3}{2}. \qquad \text{(Using Theorem 2)}$$ ∎

EXAMPLE 10   Let $f(x) = \dfrac{2x^3 - 3x^2 + 1}{x^2 + 2x + 4}$ and evaluate

a. $\lim_{x \to \infty} f(x)$ and   **b.** $\lim_{x \to -\infty} f(x)$.

Solution
a. Dividing the numerator and the denominator of the rational expression by x^2, we obtain

$$\lim_{x \to \infty} \frac{2x^3 - 3x^2 + 1}{x^2 + 2x + 4} = \lim_{x \to \infty} \frac{2x - 3 + \dfrac{1}{x^2}}{1 + \dfrac{2}{x} + \dfrac{4}{x^2}}.$$

Since the numerator becomes arbitrarily large while the denominator approaches 1 as x approaches infinity, we see that the quotient $f(x)$ gets larger and larger as x approaches infinity. In other words, the limit does not exist. We indicate this by writing

$$\lim_{x \to \infty} \frac{2x^3 - 3x^2 + 1}{x^2 + 2x + 4} = \infty.$$

b. Once again, dividing both the numerator and the denominator by x^2, we obtain

$$\lim_{x \to -\infty} \frac{2x^3 - 3x^2 + 1}{x^2 + 2x + 4} = \lim_{x \to -\infty} \frac{2x - 3 + \dfrac{1}{x^2}}{1 + \dfrac{2}{x} + \dfrac{4}{x^2}}.$$

In this case, the numerator becomes arbitrarily large in magnitude but negative in sign, while the denominator approaches 1 as x approaches negative infinity. Therefore, the quotient $f(x)$ decreases without bound and the limit does not exist. We indicate this by writing

$$\lim_{x \to -\infty} \frac{2x^3 - 3x^2 + 1}{2x^2 + 2x + 4} = -\infty.$$ ∎

Example 11 gives an application of the concept of the limit of a function at infinity.

FIGURE 2.25

As the level of production increases, the average cost approaches $100 per desk.

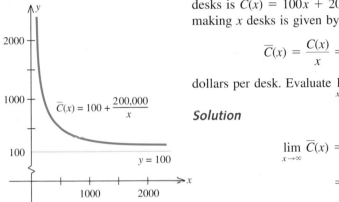

$\overline{C}(x) = 100 + \dfrac{200{,}000}{x}$

$y = 100$

EXAMPLE 11 The Custom Office Company makes a line of executive desks. It is estimated that the total cost of making x Senior Executive Model desks is $C(x) = 100x + 200{,}000$ dollars per year, so that the average cost of making x desks is given by

$$\overline{C}(x) = \frac{C(x)}{x} = \frac{100x + 200{,}000}{x} = 100 + \frac{200{,}000}{x}$$

dollars per desk. Evaluate $\lim_{x \to \infty} \overline{C}(x)$ and interpret your results.

Solution

$$\lim_{x \to \infty} \overline{C}(x) = \lim_{x \to \infty} \left(100 + \frac{200{,}000}{x} \right)$$
$$= \lim_{x \to \infty} 100 + \lim_{x \to \infty} \frac{200{,}000}{x} = 100.$$

A sketch of the graph of the function $\overline{C}(x)$ appears in Figure 2.25. The result we obtained is fully expected if we consider its economic implications. Note that as the level of production increases, the fixed cost per desk produced, represented by the term $(200{,}000/x)$, drops steadily. The average cost should approach a constant unit cost of production—$100 in this case. ∎

SELF-CHECK EXERCISES 2.4

1. Find the indicated limit if it exists.

   a. $\lim_{x \to 3} \dfrac{\sqrt{x^2 + 7} + \sqrt{3x - 5}}{x + 2}$

   b. $\lim_{x \to -1} \dfrac{x^2 - x - 2}{2x^2 - x - 3}$

2. The average cost per disc (in dollars) incurred by the Herald Record Company in pressing x compact audio discs is given by the average cost function

$$\overline{C}(x) = 1.8 + \frac{3000}{x}.$$

Evaluate $\lim_{x \to \infty} \overline{C}(x)$ and interpret your result.

Solutions to Self-Check Exercises 2.4 can be found on page 96.

2.4   EXERCISES

In Exercises 1–8, use the graph of the given function f to determine $\lim_{x \to a} f(x)$ at the indicated value of a, if it exists.

1.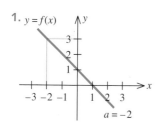

2.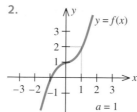

3.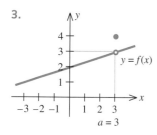

4.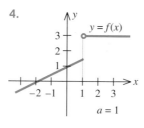

5.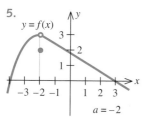

6.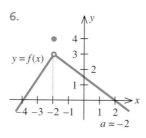

7.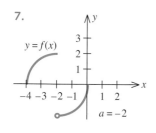

8.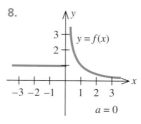

In Exercises 9–16, complete the table by computing $f(x)$ at the given values of x. Use these results to estimate the indicated limit (if it exists).

C 9. $f(x) = x^2 + 1$; $\lim_{x \to 2} f(x)$

x	1.9	1.99	1.999	2.001	2.01	2.1
f(x)						

10. $f(x) = 2x^2 - 1$; $\lim_{x \to 1} f(x)$

x	0.9	0.99	0.999	1.001	1.01	1.1
f(x)						

11. $f(x) = \dfrac{|x|}{x}$; $\lim_{x \to 0} f(x)$

x	-0.1	-0.01	-0.001	0.001	0.01	0.1
f(x)						

12. $f(x) = \dfrac{|x - 1|}{x - 1}$; $\lim_{x \to 1} f(x)$

x	0.9	0.99	0.999	1.001	1.01	1.1
f(x)						

13. $f(x) = \dfrac{1}{(x - 1)^2}$; $\lim_{x \to 1} f(x)$

x	0.9	0.99	0.999	1.001	1.01	1.1
f(x)						

14. $f(x) = \dfrac{1}{x - 2}$; $\lim_{x \to 2} f(x)$

x	1.9	1.99	1.999	2.001	2.01	2.1
f(x)						

15. $f(x) = \dfrac{x^2 + x - 2}{x - 1}$; $\lim_{x \to 1} f(x)$

x	0.9	0.99	0.999	1.001	1.01	1.1
f(x)						

16. $f(x) = \dfrac{x-1}{x-1}; \quad \lim\limits_{x \to 1} f(x)$

x	0.9	0.99	0.999	1.001	1.01	1.1
f(x)						

In Exercises 17–22, sketch the graph of the function f and evaluate $\lim\limits_{x \to a} f(x)$, if it exists, for the given values of a.

17. $f(x) = \begin{cases} x - 1 & \text{if } x \le 0 \\ -1 & \text{if } x > 0 \end{cases} \quad a = 0$

18. $f(x) = \begin{cases} x - 1 & \text{if } x \le 3 \\ -2x + 8 & \text{if } x > 3 \end{cases} \quad a = 3$

19. $f(x) = \begin{cases} x & \text{if } x < 1 \\ 0 & \text{if } x = 1 \\ -x + 2 & \text{if } x > 1 \end{cases} \quad a = 1$

20. $f(x) = \begin{cases} -2x + 4 & \text{if } x < 1 \\ 4 & \text{if } x = 1 \\ x^2 + 1 & \text{if } x > 1 \end{cases} \quad a = 1$

21. $f(x) = \begin{cases} |x| & \text{if } x \ne 0 \\ 1 & \text{if } x = 0 \end{cases} \quad a = 0$

22. $f(x) = \begin{cases} |x - 1| & \text{if } x \ne 1 \\ 0 & \text{if } x = 1 \end{cases} \quad a = 1$

In Exercises 23–40, find the indicated limit.

23. $\lim\limits_{x \to 2} 3$

24. $\lim\limits_{x \to -2} -3$

25. $\lim\limits_{x \to 3} x$

26. $\lim\limits_{x \to -2} -3x$

27. $\lim\limits_{x \to 1} (1 - 2x^2)$

28. $\lim\limits_{t \to 3} (4t^2 - 2t + 1)$

29. $\lim\limits_{x \to 1} (2x^3 - 3x^2 + x + 2)$

30. $\lim\limits_{x \to 0} (4x^5 - 20x^2 + 2x + 1)$

31. $\lim\limits_{s \to 0} (2s^2 - 1)(2s + 4)$

32. $\lim\limits_{x \to 2} (x^2 + 1)(x^2 - 4)$

33. $\lim\limits_{x \to 2} \dfrac{2x + 1}{x + 2}$

34. $\lim\limits_{x \to 1} \dfrac{x^3 + 1}{2x^3 + 2}$

35. $\lim\limits_{x \to 2} \sqrt{x + 2}$

36. $\lim\limits_{x \to -2} \sqrt[3]{5x + 2}$

37. $\lim\limits_{x \to -3} \sqrt{2x^4 + x^2}$

38. $\lim\limits_{x \to 2} \sqrt{\dfrac{2x^3 + 4}{x^2 + 1}}$

39. $\lim\limits_{x \to -1} \dfrac{\sqrt{x^2 + 8}}{2x + 4}$

40. $\lim\limits_{x \to 3} \dfrac{x\sqrt{x^2 + 7}}{2x - \sqrt{2x + 3}}$

In Exercises 41–48, find the indicated limit given that $\lim\limits_{x \to a} f(x) = 3$ and $\lim\limits_{x \to a} g(x) = 4$.

41. $\lim\limits_{x \to a} [f(x) - g(x)]$

42. $\lim\limits_{x \to a} 2f(x)$

43. $\lim\limits_{x \to a} [2f(x) - 3g(x)]$

44. $\lim\limits_{x \to a} [f(x)g(x)]$

45. $\lim\limits_{x \to a} \sqrt{g(x)}$

46. $\lim\limits_{x \to a} \sqrt[3]{5f(x) + 3g(x)}$

47. $\lim\limits_{x \to a} \dfrac{2f(x) - g(x)}{f(x)g(x)}$

48. $\lim\limits_{x \to a} \dfrac{g(x) - f(x)}{f(x) + \sqrt{g(x)}}$

In Exercises 49–62, find the indicated limit if it exists.

49. $\lim\limits_{x \to 1} \dfrac{x^2 - 1}{x - 1}$

50. $\lim\limits_{x \to -2} \dfrac{x^2 - 4}{x + 2}$

51. $\lim\limits_{x \to 0} \dfrac{x^2 - x}{x}$

52. $\lim\limits_{x \to 0} \dfrac{2x^2 - 3x}{x}$

53. $\lim\limits_{x \to -5} \dfrac{x^2 - 25}{x + 5}$

54. $\lim\limits_{b \to -3} \dfrac{b + 1}{b + 3}$

55. $\lim\limits_{x \to 1} \dfrac{x}{x - 1}$

56. $\lim\limits_{x \to 2} \dfrac{x + 2}{x - 2}$

57. $\lim\limits_{x \to -2} \dfrac{x^2 - x - 6}{x^2 + x - 2}$

58. $\lim\limits_{z \to 2} \dfrac{z^3 - 8}{z - 2}$

59. $\lim\limits_{x \to 1} \dfrac{\sqrt{x} - 1}{x - 1}$ $\left[Hint: \text{Multiply by } \dfrac{\sqrt{x} + 1}{\sqrt{x} + 1}. \right]$

60. $\lim\limits_{x \to 4} \dfrac{x - 4}{\sqrt{x} - 2}$ [Hint: See Exercise 59.]

61. $\lim\limits_{x \to 1} \dfrac{x - 1}{x^3 + x^2 - 2x}$

62. $\lim\limits_{x \to -2} \dfrac{4 - x^2}{2x^2 + x^3}$

In Exercises 63–68, use the graph of the function f to determine $\lim\limits_{x \to \infty} f(x)$ and $\lim\limits_{x \to -\infty} f(x)$, if they exist.

63.

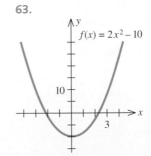

$f(x) = 2x^2 - 10$

64.

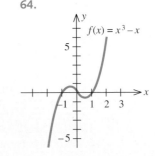

$f(x) = x^3 - x$

65.

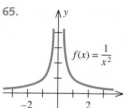

$f(x) = \dfrac{1}{x^2}$

66.

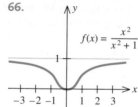

$f(x) = \dfrac{x^2}{x^2 + 1}$

67.

$f(x) = 2 - |x|$

68.

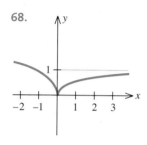

$f(x) = \begin{cases} \sqrt{-x} & \text{if } x \le 0 \\ \dfrac{x}{x+1} & \text{if } x > 0 \end{cases}$

In Exercises 69–72, complete the table by computing $f(x)$ at the given values of x. Use the results to guess at the indicated limits, if they exist.

C **69.** $f(x) = \dfrac{1}{x^2 + 1}$; $\displaystyle\lim_{x \to \infty} f(x)$ and $\displaystyle\lim_{x \to -\infty} f(x)$

x	1	10	100	1000
f(x)				

x	−1	−10	−100	−1000
f(x)				

C **70.** $f(x) = \dfrac{2x}{x + 1}$; $\displaystyle\lim_{x \to \infty} f(x)$ and $\displaystyle\lim_{x \to -\infty} f(x)$

x	1	10	100	1000
f(x)				

x	−5	−10	−100	−1000
f(x)				

C **71.** $f(x) = 3x^3 - x^2 + 10$; $\displaystyle\lim_{x \to \infty} f(x)$ and $\displaystyle\lim_{x \to -\infty} f(x)$

x	1	5	10	100	1000
f(x)					

x	−1	−5	−10	−100	−1000
f(x)					

72. $f(x) = \dfrac{|x|}{x}$; $\displaystyle\lim_{x \to \infty} f(x)$ and $\displaystyle\lim_{x \to -\infty} f(x)$

x	1	10	100	−1	−10	−100
f(x)						

In Exercises 73–80, find the indicated limits, if they exist.

73. $\displaystyle\lim_{x \to \infty} \frac{3x + 2}{x - 5}$

74. $\displaystyle\lim_{x \to -\infty} \frac{4x^2 - 1}{x + 2}$

75. $\displaystyle\lim_{x \to -\infty} \frac{3x^3 + x^2 + 1}{x^3 + 1}$

76. $\displaystyle\lim_{x \to \infty} \frac{2x^2 + 3x + 1}{x^4 - x^2}$

77. $\displaystyle\lim_{x \to -\infty} \frac{x^4 + 1}{x^3 - 1}$

78. $\displaystyle\lim_{x \to \infty} \frac{4x^4 - 3x^2 + 1}{2x^4 + x^3 + x^2 + x + 1}$

79. $\displaystyle\lim_{x \to \infty} \frac{x^5 - x^3 + x - 1}{x^6 + 2x^2 + 1}$

80. $\displaystyle\lim_{x \to \infty} \frac{2x^2 - 1}{x^3 + x^2 + 1}$

81. Toxic Waste A city's main well was recently found to be contaminated with trichloroethylene, a cancer-causing chemical, as a result of an abandoned chemical dump leaching chemicals into the water. A proposal submitted to the selectmen of the city indicates that the cost, measured in millions of dollars, of removing x percent of the toxic pollutant is given by

$$C(x) = \frac{0.5x}{100 - x} \qquad (0 < x < 100).$$

a. Find the cost of removing 50 percent of the pollutant; 60 percent; 70 percent; 80 percent; 90 percent; 95 percent.

b. Evaluate

$$\lim_{x \to 100} \frac{0.5x}{100 - x}$$

and interpret your result.

82. **Average Cost** The average cost per disc in dollars incurred by the Herald Record Company in pressing x video discs is given by the average cost function

$$\overline{C}(x) = 2.2 + \frac{2500}{x}.$$

Evaluate $\lim_{x \to \infty} \overline{C}(x)$ and interpret your result.

83. **Concentration of a Drug in the Bloodstream** The concentration of a certain drug in a patient's bloodstream t hours after injection is given by

$$C(t) = \frac{0.2t}{t^2 + 1}$$

milligrams per cubic centimeter. Evaluate $\lim_{t \to \infty} C(t)$ and interpret your result.

84. **Box Office Receipts** The total worldwide box office receipts for a long-running blockbuster movie are approximated by the function

$$T(x) = \frac{120x^2}{x^2 + 4},$$

where $T(x)$ is measured in millions of dollars and x is the number of months since the movie's release.

a. What are the total box office receipts after the first month? The second month? The third month?

b. What will the movie gross in the long run?

85. **Population Growth** A major corporation is building a 4325-acre complex of homes, offices, stores, schools, and churches in the rural community of Glen Cove. As a result of this development, the planners have estimated that Glen Cove's population t years from now (in thousands) will be given by

$$P(t) = \frac{25t^2 + 125t + 200}{t^2 + 5t + 40}.$$

a. What is the current population of Glen Cove?

b. What will the population be in the long run?

C 86. **Driving Costs** A study of driving costs of 1992 model subcompact (four-cylinder) cars found that the average cost (car payments, gas, insurance, upkeep, and depreciation), measured in cents per mile, is approximated by the function

$$C(x) = \frac{2010}{x^{2.2}} + 17.80,$$

where x denotes the number of miles (in thousands) the car is driven in a year.

a. What is the average cost of driving a subcompact car 5,000 miles a year? 10,000 miles a year? 15,000 miles a year? 20,000 miles a year? 25,000 miles a year?

b. Use (a) to help sketch the graph of the function C.

c. What happens to the average cost as the number of miles driven increases without bound?

87. **Speed of a Chemical Reaction** Certain proteins, known as enzymes, serve as catalysts for chemical reactions in living things. In 1913, Leonor Michaelis and L. M. Menten discovered the following formula giving the initial speed V (in moles per liter per second) at which the reaction begins in terms of the amount of substrate x (the substance being acted upon, measured in moles per liter):

$$V = \frac{ax}{x + b},$$

where a and b are positive constants. Evaluate

$$\lim_{x \to \infty} \frac{ax}{x + b}$$

and interpret your result.

C *In Exercises 88–90, use a computer or graphing calculator to plot the graph of the function f and find the specified limit (if it exists).*

88. $f(x) = \dfrac{x^3 - x^2 - x + 1}{x^3 - 3x + 2};\quad \lim_{x \to 1} f(x)$

89. $f(x) = \dfrac{x + 1}{\sqrt{6x^2 + 3} - 3};\quad \lim_{x \to -1} f(x)$

90. $f(x) = \dfrac{\sqrt{2x^2 + 3}}{2x + 1};\quad \lim_{x \to \infty} f(x) \;\text{ and }\; \lim_{x \to -\infty} f(x)$

SOLUTIONS TO
SELF-CHECK
EXERCISES 2.4

1. a. $\displaystyle\lim_{x \to 3} \frac{\sqrt{x^2 + 7} + \sqrt{3x - 5}}{x + 2} = \frac{\sqrt{9 + 7} + \sqrt{3(3) - 5}}{3 + 2}$

$\displaystyle = \frac{\sqrt{16} + \sqrt{4}}{5}$

$\displaystyle = \frac{6}{5}$

b. Letting x approach -1 leads to the indeterminate form $0/0$. Thus we proceed as follows:

$$\lim_{x \to -1} \frac{x^2 - x - 2}{2x^2 - x - 3} = \lim_{x \to -1} \frac{(x + 1)(x - 2)}{(x + 1)(2x - 3)}$$

$$= \lim_{x \to -1} \frac{x - 2}{2x - 3} \qquad \text{(Canceling the common factors)}$$

$$= \frac{-1 - 2}{2(-1) - 3}$$

$$= \frac{3}{5}$$

2. $\displaystyle\lim_{x \to \infty} \overline{C}(x) = \lim_{x \to \infty} \left(1.8 + \frac{3000}{x} \right)$

$$= \lim_{x \to \infty} 1.8 + \lim_{x \to \infty} \frac{3000}{x}$$

$$= 1.8$$

Our computation reveals that as the production of audio discs increases "without bound," the average cost drops and approaches a unit cost of $1.80 per disc.

2.5   One-Sided Limits and Continuity

One-Sided Limits

Consider the function f defined by

$$f(x) = \begin{cases} x - 1 & \text{if } x < 0 \\ x + 1 & \text{if } x \geq 0. \end{cases}$$

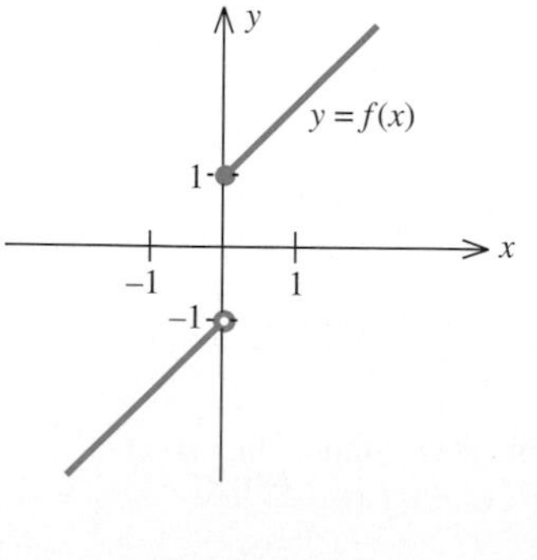

FIGURE 2.26
The function f does not have a limit as x approaches zero.

From the graph of f shown in Figure 2.26, we see that the function f does not have a limit as x approaches zero, because no matter how close x is to zero, $f(x)$ takes on values that are close to 1 if x is positive and values that are close to -1 if x is negative. Therefore, $f(x)$ cannot be close to a single number L—no matter how close x is to zero. Now if we restrict x to be greater than zero (to the right of zero), then we see that $f(x)$ can be made as close to the number 1 as we please by taking x sufficiently close to zero. In this situation, we say that the right-hand limit of f as x approaches zero (from the right) is 1, written

$$\lim_{x \to 0^+} f(x) = 1.$$

Similarly, we see that $f(x)$ can be made as close to the number -1 as we please by taking x sufficiently close to, but to the left of, zero. In this situation, we say that the left-hand limit of f as x approaches zero (from the left) is -1, written

$$\lim_{x \to 0^-} f(x) = -1.$$

These limits are called **one-sided limits.** More generally, we have the following definitions.

ONE-SIDED LIMITS

The function f has the **right-hand limit** L as x approaches a from the right, written

$$\lim_{x \to a^+} f(x) = L,$$

if the values $f(x)$ can be made as close to L as we please by taking x sufficiently close to (but not equal to) a and to the right of a.

Similarly, the function f has the **left-hand limit** M as x approaches a from the left, written

$$\lim_{x \to a^-} f(x) = M,$$

if the values $f(x)$ can be made as close to M as we please by taking x sufficiently close to (but not equal to) a and to the left of a.

The connection between one-sided limits and the two-sided limit defined earlier is given by the following theorem.

THEOREM 3

Let f be a function that is defined for all values of x close to $x = a$ with the possible exception of a itself. Then

$$\lim_{x \to a} f(x) = L \text{ if and only if } \lim_{x \to a^+} f(x) = \lim_{x \to a^-} f(x) = L.$$

Thus, the two-sided limit exists if and only if the one-sided limits exist and are equal.

EXAMPLE 1   Let

$$f(x) = \begin{cases} \sqrt{x} & \text{if } x > 0 \\ -x & \text{if } x \leq 0 \end{cases} \quad \text{and} \quad g(x) = \begin{cases} -1 & \text{if } x < 0 \\ 1 & \text{if } x \geq 0. \end{cases}$$

a. Show that $\lim_{x \to 0} f(x)$ exists by studying the one-sided limits of f as x approaches $x = 0$.

b. Show that $\lim_{x \to 0} g(x)$ does not exist.

FIGURE 2.27

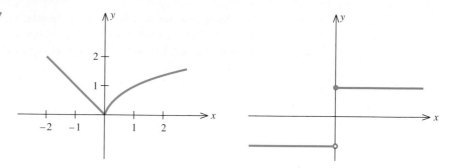

(a) $\lim\limits_{x \to 0} f(x)$ exists

(b) $\lim\limits_{x \to 0} g(x)$ does not exist

Solution

a. We find

$$\lim_{x \to 0^+} f(x) = \lim_{x \to 0^+} \sqrt{x} = 0$$

and

$$\lim_{x \to 0^-} f(x) = \lim_{x \to 0^-} (-x) = 0,$$

so

$$\lim_{x \to 0} f(x) = 0 \text{ (Figure 2.27a)}.$$

b. We have

$$\lim_{x \to 0^-} g(x) = -1 \quad \text{and} \quad \lim_{x \to 0^+} g(x) = 1,$$

and since these one-sided limits are not equal, we conclude that $\lim\limits_{x \to 0} g(x)$ does not exist (Figure 2.27b).   ∎

Continuous Functions

Continuous functions will play an important role throughout most of our study of calculus. Loosely speaking, a function is continuous at a point if the graph of the function at that point is devoid of holes, gaps, jumps, or breaks. Consider, for example, the graph of the function f depicted in Figure 2.28.

FIGURE 2.28

The graph of this function is not continuous at $x = a$, $x = b$, $x = c$, and $x = d$.

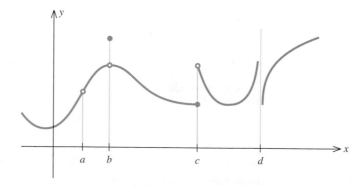

Let us take a closer look at the behavior of f at or near each of the points $x = a$, $x = b$, $x = c$, and $x = d$. First, note that f is not defined at $x = a$;

that is, the point $x = a$ is not in the domain of f, thereby resulting in a "hole" in the graph of f. Next, observe that the value of f at b, $f(b)$, is not equal to the limit of $f(x)$ as x approaches b, resulting in a "jump" in the graph of f at that point. The function f does not have a limit at $x = c$ since the left-hand and right-hand limits of $f(x)$ are not equal, also resulting in a jump in the graph of f at that point. Finally, the limit of f does not exist at $x = d$, resulting in a break in the graph of f. The function f is *discontinuous* at each of these points. It is *continuous* at all other points.

CONTINUITY AT A POINT

A function f is **continuous at the point** $x = a$ if the following conditions are satisfied:

1. $f(a)$ is defined,   **2.** $\lim\limits_{x \to a} f(x)$ exists,   **3.** $\lim\limits_{x \to a} f(x) = f(a)$.

Thus, a function f is continuous at the point $x = a$ if the limit of f at the point $x = a$ exists and has the value $f(a)$. Geometrically, f is continuous at the point $x = a$ if proximity of x to a implies the proximity of $f(x)$ to $f(a)$.

If f is not continuous at $x = a$, then f is said to be **discontinuous** at $x = a$. Also, f is **continuous on an interval** if f is continuous at every point in the interval.

FIGURE 2.29

The graph of f is continuous on the interval (a, b).

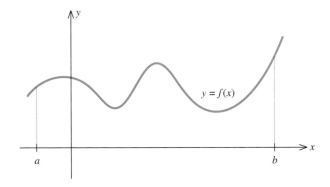

Figure 2.29 depicts the graph of a continuous function on the interval (a, b). Notice that the graph of the function over the stated interval can be sketched without lifting one's pencil from the paper.

EXAMPLE 2   Find the values of x for which each of the following functions is continuous.

a. $f(x) = x + 2$ **b.** $g(x) = \dfrac{x^2 - 4}{x - 2}$ **c.** $h(x) = \begin{cases} x + 2 & \text{if } x \neq 2 \\ 1 & \text{if } x = 2 \end{cases}$

d. $F(x) = \begin{cases} -1 & \text{if } x < 0 \\ 1 & \text{if } x \geq 0 \end{cases}$ **e.** $G(x) = \begin{cases} \dfrac{1}{x} & \text{if } x > 0 \\ -1 & \text{if } x \leq 0 \end{cases}$

The graph of each function is shown in Figure 2.30.

FIGURE 2.30

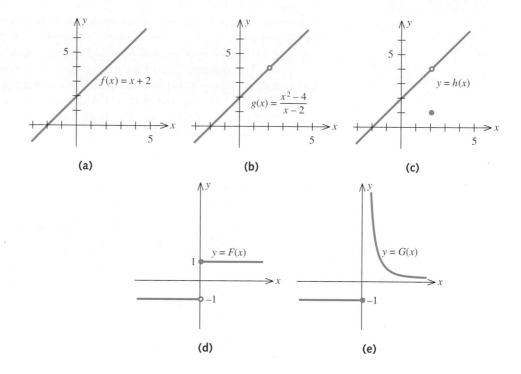

(a)   (b)   (c)

(d)   (e)

Solution

a. The function f is continuous everywhere, because the three conditions for continuity are satisfied for all values of x.

b. The function g is discontinuous at the point $x = 2$, because g is not defined at that point. It is continuous everywhere else.

c. The function h is discontinuous at $x = 2$ because the third condition for continuity is violated; the limit of $h(x)$ as x approaches 2 exists and has the value 4, but this limit is not equal to $h(2) = 1$. It is continuous for all other values of x.

d. The function F is continuous everywhere except at the point $x = 0$, where the limit of $F(x)$ fails to exist as x approaches zero (see Example 3a, Section 2.4).

e. Since the limit of $G(x)$ does not exist as x approaches zero, we conclude that G fails to be continuous at $x = 0$. The function G is continuous at all other points. ■

Properties of Continuous Functions

The following properties of continuous functions follow directly from the definition of continuity and the corresponding properties of limits. They are stated without proof.

PROPERTIES OF CONTINUOUS FUNCTIONS	**1.** The constant function $f(x) = c$ is continuous everywhere. **2.** The identity function $f(x) = x$ is continuous everywhere. *If f and g are continuous at $x = a$, then*

3. $[f(x)]^n$, where n is a real number, is continuous at $x = a$ whenever it is defined at that point.

4. $f \pm g$ is continuous at $x = a$.

5. fg is continuous at $x = a$.

6. $\dfrac{f}{g}$ is continuous at $x = a$ provided $g(a) \neq 0$.

Using these properties of continuous functions, we can prove the following results. (A proof is sketched in Exercise 75, page 107.)

CONTINUITY OF POLYNOMIAL AND RATIONAL FUNCTIONS

1. A polynomial function $y = P(x)$ is continuous at every point x.

2. A rational function $R(x) = p(x)/q(x)$ is continuous at every point x where $q(x) \neq 0$.

EXAMPLE 3 Find the values of x for which each of the following functions is continuous.

a. $f(x) = 3x^3 + 2x^2 - x + 10$ **b.** $g(x) = \dfrac{8x^{10} - 4x + 1}{x^2 + 1}$

c. $h(x) = \dfrac{4x^3 - 3x^2 + 1}{x^2 - 3x + 2}$

Solution

a. The function f is a polynomial function of degree 3, so $f(x)$ is continuous for all values of x.

b. The function g is a rational function. Observe that the denominator of g, namely $x^2 + 1$, is never equal to zero. Therefore, we conclude that g is continuous for all values of x.

c. The function h is a rational function. In this case, however, the denominator of h is equal to zero at $x = 1$ and $x = 2$, which can be seen by factoring it. Thus,

$$x^2 - 3x + 2 = (x - 2)(x - 1).$$

We conclude, therefore, that h is continuous everywhere except at $x = 1$ and $x = 2$, where it is discontinuous. ∎

Applications

Up to this point, most of the applications we have discussed involved functions that are continuous everywhere. In Example 4 we will consider an application from the field of educational psychology that involves a discontinuous function.

EXAMPLE 4 Figure 2.31 depicts the learning curve associated with a certain individual. Beginning with no knowledge of the subject being taught, the individual makes steady progress toward understanding it over the time interval $0 \le t < t_1$. In this instance, the individual's progress slows down as we approach time t_1 because he fails to grasp a particularly difficult concept. All of a sudden, a breakthrough occurs at time t_1, propelling his knowledge of the subject to a higher level. The curve is discontinuous at t_1.

FIGURE 2.31

A learning curve that is discontinuous at $t = t_1$

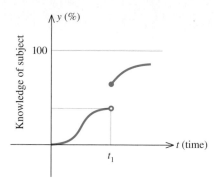

We close this section by pointing out a property of continuous functions that will play a very important role in calculus (Theorem 4).

THEOREM 4

Suppose that a continuous function f assumes the values $f(a)$ and $f(b)$ at two points $x = a$ and $x = b$ with $a < b$. If $f(a)$ and $f(b)$ have opposite signs, then there must be at least one point $x = c$, with $a < c < b$, where $f(c) = 0$ (see Figure 2.32).

FIGURE 2.32

$f(a)$ and $f(b)$ have opposite signs.

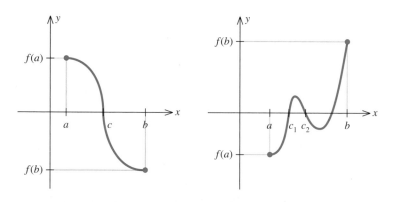

FIGURE 2.33

A discontinuous function

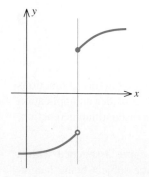

Geometrically, this property states that if the graph of a continuous function goes from above the x-axis to below the x-axis or vice versa, it must *cross* the x-axis. This, of course, is not necessarily true if the function is not continuous (Figure 2.33).

EXAMPLE 5   Let $f(x) = x^3 + x + 1$.

a. Show that f is continuous for all values of x.
b. Compute $f(-1)$ and $f(1)$ and use the results to deduce that there must be at least one point $x = c$ where c lies in the interval $(-1, 1)$ and $f(c) = 0$.

Solution

a. The function f is a polynomial function of degree 3 and is therefore continuous everywhere.

b. $f(-1) = (-1)^3 + (-1) + 1 = -1$

    $f(1) = 1^3 + 1 + 1 = 3$

Since $f(-1)$ and $f(1)$ have opposite signs, Theorem 4 tells us that there must be at least one point $x = c$ with $-1 < c < 1$ such that $f(c) = 0$. ■

**SELF-CHECK
EXERCISES 2.5**

1. Evaluate $\lim\limits_{x \to -1^-} f(x)$ and $\lim\limits_{x \to -1^+} f(x)$, where

$$f(x) = \begin{cases} 1 & \text{if } x < -1 \\ 1 + \sqrt{x+1} & \text{if } x \geq -1. \end{cases}$$

Does $\lim\limits_{x \to -1} f(x)$ exist?

2. Determine the values of x for which the given function is discontinuous. At each point of discontinuity, indicate which condition(s) for continuity are violated. Sketch the graph of the function.

a. $f(x) = \begin{cases} -x^2 + 1 & \text{if } x \leq 1 \\ x - 1 & \text{if } x > 1 \end{cases}$

b. $g(x) = \begin{cases} x + 1 & \text{if } x < -1 \\ 2 & \text{if } -1 < x \leq 1 \\ x + 3 & \text{if } x > 1 \end{cases}$

Solutions to Self-Check Exercises 2.5 can be found on page 107.

2.5   EXERCISES

In Exercises 1–8, use the graph of the function f to find $\lim\limits_{x \to a^-} f(x)$, $\lim\limits_{x \to a^+} f(x)$, *and* $\lim\limits_{x \to a} f(x)$ *at the indicated value of a, if the limit exists.*

1.

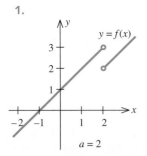

$a = 2$

2.

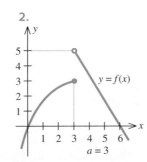

$a = 3$

3.

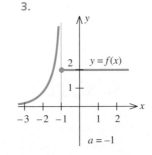

$a = -1$

4.

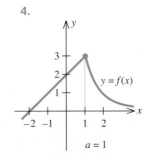

$a = 1$

5.

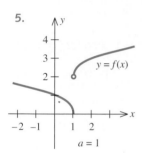

$a = 1$

6.

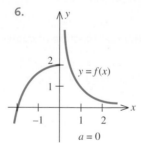

$a = 0$

7.

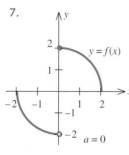

$a = 0$

8.

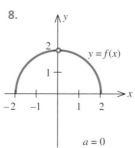

$a = 0$

28. $\lim\limits_{x \to 0^+} f(x)$ and $\lim\limits_{x \to 0^-} f(x)$, where

$$f(x) = \begin{cases} -x + 1 & \text{if } x \le 0 \\ 2x + 3 & \text{if } x > 0 \end{cases}$$

29. $\lim\limits_{x \to 1^+} f(x)$ and $\lim\limits_{x \to 1^-} f(x)$, where

$$f(x) = \begin{cases} \sqrt{x + 3} & \text{if } x \ge 1 \\ 2 + \sqrt{x} & \text{if } x < 1 \end{cases}$$

30. $\lim\limits_{x \to 1^+} f(x)$ and $\lim\limits_{x \to 1^-} f(x)$, where

$$f(x) = \begin{cases} x + 2\sqrt{x - 1} & \text{if } x \ge 1 \\ 1 - \sqrt{1 - x} & \text{if } x < 1 \end{cases}$$

In Exercises 31–38, determine the values of x, if any, at which each of the given functions is discontinuous. At each point of discontinuity, state the condition(s) for continuity that are violated.

In Exercises 9–30, find the indicated one-sided limit, if it exists.

9. $\lim\limits_{x \to 1^+} (2x + 4)$

10. $\lim\limits_{x \to 1^-} (3x - 4)$

11. $\lim\limits_{x \to 2^-} \dfrac{x - 3}{x + 2}$

12. $\lim\limits_{x \to 1^+} \dfrac{x + 2}{x + 1}$

13. $\lim\limits_{x \to 0^+} \dfrac{1}{x}$

14. $\lim\limits_{x \to 0^-} \dfrac{1}{x}$

15. $\lim\limits_{x \to 0^+} \dfrac{x - 1}{x^2 + 1}$

16. $\lim\limits_{x \to 2^+} \dfrac{x + 1}{x^2 - 2x + 3}$

17. $\lim\limits_{x \to 0^+} \sqrt{x}$

18. $\lim\limits_{x \to 2^+} 2\sqrt{x - 2}$

19. $\lim\limits_{x \to -2^+} (2x + \sqrt{2 + x})$

20. $\lim\limits_{x \to -5^+} x(1 + \sqrt{5 + x})$

21. $\lim\limits_{x \to 1^-} \dfrac{1 + x}{1 - x}$

22. $\lim\limits_{x \to 1^+} \dfrac{1 + x}{1 - x}$

23. $\lim\limits_{x \to 2^-} \dfrac{x^2 - 4}{x - 2}$

24. $\lim\limits_{x \to -3^+} \dfrac{\sqrt{x + 3}}{x^2 + 1}$

25. $\lim\limits_{x \to 3^+} \dfrac{x^2 - 9}{x + 3}$

26. $\lim\limits_{x \to 2^-} \dfrac{\sqrt[3]{x + 10}}{2x^2 + 1}$

27. $\lim\limits_{x \to 0^+} f(x)$ and $\lim\limits_{x \to 0^-} f(x)$, where

$$f(x) = \begin{cases} 2x & \text{if } x < 0 \\ x^2 & \text{if } x \ge 0 \end{cases}$$

31.

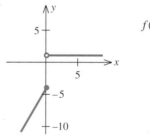

$$f(x) = \begin{cases} 2x - 4 & \text{if } x \le 0 \\ 1 & \text{if } x > 0 \end{cases}$$

32.

$$f(x) = \begin{cases} x^2 + 1 & \text{if } x \ne 0 \\ 0 & \text{if } x = 0 \end{cases}$$

33.

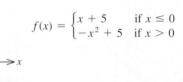

$$f(x) = \begin{cases} x + 5 & \text{if } x \le 0 \\ -x^2 + 5 & \text{if } x > 0 \end{cases}$$

34.

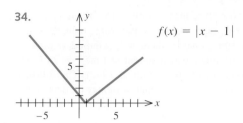

$$f(x) = |x - 1|$$

35.

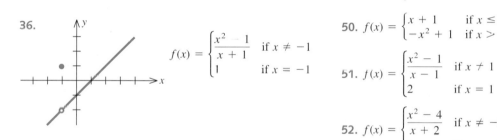

$$f(x) = \begin{cases} x + 5 & \text{if } x < 0 \\ 2 & \text{if } x = 0 \\ -x^2 + 5 & \text{if } x > 0 \end{cases}$$

36.

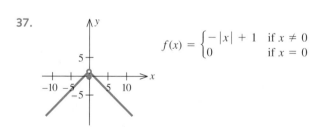

$$f(x) = \begin{cases} \dfrac{x^2 - 1}{x + 1} & \text{if } x \neq -1 \\ 1 & \text{if } x = -1 \end{cases}$$

37.

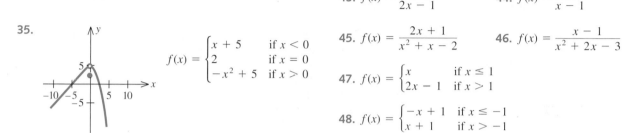

$$f(x) = \begin{cases} -|x| + 1 & \text{if } x \neq 0 \\ 0 & \text{if } x = 0 \end{cases}$$

38.

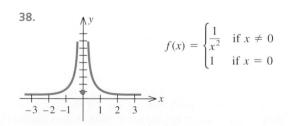

$$f(x) = \begin{cases} \dfrac{1}{x^2} & \text{if } x \neq 0 \\ 1 & \text{if } x = 0 \end{cases}$$

In Exercises 39–54, find the values of x for which each of the given functions is continuous.

39. $f(x) = 2x^2 + x - 1$

40. $f(x) = x^3 - 2x^2 + x - 1$

41. $f(x) = \dfrac{2}{x^2 + 1}$ **42.** $f(x) = \dfrac{x}{2x^2 + 1}$

43. $f(x) = \dfrac{2}{2x - 1}$ **44.** $f(x) = \dfrac{x + 1}{x - 1}$

45. $f(x) = \dfrac{2x + 1}{x^2 + x - 2}$ **46.** $f(x) = \dfrac{x - 1}{x^2 + 2x - 3}$

47. $f(x) = \begin{cases} x & \text{if } x \leq 1 \\ 2x - 1 & \text{if } x > 1 \end{cases}$

48. $f(x) = \begin{cases} -x + 1 & \text{if } x \leq -1 \\ x + 1 & \text{if } x > -1 \end{cases}$

49. $f(x) = \begin{cases} -2x + 1 & \text{if } x < 0 \\ x^2 + 1 & \text{if } x > 0 \end{cases}$

50. $f(x) = \begin{cases} x + 1 & \text{if } x \leq 1 \\ -x^2 + 1 & \text{if } x > 1 \end{cases}$

51. $f(x) = \begin{cases} \dfrac{x^2 - 1}{x - 1} & \text{if } x \neq 1 \\ 2 & \text{if } x = 1 \end{cases}$

52. $f(x) = \begin{cases} \dfrac{x^2 - 4}{x + 2} & \text{if } x \neq -2 \\ 1 & \text{if } x = -2 \end{cases}$

53. $f(x) = |x + 1|$ **54.** $f(x) = \dfrac{|x - 1|}{x - 1}$

In Exercises 55–58, determine all values of x at which the function is discontinuous.

55. $f(x) = \dfrac{2x}{x^2 - 1}$ **56.** $f(x) = \dfrac{1}{(x - 1)(x - 2)}$

57. $f(x) = \dfrac{x^2 - 2x}{x^2 - 3x + 2}$ **58.** $f(x) = \dfrac{x^2 - 3x + 2}{x^2 - 2x}$

59. The Postage Function   The graph of the "postage function"

$$f(x) = \begin{cases} 29 & \text{if } 0 < x \leq 1 \\ 52 & \text{if } 1 < x \leq 2 \\ \vdots \\ \vdots \\ 278 & \text{if } 11 < x \leq 12, \end{cases}$$

where x denotes the weight of a parcel in ounces and $f(x)$ the postage in cents, is shown in the figure. Determine the values of x for which f is discontinuous.

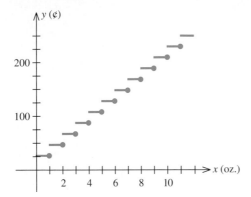

60. **Inventory Control** As part of an optimal inventory policy, the manager of an office supply company orders 500 reams of xerographic copy paper every 20 days. The accompanying graph shows the *actual* inventory level of copy paper in an office supply store during the first 60 business days of 1992. Determine the values of t for which the "inventory function" is discontinuous and give an interpretation of the graph.

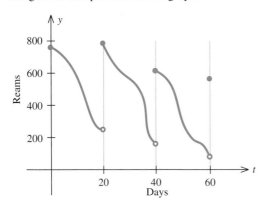

61. **Learning Curves** The following graph describes the progress Michael made in solving a problem correctly during a mathematics quiz. Here y denotes the percentage of work completed and x is measured in minutes. Give an interpretation of the graph.

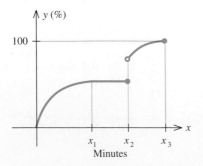

62. **Ailing Financial Institutions** The Franklin Savings and Loan Company acquired two ailing financial institutions in 1992. One of them was acquired at time $t = T_1$, and the other was acquired at time $t = T_2$ ($t = 0$ corresponds to the beginning of 1992). The following graph shows the total amount of money on deposit with Franklin.

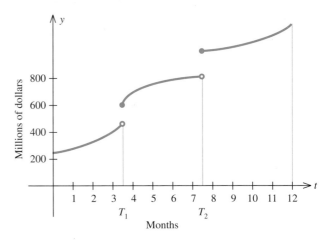

Explain the significance of the discontinuities of the function at T_1 and T_2.

63. **Energy Consumption** The following graph shows the amount of home heating oil remaining in a 200-gallon tank over a 120-day period ($t = 0$ corresponds to October 1).

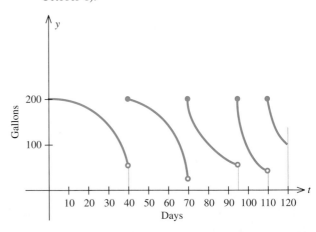

Explain why the function is discontinuous at $t = 40$, $t = 70$, $t = 95$, and $t = 110$.

64. **Administration of an Intravenous Solution** A dextrose solution is being administered to a patient intravenously. The 1-liter bottle holding the solution is removed and replaced by another as soon as the contents drop to approximately 5 percent of the initial (1-liter) amount. The rate of discharge is constant, and it takes

6 hours to discharge 95 percent of the contents of a full bottle. Draw a graph showing the amount of dextrose solution in a bottle in the IV system over a 24-hour period, assuming that we started with a full bottle.

65. **Commissions** The base salary of a salesman working on commission is $12,000. For each $50,000 of sales beyond $100,000 he is paid a $1,000 commission. Sketch a graph showing his earnings as a function of the level of his sales x. Determine the values of x for which the function f is discontinuous.

66. **Parking Fees** The fee charged per car in a downtown parking lot is $1 for the first half hour and 50 cents for each additional half hour or part thereof, subject to a maximum of $5. Derive a function f relating the parking fee to the length of time a car is left in the lot. Sketch the graph of f and determine the values of x for which the function f is discontinuous.

67. **Commodity Prices** The function that gives the cost of a certain commodity is defined by

$$C(x) = \begin{cases} 5x & \text{if } 0 < x < 10 \\ 4x & \text{if } 10 \le x < 30 \\ 3.5x & \text{if } 30 \le x < 60 \\ 3.25x & \text{if } x \ge 60, \end{cases}$$

where x is the number of pounds of a certain commodity sold and $C(x)$ is measured in dollars. Sketch the graph of the function C and determine the values of x for which the function C is discontinuous.

68. **Energy Expended by a Fish** Suppose that a fish swimming a distance of L ft at a speed of v ft/sec relative to the water and against a current flowing at the rate of u ft/sec ($u < v$) expends a total energy given by

$$E(v) = \frac{aLv^3}{v - u},$$

where E is measured in foot-pounds (ft-lb), and a is a constant.
a. Evaluate $\lim_{v \to u^+} E(v)$ and interpret your result.
b. Evaluate $\lim_{v \to \infty} E(v)$ and interpret your result.

In Exercises 69–72, (a) show that the function f is continuous for all values of x in the interval [a, b] and (b) prove that f

must have at least one zero in the interval (a, b) by showing that f(a) and f(b) have opposite signs.

69. $f(x) = x^2 - 6x + 8$ $a = 1$, $b = 3$

70. $f(x) = x^3 - 2x^2 + 3x + 2$ $a = -1$, $b = 1$

71. $f(x) = 2x^3 - 3x^2 - 36x + 14$ $a = 0$, $b = 1$

C 72. $f(x) = 2x^{5/3} - 5x^{4/3}$ $a = 14$, $b = 16$

73. Let $f(x) = x - \sqrt{1 - x^2}$.
a. Show that f is continuous for all values of x in the interval $[-1, 1]$.
b. Show that f has at least one zero in $[-1, 1]$.
c. Find the zeros of f in $[-1, 1]$ by solving the equation $f(x) = 0$.

74. Let $f(x) = \dfrac{x^2}{x^2 + 1}$.
a. Show that f is continuous for all values of x.
b. Show that $f(x)$ is nonnegative for all values of x.
c. Show that f has a zero at $x = 0$. Does this contradict Theorem 4?

75. a. Prove that a polynomial function $y = P(x)$ is continuous at every point x. Follow these steps:
   1. Use Properties 2 and 3 of continuous functions to establish that the function $g(x) - x^n$, where n is a positive integer, is continuous everywhere.
   2. Use Properties 1 and 5 to show that $f(x) = cx^n$, where c is a constant and n is a positive integer, is continuous everywhere.
   3. Use Property 4 to complete the proof of the result.
b. Prove that a rational function $R(x) = p(x)/q(x)$ is continuous at every point x where $q(x) \ne 0$. [*Hint:* Use the result of (a) and Property 6.]

C *In Exercises 76 and 77, use a computer or graphing calculator to plot the graph of the function f and find the specified limit (if it exists).*

76. $f(x) = \begin{cases} 2x^2 & \text{if } x \le 3 \\ -3x & \text{if } x > 3 \end{cases}$; $\lim_{x \to 3^+} f(x)$ and $\lim_{x \to 3^-} f(x)$

77. $f(x) = \dfrac{\sqrt{1 + x} - 1}{x}$; $\lim_{x \to 0^+} f(x)$ and $\lim_{x \to 0^-} f(x)$

SOLUTIONS TO SELF-CHECK EXERCISES 2.5

1. For $x < -1$, $f(x) = 1$, and so

$$\lim_{x \to -1^-} f(x) = \lim_{x \to -1^-} 1 = 1.$$

For $x \ge -1$, $f(x) = 1 + \sqrt{x + 1}$, and so

$$\lim_{x \to -1^+} f(x) = \lim_{x \to -1^+} (1 + \sqrt{x + 1}) = 1.$$

Since the left-hand and right-hand limits of f exist as x approaches $x = -1$ and both are equal to 1, we conclude that

$$\lim_{x \to -1} f(x) = 1.$$

2. a. The graph of f is as follows.

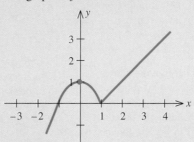

We see that f is continuous everywhere.

b. The graph of g is as follows:

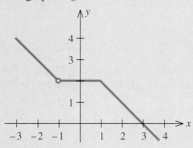

Since g is not defined at $x = -1$, it is discontinuous there. It is continuous everywhere else.

2.6   The Derivative

An Intuitive Example

We mentioned in Section 2.4 that the problem of finding the *rate of change* of one quantity with respect to another is mathematically equivalent to the problem of finding the *slope of the tangent line* to a curve at a given point on the curve. Before going on to establish this relationship, let us show its plausibility by looking at it from an intuitive point of view.

Consider the motion of the car discussed in Section 2.4. Recall that the position of the car at any time t is given by

$$s = f(t) = 2t^2 \qquad (0 \le t \le 10),$$

where s is measured in feet and t in seconds. The graph of the function f is sketched in Figure 2.34.

Observe that the graph of f rises slowly at first but more rapidly as t increases, reflecting the fact that the speed of the car is increasing with time. This observation suggests a relationship between the speed of the car at any time t and the *steepness* of the curve at the point corresponding to this value

FIGURE 2.34

Graph showing the position s of a car at time t.

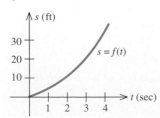

of t. Thus it would appear that we can solve the problem of finding the speed of the car at any time if we can find a way to measure the steepness of the curve at any point on the curve.

To discover a yardstick that will measure the steepness of a curve, consider the graph of a function f such as the one shown in Figure 2.35a. Think of the curve as representing a stretch of roller coaster track (Figure 2.35b). When the car is at the point P on the curve, a passenger sitting erect in the car and looking straight ahead will have a line of sight that is parallel to the line T, the tangent to the curve at P.

FIGURE 2.35

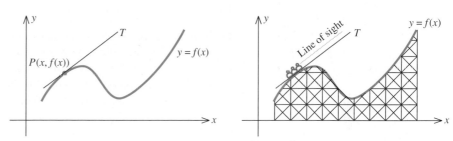

(a) T is the tangent line to the curve at P.   (b) T is parallel to the line of sight.

As Figure 2.35a suggests, the steepness of the curve—that is, the rate at which y is increasing or decreasing with respect to x—is given by the slope of the tangent line to the graph of f at the point $P(x, f(x))$. But for now we will show how this relationship can be used to estimate the rate of change of a function from its graph.

EXAMPLE 1   The graph of the function $y = N(t)$, shown in Figure 2.36, gives the number of social security beneficiaries from the beginning of 1990 ($t = 0$) through the year 2045 ($t = 55$).

FIGURE 2.36

The number of social security beneficiaries from 1990 through 2045. We can use the slope of the tangent line at the indicated points to estimate the rate at which the number of social security beneficiaries will be changing.

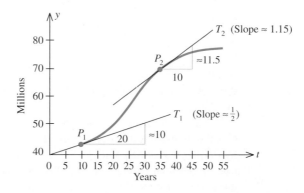

Use the graph of $y = N(t)$ to estimate the rate at which the number of social security beneficiaries will be growing at the beginning of the year 2000 ($t = 10$) and at the beginning of 2025 ($t = 35$). [Assume that the rate of change of the function N at any value of t is given by the slope of the tangent line at the point $P(t, N(t))$.]

Solution   From the figure, we see that the slope of the tangent line T_1 to the graph of $y = N(t)$ at $P_1(10, 44.7)$ is approximately 0.5. This tells us that the quantity y is increasing at the rate of $1/2$ unit per unit increase in t, when $t = 10$. In other words, at the beginning of the year 2000, the number of social security beneficiaries will be increasing at the rate of approximately .5 million, or 500,000, per year.

The slope of the tangent line T_2 at $P_2(35, 71.9)$ is approximately 1.15. This tells us that at the beginning of 2025, the number of social security beneficiaries will be growing at the rate of approximately 1.15 million, or 1,150,000, per year. ∎

Slope of a Tangent Line

In Example 1, we answered the questions raised by drawing the graph of the function N and estimating the position of the tangent lines. Ideally, however, we would like to solve a problem analytically whenever possible. In order to do this we need a precise definition of the slope of a tangent line to a curve.

To define the tangent line to a curve C at a point P on the curve, fix P and let Q be any point on C distinct from P (Figure 2.37). The straight line passing through P and Q is called a **secant line.**

FIGURE 2.37

As Q approaches P along the curve C, the secant lines approach the tangent line T.

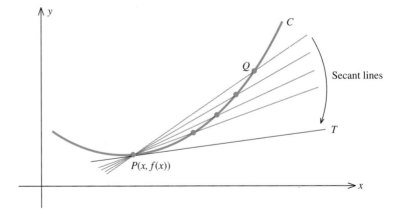

Now, as the point Q is allowed to move toward P along the curve, the secant line through P and Q rotates about the fixed point P and approaches a fixed line through P. This fixed line, which is the limiting position of the secant lines through P and Q as Q approaches P, is the **tangent line** to the graph of f at the point P.

We can describe the process more precisely as follows. Suppose the curve C is the graph of a function f defined by $y = f(x)$. Then the point P is described by $P(x, f(x))$ and the point Q by $Q(x + h, f(x + h))$, where h is some appropriate nonzero number (Figure 2.38a). Observe that we can make Q approach P along the curve C by letting h approach zero (Figure 2.38b).

Next, using the formula for the slope of a line, we can write the slope of the secant line passing through $P(x, f(x))$ and $Q(x + h, f(x + h))$ as

$$\frac{f(x + h) - f(x)}{(x + h) - x} = \frac{f(x + h) - f(x)}{h}. \qquad (5)$$

As observed earlier, Q approaches P and, therefore, the secant line through P and Q approaches the tangent line T as h approaches zero. Consequently, we

FIGURE 2.38

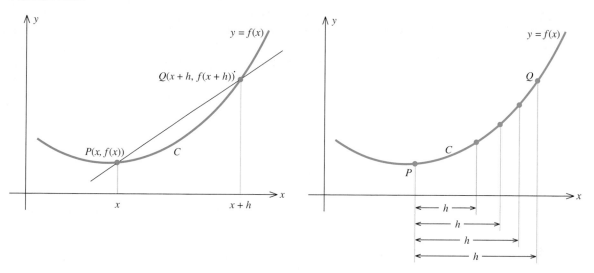

(a) The points $P(x, f(x))$ and $Q(x + h, f(x + h))$ **(b)** As h approaches 0, Q approaches P.

might expect that the slope of the secant line would approach the slope of the tangent line T as h approaches zero. This leads to the following definition.

SLOPE OF A TANGENT LINE

The slope of the tangent line to the graph f at the point $P(x, f(x))$ is given by

$$\lim_{h \to 0} \frac{f(x + h) - f(x)}{h}, \tag{6}$$

if it exists.

Rates of Change

We will now show that the problem of finding the slope of the tangent line to the graph of a function f at the point $P(x, f(x))$ is mathematically equivalent to the problem of finding the rate of change of f at x. To see this, suppose that we are given a function f that describes the relationship between the two quantities x and y:

$$y = f(x).$$

The number $f(x + h) - f(x)$ measures the change in y that corresponds to a change h in x (Figure 2.39).

Then, the **difference quotient**

$$\frac{f(x + h) - f(x)}{h} \tag{7}$$

measures the **average rate of change of y with respect to x** over the interval $[x, x + h]$. For example, if y measures the position of a car at time x, then (7) gives the average velocity of the car over the time interval $[x, x + h]$.

FIGURE 2.39
$f(x + h) - f(x)$ is the change in y that corresponds to a change h in x.

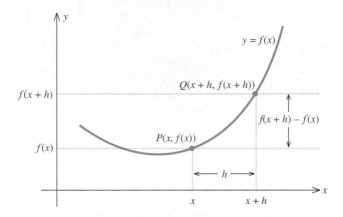

Observe that the difference quotient (7) is the same as (5). We conclude that the difference quotient (7) also measures the slope of the secant line that passes through the two points $P(x, f(x))$ and $Q(x + h, f(x + h))$ lying on the graph of $y = f(x)$. Next, by taking the limit of the difference quotient (7) as h goes to zero, that is, by evaluating

$$\lim_{h \to 0} \frac{f(x + h) - f(x)}{h}, \qquad (8)$$

we obtain the **rate of change of f at x**. For example, if y measures the position of a car at time x, then (8) gives the velocity of the car at time x. For emphasis, the rate of change of a function f at x is often called the **instantaneous rate of change of f at x**. This distinguishes it from the average rate of change of f, which is computed over an *interval* $[x, x + h]$ rather than at a *point x*.

Observe that the limit (8) is the same as (6). Therefore, the limit of the difference quotient also measures the slope of the tangent line to the graph of $y = f(x)$ at the point $(x, f(x))$.

The following summarizes this discussion.

AVERAGE AND INSTANTANEOUS RATES OF CHANGE	**Average rate of change** of f over the interval $[x, x + h]$ or **slope of the secant line** to the graph of f through the points $(x, f(x))$ and $(x + h, f(x + h))$ is $$\frac{f(x + h) - f(x)}{h}. \qquad (9)$$ **Instantaneous rate of change** of f at x or **slope of tangent line** to the graph of f at $(x, f(x))$ is $$\lim_{h \to 0} \frac{f(x + h) - f(x)}{h}. \qquad (10)$$

The Derivative

The limit (6), or (10), which measures both the slope of the tangent line to the graph of $y = f(x)$ at the point $P(x, f(x))$ and the (instantaneous) rate of change of f at x is given a special name: the **derivative of f at x**.

THE DERIVATIVE OF A FUNCTION

The derivative of a function f with respect to x is the function f' (read "f prime of x"), defined by

$$f'(x) = \lim_{h \to 0} \frac{f(x + h) - f(x)}{h}. \qquad (11)$$

The domain of f' is the set of all x where the limit exists.

Thus, the derivative of a function f is a function f' that gives the slope of the tangent line to the graph of f at *any* point $(x, f(x))$, and also the rate of change of f at x (Figure 2.40).

FIGURE 2.40

The slope of the tangent line at $P(x, f(x))$ is $f'(x)$; f changes at the rate of $f'(x)$ units per unit change in x at x.

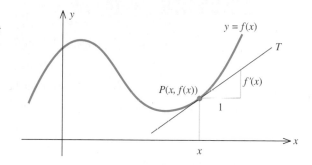

Other notations for the derivative of f include:

$$D_x f(x) \qquad \text{(Read "dee sub x of f of x")}$$

$$\frac{dy}{dx} \qquad \text{(Read "dee y dee x")}$$

and $\qquad y' \qquad$ (Read "y prime")

The last two are used when the rule for f is written in the form $y = f(x)$.

The calculation of the derivative of f is facilitated using the following Four-Step Process.

THE FOUR-STEP PROCESS FOR FINDING $f'(x)$

1. Compute $f(x + h)$.

2. Form the difference $f(x + h) - f(x)$.

3. Form the quotient $\dfrac{f(x + h) - f(x)}{h}$.

4. Compute $f'(x) = \lim\limits_{h \to 0} \dfrac{f(x + h) - f(x)}{h}$.

EXAMPLE 2 Find the slope of the tangent line to the graph of $f(x) = 3x + 5$ at any point $(x, f(x))$.

Solution The slope of the tangent line at any point on the graph of f is given by the derivative of f at x. To find the derivative, we use the Four-Step Process:

Step 1 $f(x + h) = 3(x + h) + 5 = 3x + 3h + 5$

Step 2 $f(x + h) - f(x) = (3x + 3h + 5) - (3x + 5) = 3h$

Step 3 $\dfrac{f(x + h) - f(x)}{h} = \dfrac{3h}{h} = 3$

Step 4 $f'(x) = \lim\limits_{h \to 0} \dfrac{f(x + h) - f(x)}{h} = \lim\limits_{h \to 0} 3 = 3$

We expect this result, since the tangent line to any point on a straight line must coincide with the line itself and, therefore, must have the same slope as the line. In this case, the graph of f is a straight line with slope 3. ■

EXAMPLE 3   Let $f(x) = x^2$.

a. Compute $f'(x)$.
b. Compute $f'(2)$ and interpret your result.

FIGURE 2.41
The tangent line to the graph of
$f(x) = x^2$ *at* (2, 4)

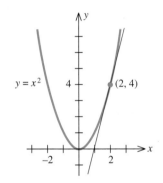

Solution
a. To find $f'(x)$ we use the Four-Step Process.

Step 1 $f(x + h) = (x + h)^2 = x^2 + 2xh + h^2$

Step 2 $f(x + h) - f(x) = x^2 + 2xh + h^2 - x^2 = 2xh + h^2 = h(2x + h)$

Step 3 $\dfrac{f(x + h) - f(x)}{h} = \dfrac{h(2x + h)}{h} = 2x + h$

Step 4 $f'(x) = \lim\limits_{h \to 0} \dfrac{f(x + h) - f(x)}{h} = \lim\limits_{h \to 0} (2x + h) = 2x$

b. $f'(2) = 2(2) = 4$. This result tells us that the slope of the tangent line to the graph of f at the point (2, 4) is 4. It also tells us that the function f is changing at the rate of 4 units per unit change in x at $x = 2$. The graph of f and the tangent line at (2, 4) are shown in Figure 2.41. ■

EXAMPLE 4   Let $f(x) = x^2 - 4x$.

a. Compute $f'(x)$.
b. Find the point on the graph of f where the tangent line to the curve is horizontal.
c. Sketch the graph of f and the tangent line to the curve at the point found in (b).
d. What is the rate of change of f at this point?

Solution
a. To find $f'(x)$, we use the Four-Step Process.

Step 1 $f(x + h) = (x + h)^2 - 4(x + h) = x^2 + 2xh + h^2 - 4x - 4h$

Step 2 $f(x + h) - f(x) = x^2 + 2xh + h^2 - 4x - 4h - (x^2 - 4x)$
$= 2xh + h^2 - 4h = h(2x + h - 4)$

Step 3 $\dfrac{f(x + h) - f(x)}{h} = \dfrac{h(2x + h - 4)}{h} = 2x + h - 4$

Step 4 $f'(x) = \lim\limits_{h \to 0} \dfrac{f(x + h) - f(x)}{h} = \lim\limits_{h \to 0} (2x + h - 4) = 2x - 4$

b. At a point on the graph of f where the tangent line to the curve is horizontal and hence has slope zero, the derivative f' of f is zero. Accordingly, to find such point(s) we set $f'(x) = 0$, which gives $2x - 4 = 0$, or $x = 2$. The corresponding value of y is given by $y = f(2) = -4$, and the required point is $(2, -4)$.

c. The graph of f and the tangent line are shown in Figure 2.42.

d. The rate of change of f at $x = 2$ is zero.

FIGURE 2.42

The tangent line to the graph of $y = x^2 - 4x$ at $(2, -4)$ is $y = -4$.

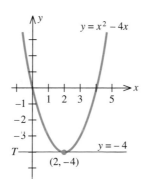

EXAMPLE 5 Let $f(x) = 1/x$.

a. Compute $f'(x)$.

b. Find the slope of the tangent line T to the graph of f at the point where $x = 1$.

c. Find an equation of the tangent line T in (b).

Solution

a. To find $f'(x)$, we use the Four-Step Process.

Step 1 $f(x + h) = \dfrac{1}{x + h}$

Step 2 $f(x + h) - f(x) = \dfrac{1}{x + h} - \dfrac{1}{x} = \dfrac{x - (x + h)}{x(x + h)} = -\dfrac{h}{x(x + h)}$

Step 3 $\dfrac{f(x + h) - f(x)}{h} = -\dfrac{h}{x(x + h)} \cdot \dfrac{1}{h} = -\dfrac{1}{x(x + h)}$

Step 4 $f'(x) = \lim\limits_{h \to 0} \dfrac{f(x + h) - f(x)}{h} = \lim\limits_{h \to 0} -\dfrac{1}{x(x + h)} = -\dfrac{1}{x^2}$

FIGURE 2.43

The tangent line to the graph of $f(x) = 1/x$ at $(1, 1)$

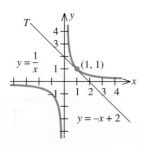

b. The slope of the tangent line T to the graph of f where $x = 1$ is given by $f'(1) = -1$.

c. When $x = 1$, $y = f(1) = 1$ and T is tangent to the graph of f at the point $(1, 1)$. From (b), we know that the slope of T is -1. Thus, an equation of T is

$$y - 1 = -1(x - 1)$$

or

$$y = -x + 2 \qquad \text{(Figure 2.43)}.$$

Applications

EXAMPLE 6   Suppose that the distance (in feet) covered by a car moving along a straight road t seconds after starting from rest is given by the function $f(t) = 2t^2 \quad (0 \le t \le 30)$.

a. Calculate the average velocity of the car over the time intervals [22, 23], [22, 22.1], and [22, 22.01].
b. Calculate the (instantaneous) velocity of the car when $t = 22$.
c. Compare the results obtained in (a) with that obtained in (b).

Solution
a. We first compute the average velocity (average rate of change of f) over the interval $[t, t + h]$ using (9). We find

$$\frac{f(t + h) - f(t)}{h} = \frac{2(t + h)^2 - 2t^2}{h}$$
$$= \frac{2t^2 + 4th + 2h^2 - 2t^2}{h}$$
$$= 4t + 2h.$$

Next, using $t = 22$ and $h = 1$, we find that the average velocity of the car over the time interval [22, 23] is

$$4(22) + 2(1) = 90,$$

or 90 feet per second. Similarly, using $t = 22$, $h = 0.1$, and $h = 0.01$, we find that its average velocities over the time intervals [22, 22.1] and [22, 22.01] are 88.2 and 88.02 feet per second, respectively.

b. Using (10), we see that the instantaneous velocity of the car at any time t is given by

$$\lim_{h \to 0} \frac{f(t + h) - f(t)}{h} = \lim_{h \to 0} (4t + 2h) \qquad \text{[Using the results from (a)]}$$
$$= 4t.$$

In particular, the velocity of the car 22 seconds from rest ($t = 22$) is given by

$$v = 4(22),$$

or 88 feet per second.

c. The computations in (a) show that as the time intervals over which the average velocity of the car are computed become smaller and smaller, the average velocities over these intervals do approach 88 ft/sec, the instantaneous velocity of the car at $t = 22$.   ■

FIGURE 2.44

The graph of the demand function $p = 144 - x^2$

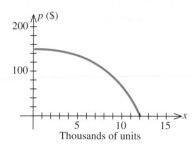

EXAMPLE 7   The management of the Titan Tire Company has determined that the weekly demand function for their Super Titan tires is given by

$$p = f(x) = 144 - x^2,$$

where p is measured in dollars and x is measured in units of a thousand (Figure 2.44).

a. Find the average rate of change in the unit price of a tire if the quantity demanded is between 5000 and 6000 tires. Between 5000 and 5100 tires. Between 5000 tires and 5010 tires.

b. What is the instantaneous rate of change of the unit price when the quantity demanded is 5000 units?

Solution

a. The average rate of change of the unit price of a tire if the quantity demanded is between x and $x + h$ is

$$\frac{f(x + h) - f(x)}{h} = \frac{[144 - (x + h)^2] - (144 - x^2)}{h}$$

$$- \frac{144 - x^2 - 2x(h) - h^2 - 144 + x^2}{h}$$

$$= -2x - h.$$

To find the average rate of change of the unit price of a tire when the quantity demanded is between 5000 and 6000 tires (that is, over the interval $[5, 6]$), we take $x = 5$ and $h = 1$, obtaining

$$-2(5) - 1 = -11,$$

or $-\$11$ per 1000 tires. (Remember, x is measured in units of a thousand.) Similarly, taking $h = 0.1$, and $h = 0.01$ with $x = 5$, we find that the average rates of change of the unit price when the quantities demanded are between 5000 and 5100 and between 5000 and 5010 are $-\$10.10$ and $-\$10.01$ per 1000 tires, respectively.

b. The instantaneous rate of change of the unit price of a tire when the quantity demanded is x units is given by

$$\lim_{h \to 0} \frac{f(x + h) - f(x)}{h} = \lim_{h \to 0} (-2x - h) \qquad \text{[Using the results from (a)]}$$

$$= -2x.$$

In particular, the instantaneous rate of change of the unit price per tire when the quantity demanded is 5000 is given by $-2(5)$, or $-\$10$ per 1000 tires. ■

The derivative of a function provides us with a tool for measuring the rate of change of one quantity with respect to another. Table 2.3 lists several other applications involving this limit.

TABLE 2.3

x stands for	y stands for	$\dfrac{f(a + h) - f(a)}{h}$ measures the	$\displaystyle\lim_{h\to 0} \dfrac{f(a + h) - f(a)}{h}$ measures the
time	**concentration of a drug** in the bloodstream at time x	average rate of change in the concentration of the drug over the time interval $[a, a + h]$	instantaneous rate of change in the concentration of the drug in the bloodstream at time $x = a$
number of items sold	**revenue** at a sales level of x units	average rate of change in the revenue when the sales level is between $x = a$ and $x = a + h$	instantaneous rate of change in the revenue when the sales level is a units
time	**volume of sales** at time x	average rate of change in the volume of sales over the time interval $[a, a + h]$	instantaneous rate of change in the volume of sales at time $x = a$
time	**population** of Drosophila (fruit flies) at time x	average rate of growth of the fruit fly population over the time interval $[a, a + h]$	instantaneous rate of change of the fruit fly population at time $x = a$
temperature in a chemical reaction	**amount of product formed in the chemical reaction** when the temperature is x degrees	average rate of formation of chemical product over the temperature range $[a, a + h]$	instantaneous rate of formation of chemical product when the temperature is a degrees

Differentiability and Continuity

In practical applications, one encounters functions that fail to be **differentiable,** that is, do not have a derivative at certain values in the domain of the function f. It can be shown that a continuous function f fails to be differentiable at a point $x = a$ when the graph of f makes an abrupt change of direction at that point. We call such a point a "corner." A function also fails to be differentiable at a point where the tangent line is vertical, since the slope of a vertical line is undefined. These cases are illustrated in Figure 2.45.

FIGURE 2.45

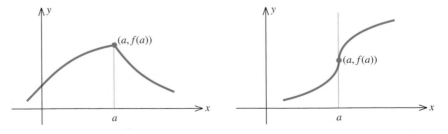

(a) The graph makes an abrupt change of direction at $x = a$.

(b) The slope at $x = a$ is undefined.

The next example illustrates a function that is not differentiable at a point.

FIGURE 2.46
The function f is not differentiable at (8, 48).

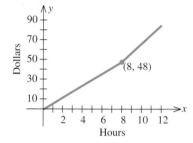

EXAMPLE 8　Mary works at the B&O department store, where, on a weekday, she is paid $6 per hour for the first 8 hours and $9 per hour for overtime. The function

$$f(x) = \begin{cases} 6x & 0 \le x \le 8 \\ 9x - 24 & 8 < x \end{cases}$$

gives Mary's earnings on a weekday in which she worked x hours. Sketch the graph of the function f and explain why it is not differentiable at $x = 8$.

Solution　The graph of f is shown in Figure 2.46. Observe that the graph of f has a corner at $x = 8$ and consequently is not differentiable at $x = 8$. ∎

We close this section by mentioning the connection between the continuity and the differentiability of a function at a given value $x = a$ in the domain of f. By re-examining the function of Example 8, it becomes clear that f is continuous everywhere and, in particular, when $x = 8$. This shows that, in general, the continuity of a function at a point $x = a$ does not necessarily imply the differentiability of the function at that point. The converse, however, is true: If a function f is differentiable at a point $x = a$, then it is continuous there.

DIFFERENTIABILITY AND CONTINUITY

If a function is differentiable at $x = a$, then it is continuous at $x = a$.

For a proof of this result, see Exercise 17, page 123.

EXAMPLE 9　Figure 2.47 depicts a portion of the graph of a function. Explain why the function fails to be differentiable at each of the points $x = a$, b, c, d, e, f, and g.

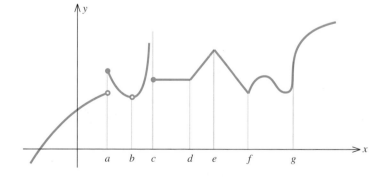

Solution　The function fails to be differentiable at the points $x = a$, b, and c because it is discontinuous at each of these points. The derivative of the function does not exist at $x = d$, e, and f because it has a kink at each of these points. Finally, the function is not differentiable at $x = g$ because the tangent line is vertical at that point. ∎

**SELF-CHECK
EXERCISES 2.6**

1. Let $f(x) = -x^2 - 2x + 3$.
   a. Find the derivative f' of f using the definition of the derivative.
   b. Find the slope of the tangent line to the graph of f at the point $(0, 3)$.
   c. Find the rate of change of f when $x = 0$.
   d. Find an equation of the tangent line to the graph of f at the point $(0, 3)$.
   e. Sketch the graph of f and the tangent line to the curve at the point $(0, 3)$.

2. The losses (in millions of dollars) due to bad loans extended chiefly in agriculture, real estate, shipping, and energy by the Franklin Bank are estimated to be

$$A = f(t) = -t^2 + 10t + 30 \qquad (0 \le t \le 10),$$

where t is the time in years ($t = 0$ corresponds to the beginning of 1988). How fast were the losses mounting at the beginning of 1991? At the beginning of 1993? How fast will the losses be mounting at the beginning of 1995? Interpret your results.

Solutions to Self-Check Exercises 2.6 can be found on page 123.

2.6 EXERCISES

1. **Average Weight of an Infant** The following graph shows the weight measurements of the average infant from the time of birth ($t = 0$) through age two ($t = 24$). By computing the slopes of the respective tangent lines, estimate the rate of change of the average infant's weight when $t = 3$ and when $t = 18$. What is the average rate of change in the average infant's weight over its first year of life?

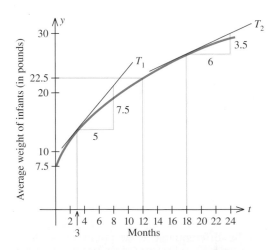

2. **Forestry** The following graph shows the volume of wood produced in a single-species forest. Here $f(t)$ is measured in cubic meters per hectare and t is measured in years. By computing the slopes of the respective tangent lines, estimate the rate at which the wood grown is changing at the beginning of the 10th year and at the beginning of the 30th year.

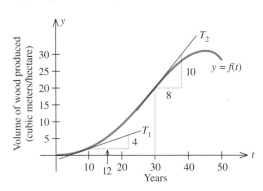

3. **TV Viewing Patterns** The following graph, based on data supplied by the A. C. Nielsen Company, shows the percentage of U.S. households watching television dur-

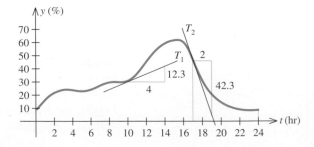

ing a 24-hr period on a weekday ($t = 0$ corresponds to 6 A.M.). By computing the slopes of the respective tangent lines, estimate the rate of change of the percentage of households watching television at 4 P.M. and 11 P.M.

4. **Crop Yield**   Productivity and yield of cultivated crops are often reduced by insect pests. The following graph shows the relationship between the yield of a certain crop, $f(x)$, as a function of the density of aphids x. (Aphids are small insects that suck plant juices.) Here $f(x)$ is measured in kg/4000 sq meters, and x is measured in hundreds of aphids per bean stem. By computing the slopes of the respective tangent lines, estimate the rate of change of the crop yield with respect to the density of aphids when that density is 200 aphids per bean stem. When it is 800 aphids per bean stem.

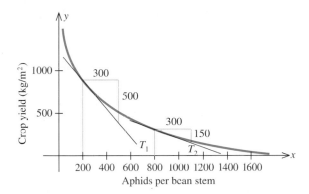

Aphids per bean stem

In Exercises 5–12, use the Four-Step Process to find the slope of the tangent line to the graph of the given function at any point.

5. $f(x) = 13$

6. $f(x) = -6$

7. $f(x) = 2x + 7$

8. $f(x) = 8 - 4x$

9. $f(x) = 3x^2$

10. $f(x) = -\frac{1}{2}x^2$

11. $f(x) = -x^2 + 3x$

12. $f(x) = 2x^2 + 5x$

In Exercises 13–18, find the slope of the tangent line to the graph of each function at the given point and determine an equation of the tangent line.

13. $f(x) = 2x + 7$ at $(2, 11)$

14. $f(x) = -3x + 4$ at $(-1, 7)$

15. $f(x) = 3x^2$ at $(1, 3)$

16. $f(x) = 3x - x^2$ at $(-2, -10)$

17. $f(x) = -\frac{1}{x}$ at $\left(3, -\frac{1}{3}\right)$

18. $f(x) = \frac{3}{2x}$ at $\left(1, \frac{3}{2}\right)$

19. Let $f(x) = 2x^2 + 1$.

   a. Find the derivative f' of f.

   b. Find an equation of the tangent line to the curve at the point $(1, 3)$.

   c. Sketch the graph of f.

20. Let $f(x) = x^2 + 6x$.

   a. Find the derivative f' of f.

   b. Find the point on the graph of f where the tangent line to the curve is horizontal. [*Hint:* Find the value of x for which $f'(x) - 0$.]

   c. Sketch the graph of f and the tangent line to the curve at the point found in (b).

21. Let $f(x) = x^2 - 2x + 1$.

   a. Find the derivative f' of f.

   b. Find the point on the graph of f where the tangent line to the curve is horizontal.

   c. Sketch the graph of f and the tangent line to the curve at the point found in (b).

   d. What is the rate of change of f at this point?

22. Let $f(x) = \dfrac{1}{x - 1}$.

   a. Find the derivative f' of f.

   b. Find an equation of the tangent line to the curve at the point $\left(-1, -\frac{1}{2}\right)$.

   c. Sketch the graph of f.

23. Let $y = f(x) = x^2 + x$.

   a. Find the average rate of change of y with respect to x in the interval from $x - 2$ to $x = 3$. In the interval from $x = 2$ to $x = 2.5$. In the interval from $x = 2$ to $x = 2.1$.

   b. Find the (instantaneous) rate of change of y at $x = 2$.

   c. Compare the results obtained in (a) with that of (b).

24. Let $y = f(x) = x^2 - 4x$.

   a. Find the average rate of change of y with respect to x in the interval from $x = 3$ to $x = 4$. In the interval from $x = 3$ to $x = 3.5$. In the interval from $x = 3$ to $x = 3.1$.

   b. Find the (instantaneous) rate of change of y at $x = 3$.

   c. Compare the results obtained in (a) with that of (b).

25. **Velocity of a Car**   Suppose that the distance s (in feet) covered by a car moving along a straight road t seconds after starting from rest is given by the function $f(t) = 2t^2 + 48t$.
    a. Calculate the average velocity of the car over the time intervals [20, 21], [20, 20.1], and [20, 20.01].
    b. Calculate the (instantaneous) velocity of the car when $t = 20$.
    c. Compare the results of (a) with that of (b).

26. **Velocity of a Ball Thrown into the Air**   A ball is thrown straight up with an initial velocity of 128 ft/sec, so that its height (in feet) after t seconds is given by $s(t) = 128t - 16t^2$.
    a. What is the average velocity of the ball over the time intervals [2, 3], [2, 2.5], and [2, 2.1]?
    b. What is the instantaneous velocity at time $t = 2$?
    c. What is the instantaneous velocity at time $t = 5$? Is the ball rising or falling at this time?
    d. When will the ball hit the ground?

27. **Cost of Producing Surfboards**   The total cost $C(x)$ [in dollars] incurred by the Aloha Company in manufacturing x surfboards a day is given by

$$C(x) = -10x^2 + 300x + 130 \qquad (0 \le x \le 15).$$

    a. Find $C'(x)$.
    b. What is the rate of change of the total cost when the level of production is ten surfboards a day?
    c. What is the average cost Aloha incurs in manufacturing ten surfboards a day?

28. **Effect of Advertising on Profit**   The quarterly profit of Cunningham Realty (in thousands of dollars) is given by

$$P(x) = -\frac{1}{3}x^2 + 7x + 30 \qquad (0 \le x \le 50),$$

    where x (in thousands of dollars) is the amount of money Cunningham spends on advertising per quarter.
    a. Find $P'(x)$.
    b. What is the rate of change of Cunningham's quarterly profit if the amount it spends on advertising is $10,000 per quarter ($x = 10$)? $30,000 per quarter ($x = 30$)?

C 29. **Demand for Tents**   The demand function for the Sportsman 5 × 7 tents is given by

$$p = f(x) = -0.1x^2 - x + 40,$$

    where p is measured in dollars and x is measured in units of a thousand.
    a. Find the average rate of change in the unit price of a tent if the quantity demanded is between 5000 and 5050 tents. Between 5000 and 5010 tents.

    b. What is the rate of change of the unit price if the quantity demanded is 5000?

30. **A Country's GNP**   The gross national product (GNP) of a certain country is projected to be

$$N(t) = t^2 + 2t + 50 \qquad (0 \le t \le 5)$$

    billion dollars t years from now. What will the rate of change of the country's GNP be two years from now? Four years from now?

31. **Growth of Bacteria**   Under a set of controlled laboratory conditions, the size of the population of a certain bacteria culture at time t (in minutes) is described by the function

$$P = f(t) = 3t^2 + 2t + 1.$$

    Find the rate of population growth at $t = 10$ minutes.

In Exercises 32–36, let x and $f(x)$ represent the given quantities. Fix $x = a$ and let h be a small positive number. Give an interpretation of the quantities

$$\frac{f(a + h) - f(a)}{h} \quad and \quad \lim_{h \to 0} \frac{f(a + h) - f(a)}{h}.$$

32. x denotes time and $f(x)$ denotes the population of seals at time x.

33. x denotes time and $f(x)$ denotes the prime interest rate at time x.

34. x denotes time and $f(x)$ denotes a country's industrial production.

35. x denotes the level of production of a certain commodity and $f(x)$ denotes the total cost incurred in producing x units of the commodity.

36. x denotes altitude and $f(x)$ denotes atmospheric pressure.

In each of Exercises 37–42, the graph of a function is shown. For each function, state whether or not (a) $f(x)$ has a limit at $x = a$, (b) $f(x)$ is continuous at $x = a$ and (c) $f(x)$ is differentiable at $x = a$. Justify your answers.

37.   38.

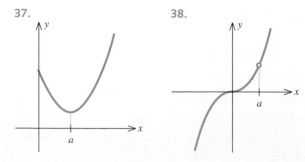

39.

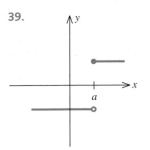

40.

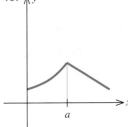

41.

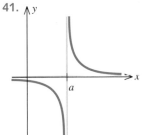

42.

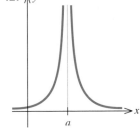

43. Sketch the graph of the function $f(x) = |x + 1|$ and show that the function does not have a derivative at $x = -1$.

44. Sketch the graph of the function $f(x) = 1/(x - 1)$ and show that the function does not have a derivative at $x = 1$.

45. Sketch the graph of the function $f(x) = x^{2/3}$. Is the function continuous at $x = 0$? Does $f'(0)$ exist? Why or why not?

46. Prove that the derivative of the function $f(x) = |x|$ for $x \neq 0$ is given by

$$f'(x) = \begin{cases} 1 & \text{if } x > 0 \\ -1 & \text{if } x < 0. \end{cases}$$

[*Hint:* Recall the definition of the absolute value of a number.]

47. Show that if a function f is differentiable at a point $x = a$, then f must be continuous at that point. [*Hint:* Write

$$f(x) - f(a) = \left[\frac{f(x) - f(a)}{x - a} \right](x - a).$$

Use the product rule for limits and the definition of the derivative to show that

$$\lim_{x \to a} [f(x) - f(a)] = 0.]$$

C *In Exercises 48 and 49, use a computer or a calculator for your calculations.*

48. The distance s (in feet) covered by a motorcycle traveling in a straight line and starting from rest in t seconds is given by the function

$$s(t) = -0.1t^3 + 2t^2 + 24t.$$

Calculate the motorcycle's average velocity over the time interval $[2, 2 + h]$ for $h = 1, 0.1, 0.01, 0.001, 0.0001$, and 0.00001, and use your results to guess at the motorcycle's instantaneous velocity at $t = 2$.

49. The daily total cost $C(x)$ incurred by Trappee and Sons, Inc., for producing x cases of Texa-Pep hot sauce is given by

$$C(x) = 0.000002x^3 + 5x + 400,$$

Calculate

$$\frac{C(100 + h) - C(100)}{h}$$

for $h = 1, 0.1, 0.01, 0.001$, and 0.0001, and use your results to estimate the rate of change of the total cost function when the level of production is 100 cases a day.

SOLUTIONS TO SELF-CHECK EXERCISES 2.6

1. a. $f'(x) = \lim_{h \to 0} \dfrac{f(x + h) - f(x)}{h}$

$= \lim_{h \to 0} \dfrac{[-(x + h)^2 - 2(x + h) + 3] - (-x^2 - 2x + 3)}{h}$

$= \lim_{h \to 0} \dfrac{-x^2 - 2xh - h^2 - 2x - 2h + 3 + x^2 + 2x - 3}{h}$

$= \lim_{h \to 0} \dfrac{h(-2x - h - 2)}{h}$

$= \lim_{h \to 0} (-2x - h - 2) = -2x - 2$

b. From the result of (a), we see that the slope of the tangent line to the graph of f at any point $(x, f(x))$ is given by

$$f'(x) = -2x - 2.$$

In particular, the slope of the tangent line to the graph of f at $(0, 3)$ is

$$f'(0) = -2.$$

c. The rate of change of f when $x = 0$ is given by $f'(0) = -2$, or -2 units per unit change in x.

d. Using the result from (b), we see that an equation of the required tangent line is

$$y - 3 = -2(x - 0)$$

or $$y = -2x + 3.$$

e.

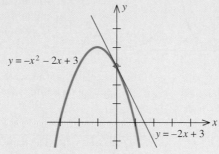

$y = -x^2 - 2x + 3$

$y = -2x + 3$

2. The rate of change of the losses at any time t is given by

$$\begin{aligned}
f'(t) &= \lim_{h \to 0} \frac{f(t + h) - f(t)}{h} \\
&= \lim_{h \to 0} \frac{[-(t + h)^2 + 10(t + h) + 30] - (-t^2 + 10t + 30)}{h} \\
&= \lim_{h \to 0} \frac{-t^2 - 2th - h^2 + 10t + 10h + 30 + t^2 - 10t - 30}{h} \\
&= \lim_{h \to 0} \frac{h(-2t - h + 10)}{h} \\
&= \lim_{h \to 0} (-2t - h + 10) \\
&= -2t + 10.
\end{aligned}$$

Therefore, the rate of change of the losses suffered by the bank at the beginning of 1991 ($t = 3$) was

$$f'(3) = -2(3) + 10 = 4;$$

that is, the losses were increasing at the rate of $4 million per year. At the beginning of 1993 ($t = 5$),

$$f'(5) = -2(5) + 10 = 0,$$

and we see that the growth in losses due to bad loans was zero at this point in time. At the beginning of 1995 ($t = 7$),

$$f'(7) = -2(7) + 10 = -4,$$

and we conclude that the losses will be decreasing at the rate of $4 million per year.

CHAPTER TWO SUMMARY OF PRINCIPAL FORMULAS AND TERMS

Formulas

1. Average rate of change of f over $[x, x + h]$
   or
   Slope of the secant line to the graph of f through $(x, f(x))$ and $(x + h, f(x + h))$
   or
   Difference quotient

$$\frac{f(x + h) - f(x)}{h}$$

2. Instantaneous rate of change of f at $(x, f(x))$
   or
   Slope of tangent line to the graph of f at $(x, f(x))$ at x
   or
   Derivative of f

$$\lim_{h \to 0} \frac{f(x + h) - f(x)}{h}$$

Terms

Function	Demand function
Domain	Supply function
Independent variable	Market equilibrium
Dependent variable	Equilibrium quantity
Graph of a function	Equilibrium price
Graph of an equation	Limit of a function
Vertical-line test	Indeterminate form
Composite function	Limit of a function at infinity
Polynomial function	Right-hand limit of a function
Linear function	Left-hand limit of a function
Quadratic function	Continuity of a function at a point
Cubic function	Secant line
Rational function	Tangent line to the graph of f
Power function	Differentiable function

CHAPTER 2 REVIEW EXERCISES

1. Find the domain of each of the following functions:

   a. $f(x) = \sqrt{9 - x}$

   b. $f(x) = \dfrac{x + 3}{2x^2 - x - 3}$

2. Let $f(x) = 3x^2 + 5x - 2$. Find
   a. $f(-2)$
   b. $f(a + 2)$
   c. $f(2a)$
   d. $f(a + h)$

3. Let $y^2 = 2x + 1$.
   a. Sketch the graph of this equation.
   b. Is y a function of x? Why?
   c. Is x a function of y? Why?

4. Sketch the graph of the function defined by

   $$f(x) = \begin{cases} x + 1 & \text{if } x < 1 \\ -x^2 + 4x - 1 & \text{if } x \geq 1. \end{cases}$$

5. Let $f(x) = 1/x$ and $g(x) = 2x + 3$. Find
   a. $f(x)g(x)$
   b. $f(x)/g(x)$
   c. $f(g(x))$
   d. $g(f(x))$

In Exercises 6–19, find the indicated limits, if they exist.

6. $\lim_{x \to 0} (5x - 3)$

7. $\lim_{x \to 1} (x^2 + 1)$

8. $\lim_{x \to -1} (3x^2 + 4)(2x - 1)$

9. $\lim_{x \to 3} \dfrac{x - 3}{x + 4}$

10. $\lim_{x \to 2} \dfrac{x + 3}{x^2 - 9}$

11. $\lim_{x \to -2} \dfrac{x^2 - 2x - 3}{x^2 + 5x + 6}$

12. $\lim_{x \to 3} \sqrt{2x^3 - 5}$

13. $\lim_{x \to 3} \dfrac{4x - 3}{\sqrt{x + 1}}$

14. $\lim_{x \to 1^+} \dfrac{x - 1}{x(x - 1)}$

15. $\lim_{x \to 1^-} \dfrac{\sqrt{x} - 1}{x - 1}$

16. $\lim_{x \to \infty} \dfrac{x^2}{x^2 - 1}$

17. $\lim_{x \to -\infty} \dfrac{x + 1}{x}$

18. $\lim_{x \to \infty} \dfrac{3x^2 + 2x + 4}{2x^2 - 3x + 1}$

19. $\lim_{x \to -\infty} \dfrac{x^2}{x + 1}$

20. Sketch the graph of the function
$$f(x) = \begin{cases} 2x - 3 & \text{if } x \le 2 \\ -x + 3 & \text{if } x > 2 \end{cases}$$
and evaluate $\lim_{x \to a^+} f(x)$, $\lim_{x \to a^-} f(x)$, and $\lim_{x \to a} f(x)$ at the point $a = 2$, if the limits exist.

21. Sketch the graph of the function
$$f(x) = \begin{cases} 4 - x & \text{if } x \le 2 \\ x + 2 & \text{if } x > 2 \end{cases}$$
and evaluate $\lim_{x \to a^+} f(x)$, $\lim_{x \to a^-} f(x)$, and $\lim_{x \to a} f(x)$ at the point $a = 2$, if the limits exist.

In Exercises 22–25, determine all values of x for which the given functions are discontinuous.

22. $g(x) = \begin{cases} x + 3 & \text{if } x \ne 2 \\ 0 & \text{if } x = 2 \end{cases}$

23. $f(x) = \dfrac{3x + 4}{4x^2 - 2x - 2}$

24. $f(x) = \begin{cases} \dfrac{1}{(x + 1)^2} & \text{if } x \ne -1 \\ 2 & \text{if } x = -1 \end{cases}$

25. $f(x) = \dfrac{|2x|}{x}$

26. Let $y = x^2 + 2$.
   a. Find the average rate of change of y with respect to x in the intervals $[1, 2]$, $[1, 1.5]$, and $[1, 1.1]$.
   b. Find the (instantaneous) rate of change of y at $x = 1$.

27. Use the definition of the derivative to find the slope of the tangent line to the graph of the function $f(x) = 3x + 5$ at any point $P(x, f(x))$ on the graph.

28. Use the definition of the derivative to find the slope of the tangent line to the graph of the function $f(x) = -1/x$ at any point $P(x, f(x))$ on the graph.

29. Use the definition of the derivative to find the slope of the tangent line to the graph of the function $f(x) = \frac{3}{2}x + 5$ at the point $(-2, 2)$ and determine an equation of the tangent line.

30. Use the definition of the derivative to find the slope of the tangent line to the graph of the function $f(x) = -x^2$ at the point $(2, -4)$ and determine an equation of the tangent line.

31. The graph of the function f is shown in the accompanying figure.

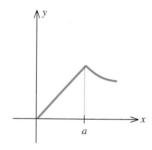

   a. Is f continuous at $x = a$? Why?
   b. Is f differentiable at $x = a$? Justify your answers.

32. Sales of a certain clock radio are approximated by the relationship $S(x) = 6{,}000x + 30{,}000$ $(0 \le x \le 5)$, where $S(x)$ denotes the number of clock radios sold in year x $(x = 0$ corresponds to the year 1992). Find the number of clock radios expected to be sold in the year 1996.

33. A company's total sales (in millions of dollars) are approximately linear as a function of time (in years). Sales in 1987 were $2.4 million, whereas sales in 1992 amounted to $7.4 million.
   a. Find an equation that gives the company's sales as a function of time.
   b. What were the sales in 1990?

34. A company has a fixed cost of $30,000 and a production cost of $6 for each unit it manufactures. A unit sells for $10.
    a. What is the cost function?
    b. What is the revenue function?
    c. What is the profit function?
    d. Compute the profit (loss) corresponding to production levels of 6,000, 8,000, and 12,000 units, respectively.

35. Find the point of intersection of the two straight lines having the equations $y = \frac{3}{4}x + 6$ and $3x - 2y + 3 = 0$.

36. The cost and revenue functions for a certain firm are given by $C(x) = 12x + 20,000$ and $R(x) = 20x$, respectively. Find the company's break-even point.

37. Given the demand equation $3x + p - 40 = 0$ and the supply equation $2x - p + 10 = 0$, where p is the unit price in dollars and x represents the quantity in units of a thousand, determine the equilibrium quantity and the equilibrium price.

38. Clark's Rule is a method for calculating pediatric drug dosages based on a child's weight. If a denotes the adult dosage (in mg) and if w is the weight of the child (in pounds), then the child's dosage is given by

$$D(w) = \frac{aw}{150}.$$

If the adult dose of a substance is 500 mg, how much should a child who weighs 35 pounds receive?

39. The monthly revenue R (in hundreds of dollars) realized in the sale of Royal electric shavers is related to the unit price p (in dollars) by the equation

$$R(p) = -\frac{1}{2}p^2 + 30p.$$

Find the revenue when an electric shaver is priced at $30.

40. The membership of the newly opened Venus Health Club is approximated by the function

$$N(x) = 200(4 + x)^{1/2} \qquad (1 \le x \le 24),$$

where $N(x)$ denotes the number of members x months after the club's grand opening. Find $N(0)$ and $N(12)$ and interpret your results.

41. Psychologist L. L. Thurstone discovered the following model for the relationship between the learning time T and the length of a list n:

$$T = f(n) = An\sqrt{n - b},$$

where A and b are constants that depend on the person and the task. Suppose that for a certain person and a certain task, $A = 4$ and $b = 4$. Compute $f(4)$, $f(5), \ldots, f(12)$ and use this information to sketch the graph of the function f. Interpret your results.

42. The monthly demand and supply functions for the Luminar desk lamp are given by

$$p = d(x) = -1.1x^2 + 1.5x + 40$$

and $\qquad p = s(x) = 0.1x^2 + 0.5x + 15,$

respectively, where p is measured in dollars and x in units of a thousand. Find the equilibrium quantity and price.

43. The Photo-Mart transfers movie films to videocassettes. The fees charged for this service are shown in the following table. Find a function C relating the cost $C(x)$ to the number of feet x of film transferred. Sketch the graph of the function C and discuss its continuity.

Length of film in feet (x)	Price ($) for conversion
$1 \le x \le 100$	5.00
$100 < x \le 200$	9.00
$200 < x \le 300$	12.50
$300 < x \le 400$	15.00
$x > 400$	$7 + 0.02x$

44. The average cost (in dollars) of producing x units of a certain commodity is given by

$$\overline{C}(x) = 20 + \frac{400}{x}.$$

Evaluate $\lim_{x \to \infty} \overline{C}(x)$ and interpret your results.

How is a pond's oxygen content affected by organic waste? In Example 7, page 144, you will see how to find the rate at which oxygen is being restored to the pond after organic waste has been dumped into it.

3

Differentiation

T his chapter gives several rules that will greatly simplify the task of finding the derivative of a function, thus enabling us to study how fast one quantity is changing with respect to another in many real-world situations. For example, we will be able to find how fast the population of an endangered species of whales will grow after certain conservation measures have been implemented, how fast an economy's consumer price index (CPI) is changing at any time, and how fast the learning time changes with respect to the length of a list. We will also see how these rules of differentiation will facilitate the study of marginal analysis, the study of the rate of change of economic quantities. Finally, we will introduce the notion of the differential of a function. Using differentials affords a relatively easy way of approximating the change in one quantity due to a small change in a related quantity.

3.1   Basic Rules of Differentiation

Four Basic Rules

The method used in Chapter 2 for computing the derivative of a function is based on a faithful interpretation of the definition of the derivative as the limit of a quotient. Thus, to find the rule for the derivative f' of a function f, we first computed the difference quotient

$$\frac{f(x + h) - f(x)}{h}$$

and then evaluated its limit as h approached zero. As you have probably observed, this method is tedious even for relatively simple functions.

   The main purpose of this chapter is to derive certain rules that will simplify the process of finding the derivative of a function. Throughout this book we will use the notation

$$\frac{d}{dx}[f(x)]$$

[read "dee, dee x of f of x"] to mean "the derivative of f with respect to x at x." In stating the rules of differentiation, we assume that the functions f and g are differentiable.

RULE 1: DERIVATIVE OF A CONSTANT

$$\frac{d}{dx}(c) = 0 \qquad (c, \text{ a constant})$$

The derivative of a constant function is equal to zero.

FIGURE 3.1

The slope of the tangent line to the graph of $f(x) = c$, where c is a constant, is zero.

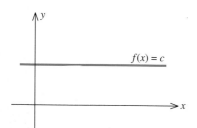

We can see this from a geometric viewpoint by recalling that the graph of a constant function is a straight line parallel to the x-axis (Figure 3.1). Since the tangent line to a straight line at any point on the line coincides with the straight line itself, its slope (as given by the derivative of $f(x) = c$) must be zero. We can also use the definition of the derivative to prove this result by computing

$$f'(x) = \lim_{h \to 0} \frac{f(x + h) - f(x)}{h}$$

$$= \lim_{h \to 0} \frac{c - c}{h}$$

$$= \lim_{h \to 0} 0 = 0.$$

EXAMPLE 1

a. If $f(x) = 28$, then

$$f'(x) = \frac{d}{dx}(28) = 0.$$

b. If $f(x) = -2$, then

$$f'(x) = \frac{d}{dx}(-2) = 0.$$ ∎

RULE 2: THE POWER RULE

If n is any real number, then $\frac{d}{dx}(x^n) = nx^{n-1}$.

Let us verify the Power Rule for the special case $n = 2$. If $f(x) = x^2$, then

$$f'(x) = \frac{d}{dx}(x^2) = \lim_{h \to 0} \frac{f(x + h) - f(x)}{h}$$

$$= \lim_{h \to 0} \frac{(x + h)^2 - x^2}{h}$$

$$= \lim_{h \to 0} \frac{x^2 + 2xh + h^2 - x^2}{h}$$

$$= \lim_{h \to 0} \frac{2xh + h^2}{h} = \lim_{h \to 0} (2x + h) = 2x,$$

as we set out to show.

The proof of the Power Rule for the general case is not easy to prove and will be omitted. However, you will be asked to prove the rule for the special case $n = 3$ in Exercise 60, page 139.

EXAMPLE 2

a. If $f(x) = x$, then

$$f'(x) = \frac{d}{dx}(x) = 1 \cdot x^{1-1} = x^0 = 1.$$

b. If $f(x) = x^8$, then

$$f'(x) = \frac{d}{dx}(x^8) = 8x^7.$$

c. If $f(x) = x^{5/2}$, then

$$f'(x) = \frac{d}{dx}(x^{5/2}) = \frac{5}{2}x^{3/2}.$$ ∎

In order to differentiate a function whose rule involves a radical, we first rewrite the rule using fractional powers. The resulting expression can then be differentiated using the Power Rule.

EXAMPLE 3   Find the derivative of the given function.

a. $f(x) = \sqrt{x}$ **b.** $g(x) = \dfrac{1}{\sqrt[3]{x}}$

Solution

a. Rewriting $\sqrt{x}$ in the form $x^{1/2}$, we obtain

$$f'(x) = \frac{d}{dx}(x^{1/2})$$

$$= \frac{1}{2}x^{-1/2} = \frac{1}{2x^{1/2}} = \frac{1}{2\sqrt{x}}.$$

b. Rewriting $\dfrac{1}{\sqrt[3]{x}}$ in the form $x^{-1/3}$, we obtain

$$g'(x) = \frac{d}{dx}(x^{-1/3})$$

$$= -\frac{1}{3}x^{-4/3} = -\frac{1}{3x^{4/3}}.$$ ∎

RULE 3: DERIVATIVE OF A CONSTANT MULTIPLE OF A FUNCTION	$\dfrac{d}{dx}[cf(x)] = c\dfrac{d}{dx}[f(x)]$ (c, a constant)

The derivative of a constant times a differentiable function is equal to the constant times the derivative of the function.

This result follows from the following computations.
    If $g(x) = cf(x)$, then

$$g'(x) = \lim_{h \to 0} \frac{g(x + h) - g(x)}{h} = \lim_{h \to 0} \frac{cf(x + h) - cf(x)}{h}$$

$$= c \lim_{h \to 0} \frac{f(x + h) - f(x)}{h}$$

$$= cf'(x).$$

EXAMPLE 4

a. If $f(x) = 5x^3$, then

$$f'(x) = \frac{d}{dx}(5x^3) = 5\frac{d}{dx}(x^3)$$

$$= 5(3x^2) = 15x^2.$$

b. If $f(x) = \dfrac{3}{\sqrt{x}}$, then

$$f'(x) = \frac{d}{dx}(3x^{-1/2})$$

$$= 3\left(-\frac{1}{2}x^{-3/2}\right) = -\frac{3}{2x^{3/2}}.$$ ■

RULE 4: THE SUM RULE

$$\frac{d}{dx}[f(x) \pm g(x)] = \frac{d}{dx}[f(x)] \pm \frac{d}{dx}[g(x)]$$

The derivative of the sum or difference of two differentiable functions is equal to the sum or difference of their derivatives.

This result may be extended to the sum and difference of any finite number of differentiable functions. Let us verify the rule for a sum of two functions.

If $s(x) = f(x) + g(x)$, then

$$s'(x) = \lim_{h \to 0} \frac{s(x + h) - s(x)}{h}$$

$$= \lim_{h \to 0} \frac{[f(x + h) + g(x + h)] - [f(x) + g(x)]}{h}$$

$$= \lim_{h \to 0} \frac{[f(x + h) - f(x)] + [g(x + h) - g(x)]}{h}$$

$$= \lim_{h \to 0} \frac{f(x + h) - f(x)}{h} + \lim_{h \to 0} \frac{g(x + h) - g(x)}{h}$$

$$= f'(x) + g'(x).$$

EXAMPLE 5   Find the derivatives of the following functions:

a. $f(x) = 4x^5 + 3x^4 - 8x^2 + x + 3$ **b.** $g(t) = \dfrac{t^2}{5} + \dfrac{5}{t^3}$

Solution

a. $f'(x) = \dfrac{d}{dx}(4x^5 + 3x^4 - 8x^2 + x + 3)$

$$= \frac{d}{dx}(4x^5) + \frac{d}{dx}(3x^4) - \frac{d}{dx}(8x^2) + \frac{d}{dx}(x) + \frac{d}{dx}(3)$$

$$= 20x^4 + 12x^3 - 16x + 1.$$

b. Here the independent variable is t instead of x, so we differentiate with respect to t. Thus,

$$g'(t) = \frac{d}{dt}\left(\frac{1}{5}t^2 + 5t^{-3}\right) \qquad \left(\text{Rewriting } \frac{1}{t^3} \text{ as } t^{-3}\right)$$

$$= \frac{2}{5}t - 15t^{-4}$$

$$= \frac{2t^5 - 75}{5t^4}. \qquad \text{(Simplifying)}$$ ■

EXAMPLE 6   Find the slope and an equation of the tangent line to the graph of $f(x) = 2x + \dfrac{1}{\sqrt{x}}$ at the point $(1, 3)$.

Solution   The slope of the tangent line at any point on the graph of f is given by

$$f'(x) = \frac{d}{dx}\left(2x + \frac{1}{\sqrt{x}}\right)$$

$$= \frac{d}{dx}(2x + x^{-1/2}) \qquad \left(\text{Rewriting } \frac{1}{\sqrt{x}} = \frac{1}{x^{1/2}} = x^{-1/2}.\right)$$

$$= 2 - \frac{1}{2}x^{-3/2} \qquad \text{(Using the Sum Rule.)}$$

$$= 2 - \frac{1}{2x^{3/2}}.$$

In particular, the slope of the tangent line to the graph of f at $(1, 3)$ [where $x = 1$] is

$$f'(1) = 2 - \frac{1}{2(1^{3/2})} = 2 - \frac{1}{2} = \frac{3}{2}.$$

Using the point-slope form of the equation of a line with slope $3/2$ and the point $(1, 3)$, we see that an equation of the tangent line is

$$y - 3 = \frac{3}{2}(x - 1) \qquad [(y - y_1) = m(x - x_1)]$$

or, upon simplification,

$$y = \frac{3}{2}x + \frac{3}{2}. \qquad \blacksquare$$

Applications

EXAMPLE 7   A group of marine biologists at the Neptune Institute of Oceanography recommended that a series of conservation measures be carried out over the next decade to save a certain species of whale from extinction. After implementing the conservation measures, the population of this species is expected to be

$$N(t) = 3t^3 + 2t^2 - 10t + 600 \qquad (0 \le t \le 10),$$

where $N(t)$ denotes the population at the end of year t. Find the rate of growth of the whale population when $t = 2$ and $t = 6$. How large will the whale population be eight years after implementing the conservation measures?

Solution   The rate of growth of the whale population at any time t is given by

$$N'(t) = 9t^2 + 4t - 10.$$

FIGURE 3.2

The whale population after year t is given by N(t).

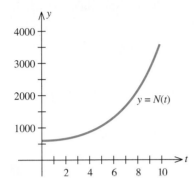

In particular, when $t = 2$ and $t = 6$, we have

$$N'(2) = 9(2)^2 + 4(2) - 10$$
$$= 34$$
$$N'(6) = 9(6)^2 + 4(6) - 10$$
$$= 338,$$

so the whale population's rate of growth will be 34 whales per year after two years and 338 per year after six years.

The whale population at the end of the eighth year will be

$$N(8) = 3(8)^3 + 2(8)^2 - 10(8) + 600$$
$$= 2184,$$

or 2184 whales. The graph of the function N appears in Figure 3.2. Note the rapid growth of the population in the later years, as the conservation measures begin to pay off, compared to the growth in the early years. ∎

EXAMPLE 8   The altitude of a rocket (in feet) t seconds into flight is given by

$$s = f(t) = -t^3 + 96t^2 + 195t + 5 \qquad (t \geq 0).$$

a. Find an expression v for the rocket's velocity at any time t.

b. Compute the rocket's velocity when $t = 0, 30, 50, 65,$ and 70. Interpret your results.

c. Using the results from the solution to (b) and the observation that at the highest point in its trajectory the rocket's velocity is zero, find the maximum altitude attained by the rocket.

Solution

a. The rocket's velocity at any time t is given by

$$v = f'(t) = -3t^2 + 192t + 195.$$

b. The rocket's velocity when $t = 0, 30, 50, 65,$ and 70 is given by

$$f'(0) = -3(0)^2 + 192(0) + 195 - 195$$
$$f'(30) = -3(30)^2 + 192(30) + 195 = 3255$$
$$f'(50) = -3(50)^2 + 192(50) + 195 = 2295$$
$$f'(65) = -3(65)^2 + 192(65) + 195 = 0$$
$$f'(70) = -3(70)^2 + 192(70) + 195 = -1065,$$

or 195, 3255, 2295, 0, and −1065 feet per second.

Thus, the rocket has an initial velocity of 195 ft/sec at $t = 0$ and accelerates to a velocity of 3255 ft/sec at $t = 30$. Fifty seconds into the flight, the rocket's velocity is 2295 ft/sec, which is less than the velocity at $t = 30$. This means that the rocket begins to decelerate after an initial period of acceleration. (Later on we will learn how to determine the rocket's maximum velocity.)

FIGURE 3.3
The rocket's altitude t seconds into flight is given by $f(t)$.

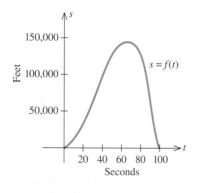

The deceleration continues: the velocity is 0 ft/sec at $t = 65$ and -1065 ft/sec when $t = 70$. This figure tells us that 70 seconds into flight the rocket is heading back to Earth with a speed of 1065 ft/sec.

c. The results of (b) show that the rocket's velocity is zero when $t = 65$. At this instant, the rocket's maximum altitude is

$$s = f(65) = -(65)^3 + 96(65)^2 + 195(65) + 5$$
$$= 143,655,$$

or 143,655 feet. A sketch of the graph of f appears in Figure 3.3.

SELF-CHECK EXERCISES 3.1

1. Find the derivative of each of the following functions using the rules of differentiation.

   a. $f(x) = 1.5x^2 + 2x^{1.5}$

   b. $g(x) = 2\sqrt{x} + \dfrac{3}{\sqrt{x}}$

2. Let $f(x) = 2x^3 - 3x^2 + 2x - 1$.

   a. Compute $f'(x)$.
   b. What is the slope of the tangent line to the graph of f when $x = 2$?
   c. What is the rate of change of the function f at $x = 2$?

3. A certain country's gross national product (GNP) [in millions of dollars] is described by the function

   $$G(t) = -2t^3 + 45t^2 + 20t + 6000 \qquad (0 \le t \le 11),$$

   where $t = 0$ corresponds to the beginning of 1984.
   a. At what rate was the GNP changing at the beginning of 1989? At the beginning of 1991? At the beginning of 1994?
   b. What was the average rate of growth of the GNP over the period 1989 to 1994?

Solutions to Self-Check Exercises 3.1 can be found on page 139.

3.1   EXERCISES

In Exercises 1–34, find the derivative of the function f by using the rules of differentiation.

1. $f(x) = -3$

2. $f(x) = 365$

3. $f(x) = x^5$

4. $f(x) = x^7$

5. $f(x) = x^{2.1}$

6. $f(x) = x^{0.8}$

7. $f(x) = 3x^2$

8. $f(x) = -2x^3$

9. $f(r) = \pi r^2$

10. $f(r) = \frac{4}{3}\pi r^3$

11. $f(x) = 9x^{1/3}$

12. $f(x) = \frac{5}{4}x^{4/5}$

13. $f(x) = 3\sqrt{x}$

14. $f(u) = \dfrac{2}{\sqrt{u}}$

15. $f(x) = 7x^{-12}$

16. $f(x) = 0.3x^{-1.2}$

17. $f(x) = 5x^2 - 3x + 7$

18. $f(x) = x^3 - 3x^2 + 1$

19. $f(x) = -x^3 + 2x^2 - 6$

20. $f(x) = x^4 - 2x^2 + 5$

21. $f(x) = 0.03x^2 - 0.4x + 10$

22. $f(x) = 0.002x^3 - 0.05x^2 + 0.1x - 20$

23. $f(x) = \dfrac{x^3 - 4x^2 + 3}{x}$

24. $f(x) = \dfrac{x^3 + 2x^2 + x - 1}{x}$

25. $f(x) = 4x^4 - 3x^{5/2} + 2$

26. $f(x) = 5x^{4/3} - \frac{2}{3}x^{3/2} + x^2 - 3x + 1$

27. $f(x) = 3x^{-1} + 4x^{-2}$

28. $f(x) = -\frac{1}{3}(x^{-3} - x^6)$

29. $f(t) = \dfrac{4}{t^4} - \dfrac{3}{t^3} + \dfrac{2}{t}$

30. $f(x) = \dfrac{5}{x^3} - \dfrac{2}{x^2} - \dfrac{1}{x} + 200$

31. $f(x) = 2x - 5\sqrt{x}$

32. $f(t) = 2t^2 + \sqrt{t^3}$

33. $f(x) = \dfrac{2}{x^2} - \dfrac{3}{x^{1/3}}$

34. $f(x) = \dfrac{3}{x^3} + \dfrac{4}{\sqrt{x}} + 1$

35. Let $f(x) = 2x^3 - 4x$. Find a. $f'(-2)$, b. $f'(0)$, and
     c. $f'(2)$.

36. Let $f(x) = 4x^{5/4} + 2x^{3/2} + x$. Find a. $f'(0)$ and
     b. $f'(16)$.

*In Exercises 37–40, find the slope and an equation of the
tangent line to the graph of the function f at the specified
point.*

37. $f(x) = 2x^2 - 3x + 4;\ (2, 6)$

38. $f(x) = -\frac{5}{3}x^2 + 2x + 2;\ \left(-1, -\frac{5}{3}\right)$

39. $f(x) = x^4 - 3x^3 + 2x^2 - x + 1;\ (1, 0)$

40. $f(x) = \sqrt{x} + \dfrac{1}{\sqrt{x}};\ \left(4, \dfrac{5}{2}\right)$

41. Let $f(x) = x^3$.
    a. Find the point on the graph of f where the tangent
       line is horizontal.
    b. Sketch the graph of f and draw the horizontal tangent
       line.

42. Let $f(x) = x^3 - 4x^2$. Find the point(s) on the graph of
    f where the tangent line is horizontal.

43. Let $f(x) = x^3 + 1$.
    a. Find the point(s) on the graph of f where the slope
       of the tangent line is equal to 12.
    b. Find the equation(s) of the tangent line(s) of (a).
    c. Sketch the graph of f showing the tangent line(s).

44. Let $f(x) = \frac{2}{3}x^3 + x^2 - 12x + 6$. Find the values of x
    for which a. $f'(x) = -12$, b. $f'(x) = 0$, and
    c. $f'(x) = 12$.

45. Let $f(x) = \frac{1}{4}x^4 - \frac{1}{3}x^3 - x^2$. Find the point(s) on the
    graph of f where the slope of the tangent line is equal
    to a. $-2x$, b. 0, and c. $10x$.

46. **Growth of a Cancerous Tumor**   The volume of a
    spherical cancer tumor is given by the function

    $$V(r) = \frac{4}{3}\pi r^3,$$

    where r is the radius of the tumor in centimeters. Find
    the rate of change in the volume of the tumor when

    a. $r = \frac{2}{3}$ cm   b. $r = \frac{5}{4}$ cm

47. **Velocity of Blood in an Artery**   The velocity (in centi-
    meters per second) of blood r centimeters from the cen-
    tral axis of an artery is given by

    $$v(r) = k(R^2 - r^2),$$

    where k is a constant and R is the radius of the artery
    (see the accompanying figure). Suppose that $k = 1000$
    and $R = 0.2$ cm. Find $v(0.1)$ and $v'(0.1)$ and interpret
    your results.

Blood vessel

48. Worker Efficiency An efficiency study conducted for the Elektra Electronics Company showed that the number of "Space Commander" walkie-talkies assembled by the average worker t hours after starting work at 8 A.M. is given by

$$N(t) = -t^3 + 6t^2 + 15t.$$

a. Find the rate at which the average worker will be assembling walkie-talkies t hours after starting work.
b. At what rate will the average worker be assembling walkie-talkies at 10 A.M.? At 11 A.M.?
c. How many walkie-talkies will the average worker assemble between 10 A.M. and 11 A.M.?

49. Consumer Price Index An economy's consumer price index (CPI) is described by the function

$$I(t) = -0.2t^3 + 3t^2 + 100 \qquad (0 \le t \le 10),$$

where $t = 0$ corresponds to 1983.

a. At what rate was the CPI changing in 1988? In 1990? In 1993?
b. What was the average rate of increase in the CPI over the period from 1988 to 1993?

50. Effect of Advertising on Sales The relationship between the amount of money x that the Cannon Precision Instruments Corporation spends on advertising and the company's total sales $S(x)$ is given by the function

$$S(x) = -0.002x^3 + 0.6x^2 + x + 500 \quad (0 \le x \le 200),$$

where x is measured in thousands of dollars. Find the rate of change of the sales with respect to the amount of money spent on advertising. Are Cannon's total sales increasing at a faster rate when the amount of money spent on advertising is (a) $100,000 or (b) $150,000?

51. Population Growth A study prepared for a Sunbelt town's chamber of commerce projected that the town's population in the next three years will grow according to the rule

$$P(t) = 50,000 + 30t^{3/2} + 20t,$$

where $P(t)$ denotes the population t months from now. How fast will the population be increasing nine months from now? Sixteen months from now?

52. Curbing Population Growth Five years ago, the government of a Pacific island state launched an extensive propaganda campaign toward curbing the country's population growth. According to the Census Department, the population (measured in thousands of people) for the following four years was

$$P(t) = -\frac{1}{3}t^3 + 64t + 3000,$$

where t is measured in years and $t = 0$ at the start of the campaign. Find the rate of change of the population at the end of years one, two, three, and four. Was the plan working?

53. Conservation of Species A certain species of turtle faces extinction because dealers collect truckloads of turtle eggs to be sold as aphrodisiacs. After severe conservation measures are implemented, it is hoped that the turtle population will grow according to the rule

$$N(t) = 2t^3 + 3t^2 - 4t + 1000 \qquad (0 \le t \le 10),$$

where $N(t)$ denotes the population at the end of year t. Find the rate of growth of the turtle population when $t = 2$ and $t = 8$. What will the population be ten years after the conservation measures are implemented?

54. Flight of a Rocket The altitude (in feet) of a rocket t seconds into flight is given by

$$s = f(t) = -2t^3 + 114t^2 + 480t + 1 \qquad (t \ge 0).$$

a. Find an expression v for the rocket's velocity at any time t.
b. Compute the rocket's velocity when $t = 0, 20, 40,$ and 60. Interpret your results.
c. Using the results from the solution to (b), find the maximum altitude attained by the rocket. [*Hint:* At its highest point, the velocity of the rocket is zero.]

55. Stopping Distance of a Racing Car During a test by the editors of an auto magazine, the stopping distance s (in feet) of the MacPherson X-2 racing car conformed to the rule

$$s = f(t) = 120t - 15t^2 \qquad (t \ge 0),$$

where t was the time (in seconds) after the brakes were applied.

a. Find an expression for the car's velocity v at any time t.
b. What was the car's velocity when the brakes were first applied?
c. What was the car's stopping distance for that particular test? [*Hint:* The stopping time is found by setting $v = 0$.]

C **56. Demand Functions** The demand function for the Luminar desk lamp is given by

$$p = f(x) = -0.1x^2 - 0.4x + 35,$$

where x is the quantity demanded (measured in thousands) and p is the unit price in dollars.
a. Find $f'(x)$.
b. What is the rate of change of the unit price when the quantity demanded is 10,000 units ($x = 10$)? What is the unit price at that level of demand?

57. Supply Functions The supply function for a certain make of transistor radio is given by

$$p = f(x) = 0.1\sqrt{x} + 10,$$

where x is the quantity supplied and p is the unit price in dollars.
a. Find $f'(x)$.
b. What is the rate of change of the unit price if the quantity supplied is 40,000 transistor radios?

C **58. Average Speed of a Vehicle on a Highway** The average speed of a vehicle on a stretch of Route 134 between 6 A.M. and 10 A.M. on a typical weekday is approximated by the function

$$f(t) = 20t - 40\sqrt{t} + 50 \qquad (0 \le t \le 4),$$

where $f(t)$ is measured in miles per hour and t is measured in hours, $t = 0$ corresponding to 6 A.M.
a. Compute $f'(t)$.
b. Compute $f(0)$, $f(1)$, and $f(2)$ and interpret your results.
c. Compute $f'(\frac{1}{2})$, $f'(1)$, and $f'(2)$, and interpret your results.

C **59. Health-Care Spending** Despite efforts at cost containment, the cost of the medicare program is increasing at a high rate. Two major reasons for this increase are an aging population and the constant development and extensive use by physicians of new technologies. Based on data from the Health Care Financing Administration and the U.S. Census Bureau, health-care spending through the year 2000 may be approximated by the function

$$S(t) = 0.02836t^3 - 0.05167t^2 + 9.60881t + 41.9 \qquad (0 \le t \le 35),$$

where $S(t)$ is the spending in billions of dollars and t is measured in years, with $t = 0$ corresponding to the beginning of 1965.
a. Find an expression for the rate of change of health-care spending at any time t.
b. How fast was health-care spending changing at the beginning of 1980? How fast will health-care spending be changing at the beginning of 2000?
c. What was the amount of health-care spending at the beginning of 1980? What will the amount of health-care spending be at the beginning of 2000?

60. Prove the Power Rule (Rule 2) for the special case $n = 3$. [*Hint:* Compute

$$\lim_{h \to 0} \left[\frac{(x + h)^3 - x^3}{h} \right].]$$

C *In Exercises 61–64, use a computer or a graphing calculator to plot the graphs of the functions f and f'. Then use the graphs to find the approximate location of the point(s) where the tangent line(s) are horizontal.*

61. $f(x) = 2x^4 - 3x^3 + 5x^2 - 20x + 40$

62. $f(x) = x^5 - 3x^3 + 2x - 4$

63. $f(x) = \dfrac{\sqrt{x} - 1}{x}$

64. $f(x) = x - 2\sqrt{x}$

SOLUTIONS TO SELF-CHECK EXERCISES 3.1

1. a. $f'(x) = \dfrac{d}{dx}(1.5x^2) + \dfrac{d}{dx}(2x^{1.5})$

$= (1.5)(2x) + (2)(1.5x^{0.5})$

$= 3x + 3\sqrt{x} = 3(x + \sqrt{x})$

b. $g'(x) = \dfrac{d}{dx}(2x^{1/2}) + \dfrac{d}{dx}(3x^{-1/2})$

$= (2)\left(\dfrac{1}{2}x^{-1/2}\right) + (3)\left(-\dfrac{1}{2}x^{-3/2}\right)$

$= x^{-1/2} - \dfrac{3}{2}x^{-3/2}$

$= \dfrac{1}{2}x^{-3/2}(2x - 3) = \dfrac{2x - 3}{2x^{3/2}}$

2. a. $f'(x) = \dfrac{d}{dx}(2x^3) - \dfrac{d}{dx}(3x^2) + \dfrac{d}{dx}(2x) - \dfrac{d}{dx}(1)$

$= (2)(3x^2) - (3)(2x) + 2$

$= 6x^2 - 6x + 2$

b. The slope of the tangent line to the graph of f when $x = 2$ is given by

$$f'(2) = 6(2)^2 - 6(2) + 2 = 14.$$

c. The rate of change of f at $x = 2$ is given by $f'(2)$. Using the results of (b), we see that $f'(2)$ is 14 units per unit change in x.

3. a. The rate at which the GNP was changing at any time t $(0 < t < 11)$ is given by

$$G'(t) = -6t^2 + 90t + 20.$$

In particular, the rates of change of the GNP at the beginning of the years 1989 $(t = 5)$, 1991 $(t = 7)$, and 1994 $(t = 10)$ are given by

$$G'(5) = 320, \quad G'(7) = 356, \quad \text{and} \quad G'(10) = 320,$$

respectively; that is, by \$320 million per year, \$356 million per year, and \$320 million per year, respectively.

b. The average rate of growth of the GNP over the period from the beginning of 1989 $(t = 5)$ to the beginning of 1994 $(t = 10)$ is given by

$$\frac{G(10) - G(5)}{10 - 5} = \frac{[-2(10)^3 + 45(10)^2 + 20(10) + 6000]}{5}$$

$$- \frac{[-2(5)^3 + 45(5)^2 + 20(5) + 6000]}{5}$$

$$= \frac{8700 - 6975}{5},$$

or \$345 million per year.

3.2   The Product and Quotient Rules

In this section, we will study two more rules of differentiation known as the **Product Rule** and the **Quotient Rule.**

The Product Rule

The derivative of the product of two differentiable functions is given by the following rule:

RULE 5: THE PRODUCT RULE	$\frac{d}{dx}[f(x)g(x)] = f(x)g'(x) + g(x)f'(x)$

The derivative of the product of two functions is the first function times the derivative of the second plus the second function times the derivative of the first.

The Product Rule may be extended to the case involving the product of any finite number of functions (see Exercise 53, p. 148). We will prove the Product Rule at the end of this section.

The derivative of the product of two functions is not given by the product of the derivatives of the functions; that is, in general

$$\frac{d}{dx}[f(x)g(x)] \neq f'(x)g'(x).$$

EXAMPLE 1   Find the derivative of the function

$$f(x) = (2x^2 - 1)(x^3 + 3).$$

Solution   By the Product Rule,

$$
\begin{aligned}
f'(x) &= (2x^2 - 1)\frac{d}{dx}(x^3 + 3) + (x^3 + 3)\frac{d}{dx}(2x^2 - 1) \\
&= (2x^2 - 1)(3x^2) + (x^3 + 3)(4x) \\
&= 6x^4 - 3x^2 + 4x^4 + 12x \\
&= 10x^4 - 3x^2 + 12x \\
&= x(10x^3 - 3x + 12).
\end{aligned}
$$

EXAMPLE 2   Differentiate (that is, find the derivative of) the function

$$f(x) = x^3(\sqrt{x} + 1).$$

Solution   First, we express the function in exponential form, obtaining

$$f(x) = x^3(x^{1/2} + 1).$$

By the Product Rule,

$$
\begin{aligned}
f'(x) &= x^3\frac{d}{dx}(x^{1/2} + 1) + (x^{1/2} + 1)\frac{d}{dx}x^3 \\
&= x^3\left(\frac{1}{2}x^{-1/2}\right) + (x^{1/2} + 1)(3x^2) \\
&= \frac{1}{2}x^{5/2} + 3x^{5/2} + 3x^2 \\
&= \frac{7}{2}x^{5/2} + 3x^2
\end{aligned}
$$

REMARK   We can also solve the problem by first expanding the product before differentiating f. Examples for which this is not possible will be considered in Section 3.3, where the true value of the Product Rule will be appreciated.   □

The Quotient Rule

The derivative of the quotient of two differentiable functions is given by the following rule:

RULE 6: THE QUOTIENT RULE	$\dfrac{d}{dx}\left[\dfrac{f(x)}{g(x)}\right] = \dfrac{g(x)f'(x) - f(x)g'(x)}{[g(x)]^2} \quad (g(x) \neq 0)$

As an aid to remembering this expression, observe that it has the following form:

$$\frac{d}{dx}\left[\frac{f(x)}{g(x)}\right] = \frac{(\text{denominator})\binom{\text{derivative of}}{\text{numerator}} - (\text{numerator})\binom{\text{derivative of}}{\text{denominator}}}{(\text{square of denominator})}.$$

For a proof of the Quotient Rule, see Exercise 54, page 148.

The derivative of a quotient is not equal to the quotient of the derivatives; that is,

$$\frac{d}{dx}\left[\frac{f(x)}{g(x)}\right] \ne \frac{f'(x)}{g'(x)}.$$

For example, if $f(x) = x^3$ and $g(x) = x^2$, then

$$\frac{d}{dx}\left[\frac{f(x)}{g(x)}\right] = \frac{d}{dx}\left(\frac{x^3}{x^2}\right) = \frac{d}{dx}(x) = 1,$$

which is not equal to

$$\frac{f'(x)}{g'(x)} = \frac{\dfrac{d}{dx}(x^3)}{\dfrac{d}{dx}(x^2)} = \frac{3x^2}{2x} = \frac{3}{2}x!$$

EXAMPLE 3   Find $f'(x)$ if $f(x) = \dfrac{x}{2x - 4}$.

Solution   Using the Quotient Rule, we obtain

$$f'(x) = \frac{(2x - 4)\dfrac{d}{dx}(x) - x\dfrac{d}{dx}(2x - 4)}{(2x - 4)^2}$$

$$= \frac{(2x - 4)(1) - x(2)}{(2x - 4)^2}$$

$$= \frac{2x - 4 - 2x}{(2x - 4)^2} = -\frac{4}{(2x - 4)^2}.$$

EXAMPLE 4   Find $f'(x)$ if $f(x) = \dfrac{x^2 + 1}{x^2 - 1}$.

Solution   By the Quotient Rule,

$$f'(x) = \frac{(x^2 - 1)\dfrac{d}{dx}(x^2 + 1) - (x^2 + 1)\dfrac{d}{dx}(x^2 - 1)}{(x^2 - 1)^2}$$

$$= \frac{(x^2 - 1)(2x) - (x^2 + 1)(2x)}{(x^2 - 1)^2}$$

$$= \frac{2x^3 - 2x - 2x^3 - 2x}{(x^2 - 1)^2}$$

$$= -\frac{4x}{(x^2 - 1)^2}.$$

EXAMPLE 5   Find $h'(x)$ if

$$h(x) = \frac{\sqrt{x}}{x^2 + 1}.$$

Solution   Rewrite $h(x)$ in the form $h(x) = \dfrac{x^{1/2}}{x^2 + 1}$. By the Quotient Rule, we find

$$h'(x) = \frac{(x^2 + 1)\dfrac{d}{dx}(x^{1/2}) - x^{1/2}\dfrac{d}{dx}(x^2 + 1)}{(x^2 + 1)^2}$$

$$= \frac{(x^2 + 1)\left(\dfrac{1}{2}x^{-1/2}\right) - x^{1/2}(2x)}{(x^2 + 1)^2}$$

$$= \frac{\dfrac{1}{2}x^{-1/2}(x^2 + 1 - 4x^2)}{(x^2 + 1)^2} \qquad \left(\begin{array}{l}\text{Factoring out } \tfrac{1}{2}x^{-1/2} \\ \text{from the numerator}\end{array}\right)$$

$$= \frac{1 - 3x^2}{2\sqrt{x}(x^2 + 1)^2}. \qquad\blacksquare$$

Applications

EXAMPLE 6   The sales (in millions of dollars) of a laser disc recording of a hit movie t years from the date of release is given by

$$S(t) = \frac{5t}{t^2 + 1}.$$

a. Find the rate at which the sales are changing at time t.
b. How fast are the sales changing at the time the laser discs are released ($t = 0$)? Two years from the date of release?

Solution
a. The rate at which the sales are changing at time t is given by $S'(t)$. Using the Quotient Rule, we obtain

$$S'(t) = \frac{d}{dt}\left[\frac{5t}{t^2 + 1}\right] = 5\frac{d}{dt}\left[\frac{t}{t^2 + 1}\right]$$

$$= 5\left[\frac{(t^2 + 1)(1) - t(2t)}{(t^2 + 1)^2}\right]$$

$$= 5\left[\frac{t^2 + 1 - 2t^2}{(t^2 + 1)^2}\right] = \frac{5(1 - t^2)}{(t^2 + 1)^2}.$$

b. The rate at which the sales are changing at the time the laser discs are released is given by

$$S'(0) = \frac{5(1 - 0)}{(0 + 1)^2} = 5;$$

that is, they are increasing at the rate of \$5 million per year.

Two years from the date of release, the sales are changing at the rate of

$$S'(2) = \frac{5(1 - 4)}{(4 + 1)^2} = -\frac{3}{5} = -0.6,$$

that is, decreasing at the rate of $600,000 per year.

The graph of the function S is shown in Figure 3.4.

FIGURE 3.4

After a spectacular rise, the sales begin to taper off.

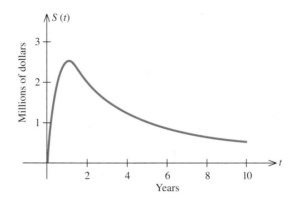

EXAMPLE 7   When organic waste is dumped into a pond, the oxidation process that takes place reduces the pond's oxygen content. However, given time, nature will restore the oxygen content to its natural level. Suppose that the oxygen content t days after organic waste has been dumped into the pond is given by

$$f(t) = 100\left[\frac{t^2 - 4t + 4}{t^2 + 4}\right] \qquad (0 \le t < \infty)$$

percent of its normal level.

a. Derive a general expression that gives the rate of change of the pond's oxygen level at any time t.

b. How fast is the pond's oxygen content changing one day after the organic waste has been dumped? Two days after? Three days after?

Solution

a. The rate of change of the pond's oxygen level at any time t is given by the derivative of the function f. Thus, the required expression is

$$
\begin{aligned}
f'(t) &= 100\frac{d}{dt}\left[\frac{t^2 - 4t + 4}{t^2 + 4}\right] \\
&= 100\left[\frac{(t^2 + 4)(2t - 4) - (t^2 - 4t + 4)(2t)}{(t^2 + 4)^2}\right] \\
&= 100\left[\frac{2t^3 - 4t^2 + 8t - 16 - 2t^3 + 8t^2 - 8t}{(t^2 + 4)^2}\right] \\
&= \frac{100(4t^2 - 16)}{(t^2 + 4)^2} = \frac{400(t^2 - 4)}{(t^2 + 4)^2}.
\end{aligned}
$$

b. The rate at which the pond's oxygen content is changing one day after the organic waste has been dumped is given by

$$f'(1) = \frac{400(1 - 4)}{(1 + 4)^2} = -48;$$

that is, it is dropping at the rate of 48 percent per day. Two days after, the rate is

$$f'(2) = \frac{400(4 - 4)}{(4 + 4)^2} = 0;$$

that is, it is neither increasing nor decreasing. Three days after, the rate is

$$f'(3) = \frac{400(3^2 - 4)}{(3^2 + 4)^2} = 11.83;$$

that is, the oxygen content is increasing at the rate of 11.83 percent per day, and the restoration process has indeed begun.   ■

Verification of the Product Rule

We will now verify the Product Rule. If $p(x) = f(x)g(x)$, then

$$p'(x) = \lim_{h \to 0} \frac{p(x + h) - p(x)}{h}$$

$$= \lim_{h \to 0} \frac{f(x + h)g(x + h) - f(x)g(x)}{h}.$$

By adding $-f(x + h)g(x) + f(x + h)g(x)$ [which is zero!] to the numerator and factoring, we have

$$p'(x) = \lim_{h \to 0} \frac{f(x + h)[g(x + h) - g(x)] + g(x)[f(x + h) - f(x)]}{h}$$

$$= \lim_{h \to 0} \left\{ f(x + h)\left[\frac{g(x + h) - g(x)}{h} \right] + g(x)\left[\frac{f(x + h) - f(x)}{h} \right] \right\}$$

$$= \lim_{h \to 0} f(x + h)\left[\frac{g(x + h) - g(x)}{h} \right]$$

$$+ \lim_{h \to 0} g(x)\left[\frac{f(x + h) - f(x)}{h} \right] \qquad \text{(By Property (iii) of limits)}$$

$$= \lim_{h \to 0} f(x + h) \cdot \lim_{h \to 0} \frac{g(x + h) - g(x)}{h}$$

$$+ \lim_{h \to 0} g(x) \cdot \lim_{h \to 0} \frac{f(x + h) - f(x)}{h} \qquad \text{(By Property (iv) of limits)}$$

$$= f(x)g'(x) + g(x)f'(x).$$

Observe that in the second from the last link in the chain of equalities, we have used the fact that $\lim_{h \to 0} f(x + h) = f(x)$ because f is continuous at x.

**SELF-CHECK
EXERCISES 3.2**

1. Find the derivative of $f(x) = \dfrac{2x + 1}{x^2 - 1}$.

2. What is the slope of the tangent line to the graph of

$$f(x) = (x^2 + 1)(2x^3 - 3x^2 + 1)$$

at the point (2, 25)? How fast is the function f changing when $x = 2$?

c 3. The total sales of the Security Products Corporation in its first two years of operation are given by

$$S = f(t) = \frac{0.3t^3}{1 + 0.4t^2} \qquad (0 \le t \le 2),$$

where S is measured in millions of dollars and $t = 0$ corresponds to the date Security Products began operations. How fast were the sales increasing at the beginning of the company's second year of operation?

Solutions to Self-Check Exercises 3.2 can be found on page 148.

3.2   EXERCISES

In Exercises 1–30, find the derivative of the given function.

1. $f(x) = 2x(x^2 + 1)$

2. $f(x) = 3x^2(x - 1)$

3. $f(t) = (t - 1)(2t + 1)$

4. $f(x) = (2x + 3)(3x - 4)$

5. $f(x) = (3x + 1)(x^2 - 2)$

6. $f(x) = (x + 1)(2x^2 - 3x + 1)$

7. $f(x) = (x^3 - 1)(x + 1)$

8. $f(x) = (x^3 - 12x)(3x^2 + 2x)$

9. $f(w) = (w^3 - w^2 + w - 1)(w^2 + 2)$

10. $f(x) = \dfrac{1}{5}x^5 + (x^2 + 1)(x^2 - x - 1) + 28$

11. $f(x) = (5x^2 + 1)(2\sqrt{x} - 1)$

12. $f(t) = (1 + \sqrt{t})(2t^2 - 3)$

13. $f(x) = (x^2 - 5x + 2)\left(x - \dfrac{2}{x}\right)$

14. $f(x) = (x^3 + 2x + 1)\left(2 + \dfrac{1}{x^2}\right)$

15. $f(x) = \dfrac{1}{x - 2}$

16. $g(x) = \dfrac{3}{2x + 4}$

17. $f(x) = \dfrac{x - 1}{2x + 1}$

18. $f(t) = \dfrac{1 - 2t}{1 + 3t}$

19. $f(x) = \dfrac{1}{x^2 + 1}$

20. $f(u) = \dfrac{u}{u^2 + 1}$

21. $f(s) = \dfrac{s^2 - 4}{s + 1}$

22. $f(x) = \dfrac{x^3 - 2}{x^2 + 1}$

23. $f(x) = \dfrac{\sqrt{x}}{x^2 + 1}$

24. $f(x) = \dfrac{x^2 + 1}{\sqrt{x}}$

25. $f(x) = \dfrac{x^2 + 2}{x^2 + x + 1}$

26. $f(x) = \dfrac{x + 1}{2x^2 + 2x + 3}$

27. $f(x) = \dfrac{(x + 1)(x^2 + 1)}{x - 2}$

28. $f(x) = (3x^2 - 1)\left(x^2 - \dfrac{1}{x}\right)$

29. $f(x) = \dfrac{x}{x^2 - 4} - \dfrac{x - 1}{x^2 + 4}$

30. $f(x) = \dfrac{x + \sqrt{3x}}{3x - 1}$

In Exercises 31–34, find the derivative of each of the given functions and evaluate $f'(x)$ at the given value of x.

31. $f(x) = (2x - 1)(x^2 + 3); \; x = 1$

32. $f(x) = \dfrac{2x+1}{2x-1}$; $x = 2$

33. $f(x) = \dfrac{x}{x^4 - 2x^2 - 1}$; $x = -1$

34. $f(x) = (\sqrt{x} + 2x)(x^{3/2} - x)$; $x = 4$

In Exercises 35–38, find the slope and an equation of the tangent line to the graph of the function f at the specified point.

35. $f(x) = (x^3 + 1)(x^2 - 2)$; (2, 18)

36. $f(x) = \dfrac{x^2}{x+1}$; $\left(2, \dfrac{4}{3}\right)$

37. $f(x) = \dfrac{x+1}{x^2+1}$; (1, 1)

38. $f(x) = \dfrac{1 + 2x^{1/2}}{1 + x^{3/2}}$; $\left(4, \dfrac{5}{9}\right)$

39. Find an equation of the tangent line to the graph of the function $f(x) = (x^3 + 1)(3x^2 - 4x + 2)$ at the point (1, 2).

40. Find an equation of the tangent line to the graph of the function $f(x) = \dfrac{3x}{x^2 - 2}$ at the point (2, 3).

41. Let $f(x) = (x^2 + 1)(2 - x)$. Find the point(s) on the graph of f where the tangent line is horizontal.

42. Let $f(x) = \dfrac{x}{x^2 + 1}$. Find the point(s) on the graph of f where the tangent line is horizontal.

43. Find the point(s) on the graph of the function $f(x) = (x^2 + 6)(x - 5)$ where the slope of the tangent line is equal to -2.

44. Find the point(s) on the graph of the function $f(x) = \dfrac{x+1}{x-1}$ where the slope of the tangent line is equal to $-\frac{1}{2}$.

45. **Concentration of a Drug in the Bloodstream**   The concentration of a certain drug in a patient's bloodstream t hours after injection is given by

$$C(t) = \dfrac{0.2t}{t^2 + 1}.$$

  a. Find the rate at which the concentration of the drug is changing with respect to time.
  b. How fast is the concentration changing $\frac{1}{2}$ hour after the injection? 1 hour after the injection? 2 hours after the injection?

46. **Cost of Removing Toxic Waste**   A city's main well was recently found to be contaminated with trichloroethylene, a cancer-causing chemical, as a result of an abandoned chemical dump leaching chemicals into the water. A proposal submitted to the city's selectmen indicates that the cost, measured in millions of dollars, of removing x percent of the toxic pollutant is given by

$$C(x) = \dfrac{0.5x}{100 - x}.$$

Find $C'(80)$, $C'(90)$, $C'(95)$, and $C'(99)$ and interpret your results.

47. **Drug Dosages**   Thomas Young has suggested the following rule for calculating the dosage of medicine for children 1 to 12 years old. If a denotes the adult dosage (in mg), and if t is the child's age (in years), then the child's dosage is given by

$$D(t) = \dfrac{at}{t + 12}.$$

Suppose that the adult dosage of a substance is 500 mg. Find an expression that gives the rate of change of a child's dosage with respect to the child's age. What is the rate of change of a child's dosage with respect to his or her age for a 6-year-old child? For a 10-year-old child?

48. **Effect of Bactericide**   The number of bacteria $N(t)$ in a certain culture t minutes after an experimental bactericide is introduced obeys the rule

$$N(t) = \dfrac{10,000}{1 + t^2} + 2,000.$$

Find the rate of change of the number of bacteria in the culture 1 minute after and 2 minutes after the bactericide is introduced. What is the population of the bacteria in the culture 1 minute after the bactericide is introduced? 2 minutes after?

49. **Demand Functions**   The demand function for the Sicard wristwatch is given by

$$d(x) = \dfrac{50}{0.01x^2 + 1} \quad (0 \le x \le 20),$$

where x (measured in units of a thousand) is the quantity demanded per week and $d(x)$ is the unit price in dollars.
  a. Find $d'(x)$.
  b. Find $d'(5)$, $d'(10)$, and $d'(15)$ and interpret your results.

50. Learning Curves From experience, the Emory Secretarial School knows that the average student taking Advanced Typing will progress according to the rule

$$N(t) = \frac{60t + 180}{t + 6} \qquad (t \geq 0),$$

where $N(t)$ measures the number of words per minute the student can type after t weeks in the course.

a. Find an expression for $N'(t)$.

b. Compute $N'(t)$ for $t = 1$, 3, 4, and 7 and interpret your results.

c. Sketch the graph of the function N. Does it confirm the results obtained in (b)?

d. What will the average student's typing speed be at the end of the 12-week course?

51. Box Office Receipts The total worldwide box office receipts for a long-running movie are approximated by the function

$$T(x) = \frac{120x^2}{x^2 + 4},$$

where $T(x)$ is measured in millions of dollars and x is the number of years since the movie's release. How fast are the total receipts changing one year after the movie's release? Three years after its release? Five years after its release?

52. Population Growth A major corporation is building a 4325-acre complex of homes, offices, stores, schools, and churches in the rural community of Glen Cove. As a result of this development, the planners have estimated that Glen Cove's population (in thousands) t years from now will be given by

$$P(t) = \frac{25t^2 + 125t + 200}{t^2 + 5t + 40}.$$

a. Find the rate at which Glen Cove's population is changing with respect to time.

b. What will the population be after ten years? At what rate will the population be increasing when $t = 10$?

53. Extend the Product Rule for differentiation to the following case involving the product of three differentiable functions: Let $h(x) = u(x)v(x)w(x)$ and show that $h'(x) = u(x)v(x)w'(x) + u(x)v'(x)w(x) + u'(x)v(x)w(x)$. [*Hint:* Let $f(x) = u(x)v(x)$, $g(x) = w(x)$, and $h(x) = f(x)g(x)$, and apply the Product Rule to the function h.]

54. Prove the Quotient Rule for differentiation (Rule 6). [*Hint:* Verify the following steps:

a. $\dfrac{k(x + h) - k(x)}{h} = \dfrac{f(x + h)g(x) - f(x)g(x + h)}{hg(x + h)g(x)}$

b. By adding $[-f(x)g(x) + f(x)g(x)]$ to the numerator and simplifying,

$$\frac{k(x + h) - k(x)}{h} = \frac{1}{g(x + h)g(x)} \times$$
$$\left\{ \left[\frac{f(x + h) - f(x)}{h} \right] \cdot g(x) - \left[\frac{g(x + h) - g(x)}{h} \right] \cdot f(x) \right\}$$

c. $k'(x) = \lim\limits_{h \to 0} \dfrac{k(x + h) - k(x)}{h}$

$$= \frac{g(x)f'(x) - f(x)g'(x)}{[g(x)]^2}.]$$

C *In Exercises 55–58, use a computer or a graphing calculator to plot the graphs of the functions f and f'. Then use the graphs to find the approximate location of the point(s) where the tangent line(s) are horizontal.*

55. $f(x) = x(x^2 - 9)$

56. $f(x) = x^3(3x^2 - 5)$

57. $f(x) = \dfrac{3x}{x^2 + 1}$

58. $f(x) = \dfrac{x^2 - 1}{x^3}$

SOLUTIONS TO SELF-CHECK EXERCISES 3.2

1. We use the Quotient Rule to obtain

$$h'(x) = \frac{(x^2 - 1)\dfrac{d}{dx}(2x + 1) - (2x + 1)\dfrac{d}{dx}(x^2 - 1)}{(x^2 - 1)^2}$$

$$= \frac{(x^2 - 1)(2) - (2x + 1)(2x)}{(x^2 - 1)^2}$$

$$= \frac{2x^2 - 2 - 4x^2 - 2x}{(x^2 - 1)^2}$$

$$= \frac{-2x^2 - 2x - 2}{(x^2 - 1)^2}$$

$$= -\frac{2(x^2 + x + 1)}{(x^2 - 1)^2}.$$

2. The slope of the tangent line to the graph of f at any point is given by

$$f'(x) = (x^2 + 1)\frac{d}{dx}(2x^3 - 3x^2 + 1)$$
$$+ (2x^3 - 3x^2 + 1)\frac{d}{dx}(x^2 + 1)$$
$$= (x^2 + 1)(6x^2 - 6x) + (2x^3 - 3x^2 + 1)(2x).$$

In particular, the slope of the tangent line to the graph of f when $x = 2$ is

$$f'(2) = (2^2 + 1)[6(2^2) - 6(2)]$$
$$+ [2(2^3) - 3(2^2) + 1][2(2)]$$
$$= 60 + 20 = 80.$$

Note that it is not necessary to simplify the expression for $f'(x)$, since we are required only to evaluate the expression at $x = 2$. We also conclude, from this result, that the function f is changing at the rate of 80 units per unit change in x when $x = 2$.

3. The rate at which the company's total sales are changing at any time t is given by

$$S'(t) = \frac{(1 + 0.4t^2)\frac{d}{dt}(0.3t^3) - (0.3t^3)\frac{d}{dt}(1 + 0.4t^2)}{(1 + 0.4t^2)^2}$$
$$= \frac{(1 + 0.4t^2)(0.9t^2) - (0.3t^3)(0.8t)}{(1 + 0.4t^2)^2}.$$

Therefore, at the beginning of the second year of operation, Security Products' sales were increasing at the rate of

$$S'(1) = \frac{(1 + 0.4)(0.9) - (0.3)(0.8)}{(1 + 0.4)^2} = 0.520408,$$

or $520,408 per year.

3.3 The Chain Rule

This section introduces another rule of differentiation called the **Chain Rule.** When used in conjunction with the rules of differentiation developed in the last two sections, the Chain Rule enables us to greatly enlarge the class of functions we are able to differentiate.

The Chain Rule

Consider the function $h(x) = (x^2 + x + 1)^2$. If we were to compute $h'(x)$ using only the rules of differentiation from the previous sections, then our approach might be to expand $h(x)$. Thus,

$$h(x) = (x^2 + x + 1)^2 = (x^2 + x + 1)(x^2 + x + 1)$$
$$= x^4 + 2x^3 + 3x^2 + 2x + 1,$$

from which we find

$$h'(x) = 4x^3 + 6x^2 + 6x + 2.$$

But what about the function $H(x) = (x^2 + x + 1)^{100}$? The same technique may be used to find the derivative of the function H, but the amount of work involved in this case would be prodigious! Consider, also, the function $G(x) = \sqrt{x^2 + 1}$. For each of the two functions H and G, the rules of differentiation of the previous sections cannot be applied directly to compute the derivatives H' and G'.

Observe that both H and G are **composite** functions; that is, each is composed of, or built up from, simpler functions. For example, the function H is composed of the two simpler functions $f(x) = x^2 + x + 1$ and $g(x) = x^{100}$ as follows:

$$H(x) = g[f(x)] = [f(x)]^{100}$$
$$= (x^2 + x + 1)^{100}.$$

In a similar manner, we see that the function G is composed of the two simpler functions $f(x) = x^2 + 1$ and $g(x) = \sqrt{x}$. Thus,

$$G(x) = g[f(x)] = \sqrt{f(x)}$$
$$= \sqrt{x^2 + 1}.$$

As a first step toward finding the derivative h' of a composite function $h = g \circ f$ defined by $h(x) = g[f(x)]$, we write

$$u = f(x) \quad \text{and} \quad y = g[f(x)] = g(u).$$

The dependency of h on g and f is illustrated in Figure 3.5.

FIGURE 3.5

The composite function
$h(x) = g[f(x)]$

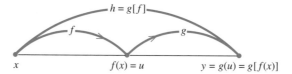

Since u is a function of x, we may compute the derivative of u with respect to x, if f is a differentiable function, obtaining $du/dx = f'(x)$. Next, if g is a differentiable function of u, we may compute the derivative of g with respect to u, obtaining $dy/du = g'(u)$. Now, since the function h is composed of the function g and the function f, we might suspect that the rule $h'(x)$ for the derivative h' of h will be given by an expression that involves the rules for the derivatives of f and g. But how do we combine these derivatives to yield h'?

This question can be answered by interpreting the derivative of each function as giving the rate of change of that function. For example, suppose that $u = f(x)$ changes three times as fast as x; that is,

$$f'(x) = \frac{du}{dx} = 3,$$

and $y = g(u)$ changes twice as fast as u; that is,

$$g'(u) = \frac{dy}{du} = 2.$$

Then we would expect $y = h(x)$ to change six times as fast as x; that is,

$$h'(x) = g'(u)f'(x) = (2)(3) = 6,$$

or equivalently,

$$\frac{dy}{dx} = \frac{dy}{du} \cdot \frac{du}{dx} = (2)(3) = 6.$$

This observation suggests the following result, which we state without proof.

RULE 7: THE CHAIN RULE	If $h(x) = g[f(x)]$, then $$h'(x) = \frac{d}{dx} g(f(x)) = g'(f(x))f'(x). \qquad (1)$$ Equivalently, if we write $y = h(x) = g(u)$, where $u = f(x)$, then $$\frac{dy}{dx} = \frac{dy}{du} \cdot \frac{du}{dx}. \qquad (2)$$

REMARKS

1. If we label the composite function h in the following manner

$$\overset{\text{inside function}}{\underset{\downarrow}{}}$$
$$h(x) = g[\underset{\uparrow}{f(x)}],$$
$$\text{outside function}$$

then $h'(x)$ is just the *derivative* of the "outside function" *evaluated at* the "inside function" times the *derivative* of the "inside function."

2. Equation (2) can be remembered by observing that if we "cancel" the du's, then

$$\frac{dy}{dx} = \frac{dy}{\cancel{du}} \cdot \frac{\cancel{du}}{dx} = \frac{dy}{dx}. \qquad \square$$

The Chain Rule for Powers of Functions

Many composite functions have the special form $h(x) = g(f(x))$, where g is defined by the rule $g(x) = x^n$ (n, a real number); that is,

$$h(x) = [f(x)]^n.$$

In other words, the function h is given by the power of a function f. The functions

$$h(x) = (x^2 + x + 1)^2, \qquad H = (x^2 + x + 1)^{100}, \quad \text{and} \quad G = \sqrt{x^2 + 1}$$

discussed earlier are examples of this type of composite function. By using the following corollary of the Chain Rule, the General Power Rule, we are able to find the derivative of this type of function much more easily than by using the Chain Rule directly.

THE GENERAL POWER RULE

If the function f is differentiable and $h(x) = [f(x)]^n$ (n, a real number), then

$$h'(x) = \frac{d}{dx}[f(x)]^n = n[f(x)]^{n-1}f'(x). \qquad (3)$$

To see this, we observe that $h(x) = g(f(x))$ where $g(x) = x^n$, so that, by virtue of the Chain Rule, we have

$$h'(x) = g'(f(x))f'(x)$$
$$= n[f(x)]^{n-1}f'(x),$$

since $g'(x) = nx^{n-1}$.

EXAMPLE 1   Find $H'(x)$, if

$$H(x) = (x^2 + x + 1)^{100}.$$

Solution   Using the General Power Rule, we obtain

$$H'(x) = 100(x^2 + x + 1)^{99} \frac{d}{dx}(x^2 + x + 1)$$
$$= 100(x^2 + x + 1)^{99}(2x + 1).$$

Observe the bonus that comes from using the Chain Rule: The answer is completely factored. ∎

EXAMPLE 2   Differentiate the function $G(x) = \sqrt{x^2 + 1}$.

Solution   We rewrite the function $G(x)$ as

$$G(x) = (x^2 + 1)^{1/2}$$

and apply the General Power Rule, obtaining

$$G'(x) = \frac{1}{2}(x^2 + 1)^{-1/2}\frac{d}{dx}(x^2 + 1)$$
$$= \frac{1}{2}(x^2 + 1)^{-1/2} \cdot 2x = \frac{x}{\sqrt{x^2 + 1}}.$$
∎

EXAMPLE 3   Differentiate the function $f(x) = x^2(2x + 3)^5$.

Solution   Applying the Product Rule followed by the General Power Rule, we obtain

$$f'(x) = x^2\frac{d}{dx}(2x + 3)^5 + (2x + 3)^5\frac{d}{dx}(x^2)$$

$$= (x^2)5(2x + 3)^4 \cdot \frac{d}{dx}(2x + 3) + (2x + 3)^5(2x)$$

$$= 5x^2(2x + 3)^4(2) + 2x(2x + 3)^5$$
$$= 2x(2x + 3)^4(5x + 2x + 3) = 2x(7x + 3)(2x + 3)^4.$$
∎

EXAMPLE 4   Find $f'(x)$ if $f(x) = \dfrac{1}{(4x^2 - 7)^2}$.

Solution   Rewriting $f(x)$ and then applying the General Power Rule, we obtain

$$f'(x) = \frac{d}{dx}\left[\frac{1}{(4x^2 - 7)^2}\right] = \frac{d}{dx}(4x^2 - 7)^{-2}$$

$$= -2(4x^2 - 7)^{-3}\frac{d}{dx}(4x^2 - 7)$$

$$= -2(4x^2 - 7)^{-3}(8x) = -\frac{16x}{(4x^2 - 7)^3}.$$ ∎

EXAMPLE 5   Find the slope of the tangent line to the graph of the function

$$f(x) = \left(\frac{2x + 1}{3x + 2}\right)^3$$

at the point $\left(0, \frac{1}{8}\right)$.

Solution   The slope of the tangent line to the graph of f at any point is given by $f'(x)$. To compute $f'(x)$, we use the General Power Rule followed by the Quotient Rule, obtaining

$$f'(x) = 3\left(\frac{2x + 1}{3x + 2}\right)^2 \frac{d}{dx}\left(\frac{2x + 1}{3x + 2}\right)$$

$$= 3\left(\frac{2x + 1}{3x + 2}\right)^2\left[\frac{(3x + 2)(2) - (2x + 1)(3)}{(3x + 2)^2}\right]$$

$$= 3\left(\frac{2x + 1}{3x + 2}\right)^2\left[\frac{6x + 4 - 6x - 3}{(3x + 2)^2}\right]$$

$$= \frac{3(2x + 1)^2}{(3x + 2)^4}.$$

In particular, the slope of the tangent line to the graph of f at $\left(0, \frac{1}{8}\right)$ is given by

$$f'(0) = \frac{3(0 + 1)^2}{(0 + 2)^4} = \frac{3}{16}.$$ ∎

Applications

EXAMPLE 6   The membership of The Fitness Center, which opened a few years ago, is approximated by the function

$$N(t) = 100(64 + 4t)^{2/3} \qquad (0 \le t \le 52),$$

where $N(t)$ gives the number of members at the beginning of week t.

a. Find $N'(t)$.
b. How fast was the center's membership increasing initially $(t = 0)$?
c. How fast was the membership increasing at the beginning of the 40th week?
d. What was the membership when the center first opened? At the beginning of the 40th week?

Solution

a. Using the General Power Rule, we obtain

$$N'(t) = \frac{d}{dt}[100(64 + 4t)^{2/3}]$$

$$= 100\frac{d}{dt}(64 + 4t)^{2/3}$$

$$= 100\left(\frac{2}{3}\right)(64 + 4t)^{-1/3}\frac{d}{dt}(64 + 4t)$$

$$= \frac{200}{3}(64 + 4t)^{-1/3}(4)$$

$$= \frac{800}{3(64 + 4t)^{1/3}}.$$

b. The rate at which the membership was increasing when the center first opened is given by

$$N'(0) = \frac{800}{3(64)^{1/3}} \approx 66.7,$$

or approximately 67 people per week.

c. The rate at which the membership was increasing at the beginning of the 40th week is given by

$$N'(40) = \frac{800}{3(64 + 160)^{1/3}} \approx 43.9,$$

or approximately 44 people per week.

d. The membership when the center first opened is given by

$$N(0) = 100(64)^{2/3} = 100(16),$$

or approximately 1600 people. The membership at the beginning of the 40th week is given by

$$N(40) = 100(64 + 160)^{2/3} \approx 3688.4,$$

or approximately 3688 people.   ∎

EXAMPLE 7   Arteriosclerosis begins during childhood when plaque (soft masses of fatty material) forms in the arterial walls, blocking the flow of blood through the arteries and leading to heart attacks, strokes, and gangrene. Suppose the idealized cross section of the aorta is circular with radius a cm, and that by year t the thickness of the plaque (assume that it is uniform) is $h = f(t)$ cm (Figure 3.6). Then the area of the opening is given by $A = \pi(a - h)^2$ square centimeters.

Suppose that the radius of an individual's artery is 1 cm ($a = 1$) and the thickness of the plaque in year t is given by

$$h = g(t) = 1 - 0.01(10,000 - t^2)^{1/2} \text{ cm.}$$

Since the area of the arterial opening is given by

$$A = f(h) = \pi(1 - h)^2,$$

the rate at which A is changing with respect to time is given by

FIGURE 3.6

Cross section of the aorta

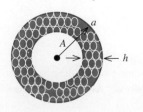

$$\frac{dA}{dt} = \frac{dA}{dh} \cdot \frac{dh}{dt} = f'(h) \cdot g'(t) \qquad \text{(By the Chain Rule)}$$

$$= 2\pi(1-h)(-1)\left[-0.01\left(\tfrac{1}{2}\right)(10{,}000 - t^2)^{-1/2}(-2t)\right] \qquad \text{(Using the Chain Rule twice)}$$

$$= -2\pi(1-h)\left[\frac{0.01t}{(10{,}000 - t^2)^{1/2}}\right]$$

$$= -\frac{0.02\pi(1-h)t}{\sqrt{10{,}000 - t^2}}.$$

For example, when $t = 50$,

$$h = g(50) = 1 - 0.01(10{,}000 - 2{,}500)^{1/2} \approx 0.134,$$

so that

$$\frac{dA}{dt} = -\frac{0.02\pi(1 - 0.134)50}{\sqrt{10{,}000 - 2{,}500}} \approx -0.03;$$

that is, the area of the arterial opening is decreasing at the rate of 0.03 cm^2/year. ∎

SELF-CHECK EXERCISES 3.3

1. Find the derivative of
$$f(x) = -\frac{1}{\sqrt{2x^2 - 1}}.$$

2. Suppose that the life expectancy at birth (in years) of a female in a certain country is described by the function
$$g(t) = 50.02(1 + 1.09t)^{0.1} \qquad (0 \le t \le 150),$$
where t is measured in years and $t = 0$ corresponds to the beginning of 1900.
   a. What is the life expectancy at birth of a female born at the beginning of 1980? At the beginning of the year 2000?
   b. How fast is the life expectancy at birth of a female born at any time t changing?

Solutions to Self-Check Exercises 3.3 can be found on page 158.

3.3 EXERCISES

In Exercises 1–46, find the derivative of the given function.

1. $f(x) = (2x - 1)^4$

2. $f(x) = (1 - x)^3$

3. $f(x) = (x^2 + 2)^5$

4. $f(t) = 2(t^3 - 1)^5$

5. $f(x) = (2x - x^2)^3$

6. $f(x) = 3(x^3 - x)^4$

7. $f(x) = (2x + 1)^{-2}$

8. $f(t) = \tfrac{1}{2}(2t^2 + t)^{-3}$

9. $f(x) = (x^2 - 4)^{3/2}$

10. $f(t) = (3t^2 - 2t + 1)^{3/2}$

11. $f(x) = \sqrt{3x - 2}$

12. $f(t) = \sqrt{3t^2 - t}$

13. $f(x) = \sqrt[3]{1 - x^2}$

14. $f(x) = \sqrt{2x^2 - 2x + 3}$

15. $f(x) = \dfrac{1}{(2x + 3)^3}$

16. $f(x) = \dfrac{2}{(x^2 - 1)^4}$

17. $f(t) = \dfrac{1}{\sqrt{2t - 3}}$

18. $f(x) = \dfrac{1}{\sqrt{2x^2 - 1}}$

19. $y = \dfrac{1}{(4x^4 + x)^{3/2}}$

20. $f(t) = \dfrac{4}{\sqrt[3]{2t^2 + t}}$

21. $f(x) = (3x^2 + 2x + 1)^{-2}$

22. $f(t) = (5t^3 + 2t^2 - t + 4)^{-3}$

23. $f(x) = (x^2 + 1)^3 - (x^3 + 1)^2$

24. $f(t) = (2t - 1)^4 + (2t + 1)^4$

25. $f(t) = (t^{-1} - t^{-2})^3$ 26. $f(v) = (v^{-3} + 4v^{-2})^3$

27. $f(x) = \sqrt{x + 1} + \sqrt{x - 1}$

28. $f(u) = (2u + 1)^{3/2} + (u^2 - 1)^{-3/2}$

29. $f(x) = 2x^2(3 - 4x)^4$ 30. $h(t) = t^2(3t + 4)^3$

31. $f(x) = (x - 1)^2(2x + 1)^4$

32. $g(u) = (1 + u^2)^5(1 - 2u^2)^8$

33. $f(x) = \left(\dfrac{x + 3}{x - 2}\right)^3$ 34. $f(x) = \left(\dfrac{x + 1}{x - 1}\right)^5$

35. $s(t) = \left(\dfrac{t}{2t + 1}\right)^{3/2}$ 36. $g(s) = \left(s^2 + \dfrac{1}{s}\right)^{3/2}$

37. $g(u) = \sqrt{\dfrac{u + 1}{3u + 2}}$ 38. $g(x) = \sqrt{\dfrac{2x + 1}{2x - 1}}$

39. $f(x) = \dfrac{x^2}{(x^2 - 1)^4}$ 40. $g(u) = \dfrac{2u^2}{(u^2 + u)^3}$

41. $h(x) = \dfrac{(3x^2 + 1)^3}{(x^2 - 1)^4}$ 42. $g(t) = \dfrac{(2t - 1)^2}{(3t + 2)^4}$

43. $f(x) = \dfrac{\sqrt{2x + 1}}{x^2 - 1}$ 44. $f(t) = \dfrac{4t^2}{\sqrt{2t^2 + 2t - 1}}$

45. $g(t) = \dfrac{\sqrt{t + 1}}{\sqrt{t^2 + 1}}$ 46. $f(x) = \dfrac{\sqrt{x^2 + 1}}{\sqrt{x^2 - 1}}$

In Exercises 47–52, find dy/du, du/dx, and dy/dx.

47. $y = u^{4/3}$ and $u = 3x^2 - 1$

48. $y = \sqrt{u}$ and $u = 7x - 2x^2$

49. $y = u^{-2/3}$ and $u = 2x^3 - x + 1$

50. $y = 2u^2 + 1$ and $u = x^2 + 1$

51. $y = \sqrt{u} + \dfrac{1}{\sqrt{u}}$ and $u = x^3 - x$

52. $y = \dfrac{1}{u}$ and $u = \sqrt{x} + 1$

In Exercises 53–56, find an equation of the tangent line to the graph of the given function at the given point.

53. $f(x) = (1 - x)(x^2 - 1)^2$; $(2, -9)$

54. $f(x) = \left(\dfrac{x + 1}{x - 1}\right)^2$; $(3, 4)$

55. $f(x) = x\sqrt{2x^2 + 7}$; $(3, 15)$

56. $f(x) = \dfrac{8}{\sqrt{x^2 + 6x}}$; $(2, 2)$

C 57. **Television Viewing**   The number of viewers of a television series introduced several years ago is approximated by the function

$$N(x) = (60 + 2x)^{2/3} \quad (1 \le x \le 26),$$

where $N(x)$ (measured in millions) denotes the number of weekly viewers of the series in the xth week. Find the rate of increase of the weekly audience at the end of the 2nd week and at the end of the 12th week. How many viewers were there in the 2nd week? In the 24th week?

C 58. **Male Life Expectancy**   Suppose that the life expectancy of a male at birth in a certain country is described by the function

$$f(t) = 46.9(1 + 1.09t)^{0.1} \quad (0 \le t \le 150),$$

where t is measured in years and $t = 0$ corresponds to the beginning of 1900. How long can a male born at the beginning of the year 2000 in that country expect to live? What is the rate of change of the life expectancy of a male born in that country at the beginning of the year 2000?

C 59. **Concentration of Carbon Monoxide in the Air**   According to a joint study conducted by Oxnard's Environmental Management Department and a state government agency, the concentration of carbon monoxide in the air due to automobile exhaust t years from now is given by

$$C(t) = 0.01(0.2t^2 + 4t + 64)^{2/3}$$

parts per million.
a. Find the rate at which the level of carbon monoxide is changing with respect to time.
b. Find the rate at which the level of carbon monoxide will be changing five years from now.

C 60. **Continuing Education Enrollment**   The registrar of Kellogg University estimates that the total student enrollment in the Continuing Education division will be given by

$$N(t) = -\dfrac{20{,}000}{\sqrt{1 + 0.2t}} + 21{,}000,$$

where $N(t)$ denotes the number of students enrolled in the division t years from now. Find an expression for $N'(t)$. How fast is the student enrollment increasing currently? How fast will it be increasing five years from now?

C **61. Air Pollution** According to the South Coast Air Quality Management District, the level of nitrogen dioxide, a brown gas that impairs breathing, present in the atmosphere on a certain May day in downtown Los Angeles is approximated by

$$A(t) = 0.03t^3(t - 7)^4 + 60.2 \qquad (0 \le t \le 7),$$

where $A(t)$ is measured in pollutant standard index and t is measured in hours, with $t = 0$ corresponding to 7 A.M.

a. Find $A'(t)$.
b. Find $A'(1)$, $A'(3)$, and $A'(4)$ and interpret your results.

C **62. Effect of Luxury Tax on Consumption** Government economists of a developing country determined that the purchase of imported perfume is related to a proposed "luxury tax" by the formula

$$N(x) = \sqrt{10,000 - 40x - 0.02x^2} \qquad (0 \le x \le 200),$$

where $N(x)$ measures the percentage of normal consumption of perfume when a "luxury tax" of x percent is imposed on it. Find the rate of change of $N(x)$ for taxes of 10 percent, 100 percent, and 150 percent.

C **63. Pulse Rate of an Athlete** The pulse rate (the number of heartbeats per minute) of a long-distance runner t seconds after leaving the starting line is given by

$$P(t) = \frac{300\sqrt{\frac{1}{2}t^2 + 2t + 25}}{t + 25} \qquad (t \ge 0).$$

Compute $P'(t)$. How fast is the athlete's pulse rate increasing 10 seconds into the run? 60 seconds into the run? 2 minutes into the run? What is her pulse rate 2 minutes into the run?

64. Thurstone Learning Model Psychologist L. L. Thurstone suggested the following relationship between the learning time T and the length of a list n:

$$T = f(n) = An\sqrt{n - b},$$

where A and b are constants that depend on the person and the task.

a. Compute dT/dn and interpret your result.
b. Suppose that for a certain person and a certain task, $A = 4$ and $b = 4$. Compute $f'(13)$ and $f'(29)$ and interpret your results.

65. Oil Spills In calm waters, the oil spilling from the ruptured hull of a grounded tanker spreads in all directions. Assuming that the area polluted is a circle and that its radius is increasing at a rate of 2 ft/sec, determine how fast the area is increasing when the radius of the circle is 40 feet.

C **66. Arteriosclerosis** Refer to Example 7, page 154. Suppose that the radius of an individual's artery is 1 cm and the thickness of the plaque (in cm) t years from now is given by

$$h = g(t) = \frac{0.5t^2}{t^2 + 10} \qquad (0 \le t \le 10).$$

How fast will the arterial opening be decreasing five years from now?

C **67. Hotel Occupancy Rates** The occupancy rate of the all-suite Wonderland Hotel, located near an amusement park, is given by the function

$$r(t) = \frac{10}{81}t^3 - \frac{10}{3}t^2 + \frac{200}{9}t + 60 \qquad (0 \le t \le 12),$$

where t is measured in months and $t = 0$ corresponds to the beginning of January. Management has estimated that the monthly revenue (in thousands of dollars per month) is approximated by the function

$$R(r) = -\frac{3}{5000}r^3 + \frac{9}{50}r^2 \qquad (0 \le r \le 100),$$

where r is the occupancy rate.

a. Find an expression that gives the rate of change of Wonderland's occupancy rate with respect to time.
b. Find an expression that gives the rate of change of Wonderland's monthly revenue with respect to the occupancy rate.
c. What is the rate of change of Wonderland's monthly revenue with respect to time at the beginning of January? At the beginning of June? [*Hint:* Use the Chain Rule to find $R'(r(0))r'(0)$ and $R'(r(6))r'(6)$.]

C **68. Effect of Housing Starts on Jobs** The president of a major housing construction firm claims that the number of construction jobs created is given by

$$N(x) = 1.42x,$$

where x denotes the number of housing starts. Suppose that the number of housing starts in the next t months is expected to be

$$x(t) = \frac{7t^2 + 140t + 700}{3t^2 + 80t + 550}$$

million units per year. Find an expression that gives the rate at which the number of construction jobs will be created t months from now. At what rate will construction jobs be created one year from now?

C **69. Demand for PCs** The quantity demanded per month, x, of a certain make of personal computer (PC) is related to the average unit price, p, (in dollars) of personal computers by the equation

$$x = f(p) = \frac{100}{9}\sqrt{810,000 - p^2}.$$

It is estimated that t months from now, the average price of a PC will be given by

$$p(t) = \frac{400}{1 + \frac{1}{8}\sqrt{t}} + 200 \qquad (0 \le t \le 60)$$

dollars. Find the rate at which the quantity demanded per month of the PCs will be changing 16 months from now.

C 70. **Cruise Ship Bookings** The management of Cruise World, operators of Caribbean luxury cruises, expects that the percentage of young adults booking passage on their cruises in the years ahead will rise dramatically. They have constructed the following model, which gives the percentage of young adult passengers in year t:

$$p = f(t) = 50\left(\frac{t^2 + 2t + 4}{t^2 + 4t + 8}\right) \qquad (0 \le t \le 5).$$

Young adults normally pick shorter cruises and generally spend less on their passage. The following model gives an approximation of the average amount of money R (in dollars) spent per passenger on a cruise when the percentage of young adults is p:

$$R(p) = 1000\left(\frac{p + 4}{p + 2}\right).$$

Find the rate at which the price of the average passage will be changing two years from now.

71. In Section 3.1, we proved that

$$\frac{d}{dx}(x^n) = nx^{n-1}$$

for the special case when $n = 2$. Use the Chain Rule to show that

$$\frac{d}{dx}(x^{1/n}) = \frac{1}{n}x^{1/n-1}$$

for any nonzero integer n, assuming that $f(x) = x^{1/n}$ is differentiable. [*Hint:* Let $f(x) = x^{1/n}$ so that $[f(x)]^n = x$. Differentiate both sides with respect to x.]

72. With the aid of Exercise 71, prove that

$$\frac{d}{dx}(x^r) = rx^{r-1}$$

for any rational number r. [*Hint:* Let $r = m/n$ where m and n are integers with $n \ne 0$, and write $x^r = (x^m)^{1/n}$.]

C *In Exercises 73–76, use a computer or a graphing calculator to plot the graphs of the functions f and f'. Then use the graphs to find the approximate location of the point(s) where the tangent line(s) are horizontal.*

73. $f(x) = \sqrt{x^2 - x^4}$

74. $f(x) = (x - 8)^2 + \left(x^2 - \frac{1}{2}\right)^2$

75. $f(x) = x - \sqrt{1 - x^2}$

76. $f(x) = x\sqrt{1 - x^2}$

SOLUTIONS TO SELF-CHECK EXERCISES 3.3

1. Rewriting, we have

$$f(x) = -(2x^2 - 1)^{-1/2}.$$

Using the General Power Rule, we find

$$\begin{aligned}
f'(x) &= -\frac{d}{dx}(2x^2 - 1)^{-1/2} \\
&= -\left(-\frac{1}{2}\right)(2x^2 - 1)^{-3/2}\frac{d}{dx}(2x^2 - 1) \\
&= \frac{1}{2}(2x^2 - 1)^{-3/2}(4x) \\
&= \frac{2x}{(2x^2 - 1)^{3/2}}.
\end{aligned}$$

2. **a.** The life expectancy at birth of a female born at the beginning of 1980 is given by

$$g(80) = 50.02[1 + 1.09(80)]^{0.1} \approx 78.29,$$

or approximately 78 years. Similarly, the life expectancy at birth of a female born at the beginning of the year 2000 is given by

$$g(100) = 50.02[1 + 1.09(100)]^{0.1} \approx 80.04,$$

or approximately 80 years.

b. The rate of change of the life expectancy at birth of a female born at any time t is given by $g'(t)$. Using the General Power Rule, we have

$$g'(t) = 50.02 \frac{d}{dt}(1 + 1.09t)^{0.1}$$

$$= (50.02)(0.1)(1 + 1.09t)^{-0.9} \frac{d}{dt}(1 + 1.09t)$$

$$= (50.02)(0.1)(1.09)(1 + 1.09t)^{-0.9}$$

$$= 5.45218(1 + 1.09t)^{-0.9}$$

$$= \frac{5.45218}{(1 + 1.09t)^{0.9}}.$$

3.4 Marginal Functions in Economics

Marginal analysis is the study of the rate of change of economic quantities. For example, an economist is not merely concerned with the value of an economy's gross national product (GNP) at a given time, but is equally concerned with the rate at which it is growing or declining. In the same vein, a manufacturer is not only interested in the total cost corresponding to a certain level of production of a commodity, but is also interested in the rate of change of the total cost with respect to the level of production, and so on. Let us begin with an example to explain the meaning of the adjective *marginal*, as used by economists.

Cost Functions

EXAMPLE 1   Suppose that the total cost in dollars incurred per week by the Polaraire Company for manufacturing x refrigerators is given by the total cost function

$$C(x) = 8000 + 200x - 0.2x^2 \qquad (0 \le x \le 400).$$

a. What is the actual cost incurred for manufacturing the 251st refrigerator?
b. Find the rate of change of the total cost with respect to x when $x = 250$.
c. Compare the results obtained in (a) and (b).

Solution

a. The actual cost incurred in producing the 251st refrigerator is the difference between the total cost incurred in producing the first 251 refrigerators and the total cost of producing the first 250 refrigerators:

$$C(251) - C(250) = [8000 + 200(251) - 0.2(251)^2]$$
$$- [8000 + 200(250) - 0.2(250)^2]$$
$$= 45{,}599.8 - 45{,}500$$
$$= 99.80, \quad \text{or} \quad \$99.80.$$

b. The rate of change of the total cost function C with respect to x is given by the derivative of C; that is, $C'(x) = 200 - 0.4x$. Thus, when the level of production is 250 refrigerators, the rate of change of the total cost with respect to x is given by

$$
\begin{aligned}
C'(250) &= 200 - 0.4(250) \\
&= 100, \quad \text{or} \quad \$100.
\end{aligned}
$$

c. From the solution to (a), we know that the actual cost for producing the 251st refrigerator is $99.80. This answer is very closely approximated by the answer to (b), $100. To see why this is so, observe that the difference $C(251) - C(250)$ may be written in the form

$$
\frac{C(251) - C(250)}{1} = \frac{C(250 + 1) - C(250)}{1} = \frac{C(250 + h) - C(250)}{h},
$$

where $h = 1$. In other words, the difference $C(251) - C(250)$ is precisely the average rate of change of the total cost function C over the interval [250, 251], or, equivalently, the slope of the secant line through the points (250, 45,500) and (251, 45,599.8). On the other hand, the number $C'(250) = 100$ is the instantaneous rate of change of the total cost function C at $x = 250$, or, equivalently, the slope of the tangent line to the graph of C at $x = 250$.

Now when h is small, the average rate of change of the function C is a good approximation to the instantaneous rate of change of the function C, or, equivalently, the slope of the secant line through the points in question is a good approximation to the slope of the tangent line through the point in question. Thus, we may expect

$$
\begin{aligned}
C(251) - C(250) &= \frac{C(251) - C(250)}{1} = \frac{C(250 + h) - C(250)}{h} \\
&\approx \lim_{h \to 0} \frac{C(250 + h) - C(250)}{h} = C'(250),
\end{aligned}
$$

which is precisely the case in this example.   ■

The actual cost incurred in producing an additional unit of a certain commodity given that a plant is already at a certain level of operation is called the **marginal cost.** Knowing this cost is very important to management in their decision-making processes. As we saw in Example 1, the marginal cost is approximated by the rate of change of the total cost function evaluated at the appropriate point. For this reason, economists have defined the **marginal cost function** to be the derivative of the corresponding total cost function. In other words, if C is a total cost function, then the marginal cost function is defined to be its derivative C'. Thus, the adjective *marginal* is synonymous with *derivative of.*

EXAMPLE 2   A subsidiary of the Elektra Electronics Company manufactures a programmable pocket calculator. Management determined that the daily total cost of producing these calculators (in dollars) is given by

$$
C(x) = 0.0001x^3 - 0.08x^2 + 40x + 5000,
$$

where x stands for the number of calculators produced.

a. Find the marginal cost function.

b. What is the marginal cost when $x = 200, 300, 400,$ and 600?

c. Interpret your results.

Solution

a. The marginal cost function C' is given by the derivative of the total cost function C. Thus,

$$C'(x) = 0.0003x^2 - 0.16x + 40.$$

b. The marginal cost when $x = 200, 300, 400,$ and 600 is given by

$$C'(200) = 0.0003(200)^2 - 0.16(200) + 40 = 20$$
$$C'(300) = 0.0003(300)^2 - 0.16(300) + 40 = 19$$
$$C'(400) = 0.0003(400)^2 - 0.16(400) + 40 = 24$$
$$C'(600) = 0.0003(600)^2 - 0.16(600) + 40 = 52,$$

or \$20, \$19, \$24, and \$52, respectively.

c. From the results of (b), we see that Elektra's actual cost for producing the 201st calculator is approximately \$20. The actual cost incurred for producing one additional calculator when the level of production is already 300 calculators is approximately \$19, and so on. Observe that when the level of production is already 600 units, the actual cost of producing one additional unit is approximately \$52. The higher cost for producing this additional unit when the level of production is 600 units may be the result of several factors, among them excessive costs incurred because of overtime or higher maintenance, production breakdown caused by greater stress and strain on the equipment, and so on. The graph of the total cost function appears in Figure 3.7. ∎

FIGURE 3.7

The cost of producing x calculators is given by C(x).

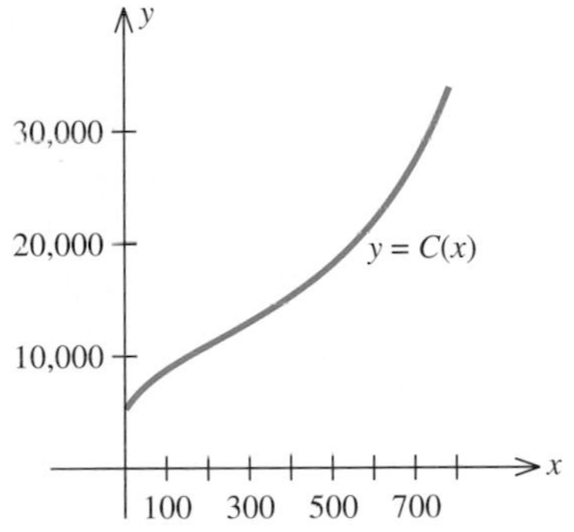

Average Cost Functions

Let us now introduce another marginal concept closely related to the marginal cost. Let $C(x)$ denote the total cost incurred in producing x units of a certain commodity. Then the **average cost** of producing x units of the commodity is obtained by dividing the total production cost by the number of units produced. Thus, if we let $\overline{C}(x)$ denote the average cost of producing x units, then

$$\overline{C}(x) = \frac{C(x)}{x}. \tag{4}$$

The function $\overline{C}$ is called the **average cost function.** The derivative $\overline{C}'$ of the average cost function, called the **marginal average cost function,** measures the rate of change of the average cost function with respect to the number of units produced.

EXAMPLE 3   The total cost of producing x units of a certain commodity is given by

$$C(x) = 400 + 20x \text{ dollars.}$$

a. Find the average cost function $\overline{C}$.

b. Find the marginal average cost function $\overline{C}'$.

c. Interpret the results obtained in (a) and (b).

Solution

a. The average cost function is given by

$$\overline{C}(x) = \frac{C(x)}{x} = \frac{400 + 20x}{x}$$

$$= 20 + \frac{400}{x}.$$

FIGURE 3.8

As the level of production increases, the average cost approaches $20.

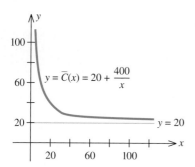

b. The marginal average cost function is

$$\overline{C}'(x) = -\frac{400}{x^2}.$$

c. Since the marginal average cost function is negative for all admissible values of x, the rate of change of the average cost function is negative for all $x > 0$; that is, $\overline{C}(x)$ decreases as x increases. However, the graph of $\overline{C}$ always lies above the horizontal line $y = 20$, but it approaches the line, since

$$\lim_{x \to \infty} \overline{C}(x) = \lim_{x \to \infty} \left(20 + \frac{400}{x} \right) = 20.$$

A sketch of the graph of the function $\overline{C}(x)$ appears in Figure 3.8. This result is fully expected if we consider the economic implications. Note that as the level of production increases, the fixed cost per unit of production, represented by the term $(400/x)$, drops steadily. The average cost approaches the constant unit of production, which is $20 in this case. ∎

EXAMPLE 4   Once again consider the subsidiary of the Elektra Electronics Company. The daily total cost for producing its programmable calculators is given by

$$C(x) = 0.0001x^3 - 0.08x^2 + 40x + 5000$$

dollars, where x stands for the number of calculators produced (see Example 2).

a. Find the average cost function $\overline{C}$.
b. Find the marginal average cost function $\overline{C}'$. Compute $\overline{C}'(500)$.
c. Sketch the graph of the function $\overline{C}$ and interpret the results obtained in (a) and (b).

Solution

a. The average cost function is given by

$$\overline{C}(x) = \frac{C(x)}{x} = 0.0001x^2 - 0.08x + 40 + \frac{5000}{x}.$$

b. The marginal average cost function is given by

$$\overline{C}'(x) = 0.0002x - 0.08 - \frac{5000}{x^2}.$$

Also, $$\overline{C}'(500) = 0.0002(500) - 0.08 - \frac{5000}{(500)^2} = 0.$$

FIGURE 3.9

The average cost reaches a minimum of $35 when 500 calculators are produced.

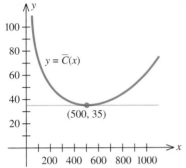

c. To sketch the graph of the function $\overline{C}$, observe that if x is a small positive number, then $\overline{C}(x) > 0$. Furthermore, $\overline{C}(x)$ becomes arbitrarily large as x approaches zero from the right, since the term $(5000/x)$ becomes arbitrarily large as x approaches zero. Next, the result $\overline{C}'(500) = 0$ obtained in (b) tells us that the tangent line to the graph of the function $\overline{C}$ is horizontal at the point $(500, 35)$ on the graph. Finally, plotting the points on the graph corresponding to, say, $x = 100, 200, 300, \ldots, 900$, we obtain the sketch in Figure 3.9. As expected, the average cost drops as the level of production increases. But in this case, as opposed to the case in Example 3, the average cost reaches a minimum value of $35, corresponding to a production level of 500, and *increases* thereafter.

This phenomenon is typical in situations where the marginal cost increases from some point on as production increases, as in Example 2. This situation is in contrast to that of Example 3, in which the marginal cost remains constant at any level of production. ∎

Revenue Functions

Another marginal concept, the *marginal revenue function,* is associated with the revenue function R, given by

$$R(x) = px, \tag{5}$$

where x is the number of units sold of a certain commodity and p is the unit selling price. In general, however, the unit selling price p of a commodity is related to the quantity x of the commodity demanded. This relationship, $p = f(x)$, is called a *demand equation* (see Section 2.3). Solving the demand equation for p in terms of x, we obtain the unit price function f, given by

$$p = f(x).$$

Thus, the revenue function R is given by

$$R(x) = px = xf(x),$$

where f is the unit price function. The derivative R' of the function R, called the **marginal revenue function,** measures the rate of change of the revenue function.

EXAMPLE 5   Suppose that the relationship between the unit price p in dollars and the quantity demanded x of the Acrosonic model F loudspeaker system is given by the equation

$$p = -0.02x + 400 \qquad (0 \le x \le 20{,}000).$$

a. Find the revenue function R.
b. Find the marginal revenue function R'.
c. Compute $R'(2000)$ and interpret your result.

Solution
a. The revenue function R is given by

$$
\begin{aligned}
R(x) &= px \\
&= x(-0.02x + 400) \\
&= -0.02x^2 + 400x \qquad (0 \le x \le 20{,}000).
\end{aligned}
$$

b. The marginal revenue function R' is given by

$$R'(x) = -0.04x + 400.$$

c. $$R'(2000) = -0.04(2000) + 400 = 320.$$

Thus, the actual revenue to be realized from the sale of the 2001st loud-speaker system is approximately $320.   ∎

Profit Functions

Our final example of a marginal function involves the profit function. The profit function P is given by

$$P(x) = R(x) - C(x), \tag{6}$$

where R and C are the revenue and cost functions and x is the number of units of a commodity produced and sold. The **marginal profit function $P'(x)$** measures the rate of change of the profit function P and provides us with a good approximation of the actual profit or loss realized from the sale of the $(x + 1)$st unit of the commodity (assuming the xth unit has been sold).

EXAMPLE 6   Refer to Example 5. Suppose that the cost of producing x units of the Acrosonic model F loudspeaker is

$$C(x) = 100x + 200,000$$

dollars.

a. Find the profit function P.
b. Find the marginal profit function P'.
c. Compute $P'(2000)$ and interpret your result.
d. Sketch the graph of the profit function P.

Solution

a. From the solution to Example 5a, we have

$$R(x) = -0.02x^2 + 400x.$$

Thus, the required profit function P is given by

$$\begin{aligned} P(x) &= R(x) - C(x) \\ &= (-0.02x^2 + 400x) - (100x + 200,000) \\ &= -0.02x^2 + 300x - 200,000. \end{aligned}$$

b. The marginal profit function P' is given by

$$P'(x) = -0.04x + 300.$$

c. $$P'(2000) = -0.04(2000) + 300 = 220.$$

Thus, the actual profit realized from the sale of the 2001st loudspeaker system is approximately $220.
d. The graph of the profit function P appears in Figure 3.10.   ∎

FIGURE 3.10

The total profit made when x loudspeakers are produced is given by P(x).

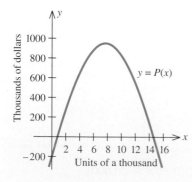

Elasticity of Demand

Finally, let us use the marginal concepts introduced in this section to derive an important criterion used by economists to analyze the demand function: *elasticity of demand.*

In what follows, it will be convenient to write the demand function f in the form $x = f(p)$; that is, we will think of the quantity demanded of a certain commodity as a function of its unit price. Since the quantity demanded of a commodity usually decreases as its unit price increases, the function f is typically a decreasing function of p (see Figure 3.11a).

FIGURE 3.11

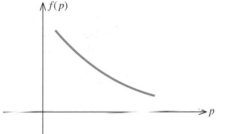

(a) A demand function

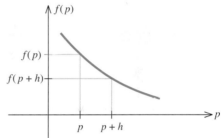

(b) $f(p + h)$ is the quantity demanded when the unit price increases from p to $p + h$ dollars.

Suppose that the unit price of a commodity is increased by h dollars from p dollars to $(p + h)$ dollars (Figure 3.11b). Then the quantity demanded drops from $f(p)$ units to $f(p + h)$ units, a change of $[f(p + h) - f(p)]$ units. The percentage change in the unit price is

$$\frac{h}{p}(100) \qquad \left(\frac{\text{change in unit price}}{\text{price } p}\right)(100),$$

and the corresponding percentage change in the quantity demanded is

$$100\left[\frac{f(p + h) - f(p)}{f(p)}\right] \qquad \left(\frac{\text{change in quantity demanded}}{\text{quantity demanded at price } p}\right)(100).$$

Now, one good way to measure the effect that a percentage change in price has on the percentage change in the quantity demanded is to look at the ratio of the latter to the former. We find

$$\frac{\text{percentage change in the quantity demanded}}{\text{percentage change in the unit price}} = \frac{100\left[\dfrac{f(p + h) - f(p)}{f(p)}\right]}{100\left(\dfrac{h}{p}\right)}$$

$$= \frac{\dfrac{f(p + h) - f(p)}{h}}{\dfrac{f(p)}{p}}.$$

If f is differentiable at p, then

$$\frac{f(p + h) - f(p)}{h} \approx f'(p)$$

when h is small. Therefore, if h is small, then the ratio is approximately equal to

$$\frac{f'(p)}{\dfrac{f(p)}{p}} = \frac{pf'(p)}{f(p)}.$$

Economists call the negative of this quantity the *elasticity of demand.*

ELASTICITY OF DEMAND

If f is a differentiable demand function defined by $x = f(p)$, then the **elasticity of demand** at price p is given by

$$E(p) = -\frac{pf'(p)}{f(p)}. \qquad (7)$$

REMARK   It will be shown later (Section 4.1) that if f is decreasing on an interval, then $f'(p) < 0$ for p in that interval. In light of this, we see that since both p and $f(p)$ are positive, the quantity $\dfrac{pf'(p)}{f(p)}$ is negative. Because economists would rather work with a positive value, the elasticity of demand $E(p)$ is defined to be the negative of this quantity. $\square$

EXAMPLE 7   Consider the demand equation

$$p = -0.02x + 400 \qquad (0 \le x \le 20{,}000),$$

which describes the relationship between the unit price in dollars and the quantity demanded x of the Acrosonic model F loudspeaker systems.

a. Find the elasticity of demand $E(p)$.
b. Compute $E(100)$ and interpret your result.
c. Compute $E(300)$ and interpret your result.

Solution
a. Solving the given demand equation for x in terms of p, we find

$$x = f(p) = -50p + 20{,}000,$$

from which we see that

$$f'(p) = -50.$$

Therefore,

$$E(p) = -\frac{pf'(p)}{f(p)} = -\frac{p(-50)}{-50p + 20{,}000}$$

$$= \frac{p}{400 - p}.$$

b. $E(100) = \dfrac{100}{400 - 100} = \dfrac{1}{3}$, which is the elasticity of demand when $p = 100$. To interpret this result, recall that $E(100)$ is the negative of the ratio of the percentage change in the quantity demanded to the percentage change in the unit price when $p = 100$. Therefore, our result tells us that when the unit price p is set at $100 per speaker, an increase of 1 percent in the unit price will cause an increase of approximately 0.33 percent in the quantity demanded.

c. $E(300) = \dfrac{300}{400 - 300} = 3$, which is the elasticity of demand when $p = 300$. It tells us that when the unit price is set at $300 per speaker, an increase of 1 percent in the unit price will cause a decrease of approximately 3 percent in the quantity demanded.   ∎

Economists often use the following terminology to describe demand in terms of elasticity.

ELASTICITY OF DEMAND

The demand is said to be **elastic** if $E(p) > 1$.
The demand is said to be **unitary** if $E(p) = 1$.
The demand is said to be **inelastic** if $E(p) < 1$.

As an illustration, our computations in Example 7 revealed that demand for Acrosonic loudspeakers is elastic when $p = 300$ but inelastic when $p = 100$. These computations confirm that when demand is elastic, a small percentage change in the unit price will result in a greater percentage change in the quantity demanded; and when demand is inelastic, a small percentage change in the unit price will cause a smaller percentage change in the quantity demanded. Finally, when demand is unitary, a small percentage change in the unit price will result in the same percentage change in the quantity demanded.

We can describe the way revenue responds to changes in the unit price using the notion of elasticity. If the quantity demanded of a certain commodity is related to its unit price by the equation $x = f(p)$, then the revenue realized through the sale of x units of the commodity at a price of p dollars each is

$$R(p) = px = pf(p).$$

The rate of change of the revenue with respect to the unit price p is given by

$$R'(p) = f(p) + pf'(p)$$
$$= f(p)\left[1 + \frac{pf'(p)}{f(p)}\right]$$
$$= f(p)[1 - E(p)].$$

Now, suppose the demand is elastic when the unit price is set at a dollars. Then $E(a) > 1$, and so $1 - E(a) < 0$. Since $f(p)$ is positive for all values of p, we see that

$$R'(a) = f(a)[1 - E(a)] < 0,$$

and so $R(p)$ is decreasing at $p = a$. This implies that a small increase in the unit price when $p = a$ results in a decrease in the revenue, whereas a small decrease in the unit price will result in an increase in the revenue. Similarly, you can show that if the demand is inelastic when the unit price is set at a dollars, then a small increase in the unit price will cause the revenue to increase, and a small decrease in the unit price will cause the revenue to decrease. Finally, if the demand is unitary when the unit price is set at a dollars, then $E(a) = 1$ and $R'(a) = 0$. This implies that a small increase or decrease in the unit price will not result in a change in the revenue. The following statements summarize this discussion.

1. If the demand is elastic at p ($E(p) > 1$), then an increase in the unit price will cause the revenue to decrease, whereas a decrease in the unit price will cause the revenue to increase.
2. If the demand is inelastic at p ($E(p) < 1$), then an increase in the unit price will cause the revenue to increase, and a decrease in the unit price will cause the revenue to decrease.
3. If the demand is unitary at p ($E(p) = 1$), then an increase in the unit price will cause the revenue to stay about the same.

REMARK   As an aid to remembering this, note that

1. if demand is elastic, then the change in revenue and the change in the unit price move in opposite directions.
2. if demand is inelastic, then they move in the same direction. ☐

EXAMPLE 8   Refer to Example 7.

a. Is demand elastic, unitary, or inelastic when $p = 100$? When $p = 300$?
b. If the price is \$100, will raising the unit price slightly cause the revenue to increase or decrease?

Solution
a. From the results of Example 7, we see that $E(100) = \frac{1}{3} < 1$ and $E(300) = 3 > 1$. We conclude accordingly that demand is inelastic when $p = 100$ and elastic when $p = 300$.
b. Since demand is inelastic when $p = 100$, raising the unit price slightly will cause the revenue to increase. ∎

SELF-CHECK EXERCISES 3.4

1. The weekly demand for Pulsar VCRs (videocassette recorders) is given by the demand equation

$$p = -0.02x + 300 \qquad (0 \le x \le 15{,}000),$$

where p denotes the wholesale unit price in dollars and x denotes the quantity demanded. The weekly total cost function associated with manufacturing these VCRs is

$$C(x) = 0.000003x^3 - 0.04x^2 + 200x + 70{,}000$$

dollars.

a. Find the revenue function R and the profit function P.
b. Find the marginal cost function C', the marginal revenue function R', and the marginal profit function P'.
c. Find the marginal average cost function $\overline{C}'$.
d. Compute $C'(3000)$, $R'(3000)$, and $P'(3000)$ and interpret your results.

2. Refer to Exercise 1. Determine whether the demand is elastic, unitary, or inelastic when $p = 100$ and when $p = 200$.

Solutions to Self-Check Exercises 3.4 can be found on page 172.

3.4 EXERCISES

C *A calculator is recommended for Exercises 1–20.*

1. **Marginal Cost** The total weekly cost in dollars incurred by the Lincoln Record Company in pressing x long-playing records is

$$C(x) = 2000 + 2x - 0.0001x^2 \qquad (0 \le x \le 6000).$$

   a. What is the actual cost incurred in producing the 1001st record? The 2001st record?
   b. What is the marginal cost when $x = 1000$? When $x = 2000$?

2. **Marginal Cost** A division of Ditton Industries manufactures the Futura model microwave oven. The daily cost (in dollars) of producing these microwave ovens is

$$C(x) = 0.0002x^3 - 0.06x^2 + 120x + 5000,$$

   where x stands for the number of units produced.
   a. What is the actual cost incurred in manufacturing the 101st oven? The 201st oven? The 301st oven?
   b. What is the marginal cost when $x = 100$? When $x = 200$? When $x = 300$?

3. **Marginal Average Cost** The Custom Office Company makes a line of executive desks. It is estimated that the total cost for making x units of their Senior Executive Model is

$$C(x) = 100x + 200,000$$

   dollars per year.
   a. Find the average cost function $\overline{C}$.
   b. Find the marginal average cost function $\overline{C}'$.
   c. What happens to $\overline{C}(x)$ when x is very large? Interpret your results.

4. **Marginal Average Cost** The management of the Thermo-Master Company, whose Mexico subsidiary manufactures an indoor-outdoor thermometer, has estimated that the total weekly cost in dollars for producing x thermometers is

$$C(x) = 5000 + 2x.$$

   a. Find the average cost function $\overline{C}$.
   b. Find the marginal average cost function $\overline{C}'$.
   c. Interpret your results.

5. Find the average cost function $\overline{C}$ and the marginal average cost function $\overline{C}'$ associated with the total cost function C of Exercise 1.

6. Find the average cost function $\overline{C}$ and the marginal average cost function $\overline{C}'$ associated with the total cost function C of Exercise 2.

7. **Marginal Revenue** The Williams Commuter Air Service realizes a monthly revenue of

$$R(x) = 8000x - 100x^2$$

   dollars when the price charged per passenger is x dollars.
   a. Find the marginal revenue R'.
   b. Compute $R'(39)$, $R'(40)$, and $R'(41)$. What do your results imply?

8. **Marginal Revenue** The management of the Acrosonic Company plans to market the Electro-Stat, an electrostatic speaker system. The marketing department has determined that the demand for these speakers is

$$p = -0.04x + 800 \qquad (0 \le x \le 20,000),$$

   where p denotes the speaker's unit price (in dollars) and x denotes the quantity demanded.
   a. Find the revenue function R.
   b. Find the marginal revenue function R'.
   c. Compute $R'(5000)$ and interpret your result.

9. **Marginal Profit**   Lynbrook West, an apartment complex, has 100 two-bedroom units. The monthly profit (in dollars) realized from renting x apartments is

$$P(x) = -10x^2 + 1760x - 50,000.$$

   a. What is the actual profit realized from renting the 51st unit, assuming that 50 units have already been rented?
   b. Compute the marginal profit when $x = 50$ and compare your results with that obtained in (a).

10. **Marginal Profit**   Refer to Exercise 8. Acrosonic's production department estimates that the total cost (in dollars) incurred in manufacturing x Electro-Stat speaker systems in the first year of production will be

$$C(x) = 200x + 300,000.$$

   a. Find the profit function P.
   b. Find the marginal profit function P'.
   c. Compute $P'(5000)$ and $P'(8000)$.
   d. Sketch the graph of the profit function and interpret your results.

11. **Marginal Cost, Revenue, and Profit**   The weekly demand for the Pulsar 25 color console television is

$$p = 600 - 0.05x \quad (0 \le x \le 12,000),$$

   where p denotes the wholesale unit price in dollars and x denotes the quantity demanded. The weekly total cost function associated with manufacturing the Pulsar 25 is given by

$$C(x) = 0.000002x^3 - 0.03x^2 + 400x + 80,000,$$

   where $C(x)$ denotes the total cost incurred in producing x sets.
   a. Find the revenue function R and the profit function P.
   b. Find the marginal cost function C', the marginal revenue function R', and the marginal profit function P'.
   c. Compute $C'(2000)$, $R'(2000)$, and $P'(2000)$ and interpret your results.
   d. Sketch the graphs of the functions C, R, and P and interpret (b) and (c) using the graphs obtained.

12. **Marginal Cost, Revenue, and Profit**   The Pulsar Corporation also manufactures a series of 19-inch color television sets. The quantity x of these sets demanded each week is related to the wholesale unit price p by the equation

$$p = -0.006x + 180.$$

The weekly total cost incurred by Pulsar for producing x sets is

$$C(x) = 0.000002x^3 - 0.02x^2 + 120x + 60,000$$

dollars. Answer the questions in Exercise 11 for these data.

13. **Marginal Average Cost**   Find the average cost function $\overline{C}$ associated with the total cost function C of Exercise 11.
   a. What is the marginal average cost function $\overline{C}'$?
   b. Compute $\overline{C}'(5,000)$ and $\overline{C}'(10,000)$ and interpret your results.
   c. Sketch the graph of $\overline{C}$.

14. **Marginal Average Cost**   Find the average cost function $\overline{C}$ associated with the total cost function C of Exercise 12.
   a. What is the marginal average cost function $\overline{C}'$?
   b. Compute $\overline{C}'(5,000)$ and $\overline{C}'(10,000)$ and interpret your results.

15. **Marginal Revenue**   The quantity of Sicard wristwatches demanded per month is related to the unit price by the equation

$$p = \frac{50}{0.01x^2 + 1} \quad (0 \le x \le 20),$$

   where p is measured in dollars and x in units of a thousand.
   a. Find the revenue function R.
   b. Find the marginal revenue function R'.
   c. Compute $R'(2)$ and interpret your result.

16. **Marginal Propensity to Consume**   The consumption function of the U.S. economy for 1929 to 1941 is

$$C(x) = 0.712x + 95.05,$$

   where $C(x)$ is the personal consumption expenditure and x is the personal income, both measured in billions of dollars. Find the rate of change of consumption with respect to income, dC/dx. This quantity is called the *marginal propensity to consume.*

17. **Marginal Propensity to Consume**   Refer to Exercise 16. Suppose that a certain economy's consumption function is

$$C(x) = 0.873x^{1.1} + 20.34,$$

   where $C(x)$ and x are measured in billions of dollars. Find the marginal propensity to consume when $x = 10$.

18. **Marginal Propensity to Save**   Suppose that $C(x)$ measures an economy's personal consumption expendi-

ture and x the personal income, both in billions of dollars. Then

$$S(x) = x - C(x) \qquad \text{(Income minus consumption)}$$

measures the economy's savings corresponding to an income of x billion dollars. Show that

$$\frac{dS}{dx} = 1 - \frac{dC}{dx}.$$

The quantity dS/dx is called the *marginal propensity to save*.

19. Refer to Exercise 18. For the consumption function of Exercise 16, find the marginal propensity to save.

20. Refer to Exercise 18. For the consumption function of Exercise 17, find the marginal propensity to save when $x = 10$.

For each demand equation in Exercises 21–26, compute the elasticity of demand and determine whether the demand is elastic, unitary, or inelastic at the indicated price.

21. $x = -\frac{3}{2}p + 9; \quad p = 2$

22. $x = -\frac{5}{4}p + 20; \quad p = 10$

23. $x + \frac{1}{3}p - 20 = 0; \quad p = 30$

24. $0.4x + p - 20 = 0; \quad p = 10$

25. $p = 144 - x^2; \quad p = 96$

26. $p = 169 - x^2; \quad p = 29$

[c] 27. **Elasticity of Demand**   The management of the Titan Tire Company has determined that the quantity demanded x of their Super Titan tires per week is related to the unit price p by the equation

$$x = \sqrt{144 - p},$$

where p is measured in dollars and x in units of a thousand.
    a. Compute the elasticity of demand when $p = 63$, when $p = 96$, and when $p = 108$.
    b. Interpret the results obtained in (a).
    c. Is the demand elastic, unitary, or inelastic when $p = 63$? When $p = 96$? When $p = 108$?

[c] 28. **Elasticity of Demand**   The demand equation for the Roland portable hair dryer is given by

$$x = \frac{1}{5}(225 - p^2) \qquad (0 \le p \le 15),$$

where x (measured in units of a hundred) is the quantity demanded per week and p is the unit price in dollars.

a. Is the demand elastic or inelastic when $p = 8$ and when $p = 10$?
b. When is the demand unitary? [*Hint:* Solve $E(p) = 1$ for p.]
c. If the unit price is lowered slightly from \$10, will the revenue increase or decrease?
d. If the unit price is increased slightly from \$8, will the revenue increase or decrease?

29. **Elasticity of Demand**   The quantity demanded per week x (in units of a hundred) of the Mikado miniature cameras is related to the unit price p (in dollars) by the demand equation

$$x = \sqrt{400 - 5p} \qquad (0 \le p \le 80).$$

a. Is the demand elastic or inelastic when $p = 40$? When $p = 60$?
b. When is the demand unitary?
c. If the unit price is lowered slightly from \$60, will the revenue increase or decrease?
d. If the unit price is increased slightly from \$40, will the revenue increase or decrease?

30. **Elasticity of Demand**   The proprietor of the Showplace, a video club, has estimated that the rental price p in dollars of prerecorded videocassette tapes is related to the quantity x rented per week by the demand equation

$$x = \frac{2}{3}\sqrt{36 - p^2} \qquad (0 \le p \le 6).$$

Currently, the rental price is \$2 per tape.
a. Is the demand elastic or inelastic at this rental price?
b. If the rental price is increased, will the revenue increase or decrease?

31. **Elasticity of Demand**   The demand function for a certain make of exercise bicycle sold exclusively through cable television is

$$p = \sqrt{9 - 0.02x} \qquad (0 \le x \le 450),$$

where p is the unit price in hundreds of dollars and x is the quantity demanded per week. Compute the elasticity of demand and determine the range of prices corresponding to inelastic, unitary, and elastic demand. [*Hint:* Solve the equation $E(p) = 1$.]

32. **Elasticity of Demand**   The demand equation for the Sicard wristwatch is given by

$$x = 10\sqrt{\frac{50 - p}{p}} \qquad (0 < p \le 50),$$

where x (measured in units of a thousand) is the quantity demanded per week and p is the unit price in dollars.

Compute the elasticity of demand and determine the range of prices corresponding to inelastic, unitary, and elastic demand.

C *Computer or graphing calculator problems.*

33. The revenue realized in selling x thousand Sicard wristwatches per month is $R(x)$ thousand dollars, where

$$R(x) = \frac{50x}{0.01x^2 + 1} \qquad (0 \le x \le 20).$$

Sketch the graphs of the functions R and R'. Then use them to determine the approximate location of the point

$(x, R(x))$, where the tangent line to the graph of R is horizontal. Interpret your results.

34. The average cost function $\overline{C}$ associated with the manufacture of a series of 19-inch color television sets is given by

$$\overline{C}(x) = 0.000002x^2 - 0.02x + 120 + \frac{60,000}{x}$$

dollars. Sketch the graphs of the functions $\overline{C}$ and $\overline{C}'$. Then use them to determine the approximate location of the point $(x, \overline{C}(x))$, where the tangent line to the graph of $\overline{C}$ is horizontal. Interpret your results.

SOLUTIONS TO SELF-CHECK EXERCISES 3.4

1. a.
$$R(x) = px$$
$$= x(-0.02x + 300)$$
$$= -0.02x^2 + 300x \qquad (0 \le x \le 15,000)$$

$$P(x) = R(x) - C(x)$$
$$= -0.02x^2 + 300x$$
$$\quad -(0.000003x^3 - 0.04x^2 + 200x + 70,000)$$
$$= -0.000003x^3 + 0.02x^2 + 100x - 70,000$$

b.
$$C'(x) = 0.000009x^2 - 0.08x + 200$$
$$R'(x) = -0.04x + 300,$$

and

$$P'(x) = -0.000009x^2 + 0.04x + 100$$

c. The average cost function is

$$\overline{C}(x) = \frac{C(x)}{x}$$
$$= \frac{0.000003x^3 - 0.04x^2 + 200x + 70,000}{x}$$
$$= 0.000003x^2 - 0.04x + 200 + \frac{70,000}{x}.$$

Therefore, the marginal average cost function is

$$\overline{C}'(x) = 0.000006x - 0.04 - \frac{70,000}{x^2}.$$

d. Using the results from (b), we find

$$C'(3000) = 0.000009(3000)^2 - 0.08(3000) + 200$$
$$= 41;$$

that is, when the level of production is already 3000 VCRs, the actual cost of producing one additional VCR is approximately \$41. Next,

$$R'(3000) = -0.04(3000) + 300 = 180;$$

that is, the actual revenue to be realized from selling the 3001st VCR is approximately \$180. Finally,

$$P'(3000) = -0.000009(3000)^2 + 0.04(3000) + 100$$
$$= 139;$$

that is, the actual profit realized from selling the 3001st VCR is approximately \$139.

2. We first solve the given demand equation for x in terms of p, obtaining

$$x = f(p) = -50p + 15,000$$

and

$$f'(p) = -50.$$

Therefore,

$$E(p) = -\frac{pf'(p)}{f(p)} = -\frac{p}{-50p + 15,000}(-50)$$

$$-\frac{p}{300 - p} \qquad (0 \le p < 300).$$

Next, we compute

$$E(100) = \frac{100}{300 - 100} = \frac{1}{2} < 1,$$

and we conclude that demand is inelastic when $p = 100$. Also,

$$E(200) = \frac{200}{300 - 200} = 2 > 1,$$

and we see that demand is elastic when $p = 200$.

3.5   Higher-Order Derivatives

Higher-Order Derivatives

The derivative f' of a function f is also a function. As such, the differentiability of f' may be considered. Thus, the function f' has a derivative f'' at a point x in the domain of f' if the limit of the quotient

$$\frac{f'(x + h) - f'(x)}{h}$$

exists as h approaches zero. In other words, it is the derivative of the first derivative.

The function f'' obtained in this manner is called the **second derivative** of the function f, just as the derivative f' of f is often called the first derivative of f. Continuing in this fashion, we are led to considering the third, fourth, and higher-order derivatives of f whenever they exist. Notations for the first, second, third, and, in general, nth derivatives of a function f at a point x are

$$f'(x), f''(x), f'''(x), \ldots, f^{(n)}(x)$$

or $\qquad D^1 f(x), D^2 f(x), D^3 f(x), \ldots, D^n f(x).$

If f is written in the form $y = f(x)$, then the notations for its derivatives are

$$y', y'', y''', \ldots, y^{(n)},$$

$$\frac{dy}{dx}, \frac{d^2y}{dx^2}, \frac{d^3y}{dx^3}, \ldots, \frac{d^n y}{dx^n},$$

or $\qquad D^1 y, D^2 y, D^3 y, \ldots, D^n y,$

respectively.

EXAMPLE 1   Find the derivatives of all orders of the polynomial function $f(x) = x^5 - 3x^4 + 4x^3 - 2x^2 + x - 8.$

Solution   We have

$$f'(x) = 5x^4 - 12x^3 + 12x^2 - 4x + 1$$

$$f''(x) = \frac{d}{dx} f'(x) = 20x^3 - 36x^2 + 24x - 4$$

$$f'''(x) = \frac{d}{dx} f''(x) = 60x^2 - 72x + 24$$

$$f^{(4)}(x) = \frac{d}{dx} f'''(x) = 120x - 72$$

$$f^{(5)}(x) = \frac{d}{dx} f^{(4)}(x) = 120$$

and, in general,

$$f^{(n)}(x) = 0 \quad \text{for } n > 5. \qquad \blacksquare$$

EXAMPLE 2   Find the third derivative of the function f defined by $y = x^{2/3}$. What is its domain?

Solution   We have

$$y' = \frac{2}{3} x^{-1/3}$$

$$y'' = \left(\frac{2}{3}\right)\left(-\frac{1}{3}\right) x^{-4/3} = -\frac{2}{9} x^{-4/3},$$

so the required derivative is

$$y''' = \left(-\frac{2}{9}\right)\left(-\frac{4}{3}\right) x^{-7/3} = \frac{8}{27} x^{-7/3} = \frac{8}{27 x^{7/3}}.$$

The common domain of the functions f', f'', and f''' is the set of all real numbers except $x = 0$. The domain of $y = x^{2/3}$ is the set of all real numbers. The graph of the function $y = x^{2/3}$ appears in Figure 3.12. $\qquad \blacksquare$

REMARK   Always simplify an expression before differentiating it to obtain the next order derivative. $\qquad \square$

FIGURE 3.12

The graph of the function $y = x^{2/3}$.

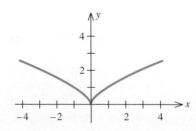

EXAMPLE 3   Find the second derivative of the function $y = (2x^2 + 3)^{3/2}$.

Solution   We have, using the General Power Rule,

$$y' = \frac{3}{2}(2x^2 + 3)^{1/2}(4x) = 6x(2x^2 + 3)^{1/2}.$$

Next, using the Product Rule and then the Chain Rule, we find

$$y'' = (6x) \cdot \frac{d}{dx}(2x^2 + 3)^{1/2} + \left[\frac{d}{dx}(6x)\right](2x^2 + 3)^{1/2}$$

$$= (6x)\left(\frac{1}{2}\right)(2x^2 + 3)^{-1/2}(4x) + 6(2x^2 + 3)^{1/2}$$

$$= 12x^2(2x^2 + 3)^{-1/2} + 6(2x^2 + 3)^{1/2}$$

$$= 6(2x^2 + 3)^{-1/2}[2x^2 + (2x^2 + 3)]$$

$$= \frac{6(4x^2 + 3)}{\sqrt{2x^2 + 3}}.$$

Applications

Just as the derivative of a function f at a point x measures the rate of change of the function f at that point, the second derivative of f (the derivative of f') measures the rate of change of the derivative f' of the function f. The third derivative of the function f, f''', measures the rate of change of f'', and so on.

In Chapter 4 we will discuss applications involving the geometric interpretation of the second derivative of a function. The following example gives an interpretation of the second derivative in a familiar role.

EXAMPLE 4   Refer to Example 6, page 116. The distance s (in feet) covered by a car moving along a straight road t seconds after starting from rest is given by the function $s = 2t^2$ ($0 \le t \le 30$). What is the car's acceleration at the end of 30 seconds?

Solution   The velocity of the car t seconds from rest is given by

$$v = \frac{ds}{dt} = \frac{d}{dt}(2t^2) = 4t.$$

The acceleration of the car t seconds from rest is given by the rate of change of the velocity of t; that is,

$$a = \frac{d}{dt}v = \frac{d}{dt}\left(\frac{ds}{dt}\right) = \frac{d^2s}{dt^2} = \frac{d}{dt}(4t) = 4,$$

or 4 feet per second per second, normally abbreviated, 4 ft/sec^2.

EXAMPLE 5   A ball is thrown straight up into the air from the roof of a building. The height of the ball as measured from the ground is given by

$$s = -16t^2 + 24t + 120,$$

where s is measured in feet and t in seconds. Find the velocity and acceleration of the ball 3 seconds after it is thrown into the air.

Solution   The velocity v and acceleration a of the ball at any time t are given by

$$v = \frac{ds}{dt} = \frac{d}{dt}(-16t^2 + 24t + 120) = -32t + 24$$

and

$$a = \frac{d^2t}{dt^2} = \frac{d}{dt}\left(\frac{ds}{dt}\right) = \frac{d}{dt}(-32t + 24) = -32.$$

Therefore, the velocity of the ball 3 seconds after it is thrown into the air is

$$v = -32(3) + 24 = -72.$$

That is, the ball is falling downward at a speed of 72 ft/sec. The acceleration of the ball is 32 ft/sec² downward at any time during the motion.   ■

FIGURE 3.13

The CPI of a certain economy from year a to year b is given by I(t).

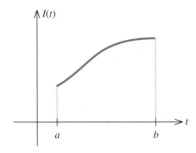

Another interpretation of the second derivative of a function—this time from the field of economics—follows. Suppose that the consumer price index (CPI) of an economy between the years a and b is described by the function $I(t)$ ($a \leq t \leq b$) [Figure 3.13]. Then the first derivative of I, $I'(t)$, gives the rate of inflation of the economy at any time t. The second derivative of I, $I''(t)$, gives the *rate of change of the inflation rate* at any time t. Thus, when the economist or politician claims that "inflation is slowing," what he or she is saying is that the rate of inflation is decreasing. Mathematically, this is equivalent to noting that the second derivative $I''(t)$ is negative at the time t under consideration. Observe that $I'(t)$ could be positive at a point in time when $I''(t)$ is negative (see Example 6). Thus, one may not draw the conclusion from the aforementioned quote that prices of goods and services are about to drop!

EXAMPLE 6   An economy's consumer price index (CPI) is described by the function

$$I(t) = -0.2t^3 + 3t^2 + 100 \qquad (0 \leq t \leq 9),$$

where $t = 0$ corresponds to the year 1988. Compute $I'(6)$ and $I''(6)$ and use these results to show that even though the CPI was rising at the beginning of 1994, "inflation was moderating" at that point in time.

FIGURE 3.14

The CPI of an economy is given by I(t).

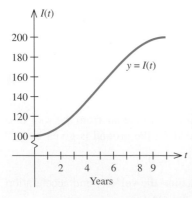

Solution   We find

$$I'(t) = -0.6t^2 + 6t$$

and

$$I''(t) = -1.2t + 6,$$

so

$$I'(6) = -0.6(6)^2 + 6(6) = 14.4$$

and

$$I''(6) = -1.2(6) + 6 = -1.2.$$

Our computations reveal that at the beginning of 1994 ($t = 6$), the CPI was increasing at the rate of 14.4 points per year, whereas the rate of the inflation rate was decreasing by 1.2 points per year. Thus, inflation was moderating at that point in time (Figure 3.14). In Section 4.3, we will see that relief actually began in early 1993.   ■

1. Find the third derivative of

$$f(x) = 2x^5 - 3x^3 + x^2 - 6x + 10.$$

2. Let

$$f(x) = \frac{1}{1 + x}.$$

Find $f'(x)$, $f''(x)$, and $f'''(x)$.

3. A certain species of turtle faces extinction because dealers collect truckloads of turtle eggs to be sold as aphrodisiacs. After severe conservation measures are implemented, it is hoped that the turtle population will grow according to the rule

$$N(t) = 2t^3 + 3t^2 - 4t + 1000 \qquad (0 \le t \le 10),$$

where $N(t)$ denotes the population at the end of year t. Compute $N''(2)$ and $N''(8)$ and interpret your results.

Solutions to Self-Check Exercises 3.5 can be found on page 178.

3.5   EXERCISES

In Exercises 1–20, find the first and second derivatives of the given function.

1. $f(x) = 4x^2 - 2x + 1$

2. $f(x) = -0.2x^2 + 0.3x + 4$

3. $f(x) = 2x^3 - 3x^2 + 1$

4. $g(x) = -3x^3 + 24x^2 + 6x - 64$

5. $h(t) = t^4 - 2t^3 + 6t^2 - 3t + 10$

6. $f(x) = x^5 - x^4 + x^3 - x^2 + x - 1$

7. $f(x) = (x^2 + 2)^5$ 8. $g(t) = t^2(3t + 1)^4$

9. $g(t) = (2t^2 - 1)^2(3t^2)$

10. $h(x) = (x^2 + 1)^2(x - 1)$

11. $f(x) = (2x^2 + 2)^{7/2}$

12. $h(w) = (w^2 + 2w + 4)^{5/2}$

13. $f(x) = x(x^2 + 1)^2$ 14. $g(u) = u(2u - 1)^3$

15. $f(x) = \dfrac{x}{2x + 1}$ 16. $g(t) = \dfrac{t^2}{t - 1}$

17. $f(s) = \dfrac{s - 1}{s + 1}$ 18. $f(u) = \dfrac{u}{u^2 + 1}$

19. $f(u) = \sqrt{4 - 3u}$ 20. $f(x) = \sqrt{2x - 1}$

In Exercises 21–28, find the third derivative of the given function.

21. $f(x) = 3x^4 - 4x^3$

22. $f(x) = 3x^5 - 6x^4 + 2x^2 - 8x + 12$

23. $f(x) = \dfrac{1}{x}$ 24. $f(x) = \dfrac{2}{x^2}$

25. $g(s) = \sqrt{3s - 2}$ 26. $g(t) = \sqrt{2t + 3}$

27. $f(x) = (2x - 3)^4$ 28. $g(t) = \left(\tfrac{1}{2}t^2 - 1\right)^5$

29. **Acceleration of a Car**   During the construction of an office building, a hammer is accidentally dropped from a height of 256 ft. The distance the hammer falls in t seconds is $s = 16t^2$. What is the hammer's velocity when it strikes the ground? What is its acceleration?

30. **A Falling Object**   The distance s (in feet) covered by a car t seconds after starting from rest is given by

$$s = 20t + 8t^2 - t^3 \qquad (0 \le t \le 6).$$

Find a general expression for the car's acceleration at any time $t (0 \le t \le 6)$. Show that the car is decelerating $2\tfrac{2}{3}$ seconds after starting from rest.

31. Crime Rates The number of major crimes committed in Bronxville between 1985 and 1992 is approximated by the function

$$N(t) = -0.1t^3 + 1.5t^2 + 100 \qquad (0 \le t \le 7),$$

where $N(t)$ denotes the number of crimes committed in year t and $t = 0$ corresponds to the year 1985. Enraged by the dramatic increase in the crime rate, Bronxville's citizens, with the help of the local police, organized "Neighborhood Crime Watch" groups in early 1989 to combat this menace.
  a. Verify that the crime rate was increasing from 1985 through 1992. [*Hint:* Compute $N'(0)$, $N'(1)$, . . . , $N'(7)$.]
  b. Show that the Neighborhood Crime Watch program was working by computing $N''(4)$, $N''(5)$, $N''(6)$, and $N''(7)$.

32. GNP of a Developing Country A developing country's gross national product (GNP) from 1986 to 1994 is approximated by the function

$$G(t) = -0.2t^3 + 2.4t^2 + 60 \qquad (0 \le t \le 8),$$

where $G(t)$ is measured in billions of dollars and $t = 0$ corresponds to the year 1986.
  a. Compute $G'(0)$, $G'(1)$, . . . , $G'(8)$.
  b. Compute $G''(0)$, $G''(1)$, . . . , $G''(8)$.
  c. Using the results obtained in (a) and (b), show that after a spectacular growth rate in the early years, the growth of the GNP cooled off.

33. Test Flight of a VTOL In a test flight of the McCord Terrier, McCord Aviation's experimental VTOL (vertical take-off and landing) aircraft, it was determined that t seconds after lift-off, when the craft was operated in the vertical take-off mode, its altitude (in feet) was

$$h(t) = \frac{1}{16}t^4 - t^3 + 4t^2 \qquad (0 \le t \le 8).$$

  a. Find an expression for the craft's velocity at time t.
  b. Find the craft's velocity when $t = 0$ (the initial velocity), $t = 4$, and $t = 8$.

  c. Find an expression for the craft's acceleration at time t.
  d. Find the craft's acceleration when $t = 0$, $t = 4$, and $t = 8$.
  e. Find the craft's height when $t = 0$, $t = 4$, and $t = 8$.

C **34. U.S. Census** According to the U.S. Census Bureau, the number of Americans aged 45 to 54 will be approximately

$$N(t) = -0.00233t^4 + 0.00633t^3 - 0.05417t^2 + 1.3467t + 25$$

million people in year t, where $t = 0$ corresponds to the beginning of 1990. Compute $N'(10)$ and $N''(10)$ and interpret your results.

C **35. Air Purification** During testing of a certain brand of air purifier, it was determined that the amount of smoke remaining t minutes after the start of the test was

$$A(t) = 100 - 17.63t + 1.915t^2 - 0.1316t^3 + 0.00468t^4 - 0.00006t^5$$

percent of the original amount. Compute $A'(10)$ and $A''(10)$ and interpret your results.

36. Let f be the function defined by the rule $f(x) = x^{7/3}$. Show that f has first- and second-order derivatives at all points x, in particular, at $x = 0$. Show also that the third derivative of f does not exist at $x = 0$.

37. Construct a function f that has derivatives of order up through and including n at a point a but fails to have the $(n + 1)$st derivative there. [*Hint:* See Exercise 36.]

38. Show that a polynomial function has derivatives of all orders. [*Hint:* Let $P(x) = a_0x^n + a_1x^{n-1} + a_2x^{n-2} + \cdots + a_n$ be a polynomial of degree n, where n is a positive integer and $a_0, a_1, \ldots, a_n$ are constants with $a_0 \ne 0$. Compute $P'(x)$, $P''(x)$, ...]

SOLUTIONS TO SELF-CHECK EXERCISES 3.5

1. $f'(x) = 10x^4 - 9x^2 + 2x - 6$

$f''(x) = 40x^3 - 18x + 2$

$f'''(x) = 120x^2 - 18$

2. We write $f(x) = (1 + x)^{-1}$ and use the General Power Rule, obtaining

$$f'(x) = (-1)(1 + x)^{-2}\frac{d}{dx}(1 + x) = -(1 + x)^{-2}(1)$$

$$= -(1 + x)^{-2} = -\frac{1}{(1 + x)^2}.$$

Continuing, we find

$$f''(x) = -(-2)(1 + x)^{-3}$$

$$= 2(1 + x)^{-3} = \frac{2}{(1 + x)^3}$$

and

$$f'''(x) = 2(-3)(1 + x)^{-4}$$

$$= -6(1 + x)^{-4}$$

$$= -\frac{6}{(1 + x)^4}.$$

3. $N'(t) = 6t^2 + 6t - 4$

$N''(t) = 12t + 6 = 6(2t + 1)$

Therefore, $N''(2) = 30$ and $N''(8) = 102$. The results of our computations reveal that at the end of year 2, the *rate* of growth of the turtle population is increasing at the rate of 30 turtles per year per year. At the end of year 8, the rate is increasing at the rate of 102 turtles per year per year. Clearly, the conservation measures are paying off handsomely.

3.6   Implicit Differentiation and Related Rates

Differentiating Implicitly

Up to now we have dealt with functions expressed in the form $y = f(x)$; that is, the dependent variable y is expressed *explicitly* in terms of the independent variable x. However, not all functions are expressed in this form. Consider, for example, the equation

$$x^2y + y - x^2 + 1 = 0. \tag{8}$$

This equation does express y *implicitly* as a function of x. In fact, solving (8) for y in terms of x, we obtain

$$(x^2 + 1)y = x^2 - 1 \qquad \text{(Implicit equation)}$$

or

$$y = f(x) = \frac{x^2 - 1}{x^2 + 1}, \qquad \text{(Explicit equation)}$$

which gives an explicit representation of f.

Next, consider the equation

$$y^4 - y^3 - y + 2x^3 - x = 8.$$

When certain restrictions are placed on x and y, this equation defines y as a function of x. But in this instance, we would be hard pressed to find y explicitly in terms of x. The following question arises naturally: How does one go about computing dy/dx in this case?

As it turns out, thanks to the Chain Rule, a method *does* exist for computing the derivative of a function directly from the implicit equation defining the function. This method is called **implicit differentiation** and is demonstrated in the next several examples.

EXAMPLE 1   Find dy/dx given the equation $y^2 = x$.

Solution   Differentiating both sides of the equation with respect to x, we obtain

$$\frac{d}{dx}(y^2) = \frac{d}{dx}(x).$$

In order to carry out the differentiation of the term $\frac{d}{dx}(y^2)$, we note that y is a function of x. Writing $y = f(x)$ to remind us of this fact, we find that

$$\frac{d}{dx}(y^2) = \frac{d}{dx}[f(x)]^2 \qquad \text{[Writing } y = f(x)]$$
$$= 2f(x)f'(x) \qquad \text{(Using the Chain Rule)}$$
$$= 2y\frac{dy}{dx}. \qquad \begin{array}{l}\text{[Returning to using } y \\ \text{instead of } f(x)]\end{array}$$

Therefore, the equation

$$\frac{d}{dx}(y^2) = \frac{d}{dx}(x)$$

is equivalent to

$$2y\frac{dy}{dx} = 1.$$

Solving for $\frac{dy}{dx}$ yields

$$\frac{dy}{dx} = \frac{1}{2y}.$$ ∎

Before considering other examples, let us summarize the important steps involved in implicit differentiation. (Here we assume that dy/dx exists.)

FINDING $\frac{dy}{dx}$ BY IMPLICIT DIFFERENTIATION

1. Differentiate both sides of the equation *with respect to x.* (Make sure that the derivative of any term involving y includes the factor dy/dx.)

2. Solve the resulting equation for $\frac{dy}{dx}$ in terms of x and y.

EXAMPLE 2   Find dy/dx given the equation

$$y^3 - y + 2x^3 - x = 8.$$

Solution Differentiating both sides of the given equation with respect to x, we obtain

$$\frac{d}{dx}(y^3 - y + 2x^3 - x) = \frac{d}{dx}(8)$$

or $$\frac{d}{dx}(y^3) - \frac{d}{dx}(y) + \frac{d}{dx}(2x^3) - \frac{d}{dx}(x) = 0.$$

Now, recalling that y is a function of x, we apply the Chain Rule to the first two terms on the left. Thus,

$$3y^2 \frac{dy}{dx} - \frac{dy}{dx} + 6x^2 - 1 = 0$$

$$(3y^2 - 1)\frac{dy}{dx} = 1 - 6x^2$$

$$\frac{dy}{dx} = \frac{1 - 6x^2}{3y^2 - 1}. \qquad ■$$

EXAMPLE 3 Consider the equation $x^2 + y^2 = 4$.

a. Find dy/dx by implicit differentiation.
b. Find the slope of the tangent line to the graph of the function $y = f(x)$ at the point $(1, \sqrt{3})$.
c. Find an equation of the tangent line of (b).

Solution
a. Differentiating both sides of the equation with respect to x, we obtain

$$\frac{d}{dx}(x^2 + y^2) = \frac{d}{dx}(4)$$

$$\frac{d}{dx}(x^2) + \frac{d}{dx}(y^2) = 0$$

$$2x + 2y\frac{dy}{dx} = 0$$

and $$\frac{dy}{dx} = -\frac{x}{y} \qquad (y \neq 0).$$

b. The slope of the tangent line to the graph of the function at the point $(1, \sqrt{3})$ is given by

$$\left.\frac{dy}{dx}\right|_{(1, \sqrt{3})} = \left.-\frac{x}{y}\right|_{(1, \sqrt{3})} = -\frac{1}{\sqrt{3}}.$$

(*Note:* This notation is read "dy/dx evaluated at the point $(1, \sqrt{3})$.")
c. An equation of the tangent line in question is found by using the point-slope form of the equation of a line with the slope $m = -1/\sqrt{3}$ and the point $(1, \sqrt{3})$. Thus,

$$y - \sqrt{3} = -\frac{1}{\sqrt{3}}(x - 1)$$

$$\sqrt{3}y - 3 = -x + 1$$

or $$x + \sqrt{3}y - 4 = 0.$$

FIGURE 3.15

The line $x + \sqrt{3}y - 4 = 0$ is tangent to the graph of the function $y = f(x)$.

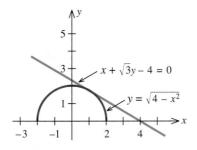

A sketch of this tangent line is shown in Figure 3.15.

We can also solve the equation $x^2 + y^2 = 4$ explicitly for y in terms of x. If we do this, we obtain

$$y = \pm\sqrt{4 - x^2}.$$

From this we see that the equation $x^2 + y^2 = 4$ defines the two functions

$$y = f(x) = \sqrt{4 - x^2}$$

and

$$y = g(x) = -\sqrt{4 - x^2}.$$

Since the point $(1, \sqrt{3})$ does not lie on the graph of $y = g(x)$, we conclude that

$$y = f(x) = \sqrt{4 - x^2}$$

is the required function. The graph of f is the upper semicircle shown in Figure 3.15. ∎

To find dy/dx at a *specific* point (a, b), differentiate the given equation implicitly with respect to x and then replace x and y by a and b, respectively, *before* solving the equation for dy/dx. This often simplifies the amount of algebra involved.

EXAMPLE 4   Find dy/dx given that x and y are related by the equation

$$x^2y^3 + 6x^2 = y + 12$$

and that $y = 2$ when $x = 1$.

Solution   Differentiating both sides of the given equation with respect to x, we obtain

$$\frac{d}{dx}(x^2y^3) + \frac{d}{dx}(6x^2) = \frac{d}{dx}(y) + \frac{d}{dx}(12)$$

$$x^2 \cdot \frac{d}{dx}(y^3) + y^3 \cdot \frac{d}{dx}(x^2) + 12x = \frac{dy}{dx} \qquad \begin{array}{l}\text{[Using the Product}\\ \text{Rule on } \frac{d}{dx}(x^2y^3)\text{]}\end{array}$$

$$3x^2y^2\frac{dy}{dx} + 2xy^3 + 12x = \frac{dy}{dx}.$$

Substituting $x = 1$ and $y = 2$ into this equation gives

$$3(1)^2(2)^2\frac{dy}{dx} + 2(1)(2)^3 + 12(1) = \frac{dy}{dx}$$

$$12\frac{dy}{dx} + 16 + 12 = \frac{dy}{dx},$$

and, solving for $\frac{dy}{dx}$,

$$\frac{dy}{dx} = -\frac{28}{11}.$$

Note that it is not necessary to find an explicit expression for dy/dx. ∎

REMARK   In Examples 3 and 4, you can verify that the points at which we evaluated dy/dx actually lie on the curve in question by showing that the coordinates of the points satisfy the given equations.   ☐

EXAMPLE 5   Find dy/dx given that x and y are related by the equation
$$\sqrt{x^2 + y^2} - x^2 = 5.$$

Solution   Differentiating both sides of the given equation with respect to x, we obtain

$$\frac{d}{dx}(x^2 + y^2)^{1/2} - \frac{d}{dx}(x^2) = \frac{d}{dx}(5)$$

(Writing $\sqrt{x^2 + y^2} = (x^2 + y^2)^{1/2}$)

$$\frac{1}{2}(x^2 + y^2)^{-1/2}\frac{d}{dx}(x^2 + y^2) - 2x = 0$$

(Using the General Power Rule on the first term.)

$$\frac{1}{2}(x^2 + y^2)^{-1/2}\left(2x + 2y\frac{dy}{dx}\right) - 2x = 0$$

$$2x + 2y\frac{dy}{dx} = 4x(x^2 + y^2)^{1/2}$$

(Transposing $2x$ and multiplying both sides by $2(x^2 + y^2)^{1/2}$)

$$2y\frac{dy}{dx} = 4x(x^2 + y^2)^{1/2} - 2x$$

$$\frac{dy}{dx} = \frac{2x\sqrt{x^2 + y^2} - x}{y}.$$ ∎

Related Rates

Implicit differentiation is a useful technique for solving a class of problems known as **related rates** problems. For example, suppose that x and y are each functions of a third variable t. Here, x might denote the mortgage rate and y the number of single-family homes sold at any time t. Further, suppose that we have an equation that gives the relationship between x and y (the number of houses sold, y, is related to the mortgage rate, x). Differentiating both sides of this equation implicitly with respect to t, we obtain an equation that gives a relationship between $\dfrac{dx}{dt}$ and $\dfrac{dy}{dt}$. In the context of our example, this equation gives us a relationship between the rate of change of the mortgage rate and the rate of change of the number of houses sold, as a function of time. Thus, knowing

$$\frac{dx}{dt}$$ (How fast the mortgage rate is changing at time t)

we can determine

$$\frac{dy}{dt}$$ (How fast the sale of houses is changing at that instant of time)

EXAMPLE 6   A study prepared for the National Association of Realtors estimates that the number of housing starts in the Southwest, $N(t)$ [in units of a million], over the next five years is related to the mortgage rate $r(t)$ [percent per year] by the equation

$$9N^2 + r = 36.$$

What is the rate of change of the number of housing starts with respect to time when the mortgage rate is 11 percent per year and is increasing at the rate of 1.5 percent per year?

Solution   We are given that

$$r = 11 \quad \text{and} \quad \frac{dr}{dt} = 1.5,$$

at a certain instant of time, and we are required to find $\dfrac{dN}{dt}$. First, by substituting $r = 11$ into the given equation, we find

$$9N^2 + 11 = 36$$
$$N^2 = \frac{25}{9},$$

or $N = 5/3$ (we reject the negative root). Next, differentiating the given equation implicitly on both sides with respect to t, we obtain

$$\frac{d}{dt}(9N^2) + \frac{d}{dt}(r) = \frac{d}{dt}(36)$$

$$18N\frac{dN}{dt} + \frac{dr}{dt} = 0. \qquad \text{(Use the Chain Rule on the first term.)}$$

Then, substituting $N = \dfrac{5}{3}$ and $\dfrac{dr}{dt} = 1.5$ into this equation gives

$$18\left(\frac{5}{3}\right)\frac{dN}{dt} + 1.5 = 0.$$

Solving this equation for dN/dt then gives

$$\frac{dN}{dt} = -\frac{1.5}{30} \approx -0.05.$$

Thus, at the instant of time under consideration, the number of housing starts is decreasing at the rate of 50,000 units per year.   ■

EXAMPLE 7   A major audio-tape manufacturer is willing to make x thousand ten-packs of metal alloy audiocassette tapes per week available in the marketplace when the wholesale price is $\$p$ per ten-pack. It is known that the relationship between x and p is governed by the supply equation

$$x^2 - 3xp + p^2 = 5.$$

How fast is the supply of tapes changing when the price per ten-pack is $11, the quantity supplied is 4000 ten-packs, and the wholesale price per ten-pack is increasing at the rate of 10 cents per ten-pack per week?

Solution   We are given that

$$p = 11, \quad x = 4, \quad \text{and} \quad \frac{dp}{dt} = 0.1$$

at a certain instant of time, and we are required to find $\frac{dx}{dt}$. Differentiating the given equation on both sides with respect to t, we obtain

$$\frac{d}{dt}(x^2) - \frac{d}{dt}(3xp) + \frac{d}{dt}(p^2) = \frac{d}{dt}(5)$$

$$2x\frac{dx}{dt} - 3\left(p\frac{dx}{dt} + x\frac{dp}{dt}\right) + 2p\frac{dp}{dt} = 0 \qquad \text{(Use the Product Rule on the second term.)}$$

Substituting the given values of p, x, and $\frac{dp}{dt}$ into the last equation, we have

$$2(4)\frac{dx}{dt} - 3[(11)\frac{dx}{dt} + 4(0.1)] + 2(11)(0.1) = 0$$

$$8\frac{dx}{dt} - 33\frac{dx}{dt} - 1.2 + 2.2 = 0$$

$$25\frac{dx}{dt} = 1,$$

or

$$\frac{dx}{dt} = 0.04.$$

Thus, at the instant of time under consideration the supply of ten-pack audiocassettes is increasing at the rate of (0.04)(1000), or 40, ten-packs per week. ■

In certain related problems, we need to formulate the problem mathematically before analyzing it. The following guidelines can be used to help solve problems of this type.

SOLVING RELATED RATES PROBLEMS	1. Assign a variable to each quantity. Draw a diagram if needed. 2. Write the *given* values of the variables and their rates of change with respect to t. 3. Find an equation giving the relationship between the variables. 4. Differentiate both sides of this equation implicitly with respect to t. 5. Replace the variables and their derivatives by the numerical data found in Step 2, and solve the equation for the required rate of change.

EXAMPLE 8   At a distance of 4000 feet from the launch site, a spectator is observing a rocket being launched. If the rocket lifts off vertically and is rising at a speed of 600 ft/sec when it is at an altitude of 3000 ft, how fast is the distance between the rocket and the spectator changing at that instant?

Solution

Step 1   Let

$$y = \text{the altitude of the rocket}$$

and $x = $ the distance between the rocket and the spectator

at any time t (see Figure 3.16).

FIGURE 3.16
The rate at which x is changing with respect to time is related to the rate of change of y with respect to time.

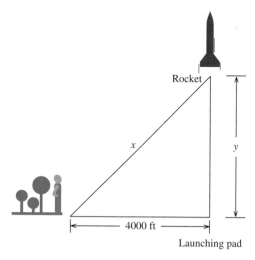

Rocket

x

y

4000 ft

Launching pad

Step 2   We are given that at a certain instant of time

$$y = 3000 \quad \text{and} \quad \frac{dy}{dt} = 600$$

and are asked to find dx/dt at that instant.

Step 3   Applying the Pythagorean Theorem to the right triangle in Figure 3.16, we find that

$$x^2 = y^2 + 4000^2.$$

Therefore, when $y = 3000$,

$$x = \sqrt{3000^2 + 4000^2} = 5000.$$

Step 4   Next, we differentiate the equation $x^2 = y^2 + 4000^2$ with respect to t, obtaining

$$2x\frac{dx}{dt} = 2y\frac{dy}{dt}.$$

(Remember that both x and y are functions of t.)

Step 5   Substituting $x = 5000$, $y = 3000$, and $dy/dt = 600$, we find

$$2(5000)\frac{dx}{dt} = 2(3000)(600)$$

or

$$\frac{dx}{dt} = 360.$$

Therefore, the distance between the rocket and the spectator is changing at a rate of 360 ft/sec.   ■

Be sure that you do *not* replace the variables in the equation found in Step 3 by their numerical values before differentiating the equation.

**SELF-CHECK
EXERCISES 3.6**

1. Given the equation $x^3 + 3xy + y^3 = 4$, find $\dfrac{dy}{dx}$ by implicit differentiation.

2. Find an equation of the tangent line to the graph of $16x^2 + 9y^2 = 144$ at the point $\left(2, -\dfrac{4\sqrt{5}}{3}\right)$.

Solutions to Self-Check Exercises 3.6 can be found on page 189.

3.6   EXERCISES

In Exercises 1–8, find the derivative dy/dx (a) by solving each of the given implicit equations for y explicitly in terms of x and (b) by differentiating each of the given equations implicitly. Show that, in each case, the results are equivalent.

1. $x + 2y = 5$

2. $3x + 4y = 6$

3. $xy = 1$

4. $xy - y - 1 = 0$

5. $x^3 - x^2 - xy = 4$

6. $x^2y - x^2 + y - 1 = 0$

7. $\dfrac{x}{y} - x^2 = 1$

8. $\dfrac{y}{x} - 2x^3 = 4$

In Exercises 9–30, find dy/dx by implicit differentiation.

9. $x^2 + y^2 = 16$

10. $2x^2 + y^2 = 16$

11. $x^2 - 2y^2 = 16$

12. $x^3 + y^3 + y - 4 = 0$

13. $x^2 - 2xy = 6$

14. $x^2 + 5xy + y^2 = 10$

15. $x^2y^2 - xy = 8$

16. $x^2y^3 - 2xy^2 = 5$

17. $x^{1/2} + y^{1/2} = 1$

18. $x^{1/3} + y^{1/3} = 1$

19. $\sqrt{x + y} = x$

20. $(2x + 3y)^{1/3} = x^2$

21. $\dfrac{1}{x^2} + \dfrac{1}{y^2} = 1$

22. $\dfrac{1}{x^3} + \dfrac{1}{y^3} = 5$

23. $\sqrt{xy} = x + y$

24. $\sqrt{xy} = 2x + y^2$

25. $\dfrac{x + y}{x - y} = 3x$

26. $\dfrac{x - y}{2x + 3y} = 2x$

27. $xy^{3/2} = x^2 + y^2$

28. $x^2y^{1/2} = x + 2y^3$

29. $(x + y)^3 + x^3 + y^3 = 0$

30. $(x + y^2)^{10} = x^2 + 25$

In Exercises 31–34, find an equation of the tangent line to the graph of the function f defined by the given equation at the indicated point.

31. $4x^2 + 9y^2 = 36$; $(0, 2)$

32. $y^2 - x^2 = 16$; $(2, 2\sqrt{5})$

33. $x^2y^3 - y^2 + xy - 1 = 0$; $(1, 1)$

34. $(x - y - 1)^3 = x$; $(1, -1)$

In Exercises 35–38, find the second derivative d^2y/dx^2 of each of the functions defined implicitly by the given equation.

35. $xy = 1$

36. $x^3 + y^3 = 28$

37. $y^2 - xy = 8$

38. $x^{1/3} + y^{1/3} = 1$

39. **Price-Demand**   Suppose that the quantity demanded weekly of the Super Titan radial tires is related to its unit price by the equation

$$p + x^2 = 144,$$

where p is measured in dollars and x is measured in units of a thousand. How fast is the quantity demanded changing when $x = 9$, $p = 63$, and the price per tire is increasing at the rate of $2 per week?

40. **Price-Supply**   Suppose the quantity, x, of Super Titan radial tires made available per week in the marketplace by the Titan Tire Company is related to the unit selling price by the equation

$$p - \frac{1}{2}x^2 = 48,$$

where x is measured in units of a thousand and p is in dollars. How fast is the weekly supply of Super Titan radial tires being introduced into the marketplace when $x = 6$, $p = 66$, and the price per tire is decreasing at the rate of $3 per week?

41. Price-Demand The demand equation for a certain brand of metal alloy audiocassette tape is

$$100x^2 + 9p^2 = 3600,$$

where x represents the number (in thousands) of ten-packs demanded per week when the unit price is $$p$. How fast is the quantity demanded increasing when the unit price per ten-pack is $14 and the selling price is dropping at the rate of 15 cents per ten-pack per week? [*Hint:* To find the value of x when $p = 14$, solve the equation $100x^2 + 9p^2 = 3600$ for x when $p = 14$.]

42. Effect of Price on Supply Suppose that the wholesale price of a certain brand of medium-size eggs p (in dollars per carton) is related to the weekly supply x (in thousands of cartons) by the equation

$$625p^2 - x^2 = 100.$$

If 25,000 cartons of eggs are available at the beginning of a certain week and the price is falling at the rate of 2 cents per carton per week, at what rate is the supply falling? [*Hint:* To find the value of p when $x = 25$, solve the supply equation for p when $x = 25$.]

43. Supply-Demand Refer to Exercise 42. If 25,000 cartons of eggs are available at the beginning of a certain week and the supply is falling at the rate of 1,000 cartons per week, at what rate is the wholesale price changing?

44. Elasticity of Demand The demand function for a certain make of cartridge typewriter ribbon is

$$p = -0.01x^2 - 0.1x + 6,$$

where p is the unit price in dollars and x is the quantity demanded each week, measured in units of a thousand. Compute the elasticity of demand and determine whether the demand is inelastic, unitary, or elastic when $x = 10$.

45. Elasticity of Demand The demand function for a certain brand of long-playing record is

$$p = -0.01x^2 - 0.2x + 8,$$

where p is the wholesale unit price in dollars and x is the quantity demanded each week, measured in units of a thousand. Compute the elasticity of demand, and determine whether the demand is inelastic, unitary, or elastic when $x = 15$.

46. The volume V of a cube with sides of length x inches is changing with respect to time. At a certain instant of time, the sides of the cube are 5 inches long and increasing at the rate of 0.1 in./sec. How fast is the volume of the cube changing at that instant of time?

47. Oil Spills In calm waters, oil spilling from the ruptured hull of a grounded tanker spreads in all directions. Assuming that the area polluted is a circle and that its radius is increasing at a rate of 2 ft/sec, determine how fast the area is increasing when the radius of the circle is 40 feet.

48. Two ships leave the same port at noon. Ship A sails north at 15 mph and Ship B sails east at 12 mph. How fast is the distance between them changing at 1 P.M.?

49. A car leaves an intersection, traveling east. Its position t seconds later is given by $x = t^2 + t$ feet. At the same time, another car leaves the same intersection, heading north, traveling $y = t^2 + 3t$ feet in t seconds. Find the rate at which the distance between the two cars will be changing 5 seconds later.

50. At a distance of 50 feet from the pad, a man observes a helicopter taking off from a heliport. If the helicopter lifts off vertically and is rising at a speed of 44 ft/sec when it is at an altitude of 120 feet, how fast is the distance between the helicopter and the man changing at that instant?

51. A spectator watches a rowing race from the edge of a river bank. The lead boat is moving in a straight line that is 120 feet from the river bank. If the boat is moving at a constant speed of 20 ft/sec, how fast is the boat moving away from the spectator when it is 50 feet past her?

52. A man 6 feet tall is walking away from a street light 18 feet high at a speed of 6 ft/sec. How fast is the tip of his shadow moving along the ground?

53. A 20-foot ladder is leaning against a wall. If the bottom of the ladder is pulled away from the wall at a rate of 2 ft/sec, how fast is the top of the ladder sliding down the wall when the bottom of the ladder is 12 feet from the wall? [*Hint:* Refer to the adjacent figure. By the Pythagorean Theorem, $x^2 + y^2 = 400$. Find dy/dt when $x = 12$ and $dx/dt = 2$.]

54. A 13-foot ladder is leaning against a wall. If the bottom of the ladder is pulled away from the wall at a rate of 2.5 ft/sec, how fast is the top of the ladder sliding down the wall when the bottom of the ladder is 12 feet from the wall?

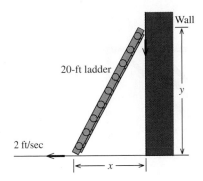

Figure for Exercise 53

SOLUTIONS TO SELF-CHECK EXERCISES 3.6

1. Differentiating both sides of the equation with respect to x, we have

$$3x^2 + 3y + 3xy' + 3y^2y' = 0$$
$$(x^2 + y) + (x + y^2)y' = 0$$

or

$$y' = -\frac{x^2 + y}{x + y^2}.$$

2. To find the slope of the tangent line to the graph of the function at any point, we differentiate the equation implicitly with respect to x, obtaining

$$32x + 18yy' = 0$$

or

$$y' = -\frac{16x}{9y}.$$

In particular, the slope of the tangent line at $\left(2, -\dfrac{4\sqrt{5}}{3}\right)$ is

$$m = -\frac{16(2)}{9\left(-\dfrac{4\sqrt{5}}{3}\right)} = \frac{8}{3\sqrt{5}}.$$

Using the point-slope form of the equation of a line, we find

$$y - \left(-\frac{4\sqrt{5}}{3}\right) = \frac{8}{3\sqrt{5}}(x - 2)$$

or

$$y = \frac{8\sqrt{5}}{15}x - \frac{36\sqrt{5}}{15}.$$

3.7 Differentials

The Millers are planning to buy a house in the near future and estimate that they will need a 30-year fixed-rate mortgage for $120,000. If the interest rate increases from the present rate of 9 percent per year to 9.4 percent per year between now and the time the Millers decide to secure the loan, approximately how much more per month will their mortgage be? (You will be asked to answer this question in Exercise 40, page 197.)

John Decker

Title: Mortgage Counselor
Institution: Shawmut Bank

Decker's job is simple: to loan money to people who want to buy a home. The hard part is deciding whether applicants qualify for one of Shawmut's 40 different mortgage plans. Sifting through income figures, current indebtedness, and credit history helps Decker determine which individuals make the best mortgage candidates.

Decker stresses that he and his colleagues "strive to grant loans. That's our job." But before allowing someone to file a formal application, Decker takes the person over several "pre-qualification hurdles" to gauge his or her ability to handle a mortgage.

To start, Decker relies on a two-tiered, debt-to-income ratio to see whether a person has sufficient gross monthly income to make payments. Under the first tier, the proposed monthly payment (principal and interest, property tax, homeowner's insurance, and, when applicable, a condo fee) cannot exceed 28 percent of an individual's gross monthly income. If Decker's initial calculations are positive, he then must determine the individual's ability to meet monthly mortgage payments while also repaying other debts, such as car and student loans, credit cards, alimony, and so on. These combined payments cannot exceed 36 percent of gross monthly income. A typical person with an $800 mortgage obligation and $550 in other payments would have to earn $3750 per month to clear these first two hurdles.

Contrary to popular belief, bankers *want* to lend money. "The idea is to grant mortgages," says Decker, which contribute substantially to a bank's profitability. But lending money means making sensible decisions about how much to lend as well as a person's ability to repay the loan.

Using a loan-to-value formula, banks might lend 80 percent of a property's value. In such cases, the applicant has to put 20 percent down to make the purchase. Or Shawmut may decide to lend up to 95 percent or as little as 75 percent of the appraised value.

Understandably, banks don't like to see bankruptcies, late payments, or liens on a personal credit history. Decker notes, however, that even this final hurdle doesn't necessarily mean failure in securing a mortgage. He works closely with each individual to overcome any stigma that might prompt a rejection.

Once Decker has put together a successful mortgage application, it is reviewed internally. Then, even though a mortgage is granted, it might come through at a slightly higher interest rate—10.4 percent instead of the expected 10 percent rate. Differentials would be used to compute the change in monthly payments, for while this might seem like a small change, on a large mortgage it could be enough of a variable to affect the new customer's ability to make monthly payments.

Using the debt-to-income ratio, Decker plugs the new variable into his formulas to determine whether a problem exists. These formulas used to compute the monthly payments involve the use of exponential functions. On a $100,000, 30-year, fixed-rate mortgage, payments will increase about $30 per month. Decker notes that such a small increase doesn't usually pose a significant problem.

If the customer can't make the new payment, however, Decker explores alternatives until he finds a solution. For Decker, it comes down to this: "If there is any possible way to give a loan, we're going to make it work."

Questions such as this one, in which one wishes to *estimate* the change in the dependent variable (monthly mortgage payment) corresponding to a small change in the independent variable (interest rate per year), occur in many real-life applications. For example:

- An economist would like to know how a small increase in a country's capital expenditure will affect the country's gross domestic output.
- A sociologist would like to know how a small increase in the amount of capital investment in a housing project will affect the crime rate.
- A businesswoman would like to know how raising a product's unit price by a small amount will affect her profit.
- A bacteriologist would like to know how a small increase in the amount of a bacteria will affect a population of bacteria.

In order to calculate these changes and estimate their effects, we use the *differential* of a function, a concept that will be introduced shortly.

Increments

Let x denote a variable quantity and suppose that x changes from x_1 to x_2. This change in x is called the **increment in x** and is denoted by the symbol Δx (read "delta x"). Thus,

$$\Delta x = x_2 - x_1. \qquad \text{(Final value minus initial value)} \qquad (9)$$

EXAMPLE 1 Find the increment in x

a. as x changes from 3 to 3.2   and   **b.** as x changes from 3 to 2.7.

Solution
a. Here $x_1 = 3$ and $x_2 = 3.2$, so

$$\Delta x = x_2 - x_1 = 3.2 - 3 = 0.2.$$

b. Here $x_1 = 3$ and $x_2 = 2.7$. Therefore,

$$\Delta x = x_2 - x_1 = 2.7 - 3 = -0.3. \qquad \blacksquare$$

Observe that Δx plays the same role that h played in Section 2.4.

Now, suppose two quantities, x and y, are related by an equation $y = f(x)$, where f is a function. If x changes from x to $x + \Delta x$, then the corresponding change in y is called the *increment in y*. It is denoted by Δy and is defined in Figure 3.17 by

$$\Delta y = f(x + \Delta x) - f(x) \qquad (10)$$

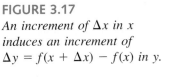

FIGURE 3.17
An increment of Δx in x induces an increment of $\Delta y = f(x + \Delta x) - f(x)$ in y.

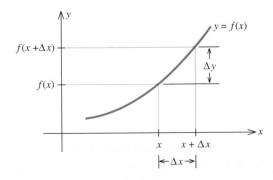

EXAMPLE 2   Let $y = x^3$. Find Δx and Δy when

a. x changes from 2 to 2.01   and   **b.** x changes from 2 to 1.98.

Solution   Let $f(x) = x^3$.

a. Here $\Delta x = 2.01 - 2 = 0.01$. Next,

$$\Delta y = f(x + \Delta x) - f(x) = f(2.01) - f(2)$$
$$= (2.01)^3 - 2^3 = 8.120601 - 8 = 0.120601.$$

b. Here $\Delta x = 1.98 - 2 = -0.02$. Next,

$$\Delta y = f(x + \Delta x) - f(x) = f(1.98) - f(2)$$
$$= (1.98)^3 - 2^3 = 7.762392 - 8 = -0.237608.$$ ∎

Differentials

We can obtain a relatively quick and simple way of approximating Δy, the change in y due to a small change Δx, by examining the graph of the function f shown in Figure 3.18.

FIGURE 3.18
If Δx is small, dy is a good approximation of Δy.

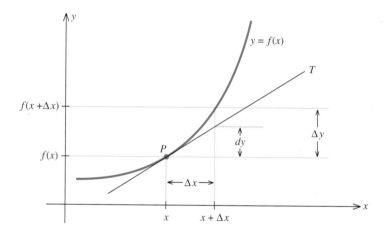

Observe that near the point of tangency P, the tangent line T is close to the graph of f. Therefore if Δx is small, then dy is a good approximation of Δy. We can find an expression for dy as follows: Notice that the slope of T is given by

$$\frac{dy}{\Delta x}. \qquad \text{(Rise divided by run)}$$

On the other hand, the slope of T is given by $f'(x)$. Therefore, we have

$$\frac{dy}{\Delta x} = f'(x),$$

or $dy = f'(x)\Delta x$. Thus, we have the approximation

$$\Delta y \approx dy = f'(x)\Delta x$$

in terms of the derivative of f at x. The quantity dy is called the *differential of y*.

THE DIFFERENTIAL	Let $y = f(x)$ define a differentiable function of x. Then,

1. the **differential dx** of the independent variable x is $dx = \Delta x$.

2. the **differential dy** of the dependent variable y is

$$dy = f'(x)\Delta x = f'(x)dx. \qquad (11)$$

REMARKS

1. For the independent variable x: There is no difference between Δx and dx—both measure the change in x from x to $x + \Delta x$.

2. For the dependent variable y: Δy measures the *actual* change in y as x changes from x to $x + \Delta x$, whereas dy measures the *approximate* change in y corresponding to the same change in x.

3. The differential dy depends on both x and dx, but for fixed x, dy is a linear function of dx.
　　　　　　　　　　　　　　　　　　　　　　　　　　　　　　　　　□

EXAMPLE 3　　Let $y = x^3$.

a. Find the differential dy of y.
b. Use dy to approximate Δy when x changes from 2 to 2.01.
c. Use dy to approximate Δy when x changes from 2 to 1.98.
d. Compare the results of (b) with those of Example 2.

Solution
a. Let $f(x) = x^3$. Then

$$dy = f'(x)dx = 3x^2dx.$$

b. Here $x = 2$ and $dx = 2.01 - 2 = 0.01$. Therefore,

$$dy = 3x^2dx = 3(2)^2(0.01) = 0.12.$$

c. Here $x = 2$ and $dx = 1.98 - 2 = -0.02$. Therefore,

$$dy = 3x^2dx = 3(2)^2(-0.02) = -0.24.$$

d. As you can see, both approximations 0.12 and -0.24 are quite close to the actual changes of Δy obtained in Example 2, 0.120601 and -0.237608.　　■

　　　Observe how much easier it is to find an approximation to the exact change in a function with the help of the differential, rather than calculating the exact change in the function itself. In the following examples, we will take advantage of this fact.

EXAMPLE 4　　Approximate the value of $\sqrt{26.5}$ using differentials. Verify your result using the $\boxed{\sqrt{\ }}$ key on your calculator.

Solution　　Since we want to compute the square root of a number, let us consider the function $y = f(x) = \sqrt{x}$. Since 25 is the number nearest 26.5 whose square root is readily recognized, let us take $x = 25$. We want to know the change in y, Δy, as x changes from $x = 25$ to $x = 26.5$, an increase of $\Delta x = 1.5$ units. Using (11), we find

$$\Delta y \approx dy = f'(x)\Delta x$$

$$= \left[\frac{1}{2\sqrt{x}}\bigg|_{x=25}\right] \cdot (1.5) = \left(\frac{1}{10}\right)(1.5) = 0.15.$$

Therefore,

$$\sqrt{26.5} - \sqrt{25} = \Delta y \approx 0.15$$

and
$$\sqrt{26.5} \approx \sqrt{25} + 0.15 = 5.15.$$

The exact value of $\sqrt{26.5}$, rounded off to five decimal places, is 5.14782. Thus, the error incurred in the approximation is 0.00218.   ■

Applications

EXAMPLE 5   The total cost incurred in operating a certain type of truck on a 500-mile trip, traveling at an average speed of v miles per hour, is estimated to be

$$C(v) = 125 + v + \frac{4500}{v}$$

dollars. Find the approximate change in the total operating cost when the average speed is increased from 55 mph to 58 mph.

Solution   With $v = 55$ and $\Delta v = dv = 3$, we find

$$\Delta C \approx dC = C'(v)dv = \left(1 - \frac{4500}{v^2}\right)\bigg|_{v=55} \cdot 3$$

$$= \left(1 - \frac{4500}{3025}\right)(3) \approx -1.46,$$

so the total operating cost is found to decrease by $1.46. This might explain why so many independent truckers often exceed the 55 mph speed limit.   ■

EXAMPLE 6   The relationship between the amount of money x spent by Cannon Precision Instruments on advertising and Cannon's total sales $S(x)$ is given by the function

$$S(x) = -0.002x^3 + 0.6x^2 + x + 500 \qquad (0 \le x \le 200),$$

where x is measured in thousands of dollars. Use differentials to estimate the change in Cannon's total sales if advertising expenditures are increased from $100,000 ($x = 100$) to $105,000 ($x = 105$).

Solution   The required change in sales is given by

$$\Delta S \approx dS = S'(100)dx$$

$$= -0.006x^2 + 1.2x + 1 \bigg|_{x=100} \cdot (5) \qquad (dx = 105 - 100 = 5)$$

$$= (-60 + 120 + 1)(5) = 305,$$

that is, an increase of $305,000.   ◼

Finally, we want to point out that if at some point in reading this section you have a sense of déjà vu, do not be surprised, because the notion of the differential was first used in Section 3.4 (see Example 1). There we took $\Delta x = 1$, since we were interested in finding the marginal cost when the level of production was increased from $x = 250$ to $x = 251$. If we had used differentials, we would have found

$$C(251) - C(250) \approx C'(250)dx,$$

so that taking $dx = \Delta x = 1$, we have $C(251) - C(250) \approx C'(250)$, which agrees with the result obtained in Example 1. Thus, in Section 3.4, we touched upon the notion of the differential, albeit in the special case in which $dx = 1$.

SELF-CHECK EXERCISES 3.7

1. Find the differential of $f(x) = \sqrt{x} + 1$.

2. A certain country's government economists have determined that the demand equation for corn in that country is given by

$$p = f(x) = \frac{125}{x^2 + 1},$$

where p is expressed in dollars per bushel and x, the quantity demanded per year, is measured in billions of bushels. The economists are forecasting a harvest of 6 billion bushels for the year. If the actual production of corn were 6.2 billion bushels for the year instead, what would be the approximate drop in the predicted price of corn per bushel?

Solutions to Self-Check Exercises 3.7 can be found on page 197.

3.7 EXERCISES

In Exercises 1–14, find the differential of the given function.

1. $f(x) = 2x^2$

2. $f(x) = 3x^2 + 1$

3. $f(x) = x^3 - x$

4. $f(x) = 2x^3 + x$

5. $f(x) = \sqrt{x + 1}$

6. $f(x) = \dfrac{3}{\sqrt{x}}$

7. $f(x) = 2x^{3/2} + x^{1/2}$

8. $f(x) = 3x^{5/6} + 7x^{2/3}$

9. $f(x) = x + \dfrac{2}{x}$

10. $f(x) = \dfrac{3}{x - 1}$

11. $f(x) = \dfrac{x - 1}{x^2 + 1}$

12. $f(x) = \dfrac{2x^2 + 1}{x + 1}$

13. $f(x) = \sqrt{3x^2 - x}$

14. $f(x) = (2x^2 + 3)^{1/3}$

15. Let f be a function defined by

$$y = f(x) = x^2 - 1.$$

a. Find the differential of f.
b. Use your result from (a) to find the approximate change in y if x changes from 1 to 1.02.
c. Find the actual change in y if x changes from 1 to 1.02 and compare your result with that obtained in (b).

16. Let f be a function defined by

    $$y = f(x) = 3x^2 - 2x + 6.$$

    a. Find the differential of f.
    b. Use your result from (a) to find the approximate change in y if x changes from 2 to 1.97.
    c. Find the actual change in y if x changes from 2 to 1.97 and compare your result with that obtained in (b).

17. Let f be a function defined by

    $$y = f(x) = \frac{1}{x}.$$

    a. Find the differential of f.
    b. Use your result from (a) to find the approximate change in y if x changes from -1 to -0.95.
    c. Find the actual change in y if x changes from -1 to -0.95 and compare your result with that obtained in (b).

18. Let f be a function defined by

    $$y = f(x) = \sqrt{2x + 1}.$$

    a. Find the differential of f.
    b. Use your result from (a) to find the approximate change in y if x changes from 4 to 4.1.
    c. Find the actual change in y if x changes from 4 to 4.1 and compare your result with that obtained in (b).

In Exercises 19–26, use differentials to approximate the given quantity.

19. $\sqrt{10}$ 20. $\sqrt{17}$ 21. $\sqrt{49.5}$

22. $\sqrt{99.7}$ 23. $\sqrt[3]{7.8}$ 24. $\sqrt[4]{81.6}$

25. $\sqrt{0.089}$ 26. $\sqrt[3]{0.00096}$

27. **Error Estimation** The length of each edge of a cube is 12 cm, with a possible error in measurement of 0.02 cm. Use differentials to estimate the error that might occur when the volume of the cube is calculated.

28. **Error Estimation** A hemisphere-shaped dome of radius 60 ft is to be coated with a layer of rust-proofer before painting. Use differentials to estimate the amount of rust-proofer needed if the coat is to be 0.01 inch thick. [*Hint:* The surface area of a hemisphere of radius r is $S = \frac{2}{3}\pi r^3$.]

29. **Growth of a Cancerous Tumor** The volume of a spherical cancer tumor is given by

    $$V(r) = \frac{4}{3}\pi r^3.$$

If the radius of a tumor is estimated at 1.1 cm, with a maximum error in measurement of 0.005 cm, determine the error that might occur when the volume of the tumor is calculated.

30. **Gross Domestic Product** An economist has determined that a certain country's gross domestic output (GDP) is approximated by the function $f(x) = 640x^{1/5}$, where $f(x)$ is measured in billions of dollars and x is the capital outlay in billions of dollars. Use differentials to estimate the change in the country's GDP if the country's capital expenditure changes from 243 billion dollars to 248 billion dollars.

31. **Learning Curves** The length of time (in seconds) a certain individual takes to learn a list of n items is approximated by

    $$f(n) = 4n\sqrt{n - 4}.$$

    Use differentials to approximate the additional time it takes the individual to learn the items on a list when n is increased from 85 to 90 items.

32. **Effect of Advertising on Profits** The relationship between Cunningham Realty's quarterly profits, $P(x)$, and the amount of money x spent on advertising per quarter is described by the function

    $$P(x) = -\frac{1}{8}x^2 + 7x + 30 \qquad (0 \le x \le 50),$$

    where both $P(x)$ and x are measured in thousands of dollars. Use differentials to estimate the increase in profits when advertising expenditure each quarter is increased from $24,000 to $26,000.

33. **Effect of Mortgage Rates on Housing Starts** A study prepared for the National Association of Realtors estimates that the number of housing starts per year over the next five years will be

    $$N(r) = \frac{7}{1 + 0.02r^2}$$

    million units, where r(percent) is the mortgage rate. Use differentials to estimate the decrease in the number of housing starts when the mortgage rate is increased from 12 to 12.5 percent.

34. **Supply-Price** The supply equation for a certain brand of transistor radio is given by

    $$p = s(x) = 0.3\sqrt{x} + 10,$$

    where x is the quantity supplied and p is the unit price in dollars. Use differentials to approximate the change in price when the quantity supplied is increased from 10,000 units to 10,500 units.

35. **Demand-Price**   The demand function for the Sentinel smoke alarm is given by

$$p = d(x) = \frac{30}{0.02x^2 + 1},$$

where x is the quantity demanded (in units of a thousand) and p is the unit price in dollars. Use differentials to estimate the change in the price p when the quantity demanded changes from 5000 to 5500 units per week.

36. **Surface Area of an Animal**   Animal physiologists use the formula

$$S = kW^{2/3}$$

to calculate an animal's surface area (in square meters) from its weight W (in kilograms), where k is a constant that depends on the animal under consideration. Suppose a physiologist calculates the surface area of a horse ($k = 0.1$). If the horse's weight is estimated at 300 kg, with a maximum error in measurement of 0.6 kg, determine the percentage error in the calculation of the horse's surface area.

37. **Forecasting Profits**   The management of Trappee and Sons, Inc., forecast that they will sell 200,000 cases of their Texa-Pep hot sauce next year. Their annual profit is described by

$$P(x) = -0.000032x^3 + 6x - 100$$

thousand dollars, where x is measured in thousands of cases. If the maximum error in the forecast is 15 percent, determine the corresponding error in Trappee's profits.

38. **Forecasting Commodity Prices**   A certain country's government economists have determined that the demand equation for soybeans in that country is given by

$$p = f(x) = \frac{55}{2x^2 + 1},$$

where p is expressed in dollars per bushel and x, the quantity demanded per year, is measured in billions of bushels. The economists are forecasting a harvest of 1.8 billion bushels for the year, with a possible error of 15 percent in their forecast. Determine the corresponding error in the predicted price per bushel of soybeans.

39. **Crime Studies**   A sociologist has found that the number of serious crimes in a certain city per year is described by the function

$$N(x) = \frac{500(400 + 20x)^{1/2}}{(5 + 0.2x)^2},$$

where x (in cents per dollar deposited) is the level of reinvestment in the area in conventional mortgages by the city's ten largest banks. Use differentials to estimate the change in the number of crimes if the level of reinvestment changes from 20 cents per dollar deposited to 22 cents per dollar deposited.

C 40. **Financing a Home**   The Millers are planning to buy a home in the near future and estimate that they will need a 30-year fixed-rate mortgage for $120,000. Their monthly payment P (in dollars) can be computed using the formula

$$P = \frac{10,000\,r}{1 - \left(1 + \dfrac{r}{12}\right)^{-360}},$$

where r is the interest rate per year.
a. Find the differential of P.
b. If the interest rate increases from the present rate of 9 percent per year to 9.2 percent per year between now and the time the Millers decide to secure the loan, approximately how much more will their monthly mortgage payment be? How much more will it be if the interest rate increases to 9.3 percent per year? To 9.4 percent per year? To 9.5 percent per year?

SOLUTIONS TO SELF-CHECK EXERCISES 3.7

1. We find

$$f'(x) = \frac{1}{2}x^{-1/2} = \frac{1}{2\sqrt{x}}.$$

Therefore, the required differential of f is

$$dy = \frac{1}{2\sqrt{x}}\,dx.$$

2. We first compute the differential

$$dp = -\frac{250x}{(x^2 + 1)^2}\,dx.$$

Next, using (11) with $x = 6$ and $dx = 0.2$, we find

$$\Delta p \approx dp = -\frac{250(6)}{(36 + 1)^2}(0.2) = -0.22,$$

or a drop in price of 22 cents per bushel.

CHAPTER THREE
SUMMARY OF
PRINCIPAL
FORMULAS AND
TERMS

Formulas

1. Derivative of a constant $\quad \dfrac{d}{dx}(c) = 0,\ c$ a constant

2. Power Rule $\quad \dfrac{d}{dx}(x^n) = nx^{n-1}$

3. Constant Multiple Rule $\quad \dfrac{d}{dx}(cu) = c\dfrac{du}{dx},\ c$ a constant

4. Sum Rule $\quad \dfrac{d}{dx}(u \pm v) = \dfrac{du}{dx} \pm \dfrac{dv}{dx}$

5. Product Rule $\quad \dfrac{d}{dx}(uv) = u\dfrac{dv}{dx} + v\dfrac{du}{dx}$

6. Quotient Rule $\quad \dfrac{d}{dx}\left(\dfrac{u}{v}\right) = \dfrac{v\dfrac{du}{dx} - u\dfrac{dv}{dx}}{v^2}$

7. Chain Rule $\quad \dfrac{dy}{dx} = \dfrac{dy}{du} \cdot \dfrac{du}{dx}$

8. General Power Rule $\quad \dfrac{d}{dx}(u^n) = nu^{n-1}\dfrac{du}{dx}$

9. Average cost function $\quad \overline{C}(x) = \dfrac{C(x)}{x}$

10. Revenue function $\quad R(x) = px$

11. Profit function $\quad P(x) = R(x) - C(x)$

12. Elasticity of demand $\quad E(p) = -\dfrac{pf'(p)}{f(p)}$

13. Differential of y $\quad dy = f'(x)\,dx$

Terms

Cost function

Marginal cost function

Marginal average cost function

Marginal revenue function

Marginal profit function

Elastic demand

Unitary demand

Inelastic demand

Higher-order derivatives

Second derivative of f

Implicit differentiation

Related rates

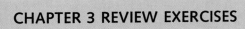

CHAPTER 3 REVIEW EXERCISES

In Exercises 1–30, find the derivative of the given function.

1. $f(x) = 3x^5 - 2x^4 + 3x^2 - 2x + 1$

2. $f(x) = 4x^6 + 2x^4 + 3x^2 - 2$

3. $g(x) = -2x^{-3} + 3x^{-1} + 2$

4. $f(t) = 2t^2 - 3t^3 - t^{-1/2}$

5. $g(t) = 2t^{-1/2} + 4t^{-3/2} + 2$

6. $h(x) = x^2 + \dfrac{2}{x}$

7. $f(t) = t + \dfrac{2}{t} + \dfrac{3}{t^2}$

8. $g(s) = 2s^2 - \dfrac{4}{s} + \dfrac{2}{\sqrt{s}}$

9. $h(x) = x^2 - \dfrac{2}{x^{3/2}}$

10. $f(x) = \dfrac{x + 1}{2x - 1}$

11. $g(t) = \dfrac{t^2}{2t^2 + 1}$

12. $h(t) = \dfrac{\sqrt{t}}{\sqrt{t} + 1}$

13. $f(x) = \dfrac{\sqrt{x} - 1}{\sqrt{x} + 1}$

14. $f(t) - \dfrac{t}{2t^2 + 1}$

15. $f(x) = \dfrac{x^2(x^2 + 1)}{x^2 - 1}$

16. $f(x) = (2x^2 + x)^3$

17. $f(x) = (3x^3 - 2)^8$

18. $h(x) = (\sqrt{x} + 2)^5$

19. $f(t) = \sqrt{2t^2 + 1}$

20. $g(t) = \sqrt[3]{1 - 2t^3}$

21. $s(t) = (3t^2 - 2t + 5)^{-2}$

22. $f(x) = (2x^3 - 3x^2 + 1)^{-3/2}$

23. $h(x) = \left(x + \dfrac{1}{x}\right)^2$

24. $h(x) = \dfrac{1 + x}{(2x^2 + 1)^2}$

25. $h(t) = (t^2 + t)^4(2t^2)$

26. $f(x) = (2x + 1)^3(x^2 + x)^2$

27. $g(x) = \sqrt{x}(x^2 - 1)^3$

28. $f(x) = \dfrac{x}{\sqrt{x^3 + 2}}$

29. $h(x) = \dfrac{\sqrt{3x + 2}}{4x - 3}$

30. $f(t) = \dfrac{\sqrt{2t + 1}}{(t + 1)^3}$

In Exercises 31–36, find the second derivative of the given function.

31. $f(x) = 2x^4 - 3x^3 + 2x^2 + x + 4$

32. $g(x) = \sqrt{x} + \dfrac{1}{\sqrt{x}}$

33. $h(t) = \dfrac{t}{t^2 + 4}$

34. $f(x) = (x^3 + x + 1)^2$

35. $f(x) = \sqrt{2x^2 + 1}$

36. $f(t) = t(t^2 + 1)^3$

In Exercises 37–42, find dy/dx by implicit differentiation.

37. $6x^2 - 3y^2 = 9$

38. $2x^3 - 3xy = 4$

39. $y^3 + 3x^2 = 3y$

40. $x^2 + 2x^2y^2 + y^2 = 10$

41. $x^2 - 4xy - y^2 = 12$

42. $3x^2y - 4xy + x - 2y = 6$

43. Let $f(x) = 2x^3 - 3x^2 - 16x + 3$.

   a. Find the points on the graph of f at which the slope of the tangent line is equal to -4.
   b. Find the equation(s) of the tangent line(s) of (a).

44. Let $f(x) = \frac{1}{3}x^3 + \frac{1}{2}x^2 - 4x + 1$.

   a. Find the points on the graph of f at which the slope of the tangent line is equal to -2.
   b. Find the equation(s) of the tangent line(s) of (a).

45. Find an equation of the tangent line to the graph of $y = \sqrt{4 - x^2}$ at the point $(1, \sqrt{3})$.

46. Find an equation of the tangent line to the graph of $y = x(x + 1)^5$ at the point $(1, 32)$.

47. Find the third derivative of the function

$$f(x) = \dfrac{1}{2x - 1}.$$

   What is its domain?

48. The number of subscribers to CNC Cable Television in the town of Randolph is approximated by the function

$$N(x) = 1000(1 + 2x)^{1/2} \qquad (1 \le x \le 30),$$

   where $N(x)$ denotes the number of subscribers to the service in the xth week. Find the rate of increase in the number of subscribers at the end of the twelfth week.

49. The total weekly cost in dollars incurred by the Herald Record Company in pressing x video discs is given by the total cost function

$$C(x) = 2500 + 2.2x \qquad (0 \le x \le 8000).$$

   a. What is the marginal cost when $x = 1000$? When $x = 2000$?

b. Find the average cost function $\overline{C}$ and the marginal average cost function $\overline{C}'$.

c. Using the results from (b), show that the average cost incurred by Herald in pressing a video disc approaches \$2.20 per disc when the level of production is high enough.

50. The marketing department of Telecon Corporation has determined that the demand for their cordless phones obeys the relationship

$$p = -0.02x + 600 \qquad (0 \le x \le 30,000),$$

where p denotes the phone's unit price (in dollars) and x denotes the quantity demanded.

a. Find the revenue function R.

b. Find the marginal revenue function R'.

c. Compute $R'(10,000)$ and interpret your result.

51. The weekly demand for the Lectro-Copy photocopying machine is given by the demand equation

$$p = 2000 - 0.04x \qquad (0 \le x \le 50,000),$$

where p denotes the wholesale unit price in dollars and x denotes the quantity demanded. The weekly total cost function for manufacturing these copiers is given by

$$C(x) = 0.000002x^3 - 0.02x^2 + 1,000x + 120,000,$$

where $C(x)$ denotes the total cost incurred in producing x units.

a. Find the revenue function R, the profit function P, and the average cost function $\overline{C}$.

b. Find the marginal cost function C', the marginal revenue function R', the marginal profit function P', and the marginal average cost function $\overline{C}'$.

c. Compute $C'(3000)$, $R'(3000)$, and $P'(3000)$.

d. Compute $\overline{C}'(5000)$ and $\overline{C}'(8000)$ and interpret your results.

When is a rocket ascending and when is it descending? What is the maximum altitude and what is the maximum velocity attained by the rocket? In Example 5, page 208, and Example 7, page 260, you will see how the techniques of calculus can be used to help answer these questions.

Applications of the Derivative

T his chapter further explores the power of the derivative, which

we use to help analyze the properties of functions. The information

obtained can then be used to accurately sketch graphs of functions. We

will also see how the derivative is used in solving a large class of

optimization problems, including finding what level of production will

yield a maximum profit for a company; finding what level of production

will result in a minimal cost to a company; finding the maximum height

attained by a rocket; finding the maximum velocity at which air is

expelled when a person coughs; and a host of other problems.

| 4.1 | **Increasing and Decreasing Functions** |

Determining the Intervals Where a Function Is Increasing or Decreasing

According to a study by the U.S. Department of Energy and the Shell Development Company, a typical car's fuel economy as a function of its speed is described by the graph shown in Figure 4.1. Observe that the fuel economy $f(x)$ in miles per gallon (mpg) improves as x, the vehicle's speed in miles per hour (mph), increases from 0 to 42, and then drops as the speed increases beyond 42 mph. We use the terms *increasing* and *decreasing* to describe the behavior of a function as we move from left to right along its graph.

FIGURE 4.1

A typical car's fuel economy improves as the speed at which it is driven increases from 0 mph to 42 mph and drops at speeds greater than 42 mph.

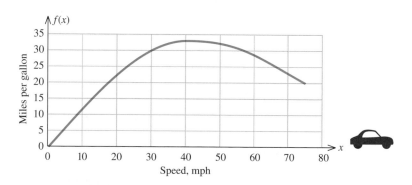

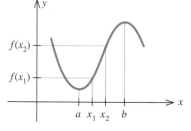

More precisely, we have the following definitions.

INCREASING AND DECREASING FUNCTIONS

A function f is **increasing** on an interval (a, b) if for any two numbers x_1 and x_2 in (a, b), $f(x_1) < f(x_2)$ whenever $x_1 < x_2$ (Figure 4.2a).

A function f is **decreasing** on an interval (a, b) if for any two numbers x_1 and x_2 in (a, b), $f(x_1) > f(x_2)$ whenever $x_1 < x_2$ (Figure 4.2b).

FIGURE 4.2

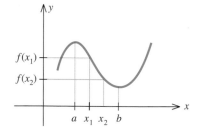

(a) f is increasing on (a, b)　　　　**(b)** f is decreasing on (a, b)

We say that f is *increasing at a point c* if there exists an interval (a, b) containing c such that f is increasing on (a, b). Similarly, we say that f is *decreasing at a point c* if there exists an interval (a, b) containing c such that f is decreasing on (a, b).

Since the rate of change of a function at a point $x = c$ is given by the derivative of the function at that point, the derivative lends itself naturally to being a tool for determining the intervals where a differentiable function is increasing or decreasing. Indeed, as we saw in Chapter 2, the derivative of a function at a point measures both the slope of the tangent line to the graph of the function at that point and the rate of change of the function at the same point. In fact, at a point where the derivative is positive, the slope of the tangent line to the graph is positive and the function is increasing. At a point where the derivative is negative, the slope of the tangent line to the graph is negative and the function is decreasing (Figure 4.3).

FIGURE 4.3

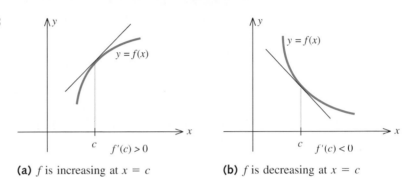

(a) f is increasing at $x = c$ **(b)** f is decreasing at $x = c$

These observations lead to the following important theorem:

THEOREM 1

a. If $f'(x) > 0$ for each value of x in an interval (a, b), then f is increasing on (a, b).
b. If $f'(x) < 0$ for each value of x in an interval (a, b), then f is decreasing on (a, b).
c. If $f'(x) = 0$ for each value of x in an interval (a, b), then f is constant on (a, b).

EXAMPLE 1   Find the interval where the function $f(x) = x^2$ is increasing and the interval where it is decreasing.

Solution   The derivative of $f(x) = x^2$ is $f'(x) = 2x$. Since

$$f'(x) = 2x > 0 \quad \text{when } x > 0$$

and $$f'(x) = 2x < 0 \quad \text{when } x < 0,$$

f is increasing on the interval $(0, \infty)$ and decreasing on the interval $(-\infty, 0)$ (Figure 4.4). ∎

FIGURE 4.4
The graph of f falls on $(-\infty, 0)$ where $f'(x) < 0$ and rises on $(0, \infty)$ where $f'(x) > 0$.

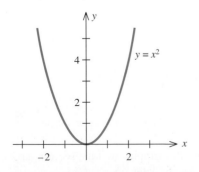

Recall that the graph of a continuous function cannot have any breaks. As a consequence, a continuous function cannot change sign unless it equals zero for some value of x. This observation suggests the following procedure for determining the sign of the derivative f' of a function f, and hence the intervals where the function f is increasing and where it is decreasing.

DETERMINING THE INTERVALS WHERE A FUNCTION IS INCREASING OR DECREASING	1. Find all values of x for which $f'(x) = 0$ or f' is discontinuous, and identify the open intervals determined by these points.
	
	2. Select a test point c in each interval found in Step 1 and determine the sign of $f'(c)$ in that interval.
	   a. If $f'(c) > 0$, f is increasing on that interval.
	   b. If $f'(c) < 0$, f is decreasing on that interval.

EXAMPLE 2   Determine the intervals where the function $f(x) = x^3 - 3x^2 - 24x + 32$ is increasing and where it is decreasing.

Solution

1. The derivative of f is

$$f'(x) = 3x^2 - 6x - 24 = 3(x + 2)(x - 4),$$

and it is continuous everywhere. The zeros of $f'(x)$ are $x = -2$ and $x = 4$.

2. To determine the sign of $f'(x)$ in the intervals $(-\infty, -2)$, $(-2, 4)$, and $(4, \infty)$, compute $f'(x)$ at a convenient test point in each interval. The results are shown in the following table.

Interval	Test point c	$f'(c)$	Sign of $f'(x)$
$(-\infty, -2)$	-3	21	$+$
$(-2, 4)$	0	-24	$-$
$(4, \infty)$	5	21	$+$

FIGURE 4.5
Sign diagram for f'

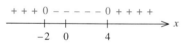

Using these results, we obtain the sign diagram shown in Figure 4.5. We conclude that f is increasing on the intervals $(-\infty, -2)$ and $(4, \infty)$ and is decreasing on the interval $(-2, 4)$. Figure 4.6 shows the graph of f.

FIGURE 4.6
The graph of f rises on $(-\infty, -2)$, falls on $(-2, 4)$, and rises again on $(4, \infty)$.

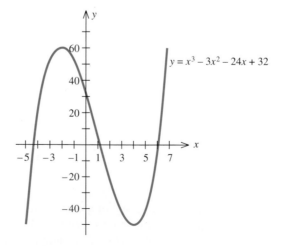

REMARK Do not be concerned with how the graphs in this section are obtained. We will learn how to sketch such graphs later on.   ☐

FIGURE 4.7
Sign diagram for f'

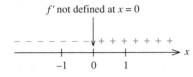

f' not defined at $x = 0$

FIGURE 4.8
f decreases on $(-\infty, 0)$ and increases on $(0, \infty)$.

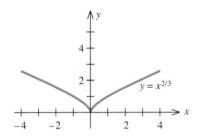

FIGURE 4.9
f' does not change sign as we move across $x = 0$.

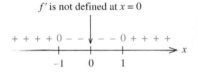

f' is not defined at $x = 0$

FIGURE 4.10
The graph of f rises on $(-\infty, -1)$, falls on $(-1, 0)$ and $(0, 1)$, and rises again on $(1, \infty)$.

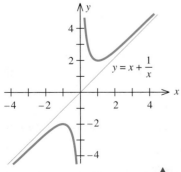

EXAMPLE 3 Find the interval where the function $f(x) = x^{2/3}$ is increasing and the interval where it is decreasing.

Solution
1. The derivative of f is

$$f'(x) = \frac{2}{3}x^{-1/3} = \frac{2}{3x^{1/3}}.$$

The function f' is not defined at $x = 0$, so f' is discontinuous there. It is continuous everywhere else. Furthermore, f' is not equal to zero anywhere. The point $x = 0$ divides the real line (the domain of f) into the intervals $(-\infty, 0)$ and $(0, \infty)$.

2. Pick a test point, say $x = -1$, in the interval $(-\infty, 0)$ and compute

$$f'(-1) = -\frac{2}{3}.$$

Since $f'(-1) < 0$, we see that $f'(x) < 0$ on $(-\infty, 0)$. Next, we pick a test point, say $x = 1$, in the interval $(0, \infty)$ and compute

$$f'(1) = \frac{2}{3}.$$

Since $f'(1) > 0$, we see that $f'(x) > 0$ on $(0, \infty)$. Figure 4.7 shows these results in the form of a sign diagram.

We conclude that f is decreasing on the interval $(-\infty, 0)$ and increasing on the interval $(0, \infty)$. The graph of f, shown in Figure 4.8, confirms these results. ∎

EXAMPLE 4 Find the intervals where the function $f(x) = x + 1/x$ is increasing and where it is decreasing.

Solution
1. The derivative of f is

$$f'(x) = 1 - \frac{1}{x^2} = \frac{x^2 - 1}{x^2}.$$

Since f' is not defined at $x = 0$, it is discontinuous there. Furthermore, $f'(x)$ is equal to zero when $x^2 - 1 = 0$ or $x = \pm 1$. These values of x partition the domain of f' into the open intervals $(-\infty, -1)$, $(-1, 0)$, $(0, 1)$, and $(1, \infty)$, where the sign of f' is different from zero.

2. To determine the sign of f' in each of these intervals, we compute $f'(x)$ at the test points $x = -2$, $-\frac{1}{2}$, $\frac{1}{2}$, and 2, respectively, obtaining $f'(-2) = \frac{3}{4}$, $f'\left(-\frac{1}{2}\right) = -3$, $f'\left(\frac{1}{2}\right) = -3$, and $f'(2) = \frac{3}{4}$. From the sign diagram for f' (Figure 4.9), we conclude that f is increasing on $(-\infty, -1)$ and $(1, \infty)$ and decreasing on $(-1, 0)$ and $(0, 1)$.

The graph of f appears in Figure 4.10. Note that f' does not change sign as we move across the point of discontinuity, $x = 0$. (Compare this with Example 3.) ∎

Example 4 reminds us that we must not automatically conclude that the derivative f' must change sign when we move across a point of discontinuity of f' or a zero of f'.

Applications

EXAMPLE 5   The altitude (in feet) of a rocket t seconds into flight is given by

$$s = f(t) = -t^3 + 96t^2 + 195t + 5 \qquad (t \geq 0).$$

a. Find the intervals where the function f is increasing and where it is decreasing.

b. Interpret the results obtained in (a).

Solution

a. The derivative of f is

$$v = f'(t) = -3t^2 + 192t + 195$$
$$= -3(t - 65)(t + 1),$$

FIGURE 4.11

Sign diagram for f'

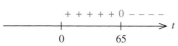

and it is equal to zero at $t = -1$ and $t = 65$. Since t must be nonnegative, we disregard the root $t = -1$. To determine the signs of $f'(t)$ in the intervals $(0, 65)$ and $(65, \infty)$, let us pick the test points $t = 1$ and $t = 70$. Since $f'(1) > 0$ and $f'(70) < 0$, we conclude that f is increasing on the interval $(0, 65)$ and decreasing on the interval $(65, \infty)$. Figure 4.11 shows the sign diagram for f', and the graph of f is sketched in Figure 4.12.

FIGURE 4.12

The altitude of the rocket after t seconds is given by $f(t)$.

FIGURE 4.13

The velocity of the rocket is positive on $(0, 65)$ when the rocket is ascending.

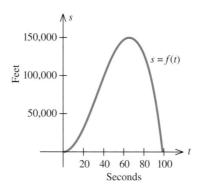

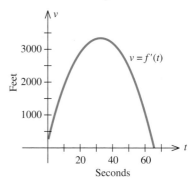

b. The derivative of the function f gives the rocket's velocity v at any time $t \geq 0$. The velocity curve is shown in Figure 4.13. Since $f'(t) > 0$ on the interval $(0, 65)$, the velocity is positive and the rocket is ascending. For $t > 65$, $f'(t) < 0$, and the rocket is descending. ∎

EXAMPLE 6   The profit function of the Acrosonic Company is given by

$$P(x) = -0.02x^2 + 300x - 200,000$$

dollars, where x is the number of Acrosonic Model F loudspeaker systems produced. Find where the function P is increasing and where it is decreasing.

FIGURE 4.14

The profit function is increasing on (0, 7500) *and decreasing on* (7500, ∞).

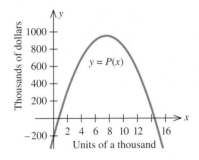

Solution The derivative P' of the function P is

$$P'(x) = -0.04x + 300 = -0.04(x - 7500).$$

Thus, $P'(x) = 0$ when $x = 7500$. Furthermore, $P'(x) > 0$ for x in the interval (0, 7500), and $P'(x) < 0$ for x in the interval (7500, ∞). This means that the profit function P is increasing on (0, 7500) and decreasing on (7500, ∞) [Figure 4.14]. ∎

EXAMPLE 7 The number of major crimes committed in the city of Bronxville from 1985 to 1992 is approximated by the function

$$N(t) = -0.1t^3 + 1.5t^2 + 100 \qquad (0 \le t \le 7),$$

where $N(t)$ denotes the number of crimes committed in year t and $t = 0$ corresponds to the beginning of 1985. Find where the function N is increasing and where it is decreasing.

Solution The derivative N' of the function N is

$$N'(t) = -0.3t^2 + 3t = -0.3t(t - 10).$$

Since $N'(t) > 0$ for t in the interval (0, 7), the function N is increasing throughout that interval (Figure 4.15).

FIGURE 4.15

The number of crimes, $N(t)$, was increasing over the seven-year interval.

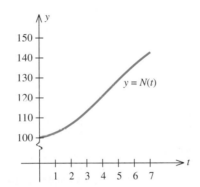

∎

SELF-CHECK EXERCISES 4.1

1. Find the intervals where the function $f(x) = \frac{2}{3}x^3 - x^2 - 12x + 3$ is increasing and the intervals where it is decreasing.

2. Certain proteins, known as enzymes, serve as catalysts for chemical reactions in living things. In 1913, Leonor Michaelis and L. M. Menten discovered the following formula giving the initial speed V (in moles per liter per second) at which the reaction begins in terms of the amount of substrate x (the substance that is being acted upon, measured in moles per liter):

$$V = \frac{ax}{x + b}$$

where a and b are positive constants. Show that the function V is always increasing and give a physical interpretation of this result.

Solutions to Self-Check Exercises 4.1 can be found on page 212.

4.1   EXERCISES

In Exercises 1–8, you are given the graph of some function f. Determine the intervals where f is increasing, constant, or decreasing.

1.

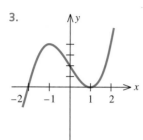

2.

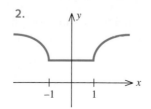

3.

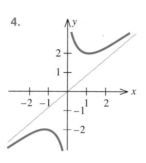

4.

5.

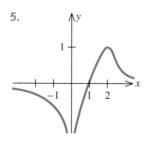

6.

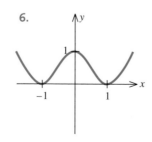

7.

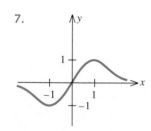

8.

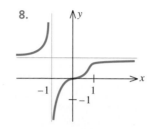

In Exercises 9–38, find the interval(s) where each of the given functions is increasing and the interval(s) where it is decreasing.

9. $f(x) = 3x + 5$

10. $f(x) = 4 - 5x$

11. $f(x) = x^2 - 3x$

12. $f(x) = 2x^2 + x + 1$

13. $g(x) = x - x^3$

14. $f(x) = x^3 - 3x^2$

15. $g(x) = x^3 + 3x^2 + 1$

16. $f(x) = x^3 - 3x + 4$

17. $g(x) = 2x^3 - 6x^2 + 7$

18. $h(x) = x^3 + 3x^2 - 9x + 8$

19. $f(x) = \frac{1}{3}x^3 - 3x^2 + 9x + 20$

20. $f(x) = \frac{2}{3}x^3 - 2x^2 - 6x - 2$

21. $h(x) = x^4 - 4x^3 + 10$

22. $g(x) = x^4 - 2x^2 + 4$

23. $f(x) = \frac{1}{x - 2}$

24. $h(x) = \frac{1}{2x + 3}$

25. $h(t) = \frac{t}{t - 1}$

26. $g(t) = \frac{2t}{t^2 + 1}$

27. $g(t) = t^2 + \frac{16}{t}$

28. $h(x) = \frac{x}{x + 1}$

29. $f(x) = x^{3/5}$

30. $f(x) = x^{2/3} + 5$

31. $f(x) = \sqrt{x + 1}$

32. $f(x) = (x - 5)^{2/3}$

33. $f(x) = \sqrt{16 - x^2}$

34. $g(x) = x\sqrt{x + 1}$

35. $f(x) = \frac{x^2 - 1}{x}$

36. $h(x) = \frac{x^2}{x - 1}$

37. $f(x) = \frac{1}{(x - 1)^2}$

38. $g(x) = \frac{x}{(x + 1)^2}$

39. A stone is thrown straight up from the roof of an 80-foot building. The distance of the stone from the ground at any time t (in seconds) is given by

$$h(t) = -16t^2 + 64t + 80$$

feet. When is the stone rising and when is it falling? If the stone were to miss the building, when would it hit the ground? [*Hint:* The stone is on the ground when $h(t) = 0$.] Sketch the graph of h.

40. Profit Functions   The Mexico subsidiary of the Thermo-Master Company manufactures an indoor-outdoor thermometer. Management estimates that the profit realizable by the company for the manufacture and sale of x units of thermometers per week is

$$P(x) = -0.001x^2 + 8x - 5000$$

dollars. Find the intervals where the profit function P is increasing and the intervals where P is decreasing.

41. **Flight of a Rocket**   The height (in feet) attained by a rocket t seconds into flight is given by the function

$$h(t) = -\frac{1}{3}t^3 + 16t^2 + 33t + 10.$$

When is the rocket rising and when is it descending?

42. **Environment of Forests**   Following the lead of the National Wildlife Federation, the Department of the Interior of a South American country began to record an index of environmental quality that measured progress and decline in the environmental quality of its forests. The index for the years 1984 through 1994 is approximated by the function

$$I(t) = \frac{1}{3}t^3 - \frac{5}{2}t^2 + 80 \qquad (0 \le t \le 10),$$

where $t = 0$ corresponds to the year 1984. Find the intervals where the function I is increasing and the intervals where it is decreasing. Interpret your results.

43. **Average Speed of a Highway Vehicle**   The average speed of a vehicle on a stretch of Route 134 between 6 A.M. and 10 A.M. on a typical weekday is approximated by the function

$$f(t) = 20t - 40\sqrt{t} + 50 \qquad (0 \le t \le 4),$$

where $f(t)$ is measured in miles per hour and t is measured in hours, with $t = 0$ corresponding to 6 A.M. Find the interval where f is increasing and the interval where f is decreasing and interpret your results.

44. **Average Cost**   The average cost in dollars incurred by the Lincoln Record Company per week in pressing x long-playing records is given by

$$\overline{C}(x) = -0.0001x + 2 + \frac{2000}{x} \qquad (0 < x \le 6000).$$

Show that $\overline{C}(x)$ is always decreasing over the interval $(0, 6000)$.

45. **Air Pollution**   According to the South Coast Air Quality Management District, the level of nitrogen dioxide, a brown gas that impairs breathing, present in the atmosphere on a certain May day in downtown Los Angeles is approximated by

$$A(t) = 0.03t^3(t - 7)^4 + 60.2 \qquad (0 \le t \le 7),$$

where $A(t)$ is measured in pollutant standard index (PSI) and t is measured in hours with $t = 0$ corresponding to 7 A.M. At what time of day is the air pollution increasing and at what time is it decreasing?

C 46. **Projected Social Security Funds**   Based on data from the Social Security Administration, the estimated cash

in the Social Security retirement and disability trust funds in the five decades beginning with the year 1990 is given by

$$A(t) = -96.6t^4 + 403.6t^3 + 660.9t^2 + 250$$
$$(0 \le t \le 5),$$

where $A(t)$ is measured in billions of dollars and t is measured in decades, with $t = 0$ corresponding to the year 1990. Find the interval where A is increasing and the interval where A is decreasing and interpret your results. [*Hint:* Use the quadratic formula.]

47. **Learning Curves**   The Emory Secretarial School finds from experience that the average student taking Advanced Typing will progress according to the rule

$$N(t) = \frac{60t + 180}{t + 6} \qquad (t \ge 0),$$

where $N(t)$ measures the number of words per minute the student can type after t weeks in the course. Compute $N'(t)$ and use this result to show that the function N is increasing on the interval $(0, \infty)$

48. **Drug Concentration in the Blood**   The concentration of a certain drug in a patient's body t hours after injection is given by

$$C(t) = \frac{t^2}{2t^3 + 1} \qquad (0 \le t \le 4)$$

milligrams per cubic centimeter. When is the concentration of the drug increasing and when is it decreasing?

49. **Air Pollution**   The amount of nitrogen dioxide, a brown gas that impairs breathing, present in the atmosphere on a certain May day in the city of Long Beach is approximated by

$$A(t) = \frac{136}{1 + 0.25(t - 4.5)^2} + 28 \qquad (0 \le t \le 11),$$

where $A(t)$ is measured in pollutant standard index (PSI) and t is measured in hours, with $t = 0$ corresponding to 7 A.M. Find the intervals where A is increasing and where A is decreasing and interpret your results.

50. **Prison Overcrowding**   The 1980s saw a trend toward old-fashioned punitive deterrence as opposed to the more liberal penal policies and community-based corrections popular in the 1960s and early 1970s. As a result, prisons became more crowded and the gap between the number of people in prison and the prison capacity widened. Based on figures from the U.S. Department of Justice, the number of prisoners (in

thousands) in federal and state prisons is approximated by the function

$$N(t) = 3.5t^2 + 26.7t + 436.2 \qquad (0 \le t \le 10),$$

where t is measured in years and $t = 0$ corresponds to 1984. The number of inmates for which prisons were designed is given by

$$C(t) = 24.3t + 365 \qquad (0 \le t \le 10),$$

where $C(t)$ is measured in thousands and t has the same meaning as before. Show that the gap between the number of prisoners and the number for which the prisons were designed has been widening at any time t. [Hint: First write a function G that gives the gap between the number of prisoners and the number for which the prisons were designed at any time t. Then show that $G'(t) > 0$ for all values of t in the interval (0, 10).]

51. **Oxygen Content of a Pond**   When organic waste is dumped into a pond, the oxidation process that takes place reduces the pond's oxygen content. However, in time, nature will restore the oxygen content to its natural level. Suppose that the oxygen content t days after organic waste has been dumped into the pond is given

by

$$f(t) = 100 \left[\frac{t^2 - 4t + 4}{t^2 + 4} \right] \qquad (0 \le t < \infty)$$

percent of its normal level. Find the intervals where the function f is increasing and the intervals where f is decreasing. Interpret your results.

52. Using Theorem 1, verify that the linear function $f(x) = mx + b$ is (a) increasing everywhere if $m > 0$, (b) decreasing everywhere if $m < 0$, and (c) constant if $m = 0$.

53. In what interval is the quadratic function

$$f(x) = ax^2 + bx + c \qquad (a \ne 0)$$

increasing? In what interval is f decreasing?

C *In Exercises 54–57, use a computer or a graphing calculator to plot the graph of f. Then use the graph to determine the intervals where f is increasing and the intervals where it is decreasing.*

54. $f(x) = 3x^5 - 10x^3$ 55. $f(x) = \dfrac{x^3}{x^2 - 3}$

56. $f(x) = x\sqrt{1 - x^2}$ 57. $f(x) = x - \sqrt{1 - x^2}$

SOLUTIONS TO SELF-CHECK EXERCISES 4.1

1. The derivative of f is

$$f'(x) = 2x^2 - 2x - 12 = 2(x + 2)(x - 3),$$

and it is continuous everywhere. The zeros of $f'(x)$ are $x = -2$ and $x = 3$. The sign diagram of f' is shown in the accompanying figure. We conclude that f is increasing on the intervals $(-\infty, -2)$ and $(3, \infty)$ and decreasing on the interval $(-2, 3)$.

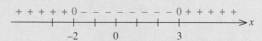

Sign diagram for f'

2. Using the Quotient Rule, we find

$$V'(x) = \frac{(x + b)\dfrac{d}{dx}(ax) - ax\dfrac{d}{dx}(x + b)}{(x + b)^2}$$

$$= \frac{(x + b)a - ax(1)}{(x + b)^2}$$

$$= \frac{ab}{(x + b)^2}.$$

Since a and b are positive constants and x is positive, we see that $V'(x) > 0$ for all x in the interval $(0, \infty)$. Therefore, V is always increasing, as we set out to show. Our results imply that the initial speed at which a chemical reaction begins increases as the amount of substrate increases.

4.2 Relative Maxima and Relative Minima

Relative Extrema

In the last section, we saw how the derivative of a function helps us determine where the graph of the function is increasing and where it is decreasing. In this section, we will see how the first derivative of a function may be used to help us locate certain "high points" and "low points" on the graph of f. Knowing these points is invaluable in sketching the graphs of functions and solving optimization problems. These "high points" and "low points" correspond to the *relative (local) maxima* and *relative minima* of a function. They are so called because they are the highest or the lowest points when compared to points nearby.

The graph shown in Figure 4.16 gives the U.S. budget deficit from 1980 ($t = 0$) through 1991. The relative maxima and the relative minima of the function f are indicated on the graph.

FIGURE 4.16
U.S. budget deficit from 1980 to 1991

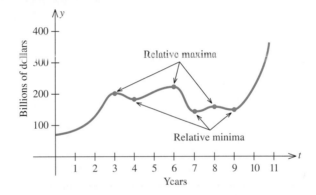

More generally, we have the following definition:

RELATIVE MAXIMUM

A function f has a **relative maximum** at $x = c$ if there exists an open interval (a, b) containing c such that $f(x) \leq f(c)$ for all x in (a, b).

Geometrically, this means that there is *some* interval containing $x = c$ such that no point on the graph of f with its x-coordinate in that interval can lie above the point $(c, f(c))$; that is, $f(c)$ is the largest value of $f(x)$ in some interval around $x = c$. Figure 4.17 depicts the graph of a function f that has a relative maximum at $x = x_1$ and another at $x = x_3$.

Observe that all the points on the graph of f with x-coordinates in the interval I_1 containing x_1 (shown in blue) lie on or below the point $(x_1, f(x_1))$. This is also true for the point $(x_3, f(x_3))$ and the interval I_3. Thus, even though there are points on the graph of f that are "higher" than the points $(x_1, f(x_1))$ and $(x_3, f(x_3))$, the latter points are "highest" relative to points in their respective neighborhoods (intervals). Points on the graph of a function f that are "highest" and "lowest" with respect to *all* points in the domain of f will be studied in Section 4.6.

FIGURE 4.17
f has a relative maximum at
$x = x_1$ *and* $x = x_3$.

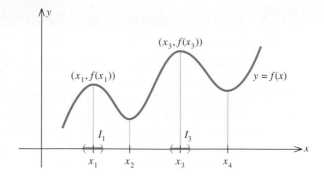

The definition of the relative minimum of a function parallels that of the relative maximum of a function.

RELATIVE MINIMUM

A function f has a **relative minimum** at $x = c$ if there exists an open interval (a, b) containing c such that $f(x) \geq f(c)$ for all x in (a, b).

The graph of the function f, depicted in Figure 4.17, has a relative minimum at $x = x_2$ and another at $x = x_4$.

Finding the Relative Extrema

We refer to the relative maximum and relative minimum of a function as the **relative extrema** of that function. As a first step in our quest for the relative extrema of a function, we will consider functions that have derivatives at such points. Suppose that f is a function that is differentiable in some interval (a, b) that contains a point $x = c$, and that f has a relative maximum at $x = c$ (Figure 4.18a).

Observe that the slope of the tangent line to the graph of f must change from positive to negative as we move across the point $x = c$ from left to right. Therefore, the tangent line to the graph of f at the point $(c, f(c))$ must be horizontal; that is, $f'(c) = 0$ (Figure 4.18a).

FIGURE 4.18

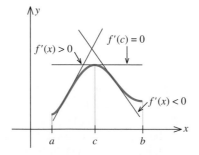

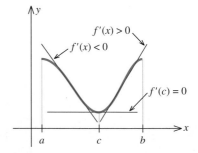

(a) f has a relative maximum at $x = c$

(b) f has a relative minimum at $x = c$

Using a similar argument, it may be shown that the derivative f' of a differentiable function f must also be equal to zero at a point $x = c$, where f has a relative minimum (Figure 4.18b).

This analysis reveals an important characteristic of the relative extrema of a differentiable function f: *At any point c where f has a relative extremum,* $f'(c) = 0$.

Before we develop a procedure for finding such points, a few words of caution are in order.

FIGURE 4.19

$f'(0) = 0$, *but f does not have a relative extremum at* $(0, 0)$.

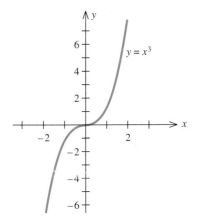

First, this result tells us that if a differentiable function f has a relative extremum at a point $x = c$, then $f'(c) = 0$. The converse of this statement—if $f'(c) = 0$ at some point $x = c$, then f must have a relative extremum at that point—is not true. Consider, for example, the function $f(x) = x^3$. Here $f'(x) = 3x^2$, so $f'(0) = 0$. Yet f has neither a relative maximum nor a relative minimum at $x = 0$ (Figure 4.19).

Second, our result assumes that the function is differentiable and thus has a derivative at a point that gives rise to a relative extremum. The functions $f(x) = |x|$ and $g(x) = x^{2/3}$ demonstrate that a relative extremum of a function may exist at a point at which the derivative does not exist. Both these functions fail to be differentiable at $x = 0$, but each has a relative minimum there. Figure 4.20 shows the graphs of these functions. Note that the slopes of the tangent lines change from negative to positive as we move across $x = 0$, just as in the case of a function that is differentiable at a value of x that gives rise to a relative minimum.

FIGURE 4.20

Each of these functions has a relative extremum at $(0, 0)$, *but the derivative does not exist there.*

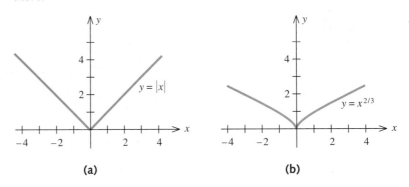

(a)　　　　　　　　　　(b)

We refer to a point in the domain of f that *may* give rise to a relative extremum as a *critical point*.

CRITICAL POINT OF f	A **critical point** of a function f is any point x in the domain of f such that $f'(x) = 0$ or $f'(x)$ does not exist.

Figure 4.21 depicts the graph of a function that has critical points at $x = a$, b, c, d, and e. Observe that $f'(x) = 0$ at $x = a$, b, and c. Next, since there is a corner at $x = d$, $f'(x)$ does not exist there. Finally, $f'(x)$ does not exist at $x = e$, because the tangent line there is vertical. Also, observe that the critical

points $x = a$, b, and d give rise to relative extrema of f, whereas the critical points $x = c$ and $x = e$ do not.

FIGURE 4.21
Critical points of f

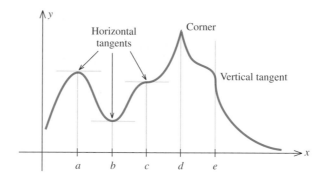

Having defined what a critical point is, we can now state a formal procedure for finding the relative extrema of a continuous function that is differentiable everywhere except at isolated values of x. Incorporated into the procedure is the so-called *First Derivative Test*, which enables us to determine whether or not a point gives rise to a relative maximum or a relative minimum of the function f.

PROCEDURE FOR FINDING THE RELATIVE EXTREMA (THE FIRST DERIVATIVE TEST)

1. Determine the critical points of f.
2. Determine the sign of $f'(x)$ to the left and right of each critical point.
   a. If $f'(x)$ changes sign from *positive* to *negative* as we move across a critical point $x = c$, then $f(c)$ is a relative maximum.
   b. If $f'(x)$ changes sign from *negative* to *positive* as we move across a critical point $x = c$, then $f(c)$ is a relative minimum.
   c. If $f'(x)$ does not change sign as we move across a critical point $x = c$, then $f(c)$ is not a relative extremum.

EXAMPLE 1   Find the relative maxima and relative minima of the function $f(x) = x^2$.

FIGURE 4.22
f has a relative minimum at $x = 0$.

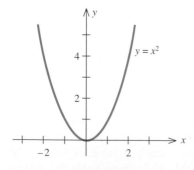

Solution   The derivative of $f(x) = x^2$ is given by $f'(x) = 2x$. Setting $f'(x) = 0$ yields $x = 0$ as the only critical point of f. Since

$$f'(x) < 0 \quad \text{if } x < 0$$

and

$$f'(x) > 0 \quad \text{if } x > 0,$$

we see that $f'(x)$ changes sign from negative to positive as move across the critical point $x = 0$. Thus we conclude that $f(0) = 0$ is a relative minimum of f (Figure 4.22). ∎

EXAMPLE 2   Find the relative maxima and relative minima of the function $f(x) = x^{2/3}$ (see Example 3, Section 4.1).

FIGURE 4.23

Sign diagram for f′

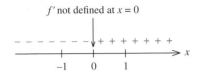

f′ not defined at x = 0

FIGURE 4.24

f has a relative minimum at x = 0.

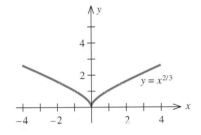

Solution The derivative of f is $f'(x) = (2/3)x^{-1/3}$. As noted in Example 3 (Section 4.1), f' is not defined at $x = 0$, is continuous everywhere else, and is not equal to zero in its domain. Thus, $x = 0$ is the only critical point of the function f.

The sign diagram obtained in Example 3 is reproduced in Figure 4.23. We can see that the sign of $f'(x)$ changes from negative to positive as we move across $x = 0$ from left to right. Thus, an application of the First Derivative Test tells us that $f(0) = 0$ is a relative minimum of f (Figure 4.24). ∎

EXAMPLE 3 Find the relative maxima and relative minima of the function

$$f(x) = x^3 - 3x^2 - 24x + 32.$$

Solution The derivative of f is

$$f'(x) = 3x^2 - 6x - 24 = 3(x + 2)(x - 4),$$

and it is continuous everywhere. The zeros of $f'(x)$, $x = -2$ and $x = 4$, are the only critical points of the function f. The sign diagram for f' is shown in Figure 4.25. Examine the two critical points $x = -2$ and $x = 4$ for a relative extremum using the First Derivative Test and the sign diagram for f':

1. *The critical point $x = -2$:* Since the function $f'(x)$ changes sign from positive to negative as we move across $x = -2$ from left to right, we conclude that a relative maximum of f occurs at $x = -2$. The value of $f(x)$ when $x = -2$ is

$$f(-2) = (-2)^3 - 3(-2)^2 - 24(-2) + 32 = 60.$$

2. *The critical point $x = 4$:* $f'(x)$ changes sign from negative to positive as we move across $x = 4$ from left to right, so $f(4) = -48$ is a relative minimum of f. The graph of f appears in Figure 4.26.

FIGURE 4.25

Sign diagram for f′

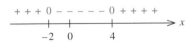

FIGURE 4.26

f has a relative maximum at x = −2 and a relative minimum at x = 4.

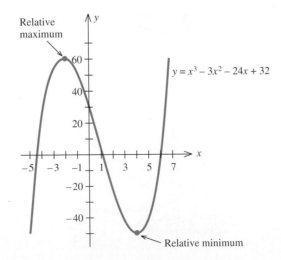

FIGURE 4.27

$x = 0$ *is not a critical point, because f is not defined at* $x = 0$.

f' is not defined at $x = 0$

FIGURE 4.28

$f(x) = x + \dfrac{1}{x}$.

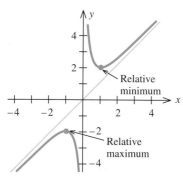

EXAMPLE 4   Find the relative maxima and the relative minima of the function

$$f(x) = x + \frac{1}{x}.$$

Solution   The derivative of f is

$$f'(x) = 1 - \frac{1}{x^2} = \frac{x^2 - 1}{x^2} = \frac{(x + 1)(x - 1)}{x^2}.$$

Since f' is equal to zero at $x = -1$ and $x = 1$, these are critical points for the function f. Next, observe that f' is discontinuous at $x = 0$. However, because *f is not defined at that point,* the point $x = 0$ does not qualify as a critical point of f. Figure 4.27 shows the sign diagram for f'.

Since $f'(x)$ changes sign from positive to negative as we move across $x = -1$ from left to right, the First Derivative Test implies that $f(-1) = -2$ is a relative maximum of the function f. Next, $f'(x)$ changes sign from negative to positive as we move across $x = 1$ from left to right, so $f(1) = 2$ is a relative minimum of the function f. The graph of f appears in Figure 4.28. Note that this function has a relative maximum that lies below its relative minimum.

■

EXAMPLE 5   Find the relative maxima and relative minima of the function

$$f(x) = 2x - 3x^{2/3}.$$

Solution   The derivative of f is

$$f'(x) = 2 - 3\left(\frac{2}{3}x^{-1/3}\right) = 2 - \frac{2}{x^{1/3}}$$

$$= \frac{2x^{1/3} - 2}{x^{1/3}} = \frac{2(x^{1/3} - 1)}{x^{1/3}}$$

FIGURE 4.29

Sign diagram for f'

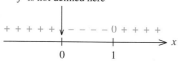

f' is not defined here

and is continuous everywhere except at $x = 0$. Furthermore, $f'(x) = 0$ if $x = 1$. Therefore, the critical points of f are $x = 0$ and $x = 1$. The sign diagram of f is shown in Figure 4.29.

Since $f'(x)$ changes sign from positive to negative as we move across $x = 0$, we see that a relative maximum of f occurs at the point $x = 0$. Its value is $f(0) = 0$. Next, observe that $f'(x)$ changes sign from negative to positive as we move across $x = 1$. So f has a relative minimum at $x = 1$ with value $f(1) = 2 - 3 = -1$. The graph of f is shown in Figure 4.30.

FIGURE 4.30

The graph of $f(x) = 2x - 3x^{2/3}$ *has a relative maximum at* $x = 0$ *and a relative minimum at* $x = 1$.

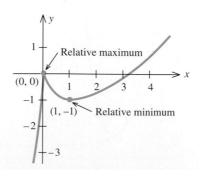

■

SELF-CHECK EXERCISES 4.2

1. Find the relative extrema of the function $f(x) = x^3 + \frac{7}{2}x^2 - 6x - 3$.

2. Find the relative extrema of $f(x) = \dfrac{x^2}{1 - x^2}$.

Solutions to Self-Check Exercises 4.2 can be found on page 220.

4.2   EXERCISES

In Exercises 1–8, you are given the graph of a function f. Determine the relative maxima and relative minima, if any.

1.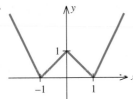

2.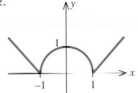

3.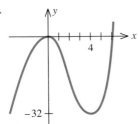

4.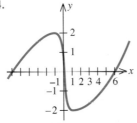

5.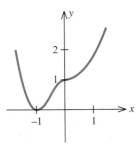

6.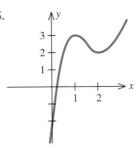

7.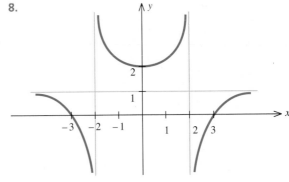

8.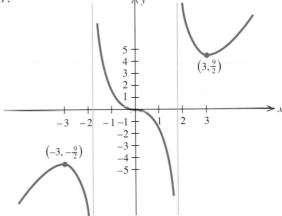

In Exercises 9–34, find the relative maxima and relative minima, if any, of each function.

9. $f(x) = x^2 - 4x$

10. $g(x) = x^2 + 3x + 8$

11. $f(x) = \frac{1}{2}x^2 - 2x + 4$

12. $h(t) = -t^2 + 6t + 6$

13. $f(x) = x^{2/3} + 2$

14. $f(x) = x^{5/3}$

15. $g(x) = x^3 - 3x^2 + 4$

16. $f(x) = x^3 - 3x + 6$

17. $F(x) = \frac{1}{3}x^3 - x^2 - 3x + 4$

18. $g(x) = 2x^3 - \frac{1}{2}x^2 - 2x + 4$

19. $f(x) = 2x^3 - \frac{1}{2}x^2 - x + 2$

20. $f(x) = \frac{1}{2}x^4 - x^2$

21. $g(x) = x^4 - 4x^3 + 8$

22. $h(x) = \frac{1}{2}x^4 - 3x^2 + 4x - 8$

23. $f(x) = 3x^4 - 2x^3 + 4$

24. $F(t) = 3t^5 - 20t^3 + 20$

25. $g(x) = \dfrac{x + 1}{x}$

26. $h(x) = \dfrac{x}{x + 1}$

27. $f(x) = x + \dfrac{9}{x} + 2$ **28.** $g(x) = 2x^2 + \dfrac{4000}{x} + 10$

29. $f(x) = \dfrac{x}{1 + x^2}$ **30.** $g(x) = \dfrac{x}{x^2 - 1}$

31. $f(x) = \dfrac{x^2}{x^2 - 4}$ **32.** $g(t) = \dfrac{t^2}{1 + t^2}$

33. $f(x) = (x - 1)^{2/3}$ **34.** $g(x) = x\sqrt{x - 4}$

35. Show that the quadratic function

$$f(x) = ax^2 + bx + c \qquad (a \neq 0)$$

has a relative extremum when $x = -b/2a$. Also, show that the relative extremum is a relative maximum if $a < 0$ and a relative minimum if $a > 0$.

36. Show that the cubic function

$$f(x) = ax^3 + bx^2 + cx + d \qquad (a \neq 0)$$

has no relative extremum if and only if $b^2 - 3ac \leq 0$.

C *In Exercises 37–40, use a computer or a graphing calculator to plot the graph of f. Then use the graph to find the approximate locations of the relative extrema of f.*

37. $f(x) = x^4 - x^3 + 3x^2 - 2x + 2$

38. $f(x) = x^5 - 2x^4 + 3x^2 - 2x + 1$

39. $f(x) = x - \sqrt{1 - x^2}$

40. $f(x) = \dfrac{\sqrt{x}(x^2 - 1)^2}{x - 2}$

SOLUTIONS TO SELF-CHECK EXERCISES 4.2

1. The derivative of f is

$$f'(x) = 3x^2 + 7x - 6 = (3x - 2)(x + 3),$$

and it is continuous everywhere. The zeros of $f'(x)$ are $x = -3$ and $x = \frac{2}{3}$, both of which are critical points of f. From the sign diagram for f', we see that a relative maximum of f occurs at $x = -3$ and a relative minimum of f occurs at $x = \frac{2}{3}$. Since

$$+ + + + 0 - - - - - - - - 0 + + + + + + + + +$$

Sign diagram for f'

$$f(-3) = (-3)^3 + \frac{7}{2}(-3)^2 - 6(-3) - 3$$

$$= \frac{39}{2}$$

and

$$f\left(\frac{2}{3}\right) = \left(\frac{2}{3}\right)^3 + \frac{7}{2}\left(\frac{2}{3}\right)^2 - 6\left(\frac{2}{3}\right) - 3$$

$$= -\frac{139}{27},$$

we conclude that $f(-3) = \frac{39}{2}$ is a relative maximum of f and $f\left(\frac{2}{3}\right) = -\frac{139}{27}$ is a relative minimum of f.

2. The derivative of f is

$$f'(x) = \frac{(1 - x^2)\dfrac{d}{dx}(x^2) - x^2\dfrac{d}{dx}(1 - x^2)}{(1 - x^2)^2}$$

$$= \frac{(1 - x^2)(2x) - x^2(-2x)}{(1 - x^2)^2} = \frac{2x}{(1 - x^2)^2},$$

and it is continuous everywhere except at $x = \pm 1$. Since $f'(x)$ is equal to zero at $x = 0$, $x = 0$ is a critical point of f. Next, observe that $f'(x)$ is discontinuous at $x = \pm 1$, but since these points are not in the domain of f, they do not qualify as critical points of f. Finally, from the sign diagram of f' shown in the accompanying figure, we conclude that $f(0) = 0$ is a relative minimum of f.

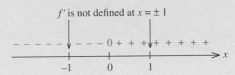

Sign diagram for f'

4.3   Concavity and Points of Inflection

Determining the Intervals of Concavity

Consider the graphs shown in Figure 4.31, which give the estimated population of the world and of the United States through the year 2000. Both graphs are rising, indicating that both the U.S. population and the world population will continue to increase through the year 2000. But observe that the graph in Figure 4.31a opens upward, whereas the graph in Figure 4.31b opens downward. What is the significance of this? To answer this question, let us look at the slopes of the tangent lines to various points on each graph (Figure 4.32).

In Figure 4.32a we see that the slopes of the tangent lines to the graph are increasing as we move from left to right. Since the slope of the tangent line to the graph at a point on the graph measures the rate of change of the function at that point, we conclude that the world population is not only increasing through the year 2000 but is increasing at an *increasing* pace. A similar analysis of Figure 4.32b reveals that the U.S. population is increasing, but at a *decreasing* pace.

FIGURE 4.31

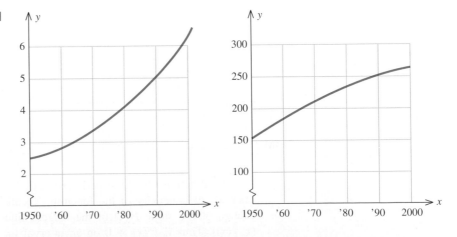

(a) World population in billions   **(b)** U.S. population in millions

FIGURE 4.32

(a) Slopes of tangent lines are increasing

(b) Slopes of tangent lines are decreasing

The shape of a curve can be described using the notion of *concavity:*

CONCAVITY OF A FUNCTION *f*

Let the function *f* be differentiable on an interval (a, b). Then

1. *f* is **concave upward** on (a, b) if f' is increasing on (a, b).

2. *f* is **concave downward** on (a, b) if f' is decreasing on (a, b).

Geometrically, a curve is concave upward if it lies above its tangent lines (Figure 4.33a). Similarly, a curve is concave downward if it lies below its tangent lines (Figure 4.33b).

We also say that *f* is **concave upward at a point** $x = c$ if there exists an interval (a, b) containing *c* in which *f* is concave upward. Similarly, we say that *f* is **concave downward at a point** $x = c$ if there exists an interval (a, b) containing *c* in which *f* is concave downward.

FIGURE 4.33

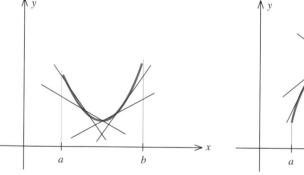

(a) *f* is concave upward on (a, b)

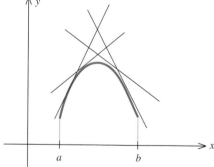

(b) *f* is concave downward on (a, b)

If a function *f* has a second derivative f'', we can use f'' to determine the intervals of concavity of the function. Recall that $f''(x)$ measures the rate of change of the slope $f'(x)$ of the tangent line to the graph of *f* at the point $(x, f(x))$. Thus if $f''(x) > 0$ on an interval (a, b), then the slopes of the tangent lines to the graph of *f* are increasing on (a, b) and so *f* is concave upward on (a, b). Similarly, if $f''(x) < 0$ on (a, b), then f is concave downward on (a, b). These observations suggest the following theorem.

THEOREM 2

a. If $f''(x) > 0$ for each value of x in (a, b), then f is concave upward on (a, b).

b. If $f''(x) < 0$ for each value of x in (a, b), then f is concave downward on (a, b).

The following procedure, based on the conclusions of Theorem 2, may be used to determine the intervals of concavity of a function.

DETERMINING THE INTERVALS OF CONCAVITY OF f

1. Determine the values of x for which f'' is zero or where f'' is not defined, and identify the open intervals determined by these points.

2. Determine the sign of f'' in each interval found in Step 1. To do this, compute $f''(c)$, where c is any conveniently chosen test point in the interval.

   a. If $f''(c) > 0$, f is concave upward on that interval.

   b. If $f''(c) < 0$, f is concave downward on that interval.

EXAMPLE 1   Determine where the function $f(x) = x^3 - 3x^2 - 24x + 32$ is concave upward and where it is concave downward.

Solution   Here,

$$f'(x) - 3x^2 - 6x - 24,$$

so

$$f''(x) = 6x - 6 = 6(x - 1),$$

and f'' is defined everywhere. Setting $f''(x) = 0$ gives $x = 1$. The sign diagram of f'' appears in Figure 4.34. We conclude that f is concave downward on the interval $(-\infty, 1)$ and is concave upward on the interval $(1, \infty)$. Figure 4.35 shows the graph of f.

FIGURE 4.34
Sign diagram for f''

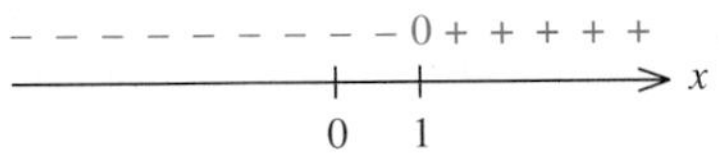

```
- - - - - - - - - -0+ + + + +
————————————+———+————————→ x
             0   1
```

FIGURE 4.35
f is concave down on $(-\infty, 1)$ and concave upward on $(1, \infty)$.

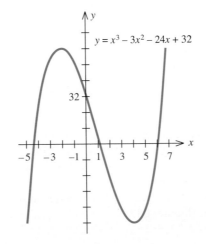

$y = x^3 - 3x^2 - 24x + 32$

EXAMPLE 2   Determine the intervals where the function $f(x) = x + 1/x$ is concave upward and where it is concave downward.

FIGURE 4.36

The sign diagram for f''

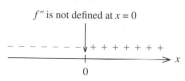

f'' is not defined at $x = 0$

Solution   We have

$$f'(x) = 1 - \frac{1}{x^2}$$

and

$$f''(x) = \frac{2}{x^3}.$$

We deduce from the sign diagram for f'' (Figure 4.36) that the function f is concave downward on the interval $(-\infty, 0)$ and concave upward on the interval $(0, \infty)$. The graph of f is sketched in Figure 4.37.

FIGURE 4.37

f is concave down on $(-\infty, 0)$ and concave up on $(0, \infty)$.

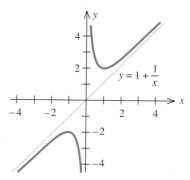

$y = 1 + \dfrac{1}{x}$

Inflection Points

FIGURE 4.38

The graph of S has a point of inflection at $(50, 2700)$.

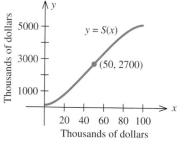

Thousands of dollars

Figure 4.38 shows the total sales S of a manufacturer of automobile air conditioners versus the amount of money x that the company spends on advertising its product. Notice that the graph of the continuous function $y = S(x)$ changes concavity—from upward to downward—at the point $(50, 2700)$. This point is called an *inflection point* of S. To understand the significance of this inflection point, observe that the total sales increase rather slowly at first, but as more money is spent on advertising, the total sales increase rapidly. This rapid increase reflects the effectiveness of the company's ads. However, a point is soon reached after which any additional advertising expenditure results in increased sales but at a slower rate of increase. This point, commonly known as the **point of diminishing returns,** is the point of inflection of the function S. We will return to this example later.

Let us now state formally the definition of an inflection point.

INFLECTION POINT   A point on the graph of a differentiable function f at which the concavity changes is called an **inflection point.**

Observe that the graph of a function crosses its tangent line at a point of inflection (Figure 4.39).

FIGURE 4.39
At each point of inflection, the graph of a function crosses its tangent line.

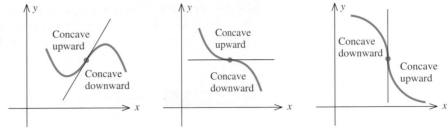

The following procedure may be used to find inflection points.

FINDING INFLECTION POINTS

1. Compute $f''(x)$.
2. Determine the points in the domain of f for which $f''(x) = 0$ or $f''(x)$ does not exist.
3. Determine the sign of $f''(x)$ to the left and right of each point $x = c$ found in Step 2. If there is a change in the sign of $f''(x)$ as we move across the point $x = c$, then $(c, f(c))$ is an inflection point of f.

The points determined in Step 2 are only *candidates* for the inflection points of f. For example, you can easily verify that $f''(0) = 0$ if $f(x) = x^4$, but a sketch of the graph of f will show that $(0, 0)$ is not an inflection point of f.

FIGURE 4.40
Sign diagram for f''

FIGURE 4.41
f has an inflection point at $(0, 0)$.

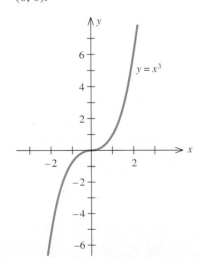

EXAMPLE 3   Find the points of inflection of the function $f(x) = x^3$.

Solution

$$f'(x) = 3x^2,$$

so

$$f''(x) = 6x.$$

Observe that f'' is continuous everywhere and is zero if $x = 0$. The sign diagram of f'' is shown in Figure 4.40. From this diagram, we see that $f''(x)$ changes sign as we move across $x = 0$. Thus the point $(0, 0)$ is an inflection point of the function f (Figure 4.41). ∎

EXAMPLE 4   Determine the intervals where the function $f(x) = (x - 1)^{5/3}$ is concave upward and where it is concave downward and find the inflection points of f.

Solution   The first derivative of f is

$$f'(x) = \frac{5}{3}(x - 1)^{2/3},$$

and the second derivative of f is

$$f''(x) = \frac{10}{9}(x - 1)^{-1/3} = \frac{10}{9(x - 1)^{1/3}}.$$

We see that f'' is not defined at $x = 1$. Furthermore, $f''(x)$ is not equal to zero anywhere. The sign diagram of f'' is shown in Figure 4.42. From the sign

FIGURE 4.42
Sign diagram for f″

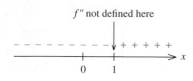

f″ not defined here

FIGURE 4.43
*f has an inflection point at
(1, 0).*

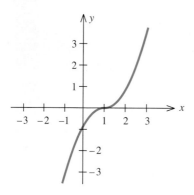

FIGURE 4.44
Sign diagram for f″

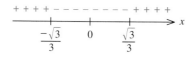

FIGURE 4.45
*The graph of f(x) = 1/(x² + 1)
is concave upward on
(−∞, −√3/3) ∪ (√3/3, ∞)
and concave downward on
(−√3/3, √3/3).*

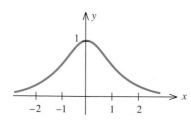

diagram, we see that f is concave downward on $(-\infty, 1)$ and concave upward on $(1, \infty)$. Next, since $x = 1$ does lie in the domain of f, our computations also reveal that the point $(1, 0)$ is an inflection point of f (Figure 4.43).   ■

EXAMPLE 5   Determine the intervals where the function

$$f(x) = \frac{1}{x^2 + 1}$$

is concave upward and where it is concave downward and find the inflection points of f.

Solution   The first derivative of f is

$$f'(x) = \frac{d}{dx}(x^2 + 1)^{-1} = -2x(x^2 + 1)^{-2} \quad \text{(Using the General Power Rule)}$$

$$= -\frac{2x}{(x^2 + 1)^2}.$$

Next, using the Quotient Rule, we find

$$f''(x) = \frac{(x^2 + 1)^2(-2) + (2x)2(x^2 + 1)(2x)}{(x^2 + 1)^4}$$

$$= \frac{(x^2 + 1)[-2(x^2 + 1) + 8x^2]}{(x^2 + 1)^4} = \frac{(x^2 + 1)(6x^2 - 2)}{(x^2 + 1)^4}$$

$$= \frac{2(3x^2 - 1)}{(x^2 + 1)^3}. \quad \text{(Canceling the common factors)}$$

Observe that f'' is continuous everywhere and is zero if

$$3x^2 - 1 = 0$$

$$x^2 = \frac{1}{3},$$

or $x = \pm\sqrt{3}/3$. The sign diagram for f'' is shown in Figure 4.44. From the sign diagram for f'', we see that f is concave upward on $(-\infty, -\sqrt{3}/3) \cup (\sqrt{3}/3, \infty)$ and concave downward on $(-\sqrt{3}/3, \sqrt{3}/3)$. Also, observe that $f''(x)$ changes sign as we move across the points $x = -\sqrt{3}/3$ and $x = \sqrt{3}/3$. Since

$$f\left(-\frac{\sqrt{3}}{3}\right) = \frac{1}{\frac{1}{3} + 1} = \frac{3}{4} \quad \text{and} \quad f\left(\frac{\sqrt{3}}{3}\right) = \frac{3}{4},$$

we see that the points $(-\sqrt{3}/3, 3/4)$ and $(\sqrt{3}/3, 3/4)$ are inflection points of f. The graph of f is shown in Figure 4.45.   ■

Applications

Examples 6 and 7 illustrate familiar interpretations of the significance of the inflection point of a function.

EXAMPLE 6   The total sales S, in thousands of dollars, of the Arctic Air Corporation, a manufacturer of automobile air conditioners, is related to the amount of money x the company spends on advertising its products by the formula

$$S = -0.01x^3 + 1.5x^2 + 200 \qquad (0 \le x \le 100),$$

where x is measured in thousands of dollars. Find the inflection point of the function S.

Solution   The first two derivatives of S are given by

$$S' = -0.03x^2 + 3x$$

and

$$S'' = -0.06x + 3.$$

Setting $S'' = 0$ gives $x = 50$ as the only candidate for an inflection point of S. Moreover, since

$$S'' > 0 \quad \text{for} \quad x < 50$$

and

$$S'' < 0 \quad \text{for} \quad x > 50,$$

the point $(50, 2700)$ is an inflection point of the function S. The graph of S appears in Figure 4.46. Notice that this is the graph of the function we discussed earlier. ∎

FIGURE 4.46

The graph of $S(x)$ has a point of inflection at $(50, 2700)$.

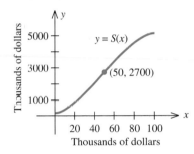

EXAMPLE 7   An economy's consumer price index (CPI) is described by the function

$$I(t) = -0.2t^3 + 3t^2 + 100 \qquad (0 \le t \le 9),$$

where $t = 0$ corresponds to the year 1984. Find the point of inflection of the function I and discuss its significance.

Solution   The first two derivatives of I are given by

$$I'(t) = -0.6t^2 + 6t$$

and

$$I''(t) = -1.2t + 6 = -1.2(t - 5).$$

Setting $I''(t) = 0$ gives $t = 5$ as the only candidate for an inflection point of I. Next, we observe that

$$I'' > 0 \quad \text{for} \quad t < 5$$

and

$$I'' < 0 \quad \text{for} \quad t > 5,$$

so the point $(5, 150)$ is an inflection point of I. The graph of I is sketched in Figure 4.47.

Since the second derivative of I measures the rate of change of the inflation rate, our computations reveal that the rate of inflation had, in fact, peaked at $t = 5$. Thus, relief actually began at the beginning of 1989. ∎

FIGURE 4.47

The graph of $I(t)$ has a point of inflection at $(5, 150)$.

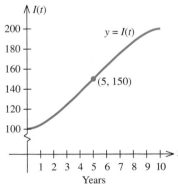

The Second Derivative Test

We will now show how the second derivative f'' of a function f can be used to help us determine whether a critical point of f is a relative extremum of f.

Figure 4.48a shows the graph of a function that has a relative maximum at $x = c$. Observe that f is concave downward at that point. Similarly, Figure 4.48b shows that at a relative minimum of f the graph is concave upward.

FIGURE 4.48

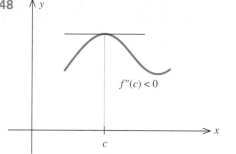

(a) f has a relative maximum at $x = c$ (b) f has a relative minimum at $x = c$

But from our previous work we know that f is concave downward at $x = c$ if $f''(c) < 0$ and f is concave upward at $x = c$ if $f''(c) > 0$. These observations suggest the following alternative procedure for determining whether a critical point of f gives rise to a relative extremum of f. This result is called the **Second Derivative Test** and is applicable when f'' exists.

THE SECOND DERIVATIVE TEST

1. Compute $f'(x)$ and $f''(x)$.
2. Find all the critical points of f at which $f'(x) = 0$.
3. Compute $f''(c)$ for each such critical point c.
   a. If $f''(c) < 0$, then f has a relative maximum at c.
   b. If $f''(c) > 0$, then f has a relative minimum at c.
   c. If $f''(c) = 0$, the test fails; that is, it is inconclusive.

REMARK   As stated in 3(c), the Second Derivative Test does not yield a conclusion if $f''(c) = 0$ or if $f''(c)$ does not exist. In other words, $x = c$ may give rise to a relative extremum or an inflection point (see Exercise 83, page 234). In such cases, you should revert to the First Derivative Test.   □

EXAMPLE 8   Determine the relative extrema of the function

$$f(x) = x^3 - 3x^2 - 24x + 32$$

using the Second Derivative Test. (See Example 3, Section 4.2.)

Solution   We have

$$f'(x) = 3x^2 - 6x - 24 = 3(x + 2)(x - 4),$$

so $f'(x) = 0$ gives $x = -2$ and $x = 4$, the critical points of f, as in Example 3. Next, we compute

$$f''(x) = 6x - 6 = 6(x - 1).$$

Since

$$f''(-2) = 6(-2 - 1) = -18 < 0,$$

the Second Derivative Test implies that $f(-2) = 60$ is a relative maximum of f. Also,

$$f''(4) = 6(4 - 1) = 18 > 0,$$

and the Second Derivative Test implies that $f(4) = -48$ is a relative minimum of f, which confirms the results obtained earlier.   ∎

Comparing the First and Second Derivative Tests

Notice that both the First Derivative Test and the Second Derivative Test are used to classify the critical points of f. What are the pros and cons of the two tests? Since the Second Derivative Test is applicable only when f'' exists, it is less versatile than the First Derivative Test. For example, it cannot be used to locate the relative minimum $f(0) = 0$ of the function $f(x) = x^{2/3}$.

Furthermore, the Second Derivative Test is inconclusive when f'' is equal to zero at a critical point of f, whereas the First Derivative Test always yields positive conclusions. The Second Derivative Test is also inconvenient to use when f'' is difficult to compute. On the plus side, if it is easy to compute f'', then we use the Second Derivative Test, since it involves just the evaluation of f'' at the critical point(s) of f. Also, the conclusions of the Second Derivative Test are important in theoretical work.

We close this section by summarizing the different roles played by the first derivative f' and the second derivative f'' of a function f in determining the properties of the graph of f. The first derivative f' tells us where f is increasing and where f is decreasing, whereas the second derivative f'' tells us where f is concave upward and where f is concave downward. These different properties of f are reflected by the signs of f' and f'' in the interval of interest. The following table shows the general characteristics of the function f for various possible combinations of the signs of f' and f'' in the interval (a, b).

Signs of f' and f''	Properties of the graph of f	General shape of the graph of f
$f'(x) > 0$ $f''(x) > 0$	f increasing f concave upward	
$f'(x) > 0$ $f''(x) < 0$	f increasing f concave downward	
$f'(x) < 0$ $f''(x) > 0$	f decreasing f concave upward	
$f'(x) < 0$ $f''(x) < 0$	f decreasing f concave downward	

**SELF-CHECK
EXERCISES 4.3**

1. Determine where the function $f(x) = 4x^3 - 3x^2 + 6$ is concave upward and where it is concave downward.

2. Using the Second Derivative Test, if applicable, find the relative extrema of the function $f(x) = 2x^3 - \frac{1}{2}x^2 - 12x - 10$.

3. A certain country's gross national product (GNP) [in millions of dollars] in year t is described by the function

$$G(t) = -2t^3 + 45t^2 + 20t + 6000 \qquad (0 \le t \le 11),$$

where $t = 0$ corresponds to the beginning of the year 1985. Find the inflection point of the function G and discuss its significance.

Solutions to Self-Check Exercises 4.3 can be found on page 234.

4.3 EXERCISES

In Exercises 1–8, you are given the graph of some function f. Determine the intervals where f is concave upward and where it is concave downward. Also, find all inflection points of f, if any.

1.
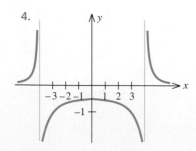

2.
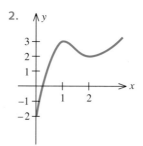

3.
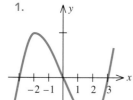

4.
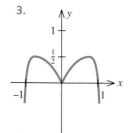

5.
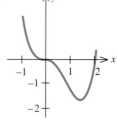

6.
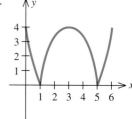

7.
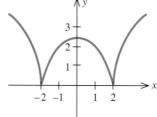

8.
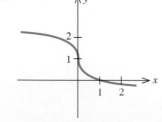

In Exercises 9–12, determine which graph—a, b, or c—is the graph of the function f with the specified properties.

9. $f(2) = 1, f'(2) > 0,$ and $f''(2) < 0$

a.

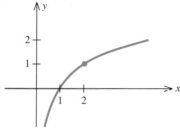

b.

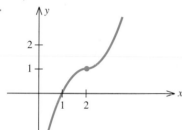

c.

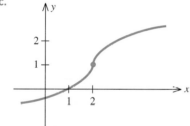

10. $f(1) = 2, f'(x) > 0$ on $(-\infty, 1) \cup (1, \infty),$ and $f''(1) = 0$

a.

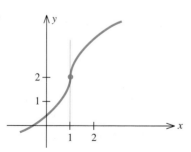

b.

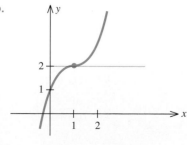

c.

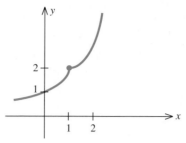

11. $f'(0)$ is undefined, f is decreasing on $(-\infty, 0),$ f is concave downward on $(0, 3),$ and f has an inflection point at $x = 3$

a.

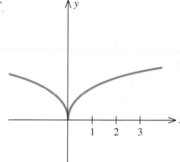

b.

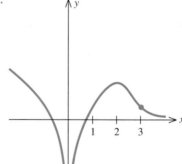

c.

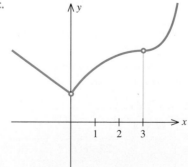

12. f is decreasing on $(-\infty, 2)$, increasing on $(2, \infty)$, f is concave upward on $(1, \infty)$, and has inflection points at $x = 0$ and $x = 1$

a.

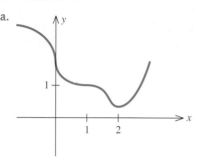

b.

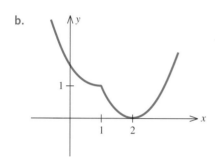

c.

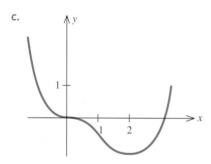

In Exercises 13–18, show that the given function is concave upward wherever it is defined.

13. $f(x) = 4x^2 - 12x + 7$

14. $g(x) = x^4 + \frac{1}{2}x^2 + 6x + 10$

15. $h(x) = \dfrac{1}{x^2}$ **16.** $f(x) = \dfrac{1}{x^4}$

17. $h(x) = \sqrt{x^2 + 4}$ **18.** $g(x) = -\sqrt{4 - x^2}$

In Exercises 19–42, determine where the given function is concave upward and where it is concave downward.

19. $f(x) = 2x^2 - 3x + 4$

20. $g(x) = -x^2 + 3x + 4$

21. $f(x) = x^3 - 1$ **22.** $g(x) = x^3 - x$

23. $f(x) = x^3 - 6x^2 + 2x + 1$

24. $g(t) = -5t^3 + 15t^2 + 24t - 192$

25. $f(x) = x^4 - 6x^3 + 2x + 8$

26. $f(x) = 3x^4 - 6x^3 + x - 8$

27. $f(x) = x^{4/7}$ **28.** $f(x) = \sqrt[3]{x}$

29. $f(x) = \sqrt{4 - x}$ **30.** $g(x) = \sqrt{x - 2}$

31. $f(x) = \dfrac{1}{x - 2}$ **32.** $g(x) = \dfrac{x}{x + 1}$

33. $f(x) = \dfrac{1}{2 + x^2}$ **34.** $g(x) = \dfrac{x}{1 + x^2}$

35. $h(t) = \dfrac{t^2}{t - 1}$ **36.** $f(x) = \dfrac{x + 1}{x - 1}$

37. $g(x) = x + \dfrac{1}{x^2}$ **38.** $h(r) = -\dfrac{1}{(r - 2)^2}$

39. $f(x) = \dfrac{x^2}{x^2 - 1}$ **40.** $f(x) = \dfrac{x^2 - 2}{x^3}$

41. $g(t) = (2t - 4)^{1/3}$ **42.** $f(x) = (x - 2)^{2/3}$

In Exercises 43–54, find the inflection points, if any, of each of the given functions.

43. $f(x) = x^3 - 2$ **44.** $g(x) = x^3 - 6x$

45. $f(x) = 6x^3 - 18x^2 + 12x - 15$

46. $g(x) = 2x^3 - 3x^2 + 18x - 8$

47. $f(x) = 3x^4 - 4x^3 + 1$

48. $f(x) = x^4 - 2x^3 + 6$

49. $g(t) = \sqrt[3]{t}$ **50.** $f(x) = \sqrt[5]{x}$

51. $f(x) = (x - 1)^3 + 2$ **52.** $f(x) = (x - 2)^{4/3}$

53. $f(x) = \dfrac{2}{1 + x^2}$ **54.** $f(x) = 2 + \dfrac{3}{x}$

In Exercises 55–72, find the relative extrema, if any, of each function. Use the Second Derivative Test, if applicable.

55. $f(x) = -x^2 + 2x + 4$ **56.** $g(x) = 2x^2 + 3x + 7$

57. $f(x) = 2x^3 + 1$ **58.** $g(x) = x^3 - 6x$

59. $f(x) = \frac{1}{3}x^3 - 2x^2 - 5x - 10$

60. $f(x) = 2x^3 + 3x^2 - 12x - 4$

61. $g(t) = t + \dfrac{9}{t}$ **62.** $f(t) = 2t + \dfrac{3}{t}$

63. $f(x) = \dfrac{x}{1 - x}$ **64.** $f(x) = \dfrac{2x}{x^2 + 1}$

65. $f(t) = t^2 - \dfrac{16}{t}$ **66.** $g(x) = x^2 + \dfrac{2}{x}$

67. $g(s) = \dfrac{s}{1 + s^2}$ **68.** $g(x) = \dfrac{1}{1 + x^2}$

69. $f(x) = \dfrac{x^4}{x - 1}$ **70.** $f(x) = \dfrac{x^2}{x^2 + 1}$

71. $g(x) = \dfrac{2 - x}{(x + 2)^3}$ **72.** $f(x) = \dfrac{x^2 + 4}{x^2 - 1}$

73. Effect of Budget Cuts on Drug-Related Crimes   The following graphs were used by a police commissioner to illustrate what effect a budget cut would have on crime in the city. The number $N_1(t)$ gives the projected number of drug-related crimes in the next twelve months. The number $N_2(t)$ gives the projected number of drug-related crimes in the same time frame if next year's budget is cut.

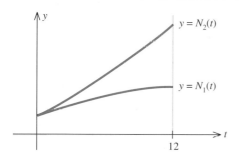

a. Explain why $N'_1(t)$ and $N'_2(t)$ are both positive on the interval $(0, 12)$.
b. What are the signs of $N''_1(t)$ and $N''_2(t)$ on the interval $(0, 12)$?
c. Interpret the results of (b).

74. Demand for RNs   The following graph gives the total number of "help wanted" ads for RNs (registered nurses) in 22 cities over the last 12 months as a function of time t (t measured in months).
a. Explain why $N'(t)$ is positive on the interval $(0, 12)$.
b. Determine the signs of $N''(t)$ on the interval $(0, 6)$ and the interval $(6, 12)$.
c. Interpret the results of (b).

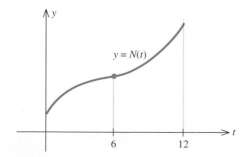

75. Effect of Advertising on Sales   The total sales S, in thousands of dollars, of the Cannon Precision Instruments Corporation is related to the amount of money x that Cannon spends on advertising its products by the function

$$S(x) = -0.002x^3 + 0.6x^2 + x + 500$$
$$(0 \le x \le 200),$$

where x is measured in thousands of dollars. Find the inflection point of the function S and discuss its significance.

76. Forecasting Profits   As a result of increasing energy costs, the growth rate of the profit of the four-year-old Venice Glassblowing Company has begun to decline. Venice's management, after consulting with energy experts, decides to implement certain energy-conservation measures aimed at cutting energy bills. The general manager reports that, according to his calculations, the growth rate of Venice's profit should once again be on the increase within four years. If Venice's profit (in hundreds of dollars) x years from now is given by the function

$$P(x) = x^3 - 9x^2 + 40x + 50 \qquad (0 \le x \le 8),$$

determine whether the general manager's forecast will be accurate. [*Hint:* Find the inflection point of the function P and study the concavity of P.]

77. Worker Efficiency   An efficiency study conducted for the Elektra Electronics Company showed that the number of Space Commander walkie-talkies assembled by the average worker t hours after starting work at 8 A.M. is given by

$$N(t) = -t^3 + 6t^2 + 15t \qquad (0 \le t \le 4).$$

At what time during the morning shift is the average worker performing at peak efficiency?

C 78. Cost of Producing Calculators   A subsidiary of Elektra Electronics manufactures a programmable electronic pocket calculator. Management determines that the daily cost $C(x)$ of producing these calculators (in dollars) is

$$C(x) = 0.0001x^3 - 0.08x^2 + 40x + 5000,$$

where x stands for the number of calculators produced. Find the inflection point of the function C and interpret your result.

C 79. Flight of a Rocket   The altitude (in feet) of a rocket t seconds into flight is given by

$$s = f(t) = -t^3 + 54t^2 + 480t + 6.$$

Find the point of inflection of the function f and interpret your result. What is the maximum velocity attained by the rocket?

C **80. Air Pollution**   The level of ozone, an invisible gas that irritates and impairs breathing, present in the atmosphere on a certain May day in the city of Riverside was approximated by

$$A(t) = 1.0974t^3 - 0.0915t^4 \qquad (0 \le t \le 11),$$

where $A(t)$ is measured in pollutant standard index (PSI) and t is measured in hours, with $t = 0$ corresponding to 7 A.M. Use the Second Derivative Test to show that the function A has a relative maximum at approximately $t = 9$. Interpret your results.

81. Show that the quadratic function

$$f(x) = ax^2 + bx + c \qquad (a \ne 0)$$

is concave upward if $a > 0$ and concave downward if $a < 0$. Thus, by examining the sign of the coefficient of x^2, one can tell immediately whether the parabola opens upward or downward.

82. Show that the cubic function

$$f(x) = ax^3 + bx^2 + cx + d \qquad (a \ne 0)$$

has one and only one inflection point. Find the coordinates of this point.

83. Consider the functions $f(x) = x^3$, $g(x) = x^4$, and $h(x) = -x^4$.
   a. Show that $x = 0$ is a critical point of each of the functions f, g, and h.
   b. Show that the second derivative of each of the functions f, g, and h equals zero at $x = 0$.
   c. Show that f has neither a relative maximum nor a relative minimum at $x = 0$, that g has a relative minimum at $x = 0$, and that h has a relative maximum at $x = 0$.

C *In Exercises 84–87, use a computer or a graphing calculator to plot the graph of f. Then use the graph to find the intervals where f is concave upward and where it is concave downward, and the inflection points of f, if any.*

84. $f(x) = x^4 - x^3 + 3x^2 - 2x + 2$

85. $f(x) = x^3(x - 7)^4$ **86.** $f(x) = \dfrac{x^2 - 1}{x^3}$

87. $f(x) = \dfrac{x + 1}{\sqrt{x}}$

SOLUTIONS TO SELF-CHECK EXERCISES 4.3

1. We first compute

$$f'(x) = 12x^2 - 6x$$

and

$$f''(x) = 24x - 6 = 6(4x - 1).$$

Observe that f'' is continuous everywhere and has a zero at $x = 1/4$. The sign diagram of f'' is shown in the accompanying figure.

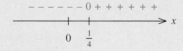

Sign diagram for f''

From the sign diagram for f'' we see that f is concave upward on $\left(\frac{1}{4}, \infty\right)$ and concave downward on $\left(-\infty, \frac{1}{4}\right)$.

2. First, we find the critical points of f by solving the equation

$$f'(x) = 6x^2 - x - 12 = 0;$$

that is,

$$(3x + 4)(2x - 3) = 0,$$

giving $x = -\frac{4}{3}$ and $x = \frac{3}{2}$. Next, we compute

$$f''(x) = 12x - 1.$$

Since

$$f''\left(-\frac{4}{3}\right) = 12\left(-\frac{4}{3}\right) - 1 = -17 < 0,$$

the Second Derivative Test implies that $f\left(-\frac{4}{3}\right) = \frac{10}{27}$ is a relative maximum of f. Also,

$$f''\left(\frac{3}{2}\right) = 12\left(\frac{3}{2}\right) - 1 = 17 > 0,$$

and we see that $f\left(\frac{3}{2}\right) = -\frac{179}{8}$ is a relative minimum.

3. We compute the second derivative of G. Thus,

$$G'(t) = -6t^2 + 90t + 20$$

$$G''(t) = -12t + 90.$$

Now G'' is continuous everywhere and $G''(t) = 0$ when $t = \frac{15}{2}$, giving $t = \frac{15}{2}$ as the only candidate for an inflection point of G. Since $G''(t) > 0$ for $t < \frac{15}{2}$ and $G''(t) < 0$ for $t > \frac{15}{2}$, we see that the point $\left(\frac{15}{2}, \frac{15675}{2}\right)$ is an inflection point of G. The results of our computations tell us that the country's GNP was increasing most rapidly at the beginning of July 1992.

4.4   Curve Sketching I

A Real-Life Example

As we have seen on numerous occasions, the graph of a function is a useful aid for visualizing the function's properties. From a practical point of view, the graph of a function also gives, at one glance, a complete summary of all the information captured by the function.

Consider, for example, the graph of the function giving the Dow Jones Industrial Average (DJIA) on Black Monday, October 19, 1987 (Figure 4.49). Here $t = 0$ corresponds to 8:30 A.M., when the market was open for business, and $t = 7.5$ corresponds to 4 P.M., the closing time. The following information may be gleaned from studying the graph.

FIGURE 4.49
The Dow Jones Industrial Average on Black Monday

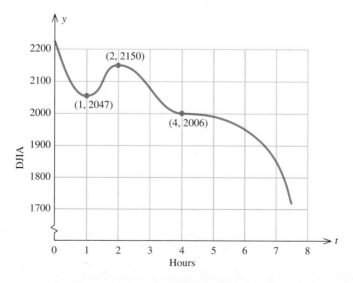

The graph is *decreasing* rapidly from $t = 0$ to $t = 1$, reflecting the sharp drop in the index in the first hour of trading. The point (1, 2047) is a *relative*

minimum point of the function, and this turning point coincides with the start of an aborted recovery. The short-lived rally, represented by the portion of the graph that is *increasing* on the interval (1, 2), quickly fizzled out at $t = 2$ (10:30 A.M.). The *relative maximum* point (2, 2150) marks the highest point of the recovery. The function is decreasing in the rest of the interval. The point (4, 2006) is an *inflection point* of the function; it shows that there was a temporary respite at $t = 4$ (12:30 P.M.). However, selling pressure continued unabated, and the Dow Jones Industrial Index continued to fall until the closing bell. Finally, the graph also shows that the index opened at the high of the day [$f(0) = 2247$ is the *absolute maximum* of the function] and closed at the low of the day [$f\left(\frac{15}{2}\right) = 1739$ is the *absolute minimum* of the function], a drop of 508 points![†]

Before we turn our attention to the actual task of sketching the graph of a function, let us look at some properties of graphs that will be helpful in this connection.

Symmetry

Consider the graph of the function f defined by $f(x) = x^2$, shown in Figure 4.50a. Note that the portion of the graph of f on one side of the y-axis is the mirror reflection of the portion of the graph on the other side. In this situation, we say that the graph of f is symmetric with respect to the y-axis.

To devise a test for symmetry with respect to the y-axis, observe that whenever the point (x, y) lies on the graph of f, so does the point $(-x, y)$, and vice versa (Figure 4.50b). But the point $(-x, y)$ lies on the graph of f if and only if it satisfies the equation $y = f(-x)$. Thus, the graph of a function f is symmetric with respect to the y-axis if and only if $f(-x) = f(x)$. For example, for the function $f(x) = x^2$, we see that

$$f(-x) = (-x)^2 = x^2 = f(x),$$

and so the graph of f is symmetric with respect to the y-axis, as we saw earlier.

FIGURE 4.50

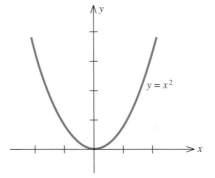

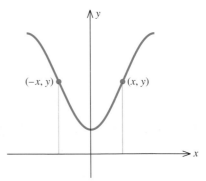

(a) The graph of $y = x^2$ is symmetric with respect to the y-axis.

(b) Symmetry with respect to the y-axis

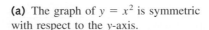

[†] Absolute maxima and absolute minima of functions are covered in Section 4.6.

Next, consider the graph of the function f defined by $f(x) = x^3$, shown in Figure 4.51a. Note that the portion of the graph lying on one side of the *origin* can be obtained by reflecting the other portion of the curve first with respect to one axis followed by a reflection with respect to the second axis. Referring to Figure 4.51b, we see that whenever the point (x, y) lies on the graph of f, so does the point $(-x, -y)$. But the point $(-x, -y)$ lies on the graph of f if and only if $-y = f(-x)$; that is, $y = -f(-x)$. Thus, the graph of f is symmetric with respect to the origin if and only if $f(-x) = -f(x)$.

FIGURE 4.51

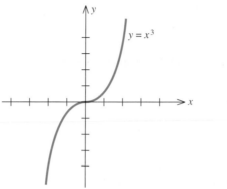

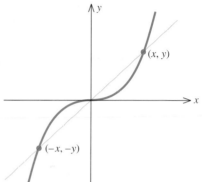

(a) The graph of $y = x^3$ is symmetric with respect to the origin.

(b) Symmetry with respect to the origin

For the function $f(x) = x^3$, we find

$$f(-x) = (-x)^3 = -x^3 = -f(x),$$

and so the graph of f is symmetric with respect to the origin, as we noted earlier. The following is a summary of these results.

TESTS FOR SYMMETRY

Let f be a function defined by the equation $y = f(x)$.

a. If $f(-x) = f(x)$, then the graph of f is symmetric with respect to the y-axis.

b. If $f(-x) = -f(x)$, then the graph of f is symmetric with respect to the origin.

EXAMPLE 1   Test the graphs of the following functions for symmetry with respect to the y-axis and with respect to the origin.

a. $f(x) = \dfrac{1}{x^2 + 1}$ **b.** $g(x) = x^3 - x$ **c.** $h(x) = 1 + \dfrac{1}{x}$

Solution

a. We compute $f(-x)$ by replacing x in $f(x)$ by $-x$, obtaining

$$f(-x) = \frac{1}{(-x)^2 + 1} = \frac{1}{x^2 + 1}.$$

FIGURE 4.52

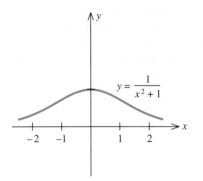

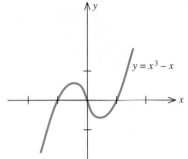

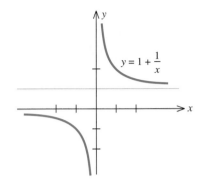

(a) The graph is symmetric with respect to the y-axis.

(b) The graph is symmetric with respect to the origin.

(c) The graph is not symmetric with respect to the y-axis or the origin.

Since this is equal to $f(x)$, we conclude that the graph of f is symmetric with respect to the y-axis (Figure 4.52a).

b. Here

$$g(-x) = (-x)^3 - (-x) = -x^3 + x$$
$$= -(x^3 - x) = -g(x),$$

and we see that the graph of g is symmetric with respect to the origin (Figure 4.52b).

c. We find

$$h(-x) = 1 + \frac{1}{-x} = 1 - \frac{1}{x},$$

which is neither equal to $h(x)$ nor $-h(x)$. Therefore, the graph of h is neither symmetric with respect to the y-axis nor symmetric with respect to the origin (Figure 4.52c). ∎

Vertical Asymptotes

Before going on, you might want to review the material on one-sided limits and the limit at infinity of a function (Sections 2.4 and 2.5).

Consider the graph of the function

$$f(x) = \frac{x + 1}{x - 1},$$

FIGURE 4.53

The graph of f has a vertical asymptote at $x = 1$.

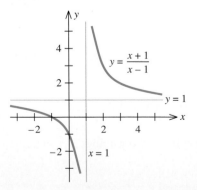

shown in Figure 4.53. Observe that $f(x)$ increases without bound (tends to infinity) as x approaches $x = 1$ from the right; that is,

$$\lim_{x \to 1^+} \frac{x + 1}{x - 1} = \infty.$$

You can verify this by taking a sequence of values of x approaching $x = 1$ from the right and looking at the corresponding values of $f(x)$.

Here is another way of looking at the situation: Observe that if x is a number that is a little larger than 1, then both $(x + 1)$ and $(x - 1)$ are positive, so that $(x + 1)/(x - 1)$ is also positive. As x approaches $x = 1$, the numerator

$(x + 1)$ approaches the number 2, but the denominator $(x - 1)$ approaches zero, so the quotient $(x + 1)/(x - 1)$ approaches infinity, as observed earlier. The line $x = 1$ is called a *vertical asymptote* of the graph of f.

For the function $f(x) = (x + 1)/(x - 1)$, you can show that

$$\lim_{x \to 1^-} \frac{x + 1}{x - 1} = -\infty,$$

and this tells us how $f(x)$ approaches the asymptote $x = 1$ from the left. More generally, we have the following definition.

VERTICAL ASYMPTOTES

The line $x = a$ is a **vertical asymptote** of the graph of a function f if either

$$\lim_{x \to a^+} f(x) = \infty \qquad (\text{or } -\infty)$$

or

$$\lim_{x \to a^-} f(x) = \infty \qquad (\text{or } -\infty).$$

REMARK Although a vertical asymptote of a graph is not part of the graph, it serves as a useful aid for sketching the graph. □

For rational functions

$$f(x) = \frac{P(x)}{Q(x)},$$

there is a simple criterion for determining whether the graph of f has any vertical asymptotes.

FINDING VERTICAL ASYMPTOTES OF RATIONAL FUNCTIONS

Suppose that f is a rational function

$$f(x) = \frac{P(x)}{Q(x)},$$

where P and Q are polynomial functions. Then the line $x = a$ is a vertical asymptote of the graph of f if $Q(a) = 0$ but $P(a) \neq 0$.

For the function

$$f(x) = \frac{x + 1}{x - 1}$$

considered earlier, $P(x) = x + 1$ and $Q(x) = x - 1$. Observe that $Q(1) = 0$ but $P(1) = 2 \neq 0$, so $x = 1$ is a vertical asymptote of the graph of f.

EXAMPLE 2 Find the vertical asymptotes of the graph of the function

$$f(x) = \frac{x^2}{4 - x^2}.$$

Solution   The function f is a rational function with $P(x) = x^2$ and $Q(x) = 4 - x^2$. The zeros of Q are found by solving

$$4 - x^2 = 0;$$

that is,

$$(2 - x)(2 + x) = 0,$$

giving $x = -2$ and $x = 2$. These are candidates for the vertical asymptotes of the graph of f. Examining $x = -2$, we compute $P(-2) = (-2)^2 = 4 \neq 0$, and we see that $x = -2$ is indeed a vertical asymptote of the graph of f. Similarly, we find $P(2) = 2^2 = 4 \neq 0$, and so $x = 2$ is also a vertical asymptote of the graph of f. The graph of f sketched in Figure 4.54 confirms these results.

FIGURE 4.54
$x = -2$ and $x = 2$ are vertical asymptotes of the graph of f.

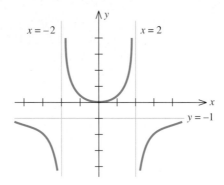

Recall that in order for the line $x = a$ to be a vertical asymptote of the graph of a rational function f, *only* the denominator of $f(x)$ must be equal to zero at $x = 0$. If *both* $P(a)$ and $Q(a)$ are equal to zero, then $x = a$ need *not* be a vertical asymptote. For example, look at the function $f(x) = \dfrac{2(x^2 - 4)}{x - 2}$, whose graph appears in Figure 2.19, page 83.

FIGURE 4.55
The graph of f has a horizontal asymptote at $y = 1$.

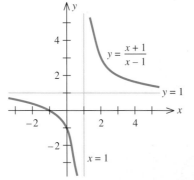

Horizontal Asymptotes

Let us return to the function f defined by

$$f(x) = \frac{x + 1}{x - 1}$$

(Figure 4.55).

Observe that $f(x)$ approaches the horizontal line $y = 1$ as x approaches infinity, and, in this case, $f(x)$ approaches $y = 1$ as x approaches minus infinity as well. The line $y = 1$ is called a *horizontal asymptote* of the graph of f. More generally, we have the following definition.

HORIZONTAL ASYMPTOTES	The line $y = b$ is a **horizontal asymptote** of the graph of a function f if either $$\lim_{x \to \infty} f(x) = b \quad \text{or} \quad \lim_{x \to -\infty} f(x) = b.$$

For the function

$$f(x) = \frac{x + 1}{x - 1},$$

we see that

$$\lim_{x \to \infty} \frac{x + 1}{x - 1} = \lim_{x \to \infty} \frac{1 + \dfrac{1}{x}}{1 - \dfrac{1}{x}} \qquad \text{(Divide numerator and denominator by } x.\text{)}$$

$$= 1.$$

Also,

$$\lim_{x \to -\infty} \frac{x + 1}{x - 1} = \lim_{x \to -\infty} \frac{1 + \dfrac{1}{x}}{1 - \dfrac{1}{x}}$$

$$= 1.$$

In either case, we conclude that $y = 1$ is a horizontal asymptote of the graph of f, as observed earlier.

EXAMPLE 3   Find the horizontal asymptotes of the graph of the function

$$f(x) = \frac{x^2}{4 - x^2}.$$

Solution   We compute

$$\lim_{x \to \infty} \frac{x^2}{4 - x^2} = \lim_{x \to \infty} \frac{1}{\dfrac{4}{x^2} - 1} \qquad \text{(Dividing numerator and denominator by } x^2\text{)}$$

$$= -1,$$

and so $y = -1$ is a horizontal asymptote, as before. The graph of f sketched in Figure 4.56 confirms this result.

FIGURE 4.56

The graph of f has a horizontal asymptote at $y = -1$.

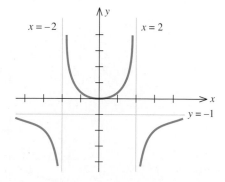

We close this section by pointing out the following property of polynomial functions.

A polynomial function has no vertical or horizontal asymptotes.

To see this, note that a polynomial function, $P(x)$ can be written as a rational function with denominator equal to 1. Thus,

$$P(x) = \frac{P(x)}{1}.$$

Since the denominator is never equal to zero, P has no vertical asymptotes. Next, if P is a polynomial of degree greater than or equal to 1, then

$$\lim_{x \to \infty} P(x) \quad \text{and} \quad \lim_{x \to -\infty} P(x)$$

are either infinity or minus infinity; that is, they do not exist. Therefore, P has no horizontal asymptotes.

**SELF-CHECK
EXERCISES 4.4**

1. Test the graphs of the following functions for symmetry with respect to the y-axis and with respect to the origin.

   a. $f(x) = \dfrac{x}{x^2 - 1}$ **b.** $f(x) = \dfrac{x^4 + x^2 + 1}{x^2}$

2. Find the horizontal and vertical asymptotes of the graph of the function

   $$f(x) = \frac{2x^2}{x^2 - 1}.$$

Solutions to Self-Check Exercises 4.4 can be found on page 245.

 4.4   EXERCISES

In Exercises 1–6, determine whether the graph of the given function is symmetric with respect to the y-axis, the origin, or neither.

1.

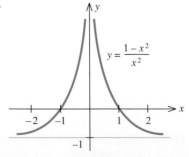

$$y = \frac{1 - x^2}{x^2}$$

2.

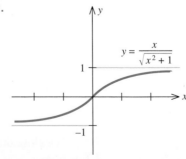

$$y = \frac{x}{\sqrt{x^2 + 1}}$$

3.

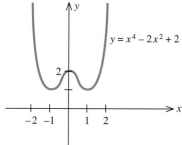

$y = x^4 - 2x^2 + 2$

9. $g(x) = x^2 - 4$

10. $f(x) = 2x^2 + 1$

11. $h(t) = t^4 - t^2 + 1$

12. $f(x) = x^5 - x^3 + x$

13. $f(x) = x^3 + x^2$

14. $g(r) = r^4 - r^2 - r$

15. $f(t) = 2t^{-2}$

16. $h(x) = 3x^{-3}$

17. $f(x) = \dfrac{1}{x^2 - 4}$

18. $g(t) = \dfrac{t}{t^2 + 4}$

19. $f(x) = \dfrac{x^3}{x^3 - 1}$

20. $f(x) = 1 + \dfrac{x}{x^2 + 1}$

21. $f(x) = \sqrt{x^2 - 9}$

22. $g(t) = t\sqrt{9 - t^2}$

In Exercises 23–30, find the horizontal and vertical asymptotes of the given graph.

23.

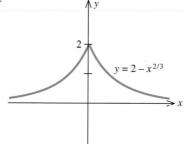

$y = 2 - x^{2/3}$

4.

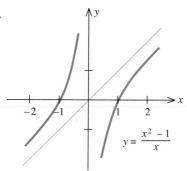

$y = \dfrac{x^2 - 1}{x}$

5.

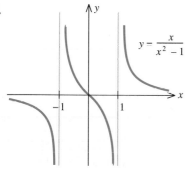

$y = \dfrac{x}{x^2 - 1}$

24.

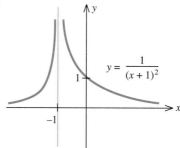

$y = \dfrac{1}{(x + 1)^2}$

6.

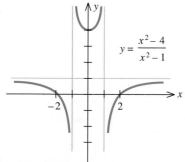

$y = \dfrac{x^2 - 4}{x^2 - 1}$

25.

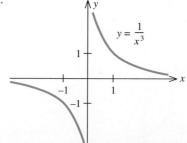

$y = \dfrac{1}{x^3}$

In Exercises 7–22, test the graph of the given function for symmetry with respect to the y-axis and with respect to the origin.

7. $f(x) = -2$

8. $f(x) = 2x$

26.

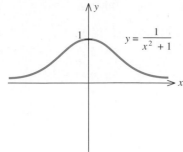

$$y = \frac{1}{x^2 + 1}$$

27.

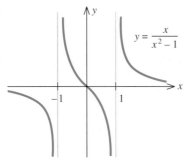

$$y = \frac{x}{x^2 - 1}$$

28.

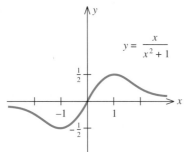

$$y = \frac{x}{x^2 + 1}$$

29.

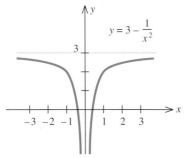

$$y = 3 - \frac{1}{x^2}$$

30.

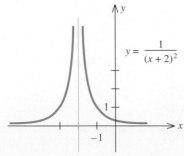

$$y = \frac{1}{(x + 2)^2}$$

In Exercises 31–48, find the horizontal and vertical asymptotes of the graph of the given function. (You need not sketch the graph.)

31. $f(x) = \dfrac{1}{x}$ **32.** $f(x) = \dfrac{1}{x + 2}$

33. $f(x) = -\dfrac{2}{x^2}$ **34.** $g(x) = \dfrac{1}{1 + 2x^2}$

35. $f(x) = \dfrac{x - 1}{x + 1}$ **36.** $g(t) = \dfrac{t + 1}{2t - 1}$

37. $h(x) = x^3 - 3x^2 + x + 1$

38. $g(x) = 2x^3 + x^2 + 1$

39. $f(t) = \dfrac{t^2}{t^2 - 9}$ **40.** $g(x) = \dfrac{x^3}{x^2 - 4}$

41. $f(x) = \dfrac{3x}{x^2 - x - 6}$ **42.** $g(x) = \dfrac{2x}{x^2 + x - 2}$

43. $g(t) = 2 + \dfrac{5}{(t - 2)^2}$ **44.** $f(x) = 1 + \dfrac{2}{x - 3}$

45. $f(x) = \dfrac{x^2 - 2}{x^2 - 4}$ **46.** $h(x) = \dfrac{2 - x^2}{x^2 + x}$

47. $g(x) = \dfrac{x^3 - x}{x(x + 1)}$

48. $f(x) = \dfrac{x^4 - x^2}{x(x - 1)(x + 2)}$

49. Cost of Removing Toxic Pollutants   A city's main well was recently found to be contaminated with tri-chloroethylene (a cancer-causing chemical) as a result of an abandoned chemical dump leaching chemicals into the water. A proposal submitted to the city's selectmen indicated that the cost, measured in millions of dollars, of removing x percent of the toxic pollutants is given by

$$C(x) = \frac{0.5x}{100 - x}.$$

a. Find the vertical asymptote of $C(x)$.
b. Is it possible to remove 100 percent of the toxic pollutant from the water?

50. Average Cost of Producing Video Discs   The average cost per disc (in dollars) incurred by the Herald Record Company in pressing x video discs is given by the average cost function

$$\overline{C}(x) = 2.2 + \frac{2500}{x}.$$

a. Find the horizontal asymptote of $\overline{C}(x)$.
b. What is the limiting value of the average cost?

51. **Concentration of a Drug in the Bloodstream** The concentration of a certain drug in a patient's bloodstream t hours after injection is given by

$$C(t) = \frac{0.2t}{t^2 + 1}$$

milligrams per cubic centimeter.
a. Find the horizontal asymptote of $C(t)$.
b. Interpret your result.

52. **Effect of Enzymes on Chemical Reactions** Certain proteins, known as enzymes, serve as catalysts for chemical reactions in living things. In 1913, Leonor Michaelis and L. M. Menten discovered the following formula giving the initial speed V (in moles per liter per second) at which the reaction begins in terms of the amount of substrate x (the substance that is being acted upon, measured in moles per liter):

$$V = \frac{ax}{x + b},$$

where a and b are positive constants.
a. Find the horizontal asymptote of V.
b. Interpret your result.

SOLUTIONS TO SELF-CHECK EXERCISES 4.4

1. a. $f(-x) = \dfrac{-x}{(-x)^2 - 1} = -\dfrac{x}{x^2 - 1} = -f(x)$,

   and so the graph of f is symmetric with respect to the origin.

   b. $f(-x) = \dfrac{(-x)^4 + (-x)^2 + 1}{(-x)^2}$

   $= \dfrac{x^4 + x^2 + 1}{x^2}$

   $= f(x)$,

   and so the graph of f is symmetric with respect to the y-axis.

2. Since

$$\lim_{x \to \infty} \frac{2x^2}{x^2 - 1} = \lim_{x \to \infty} \frac{2}{1 - \dfrac{1}{x^2}} \qquad \text{(Divide numerator and denominator by } x^2 \text{.)}$$

$$= 2,$$

   we see that $y = 2$ is a horizontal asymptote. Next, since

$$x^2 - 1 = (x + 1)(x - 1) = 0$$

   implies $x = -1$ or $x = 1$, these are candidates for the vertical asymptotes of f. Since the numerator of f is not equal to zero for $x = -1$ or $x = 1$, we conclude that $x = -1$ and $x = 1$ are vertical asymptotes of the graph of f.

4.5 Curve Sketching II

A Guide to Curve Sketching

In the last several sections we saw how the first and second derivatives of a function are used to reveal various properties of the graph of a function f. In this section we will show how this information can be used to help us sketch the graph of f. We begin by giving a general procedure for curve sketching.

A GUIDE TO CURVE SKETCHING

1. Determine the domain of f.
2. Find the x- and y-intercepts of f.[†]
3. Determine whether the graph of f is symmetric with respect to the y-axis or the origin.
4. Determine the behavior of f for large absolute values of f.
5. Find all horizontal and vertical asymptotes of f.
6. Determine the intervals where f is increasing and where f is decreasing.
7. Find the relative extrema of f.
8. Determine the concavity of f.
9. Find the inflection points of f.
10. Plot a few additional points to help in further identifying the shape of the graph of f and sketch the graph.

We will now illustrate the techniques of curve sketching with several examples.

Three Step-by-Step Examples

EXAMPLE 1   Sketch the graph of the function

$$y = f(x) = x^3 - 6x^2 + 9x + 2.$$

Solution   Obtain the following information on the graph of f.

1. The domain of f is the interval $(-\infty, \infty)$.
2. By setting $x = 0$, we find that the y-intercept is 2. The x-intercept is found by setting $y = 0$, which, in this case, leads to a cubic equation. Since the solution is not readily found, we will not make use of this information.
3. There is no symmetry of f with respect to the y-axis or the origin.
4. Since

$$\lim_{x \to -\infty} f(x) = \lim_{x \to -\infty} (x^3 - 6x^2 + 9x + 2) = -\infty$$

and $$\lim_{x \to \infty} f(x) = \lim_{x \to \infty} (x^3 - 6x^2 + 9x + 2) = \infty,$$

we see that f decreases without bound as x decreases without bound, and f increases without bound as x increases without bound.

5. Since f is a polynomial function, there are no asymptotes.

6. $$f'(x) = 3x^2 - 12x + 9 = 3(x^2 - 4x + 3)$$
$$= 3(x - 3)(x - 1)$$

Setting $f'(x) = 0$ gives $x = 1$ or $x = 3$. The sign diagram for f' shows that f is increasing on the intervals $(-\infty, 1)$ and $(3, \infty)$ and decreasing on the interval $(1, 3)$ [Figure 4.57].

FIGURE 4.57
Sign diagram for f'

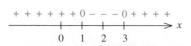

[†]The equation $f(x) = 0$ may be difficult to solve, in which case one may decide against finding the x-intercepts.

7. From the results of Step 6 we see that $x = 1$ and $x = 3$ are critical points of f. Furthermore, f' changes sign from positive to negative as we move across $x = 1$, so a relative maximum of f occurs at $x = 1$. Similarly, we see that a relative minimum of f occurs at $x = 3$. Now,

$$f(1) = 1 - 6 + 9 + 2 = 6$$

and $$f(3) = 3^3 - 6(3)^2 + 9(3) + 2 = 2,$$

so $f(1) = 6$ is a relative maximum of f and $f(3) = 2$ is a relative minimum of f.

FIGURE 4.58

Sign diagram for f''

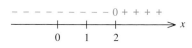

FIGURE 4.59

We first plot the intercept, the relative extrema, and the inflection point.

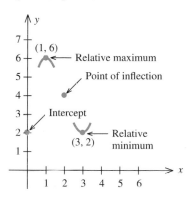

8. $$f''(x) = 6x - 12 = 6(x - 2),$$

which is equal to zero when $x = 2$. The sign diagram of f'' shows that f is concave downward on the interval $(-\infty, 2)$ and concave upward on the interval $(2, \infty)$ [Figure 4.58].

9. From the results of Step 8, we see that f'' changes sign as we move across the point $x = 2$. Next,

$$f(2) = 2^3 - 6(2)^2 + 9(2) + 2 = 4,$$

and so the required inflection point of f is $(2, 4)$.

10. Summarizing, we have

Domain	$(-\infty, \infty)$
Intercept	$(0, 2)$
Symmetry	None
$\lim\limits_{x \to -\infty} f(x); \lim\limits_{x \to \infty} f(x)$	$-\infty; \infty$
Asymptotes	None
Intervals where f is $\nearrow$ or $\searrow$	$\nearrow$ on $(-\infty, 1) \cup (3, \infty)$; $\searrow$ on $(1, 3)$
Relative extrema	Rel. max. at $(1, 6)$; rel. min. at $(3, 2)$
Concavity	Downward on $(-\infty, 2)$; upward on $(2, \infty)$
Point of inflection	$(2, 4)$

FIGURE 4.60

The graph of $y = x^3 - 6x^2 + 9x + 2$

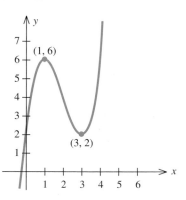

In general, it is a good idea to start graphing by plotting the intercept, relative extrema, and inflection point (Figure 4.59). Then, using the rest of the information, we complete the graph of f as sketched in Figure 4.60. ∎

EXAMPLE 2   Sketch the graph of the function

$$y = f(x) = \frac{x + 1}{x - 1}.$$

Solution Obtain the following information:

1. f is undefined when $x = 1$, so the domain of f is the set of all real numbers other than $x = 1$.

2. Setting $y = 0$ gives -1, the x-intercept of f. Next, setting $x = 0$ gives -1 as the y-intercept of f.

3. y is not symmetric with respect to the y-axis or the origin.

4. Earlier we found that

$$\lim_{x \to \infty} \frac{x + 1}{x - 1} = 1 \quad \text{and} \quad \lim_{x \to -\infty} \frac{x + 1}{x - 1} = 1$$

(see page 241). Consequently, we see that $f(x)$ approaches the line $y = 1$ as $|x|$ becomes arbitrarily large. For $x > 1$, $f(x) > 1$ and $f(x)$ approaches the line $y = 1$ from above. For $x < 1$, $f(x) < 1$, so $f(x)$ approaches the line $y = 1$ from below.

5. The straight line $x = 1$ is a vertical asymptote of the graph of f. Also, from the results of Step 4, we conclude that $y = 1$ is a horizontal asymptote of the graph of f.

6.
$$f'(x) = \frac{(x - 1)(1) - (x + 1)(1)}{(x - 1)^2} = -\frac{2}{(x - 1)^2}$$

and is discontinuous at $x = 1$. The sign diagram of f' shows that $f'(x) < 0$ whenever it is defined. Thus, f is decreasing on the intervals $(-\infty, 1)$ and $(1, \infty)$ [Figure 4.61].

FIGURE 4.61
The sign diagram for f'

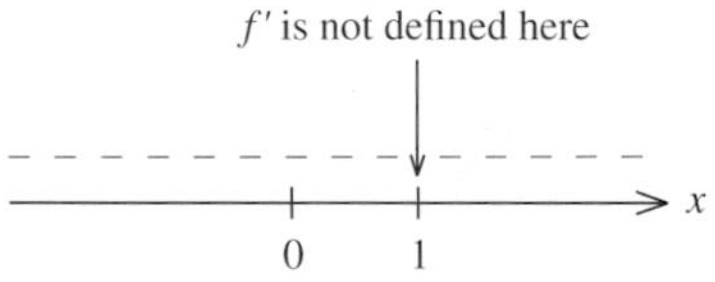

f' is not defined here

7. From the results of Step 6 we see that there are no critical points of f, since $f'(x)$ is never equal to zero for any value of x in the domain of f.

8.
$$f''(x) = \frac{d}{dx}[-2(x - 1)^{-2}] = 4(x - 1)^{-3} = \frac{4}{(x - 1)^3}.$$

The sign diagram of f'' shows immediately that f is concave downward on the interval $(-\infty, 1)$ and concave upward on the interval $(1, \infty)$ [Figure 4.62].

FIGURE 4.62
The sign diagram for f''

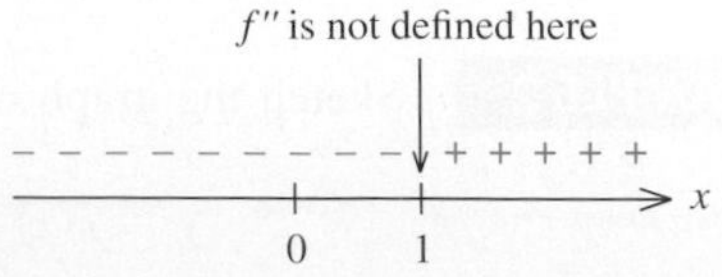

f'' is not defined here

9. From the results of Step 8 we see that there are no candidates for inflection points of f, since $f''(x)$ is never equal to zero for any value of x in the domain of f. Hence f has no inflection points.

10. Summarizing, we have

Domain	$(-\infty, 1) \cup (1, \infty)$
Intercepts	$(-1, 0)$; $(0, -1)$
Symmetry	None
$\lim\limits_{x \to -\infty} f(x)$; $\lim\limits_{x \to \infty} f(x)$	1; 1
Asymptotes	$x = 1$ is a vertical asymptote; $y = 1$ is a horizontal asymptote
Intervals where f is ↗ or ↘	↘ on $(-\infty, 1) \cup (1, \infty)$
Relative extrema	None
Concavity	Downward on $(-\infty, 1)$; upward on $(1, \infty)$
Points of inflection	None

FIGURE 4.63

The graph of f has a horizontal asymptote at $y = 1$ and a vertical asymptote at $x = 1$.

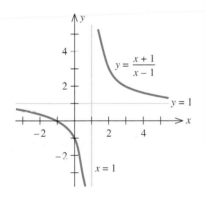

The graph of f is sketched in Figure 4.63.

EXAMPLE 3   Sketch the graph of the function

$$f(x) = \frac{x}{1 + x^2}.$$

Solution   First, obtain the following data on f:

1. Since $1 + x^2 \geq 1$ for all real values of x, the denominator of f is not equal to zero for any real value of x. Therefore, the domain of f is the set of all real numbers.

2. Setting $y = f(x) = 0$ gives 0 as the x-intercept of f. Setting $x = 0$ gives 0 as the y-intercept of f.

3. Since $f(-x) = -f(x)$, we see that f is symmetric with respect to the origin.

4. Since

$$\lim_{x \to \infty} \frac{x}{1 + x^2} = 0 \quad \text{and} \quad \lim_{x \to -\infty} \frac{x}{1 + x^2} = 0,$$

we see that $f(x)$ approaches zero as x becomes arbitrarily large in absolute value. When $x > 0$, $f(x) > 0$, and $f(x)$ approaches the line $y = 0$ from above. Similarly, when $x < 0$, $f(x) < 0$, and $f(x)$ approaches the line $y = 0$ from below.

5. There are no vertical asymptotes of f. However, $y = 0$ is a horizontal asymptote of f (see Step 4).

FIGURE 4.64
Sign diagram for f'

6.
$$f'(x) = \frac{(1 + x^2)(1) - x(2x)}{(1 + x^2)^2} = \frac{1 - x^2}{(1 + x^2)^2}$$

Setting $f'(x) = 0$ gives $x = \pm 1$. The sign diagram of f' shows that f is decreasing on the intervals $(-\infty, -1)$ and $(1, \infty)$ and increasing on the interval $(-1, 1)$ [Figure 4.64].

7. From the results of Step 6 we see that f has critical points at $x = -1$ and $x = 1$. The First Derivative Test confirms that $f(-1) = -\frac{1}{2}$ is a relative minimum of f, and $f(1) = \frac{1}{2}$ is a relative maximum of f.

8.
$$f''(x) = \frac{(1 + x^2)^2(-2x) - (1 - x^2)2(1 + x^2)(2x)}{(1 + x^2)^4}$$
$$= \frac{(1 + x^2)[-2x - 2x^3 - 4x + 4x^3]}{(1 + x^2)^4} = \frac{2x(x^2 - 3)}{(1 + x^2)^3}$$

FIGURE 4.65
Sign diagram for f''

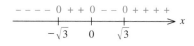

Setting $f''(x) = 0$ gives $x = 0$, $-\sqrt{3}$, or $\sqrt{3}$. The sign diagram of f'' shows that f is concave downward on the intervals $(-\infty, -\sqrt{3})$ and $(0, \sqrt{3})$ and concave upward on the intervals $(-\sqrt{3}, 0)$ and $(\sqrt{3}, \infty)$ [Figure 4.65].

9. From the results of Step 8 we see that $x = \pm\sqrt{3}$ and $x = 0$ are candidates for inflection points of f. The sign diagram of f'' verifies that the points $(-\sqrt{3}, -\sqrt{3}/4)$, $(0, 0)$, and $(\sqrt{3}, \sqrt{3}/4)$ are, in fact, inflection points of f.

10. Summarizing, we have

Domain	$(-\infty, \infty)$
Intercepts	$(0, 0)$
Symmetry	With respect to the origin
$\lim\limits_{x \to -\infty} f(x)$; $\lim\limits_{x \to \infty} f(x)$	0; 0
Asymptotes	$y = 0$ is a horizontal asymptote.
Intervals where f is ↗ or ↘	↘ on $(-\infty, -1) \cup (1, \infty)$; ↗ on $(-1, 1)$
Relative extrema	Rel. min. at $\left(-1, -\frac{1}{2}\right)$; rel. max. at $\left(1, \frac{1}{2}\right)$
Concavity	Downward on $(-\infty, -\sqrt{3}) \cup (0, \sqrt{3})$; upward on $(-\sqrt{3}, 0) \cup (\sqrt{3}, \infty)$
Points of inflection	$\left(-\sqrt{3}, -\frac{\sqrt{3}}{4}\right)$, $(0, 0)$, and $\left(\sqrt{3}, \frac{\sqrt{3}}{4}\right)$

The graph of *f* is sketched in Figure 4.66.

FIGURE 4.66

The graph of $y = \dfrac{x}{1 + x^2}$

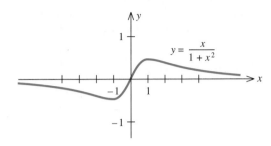

REMARK   Example 3 shows that it is possible for the graph of a function to intersect one of its asymptotes. In this case, the graph of *f* intersects the asymptote $y = 0$ of *f*.   □

SELF-CHECK EXERCISES 4.5

1. Sketch the graph of the function

$$f(x) = \frac{2}{3}x^3 - 2x^2 - 6x + 4.$$

Solutions to Self-Check Exercises 4.5 can be found on page 252.

4.5   EXERCISES

In Exercises 1–26, sketch the graph of the given function using the curve-sketching guide of this section.

1. $f(x) = x^2 - 2x + 3$
2. $g(x) = 4 - 3x - 2x^3$
3. $f(x) = 2x^3 + 1$
4. $h(x) = x^3 - 3x + 1$
5. $f(t) = 2t^3 - 15t^2 + 36t - 20$
6. $f(x) = -2x^3 + 3x^2 + 12x + 2$
7. $f(x) = x^3 - 6x^2 + 9x + 3$
8. $g(x) = \frac{1}{3}x^3 - 2x^2 + 2x + 1$
9. $f(t) = 3t^4 + 4t^3$
10. $h(x) = \frac{3}{2}x^4 - 2x^3 - 6x^2 + 8$
11. $f(x) = \sqrt{x^2 + 5}$
12. $f(t) = \sqrt{t^2 - 4}$
13. $f(x) = \sqrt[3]{x^2}$
14. $g(x) = \frac{1}{2}x - \sqrt{x}$
15. $f(x) = \dfrac{1}{x + 1}$
16. $g(x) = \dfrac{2}{x - 1}$
17. $g(x) = \dfrac{x}{x - 1}$
18. $h(x) = \dfrac{x + 2}{x - 2}$
19. $g(x) = \dfrac{x}{x^2 - 4}$
20. $f(t) = \dfrac{t^2}{1 + t^2}$

21. $f(x) = \dfrac{x^2 - 9}{x^2 - 4}$
22. $g(t) = -\dfrac{t^2 - 2}{t - 1}$
23. $h(x) = \dfrac{1}{x^2 - x - 2}$
24. $g(t) = \dfrac{t + 1}{t^2 - 2t - 1}$
25. $g(x) = (x + 2)^{3/2} + 1$
26. $h(x) = (x - 1)^{2/3} + 1$

27. **Worker Efficiency**   An efficiency study showed that the total number of cordless telephones assembled by an average worker at Delphi Electronics *t* hours after starting work at 8 A.M. is given by

$$N(t) = -\frac{1}{2}t^3 + 3t^2 + 10t \qquad (0 \le t \le 4).$$

Sketch the graph of the function *N* and interpret your results.

28. **GNP of a Developing Country**   A developing country's gross national product (GNP) from 1986 to 1994 is approximated by the function

$$G(t) = -0.2t^3 + 2.4t^2 + 60 \qquad (0 \le t \le 8),$$

where $G(t)$ is measured in billions of dollars and $t = 0$ corresponds to the year 1986. Sketch the graph of the function *G* and interpret your results.

29. **Concentration of a Drug in the Bloodstream**   The concentration of a certain drug in a patient's bloodstream t hours after injection is given by

$$C(t) = \frac{0.2t}{t^2 + 1}$$

milligrams per cubic centimeter. Sketch the graph of the function C and interpret your results.

30. **Oxygen Content of a Pond**   When organic waste is dumped into a pond, the oxidation process that takes place reduces the pond's oxygen content. However, given time, nature will restore the oxygen content to its natural level. Suppose that the oxygen content t days after organic waste has been dumped into the pond is given by

$$f(t) = 100\left[\frac{t^2 - 4t + 4}{t^2 + 4}\right] \qquad (0 \leq t < \infty)$$

percent of its normal level. Sketch the graph of the function f and interpret your results.

31. **Box Office Receipts**   The total worldwide box office receipts for a long-running movie are approximated by the function

$$T(x) = \frac{120x^2}{x^2 + 4},$$

where $T(x)$ is measured in millions of dollars and x is the number of years since the movie's release. Sketch the graph of the function T and interpret your results.

32. **Cost of Removing Toxic Pollutants**   A city's main well was recently found to be contaminated with trichloroethylene, a cancer-causing chemical, as a result of an abandoned chemical dump leaching chemicals into the water. A proposal submitted to the city's selectmen indicates that the cost, measured in millions of dollars, of removing x percent of the toxic pollutant is given by

$$C(x) = \frac{0.5x}{100 - x}.$$

Sketch the graph of the function C and interpret your results.

C   *In Exercises 33–36, use a computer or a graphing calculator to plot the graph of f.*

33. $f(x) = x^5 - 4x^3 + 2x + 1$

34. $f(x) = \dfrac{x^3 - 2x + 1}{x^2 - 2x + 3}$ 35. $f(x) = \dfrac{x + 1}{\sqrt{2x^2 + 1}}$

36. $f(x) = \dfrac{\sqrt{x} + 1}{x^2 - 3x + 1}$

SOLUTIONS TO SELF-CHECK EXERCISES 4.5

1. We obtain the following information on the graph of f.
   (1) The domain of f is the interval $(-\infty, \infty)$.
   (2) By setting $x = 0$, we find the y-intercept is 4.
   (3) There is no symmetry of f with respect to the y-axis or the origin since $f(-x) \neq f(x)$ and $f(-x) \neq -f(x)$, as you may verify.
   (4) Since

   $$\lim_{x \to -\infty} f(x) = \lim_{x \to -\infty}\left(\frac{2}{3}x^3 - 2x^2 - 6x + 4\right) = -\infty$$

   and

   $$\lim_{x \to \infty} f(x) = \lim_{x \to \infty}\left(\frac{2}{3}x^3 - 2x^2 - 6x + 4\right) = \infty,$$

   we see that $f(x)$ decreases without bound as x decreases without bound, and $f(x)$ increases without bound as x increases without bound.
   (5) Since f is a polynomial function, there are no asymptotes.
   (6) $f'(x) = 2x^2 - 4x - 6 = 2(x^2 - 2x - 3)$
   $$= 2(x + 1)(x - 3)$$

   Setting $f'(x) = 0$ gives $x = -1$ or $x = 3$. The sign diagram for f' shows that f is increasing on the intervals $(-\infty, -1)$ and $(3, \infty)$ and decreasing on $(-1, 3)$.

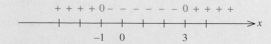

   Sign diagram for f'

(7) From the results of Step 6 we see that $x = -1$ and $x = 3$ are critical points of f. Furthermore, the sign diagram of f' tells us that $x = -1$ gives rise to a relative maximum of f and $x = 3$ gives rise to a relative minimum of f. Now,

$$f(-1) = \frac{2}{3}(-1)^3 - 2(-1)^2 - 6(-1) + 4 = \frac{22}{3}$$

and

$$f(3) = \frac{2}{3}(3)^3 - 2(3)^2 - 6(3) + 4 = -14,$$

so $f(-1) = \frac{22}{3}$ is a relative maximum of f and $f(3) = -14$ is a relative minimum of f.

(8) $$f''(x) = 4x - 4 = 4(x - 1),$$

which is equal to zero when $x = 1$. The sign diagram of f'' shows that f is concave downward on the interval $(-\infty, 1)$ and concave upward on the interval $(1, \infty)$.

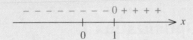

Sign diagram for f''

(9) From the results of Step 8 we see that $x = 1$ is the only candidate for an inflection point of f. Since $f''(x)$ changes sign as we move across the point $x = 1$ and

$$f(1) = \frac{2}{3}(1)^3 - 2(1)^2 - 6(1) + 4 = -\frac{10}{3},$$

we see that the required inflection point is $\left(1, -\frac{10}{3}\right)$.

(10) Summarizing this information, we have

Domain	$(-\infty, \infty)$
Intercept	$(0, 4)$
Symmetry	None
Intervals where f is ↗ or ↘	↗ on $(-\infty, -1) \cup (3, \infty)$; ↘ on $(-1, 3)$
Relative extrema	Rel. max. at $\left(-1, \frac{22}{3}\right)$; rel. min. at $(3, -14)$
Concavity	Downward on $(-\infty, 1)$; upward on $(1, \infty)$
Point of inflection	$\left(1, -\frac{10}{3}\right)$
$\lim\limits_{x \to -\infty} f(x)$; $\lim\limits_{x \to \infty} f(x)$	$-\infty$; ∞
Asymptotes	None

The graph of f is sketched in the accompanying figure.

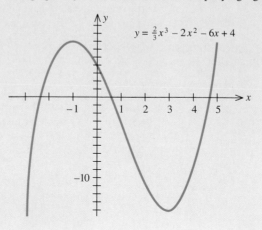

$$y = \tfrac{2}{3}x^3 - 2x^2 - 6x + 4$$

4.6   Optimization I

Absolute Extrema

The graph of the function f in Figure 4.67 shows the average age of cars in use in the United States from the beginning of 1946 ($t = 0$) to the beginning of 1990 ($t = 44$). Observe that the highest average age of cars in use during this period is 9 years, whereas the lowest average age of cars in use during the same period is $5\frac{1}{2}$ years. The number 9, the largest value of $f(t)$ for all values of t in the interval $[0, 44]$ (the domain of f) is called the *absolute maximum value of f* on that interval. The number $5\frac{1}{2}$, the smallest value of $f(t)$ for all values of t in $[0, 44]$, is called the *absolute minimum value of f* on that interval. Notice, too, that the absolute maximum value of f is attained at the endpoint $t = 0$ of the interval, whereas the absolute minimum value of f is attained at the two interior points $t = 12$ (corresponding to 1956) and $t = 23$ (corresponding to 1969).

FIGURE 4.67

$f(t)$ gives the average age of cars in use in year t, t in [0, 44].

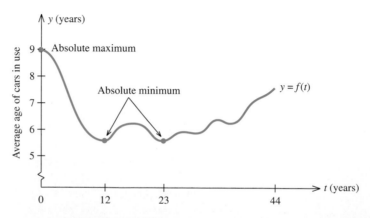

A precise definition of the absolute extrema (absolute maximum or absolute minimum) of a function follows.

THE ABSOLUTE EXTREMA OF A FUNCTION *f*

If $f(x) \leq f(c)$ for all x in the domain of f, then $f(c)$ is called the **absolute maximum value** of f.

If $f(x) \geq f(c)$ for all x in the domain of f, then $f(c)$ is called the **absolute minimum value** of f.

Figure 4.68 shows the graphs of several functions and gives the absolute maximum and absolute minimum of each function, if they exist.

FIGURE 4.68

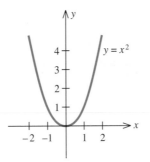

(a) $f(0) = 0$ is the absolute minimum of f; f has no absolute maximum

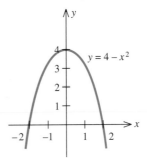

(b) $f(0) = 4$ is the absolute maximum of f; f has no absolute minimum

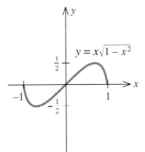

(c) $f\left(\dfrac{\sqrt{2}}{2}\right) = \dfrac{1}{2}$ is the absolute maximum of f.

$f\left(-\dfrac{\sqrt{2}}{2}\right) = -\dfrac{1}{2}$ is the absolute minimum of f.

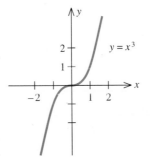

(d) f has no absolute extrema.

Absolute Extrema on a Closed Interval

As the preceding examples show, a continuous function defined on an arbitrary interval does not always have an absolute maximum or an absolute minimum. But there is an important case that arises often in practical applications in which both the absolute maximum and the absolute minimum of a function are guaranteed to exist. This occurs when a continuous function is defined on a *closed* interval. Let us state this important result in the form of a theorem, whose proof we will omit.

THEOREM 3

If a function f is continuous on a closed interval $[a, b]$, then f has both an absolute maximum value and an absolute minimum value on $[a, b]$.

Observe that if an absolute extremum of a continuous function f occurs at a point in an open interval (a, b), then it must be a relative extremum of f and hence its x-coordinate must be a critical point of f. Otherwise, the absolute extremum of f must occur at one or both of the endpoints of the interval $[a, b]$. A typical situation is illustrated in Figure 4.69.

FIGURE 4.69

The relative minimum of f at x_3 is the absolute minimum of f. The right endpoint b gives rise to the absolute maximum value f(b) of f.

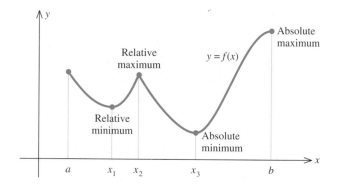

Here x_1, x_2, and x_3 are critical points of f. The absolute minimum of f occurs at x_3, which lies in the open interval (a, b) and is a critical point of f. The absolute maximum of f occurs at b, an endpoint. This observation suggests the following procedure for finding the absolute extrema of a continuous function on a closed interval.

FINDING THE ABSOLUTE EXTREMA OF f ON A CLOSED INTERVAL

1. Find the critical points of f that lie in (a, b).
2. Compute the value of f at each of these critical points of f and compute $f(a)$ and $f(b)$.
3. The absolute maximum value and absolute minimum value of f will correspond to the largest and smallest numbers, respectively, found in Step 2.

EXAMPLE 1   Find the absolute extrema of the function $F(x) = x^2$ defined on the interval $[-1, 2]$.

Solution   The function F is continuous on the closed interval $[-1, 2]$ and differentiable on the open interval $(-1, 2)$. The derivative of F is

$$F'(x) = 2x,$$

so $x = 0$ is the only critical point of F. Next, evaluate $F(x)$ at $x = -1$, $x = 0$, and $x = 2$. Thus,

$$F(-1) = 1, \qquad F(0) = 0, \quad \text{and} \quad F(2) = 4.$$

FIGURE 4.70

FIGURE 4.70

F has an absolute minimum value of 0 and an absolute maximum value of 4.

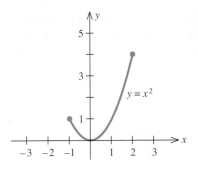

It follows that 0 is the absolute minimum value of F and 4 is the absolute maximum value of F. The graph of F, which appears in Figure 4.70, confirms our results. ■

EXAMPLE 2   Find the absolute extrema of the function

$$f(x) = x^3 - 2x^2 - 4x + 4$$

defined on the interval $[0, 3]$.

Solution   The function f is continuous on the closed interval $[0, 3]$ and differentiable on the open interval $(0, 3)$. The derivative of f is

$$f'(x) = 3x^2 - 4x - 4 = (3x + 2)(x - 2),$$

and it is equal to zero when $x = -\frac{2}{3}$ and $x = 2$. Since the point $x = -\frac{2}{3}$ lies outside the interval $[0, 3]$, it is dropped from further consideration and $x = 2$ is seen to be the sole critical point of f. Next, we evaluate $f(x)$ at the critical point of f as well as at the endpoints of f, obtaining

$$f(0) = 4, \quad f(2) = -4, \quad \text{and} \quad f(3) = 1.$$

From these results, we conclude that -4 is the absolute minimum value of f and 4 is the absolute maximum value of f. The graph of f, which appears in Figure 4.71, confirms our results. Observe that the absolute maximum of f occurs at the endpoint $x = 0$ of the interval $[0, 3]$, while the absolute minimum of f occurs at $x = 2$, which is a point in the interval $(0, 3)$.

FIGURE 4.71

f has an absolute maximum value of 4 and an absolute minimum value of -4.

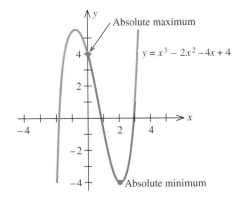

EXAMPLE 3   Find the absolute maximum and absolute minimum values of the function $f(x) = x^{2/3}$ on the interval $[-1, 8]$.

Solution   The derivative of f is

$$f'(x) = \frac{2}{3}x^{-1/3} = \frac{2}{3x^{1/3}}.$$

Note that f' is not defined at $x = 0$, is continuous everywhere else, and is not equal to zero for all x. Therefore, $x = 0$ is the only critical point of f. Evaluating $f(x)$ at $x = -1, 0,$ and 8, we obtain

$$f(-1) = 1, \quad f(0) = 0, \quad \text{and} \quad f(8) = 4.$$

FIGURE 4.72
*f has an absolute minimum
value of f(0) = 0 and an
absolute maximum value of
f(8) = 4.*

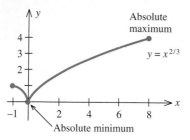

We conclude that the absolute minimum value of f is 0, attained at $x = 0$, and the absolute maximum value of f is 4, attained at $x = 8$ (Figure 4.72). ∎

Applications

Many real-world applications call for finding the absolute maximum value or the absolute minimum value of a given function. For example, management is interested in finding what level of production will yield the maximum profit for a company; a farmer is interested in finding the right amount of fertilizer to maximize crop yield; a doctor is interested in finding the maximum concentration of a drug in a patient's body and the time at which it occurs; and an engineer is interested in finding the dimensions of a container with a specified shape and volume that can be constructed at a minimum cost.

EXAMPLE 4   The Acrosonic Company's total profit from manufacturing and selling x units of their Model F loudspeaker systems is given by

$$P(x) = -0.02x^2 + 300x - 200{,}000 \qquad (0 \le x \le 20{,}000)$$

dollars. How many units of the loudspeaker system must Acrosonic produce to maximize its profits?

Solution   To find the absolute maximum of P on [0, 20,000], first find the critical points of P on the interval (0, 20,000). To do this, compute

$$P'(x) = -0.04x + 300.$$

Solving the equation $P'(x) = 0$ gives $x = 7{,}500$. Next, evaluate $P(x)$ at $x = 7{,}500$ as well as at the endpoints $x = 0$ and $x = 20{,}000$ of the interval [0, 20,000], obtaining

$$P(0) = -200{,}000$$
$$P(7{,}500) = 925{,}000$$
and
$$P(20{,}000) = -2{,}200{,}000.$$

From these computations we see that the absolute maximum value of the function P is 925,000. Thus, by producing 7,500 units, Acrosonic will realize a maximum profit of $925,000. The graph of P is sketched in Figure 4.73. ∎

FIGURE 4.73
*P has an absolute maximum at
(7,500, 925,000).*

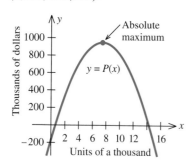

EXAMPLE 5   When a person coughs, the trachea, or windpipe, contracts, allowing air to be expelled at a maximum velocity. It can be shown that, during a cough, the velocity v of airflow is given by the function

$$v = f(r) = kr^2(R - r),$$

where r is the trachea's radius (in centimeters) during a cough, R is the trachea's normal radius (in centimeters), and k is a positive constant that depends on the length of the trachea. Find the radius r for which the velocity of airflow is greatest.

Solution   To find the absolute maximum of f on $[0, R]$, first find the critical points of f on the interval $(0, R)$. We compute

$$f'(r) = 2kr(R - r) - kr^2 \qquad \text{(Using the Product Rule)}$$
$$= -3kr^2 + 2kRr = kr(-3r + 2R).$$

Setting $f'(r) = 0$ gives $r = 0$ or $r = \frac{2}{3}R$, and so $r = \frac{2}{3}R$ is the sole critical point of f ($r = 0$ is an endpoint). Evaluating $f(r)$ at $r = \frac{2}{3}R$, as well as at the endpoints $r = 0$ and $r = R$, we obtain

$$f(0) = 0$$
$$f\left(\frac{2}{3}R\right) = \frac{4k}{27}R^3$$

and
$$f(R) = 0,$$

from which we deduce that the velocity of airflow is greatest when the radius of the contracted trachea is $\frac{2}{3}R$, that is, when the radius is contracted by approximately 33 percent. The graph of the function f is shown in Figure 4.74.

FIGURE 4.74
The velocity of airflow is greatest when the radius of the contracted trachea is $\frac{2}{3}R$.

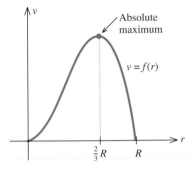

 EXAMPLE 6   The daily average cost function (in dollars per unit) of the Elektra Electronics Company is given by

$$\overline{C}(x) = 0.0001x^2 - 0.08x + 40 + \frac{5,000}{x} \qquad (x > 0),$$

where x stands for the number of programmable electronic calculators Elektra produces. Show that a production level of 500 units per day results in a minimum average cost for the company.

Solution   The domain of the function $\overline{C}$ is the interval $(0, \infty)$, which is not closed. In order to solve the problem, we resort to the graphic method. Using the techniques of graphing from the last section, we sketch the graph of $\overline{C}$ (Figure 4.75).

Now,

$$\overline{C}'(x) = 0.0002x - 0.08 - \frac{5,000}{x^2}.$$

Substituting the given value of x, 500, into $\overline{C}'(x)$ gives $\overline{C}'(500) = 0$, so $x = 500$ is a critical point of $\overline{C}$. Next,

$$\overline{C}''(x) = 0.0002 + \frac{10,000}{x^3}.$$

Thus,
$$\overline{C}''(500) = 0.0002 + \frac{10,000}{(500)^3} > 0,$$

and by the Second Derivative Test, a relative minimum of the function $\overline{C}$ occurs at the point $x = 500$. Furthermore, $\overline{C}''(x) > 0$ for $x > 0$, which implies that the graph of $\overline{C}$ is concave upward everywhere, so the relative minimum of $\overline{C}$ must be the absolute minimum of $\overline{C}$. The minimum average cost is given by

FIGURE 4.75
The minimum average cost is $35/unit.

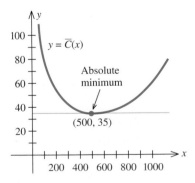

$$\overline{C}(500) = 0.0001(500)^2 - 0.08(500) + 40 + \frac{5,000}{500}$$

$$= 35,$$

or $35 per unit.   ∎

EXAMPLE 7   The altitude (in feet) of a rocket t seconds into flight is given by

$$s = f(t) = -t^3 + 96t^2 + 195t + 5 \qquad (t \geq 0).$$

a. Find the maximum altitude attained by the rocket.
b. Find the maximum velocity attained by the rocket.

Solution

a. The maximum altitude attained by the rocket is given by the largest value of the function f in the closed interval $[0, T]$, where T denotes the time the rocket impacts the Earth. We know that such a number exists because the dominant term in the expression for the continuous function f is $-t^3$. So for t large enough, the value of $f(t)$ must change from positive to negative and, in particular, it must attain the value 0 for some T.

To find the absolute maximum of f, compute

$$f'(t) = -3t^2 + 192t + 195$$
$$= -3(t - 65)(t + 1)$$

and solve the equation $f'(t) = 0$, obtaining $t = -1$ and $t = 65$. Ignore $t = -1$, since it lies outside the interval $[0, T]$. This leaves the critical point $t = 65$ of f. Continuing, we compute

$$f(0) = 5, \qquad f(65) = 143,655, \quad \text{and} \quad f(T) = 0$$

and conclude, accordingly, that the absolute maximum value of f is 143,655. Thus, the maximum altitude of the rocket is 143,655 feet, attained 65 seconds into flight. The graph of f is sketched in Figure 4.76.

b. To find the maximum velocity attained by the rocket, find the largest value of the function that describes the rocket's velocity at any time t, namely,

$$v = f'(t) = -3t^2 + 192t + 195 \qquad (t \geq 0).$$

We find the critical point of v by setting $v' = 0$. But

$$v' = -6t + 192$$

and the critical point of v is $t = 32$. Since

$$v'' = -6 < 0,$$

the Second Derivative Test implies that a relative maximum of v occurs at $t = 32$. Our computation has, in fact, clarified the property of the "velocity curve." Since $v'' < 0$ everywhere, the velocity curve is concave downward everywhere. With this observation, we assert that the relative maximum must, in fact, be the absolute maximum of v. The maximum velocity of the rocket is given by evaluating v at $t = 32$,

$$f'(32) = -3(32)^2 + 192(32) + 195,$$

FIGURE 4.76
The maximum altitude of the rocket is 143,655 feet.

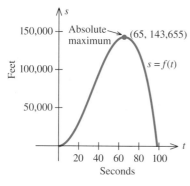

or 3267 ft/sec. The graph of the velocity function v is sketched in Figure 4.77.

FIGURE 4.77

The maximum velocity of the rocket is 3267 ft/sec.

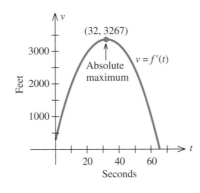

SELF-CHECK EXERCISES 4.6

1. Let $f(x) = x - 2\sqrt{x}$.

   a. Find the absolute extrema of f on the interval $[0, 9]$.
   b. Find the absolute extrema of f.

2. Find the absolute extrema of $f(x) = 3x^4 + 4x^3 + 1$ on $[-2, 1]$.

c 3. The operating rate (expressed as a percentage) of factories, mines, and utilities in a certain region of the country on the tth day of the year 1994 is given by the function

$$f(t) = 80 + \frac{1{,}200\,t}{t^2 + 40{,}000} \qquad (0 \le t \le 250).$$

   On which day of the first 250 days of 1994 was the manufacturing capacity operating rate highest?

Solutions to Self-Check Exercises 4.6 can be found on page 265.

4.6   EXERCISES

In Exercises 1–8, you are given the graph of some function f defined on the indicated interval. Find the absolute maximum and the absolute minimum of f, if they exist.

1.
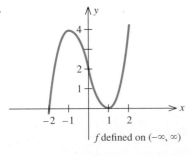

f defined on $(-\infty, \infty)$

2.
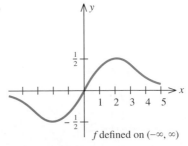

f defined on $(-\infty, \infty)$

3.

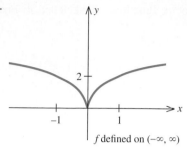

f defined on $(-\infty, \infty)$

4.

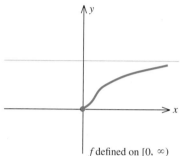

f defined on $[0, \infty)$

5.

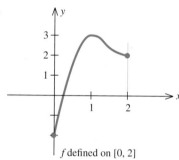

f defined on $[0, 2]$

6.

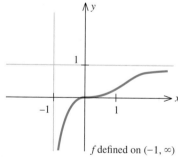

f defined on $(-1, \infty)$

7.

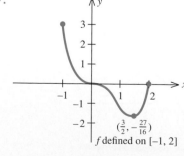

$\left(\frac{3}{2}, -\frac{27}{16}\right)$

f defined on $[-1, 2]$

8.

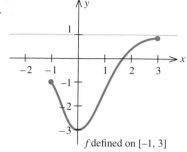

f defined on $[-1, 3]$

In Exercises 9–38, find the absolute maximum value and the absolute minimum value, if any, of the given function.

9. $f(x) = 2x^2 + 3x - 4$ **10.** $g(x) = -x^2 + 4x + 3$

11. $h(x) = x^{1/3}$ **12.** $f(x) = x^{2/3}$

13. $f(x) = \dfrac{1}{1 + x^2}$ **14.** $f(x) = \dfrac{x}{1 + x^2}$

15. $f(x) = x^2 - 2x - 3$ on $[-2, 3]$

16. $g(x) = x^2 - 2x - 3$ on $[0, 4]$

17. $f(x) = -x^2 + 4x + 6$ on $[0, 5]$

18. $f(x) = -x^2 + 4x + 6$ on $[3, 6]$

19. $f(x) = x^3 + 3x^2 - 1$ on $[-3, 2]$

20. $g(x) = x^3 + 3x^2 - 1$ on $[-3, 1]$

21. $g(x) = 3x^4 + 4x^3$ on $[-2, 1]$

22. $f(x) = \frac{1}{2}x^4 - \frac{2}{3}x^3 - 2x^2 + 3$ on $[-2, 3]$

23. $f(x) = \dfrac{x + 1}{x - 1}$ on $[0, 3]$ **24.** $g(t) = \dfrac{t}{t - 1}$ on $[2, 4]$

25. $f(x) = 4x + \dfrac{1}{x}$ on $[1, 3]$ **26.** $f(x) = 9x - \dfrac{1}{x}$ on $[1, 3]$

27. $f(x) = \frac{1}{2}x^2 - 2\sqrt{x}$ on $[0, 3]$

28. $g(x) = \frac{1}{8}x^2 - 4\sqrt{x}$ on $[0, 9]$

29. $f(x) = \dfrac{1}{x}$ on $(0, \infty)$ **30.** $g(x) = \dfrac{1}{x + 1}$ on $(0, \infty)$

31. $f(x) = 3x^{2/3} - 2x$ on $[0, 3]$

32. $g(x) = x^2 + 2x^{2/3}$ on $[-2, 2]$

33. $f(x) = x^{2/3}(x^2 - 4)$ on $[-1, 2]$

34. $f(x) = x^{2/3}(x^2 - 4)$ on $[-1, 3]$

35. $f(x) = \dfrac{x}{x^2 + 2}$ on $[-1, 2]$

36. $f(x) = \dfrac{1}{x^2 + 2x + 5}$ on $[-2, 1]$

37. $f(x) = \dfrac{x}{\sqrt{x^2 + 1}}$ on $[-1, 1]$

38. $g(x) = x\sqrt{4 - x^2}$ on $[0, 2]$

39. A stone is thrown straight up from the roof of an 80-foot building. The height of the stone at any time t (in seconds), measured from the ground, is given by

$$h(t) = -16t^2 + 64t + 80$$

feet. What is the maximum height the stone reaches?

40. **Maximizing Profits** Lynbrook West, an apartment complex, has 100 two-bedroom units. The monthly profit realized from renting out x apartments is given by

$$P(x) = -10x^2 + 1{,}760x - 50{,}000$$

dollars. Find how many units should be rented out in order to maximize the monthly rental profit. What is the maximum monthly profit realizable?

41. **Maximizing Profits** The estimated monthly profit realizable by the Cannon Precision Instruments Corporation for manufacturing and selling x units of its Model M1 camera is

$$P(x) = -0.04x^2 + 240x - 10{,}000$$

dollars. Determine how many cameras Cannon should produce per month in order to maximize its profits.

42. **Flight of a Rocket** The altitude (in feet) attained by a model rocket t seconds into flight is given by the function

$$h(t) = -\frac{1}{3}t^3 + 4t^2 + 20t + 2.$$

Find the maximum altitude attained by the rocket.

43. **Maximizing Profits** The management of Trappee and Sons, Inc., producers of the famous Texa-Pep hot sauce, estimate that their profit from the daily production and sale of x cases (each case consisting of 24 bottles) of the hot sauce is given by

$$P(x) = -0.000002x^3 + 6x - 400$$

dollars. What is the largest possible profit Trappee can make in one day?

44. **Maximizing Profits** The quantity demanded per month of the Walter Serkin recording of Beethoven's *Moonlight Sonata*, manufactured by Phonola Record Industries, is related to the price per record. The equation

$$p = -0.00042x + 6 \qquad (0 \le x \le 12{,}000),$$

where p denotes the unit price in dollars and x is the number of records demanded, relates the demand to the price. The total monthly cost for pressing and packaging x copies of this classical recording is given by

$$C(x) = 600 + 2x - 0.00002x^2 \qquad (0 \le x \le 20{,}000)$$

dollars. Determine how many copies Phonola should produce per month in order to maximize its profits. [*Hint:* The revenue is $R(x) = px$ and the profit is $P(x) = R(x) - C(x)$.]

45. **Maximizing Profit** A manufacturer of tennis rackets finds that the total cost $C(x)$ [in dollars] of manufacturing x rackets per day is given by $C(x) = 400 + 4x + 0.0001x^2$. Each racket can be sold at a price of p dollars, where p is related to x by the demand equation $p = 10 - 0.0004x$. If all the rackets that are manufactured can be sold, find the daily level of production that will yield a maximum profit for the manufacturer.

C 46. **Maximizing Profit** The weekly demand for the Pulsar 25-inch color console television is given by the demand equation

$$p = -0.05x + 600 \qquad (0 \le x \le 12{,}000),$$

where p denotes the wholesale unit price in dollars and x denotes the quantity demanded. The weekly total cost function associated with manufacturing these sets is given by

$$C(x) = 0.000002x^3 - 0.03x^2 + 400x + 80{,}000,$$

where $C(x)$ denotes the total cost incurred in producing x sets. Find the level of production that will yield a maximum profit for the manufacturer. [*Hint:* Use the quadratic formula.]

C 47. **Minimizing Average Costs** Suppose that the total cost function for manufacturing a certain product is $C(x) = 0.2(0.01x^2 + 120)$ dollars, where x represents the number of units produced. Find the level of production that will minimize the average cost.

48. **Minimizing Production Costs** The total monthly cost in dollars incurred by Cannon Precision Instruments Corporation for manufacturing x units of the Model M1 camera is given by the function

$$C(x) = 0.0025x^2 + 80x + 10{,}000.$$

a. Find the average cost function $\overline{C}$.
b. Find the level of production that results in the smallest average production cost.
c. Find the level of production for which the average cost is equal to the marginal cost.
d. Compare the result of (c) with that of (b).

49. **Minimizing Production Costs**   The daily total cost in dollars incurred by Trappee and Sons, Inc., for producing x cases of Texa-Pep hot sauce is given by the function

$$C(x) = 0.000002x^3 + 5x + 400.$$

Using this function, answer the questions posed in Exercise 48.

50. **Maximizing Revenue**   Suppose that the quantity demanded per week of a certain dress is related to the unit price p by the demand equation $p = \sqrt{800 - x}$, where p is in dollars and x is the number of dresses made. How many dresses should be made and sold per week in order to maximize the revenue? [Hint: $R(x) = px$.]

51. **Maximizing Revenue**   The quantity demanded per month of the Sicard wristwatch is related to the unit price by the equation

$$p = \frac{50}{0.01x^2 + 1} \qquad (0 \le x \le 20),$$

where p is measured in dollars and x is measured in units of a thousand. How many watches must be sold to yield a maximum revenue?

52. **Oxygen Content of a Pond**   When organic waste is dumped into a pond, the oxidation process that takes place reduces the pond's oxygen content. However, given time, nature will restore the oxygen content to its natural level. Suppose that the oxygen content t days after organic waste has been dumped into the pond is given by

$$f(t) = 100 \left[\frac{t^2 - 4t + 4}{t^2 + 4} \right] \qquad (0 \le t < \infty)$$

percent of its normal level.
a. When is the level of oxygen content lowest?
b. When is the rate of oxygen regeneration greatest?

c 53. **Air Pollution**   The amount of nitrogen dioxide, a brown gas that impairs breathing, present in the atmosphere on a certain May day in the city of Long Beach is approximated by

$$A(t) = \frac{136}{1 + 0.25(t - 4.5)^2} + 28 \qquad (0 \le t \le 11),$$

where $A(t)$ is measured in pollutant standard index (PSI) and t is measured in hours, with $t = 0$ corresponding to 7 A.M. Determine the time of day when the pollution is at its highest level.

54. **Minimizing Production Costs**   Prove that if a cost function $C(x)$ is concave upward ($C''(x) > 0$), then the

level of production that will result in the smallest average production cost occurs when

$$\overline{C}(x) = C'(x),$$

that is, when the average cost $\overline{C}(x)$ is equal to the marginal cost $C'(x)$. [Hints: 1. Show that

$$\overline{C}'(x) = \frac{xC'(x) - C(x)}{x^2},$$

so that the critical point of the function $\overline{C}$ occurs when

$$xC'(x) - C(x) = 0.$$

2. Show that

$$\overline{C}''(x) = \frac{C''(x)}{x}.$$

Use the Second Derivative Test to reach the desired conclusion.]

55. **Maximizing Revenue**   The average revenue is defined as the function

$$\overline{R}(x) = \frac{R(x)}{x} \qquad (x > 0).$$

Prove that if a revenue function $R(x)$ is concave downward ($R''(x) < 0$), then the level of sales that will result in the largest average revenue occurs when $\overline{R}(x) = R'(x)$. [Hint: Use an approach similar to the one given in the hint for Exercise 54.]

56. **Velocity of Blood**   According to a law discovered by the nineteenth-century physician Jean Louis Marie Poiseuille, the velocity (in centimeters per second) of blood r centimeters from the central axis of an artery is given by

$$v(r) = k(R^2 - r^2),$$

where k is a constant and R is the radius of the artery. Show that the velocity of blood is greatest along the central axis.

57. **GNP of a Developing Country**   A developing country's gross national product (GNP) from 1986 to 1994 is approximated by the function

$$G(t) = -0.2t^3 + 2.4t^2 + 60 \qquad (0 \le t \le 8),$$

where $G(t)$ is measured in billions of dollars and $t = 0$ corresponds to the year 1986. Show that the growth rate of the country's GNP was maximal in 1990.

58. **Crime Rates**   The number of major crimes committed in the city of Bronxville between 1987 and 1994 is approximated by the function

$$N(t) = -0.1t^3 + 1.5t^2 + 100 \qquad (0 \le t \le 7),$$

where $N(t)$ denotes the number of crimes committed in year t ($t = 0$ corresponds to the year 1987). Enraged by the dramatic increase in the crime rate, the citizens of Bronxville, with the help of the local police, organized "Neighborhood Crime Watch" groups in early 1991 to combat this menace. Show that the growth in the crime rate was maximal in 1992, giving credence to the claim that the Neighborhood Crime Watch program was working.

C **59. Social Security Surplus** Based on data from the Social Security Administration, the estimated cash in the Social Security retirement and disability trust funds in the five decades beginning with the year 1990 is given by

$$A(t) = -96.6t^4 + 403.6t^3 + 660.9t^2 + 250$$
$$(0 \le t \le 5),$$

where $A(t)$ is measured in billions of dollars and t is measured in decades, with $t = 0$ corresponding to the year 1990. Show that the social security surplus will be at its highest level at approximately the beginning of the year 2030. [*Hint:* Use the quadratic formula.]

60. Energy Expended by a Fish It has been conjectured that a fish swimming a distance of L ft at a speed of v ft/sec relative to the water and against a current flowing at the rate of u ft/sec ($u < v$) expends a total energy given by

$$E(v) = \frac{aLv^3}{v - u},$$

where E is measured in foot-pounds (ft-lbs), and a is a constant. Find the speed v at which the fish must swim in order to minimize the total energy expended. (*Note:* This result has been verified by biologists.)

61. Let f be a constant function; that is, $f(x) = c$ where c is some real number. Show that every point $x = a$ is an absolute maximum and, at the same time, an absolute minimum of f.

62. Show that a polynomial function defined on the interval $(-\infty, \infty)$ cannot have both an absolute maximum and an absolute minimum unless it is a constant function.

63. One condition that must be satisfied before Theorem 3 (page 256) is applicable is that the function f must be continuous on the closed interval $[a, b]$. Define a function f on the closed interval $[-1, 1]$ by

$$f(x) = \begin{cases} \dfrac{1}{x} & \text{if } x \in [-1, 1] \quad (x \ne 0) \\ 0 & \text{if } x = 0. \end{cases}$$

a. Show that f is not continuous at $x = 0$.
b. Show that $f(x)$ does not attain an absolute maximum or an absolute minimum on the interval $[-1, 1]$.
c. Confirm your results by sketching the function f.

C *In Exercises 64–67, use a computer or a graphing calculator to plot the graph of f. Then zoom in on the graph to find the absolute maximum value and the absolute minimum value (if any) of f on the indicated interval.*

64. $f(x) = x^5 - 3x^4 + 2x + 1$ on $[-2, 3]$

65. $f(x) = \dfrac{x^3 - 1}{x^2}$ on $[1, 3]$

66. $f(x) = \dfrac{x^3 - x^2 + 1}{x - 2}$ on $[1, 3]$

67. $f(x) = \dfrac{2x - 3}{x^2 + 1}$ on $[3, \infty)$

SOLUTIONS TO SELF-CHECK EXERCISES 4.6

1. a. The function f is continuous in its domain and differentiable in the interval $(0, 9)$. The derivative of f is

$$f'(x) = 1 - x^{-1/2} = \frac{x^{1/2} - 1}{x^{1/2}},$$

and it is equal to zero when $x = 1$. Evaluating $f(x)$ at the endpoints $x = 0$ and $x = 9$ and at the critical point $x = 1$ of f, we have

$$f(0) = 0, \quad f(1) = -1, \quad \text{and} \quad f(9) = 3.$$

From these results, we see that -1 is the absolute minimum value of f and 3 is the absolute maximum value of f.

b. In this case the domain of f is the interval $[0, \infty)$, which is not closed. Therefore, we resort to the graphic method. Using the techniques of graphing, we sketch the graph of f depicted in the accompanying figure.

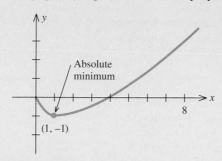

The graph of f shows that -1 is the absolute minimum value of f, but f has no absolute maximum since $f(x)$ increases without bound as x increases without bound.

2. The function f is continuous on the interval $[-2, 1]$. It is also differentiable on the open interval $(-2, 1)$. The derivative of f is

$$f'(x) = 12x^3 + 12x^2 = 12x^2(x+1),$$

and it is continuous on $(-2, 1)$. Setting $f'(x) = 0$ gives $x = -1$ and $x = 0$ as critical points of f. Evaluating $f(x)$ at these critical points of f as well as at the endpoints of the interval $[-2, 1]$, we obtain

$$f(-2) = 17, \quad f(-1) = 0, \quad f(0) = 1, \quad \text{and} \quad f(1) = 8.$$

From these results, we see that 0 is the absolute minimum value of f and 17 is the absolute maximum value of f.

3. The problem is solved by finding the absolute maximum of the function f on $[0, 250]$. Differentiating $f(t)$, we obtain

$$f'(t) = \frac{(t^2 + 40{,}000)(1{,}200) - 1{,}200t(2t)}{(t^2 + 40{,}000)^2}$$

$$= \frac{-1{,}200\,(t^2 - 40{,}000)}{(t^2 + 40{,}000)^2}.$$

Upon setting $f'(t) = 0$ and solving the resulting equation, we obtain $t = -200$ or 200. Since -200 lies outside the interval $[0, 250]$, we are interested only in the critical point $t = 200$ of f. Evaluating $f(t)$ at $t = 0$, $t = 200$, and $t = 250$, we find

$$f(0) = 80, \quad f(200) = 83, \quad \text{and} \quad f(250) = 82.93.$$

We conclude that the manufacturing capacity operating rate was the highest on the 200th day of 1994, that is, a little past the middle of July of 1994.

4.7   # Optimization II

Section 4.6 outlined how to find the solution to certain optimization problems in which the objective function is given. In this section, we will consider problems in which we are required to first find the appropriate function to be optimized. The following guidelines will be useful for solving these problems.

GUIDELINES FOR SOLVING OPTIMIZATION PROBLEMS

1. Assign a letter to each variable mentioned in the problem. If appropriate, draw and label a figure.
2. Find an expression for the quantity to be optimized.
3. Use the conditions given in the problem to write the quantity to be optimized as a function f of *one* variable. Note any restrictions to be placed on the domain of f from physical considerations of the problem.
4. Optimize the function f over its domain using the methods of Section 4.6.

REMARK   In carrying out Step 4, remember that if the function f to be optimized is continuous on a closed interval, then the absolute maximum and absolute minimum of f are, respectively, the largest and smallest values of $f(x)$ on the set comprising the critical points of f and the endpoints of the interval. If the domain of f is not a closed interval, then we resort to the graphical method. ☐

Maximization Problems

EXAMPLE 1   A man wishes to have a rectangular-shaped garden in his backyard. He has 50 feet of fencing material with which to enclose his garden. Find the dimensions for the largest garden he can have if he uses all of the fencing material.

Solution

Step 1   Let x and y denote the dimensions (in feet) of two adjacent sides of the garden (Figure 4.78), and let A denote its area.

Step 2   The area of the garden,

$$A = xy, \qquad (1)$$

is the quantity to be maximized.

Step 3   The perimeter of the rectangle, $(2x + 2y)$ feet, must equal 50 feet. Therefore, we have the equation

$$2x + 2y = 50.$$

Next, solving this equation for y in terms of x yields

$$y = 25 - x, \qquad (2)$$

which, when substituted into (1), gives

$$A = x(25 - x)$$
$$= -x^2 + 25x.$$

(Remember, the function to be optimized must involve just one variable.) Since the sides of the rectangle must be nonnegative, we must have $x \geq 0$ and $y = 25 - x \geq 0$; that is, we must have $0 \leq x \leq 25$. Thus the problem is reduced to that of finding the absolute maximum of $A = f(x) = -x^2 + 25x$ on the closed interval $[0, 25]$.

FIGURE 4.78

What is the maximum rectangular area that can be enclosed with 50 feet of fencing?

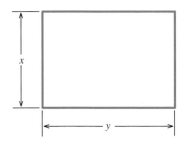

Step 4   Observe that f is continuous on $[0, 25]$. So the absolute maximum value of f must occur at the endpoint(s) or at the critical point(s) of f. The derivative of the function A is given by

$$A' = f'(x) = -2x + 25.$$

Setting $A' = 0$ gives
$$-2x + 25 = 0,$$

or $x = 12.5$, as the critical point of A. Next, we evaluate the function $A = f(x)$ at $x = 12.5$ and at the endpoints $x = 0$ and $x = 25$ of the interval $[0, 25]$, obtaining

$$f(0) = 0, \quad f(12.5) = 156.25, \quad \text{and} \quad f(25) = 0.$$

We see that the absolute maximum value of the function f is 156.25. From (2) we see that when $x = 12.5$, the value of y is given by $y = 12.5$. Thus, the garden would be of maximum area (156.25 square feet) if it were in the form of a square with sides 12.5 feet long.   ∎

EXAMPLE 2   By cutting away identical squares from each corner of a rectangular piece of cardboard and folding up the resulting flaps, the cardboard may be turned into an open box. If the cardboard is 16 inches long and 10 inches wide, find the dimensions of the box that will yield the maximum volume.

Solution

Step 1   Let x denote the length (in inches) of one side of each of the identical squares to be cut out of the cardboard (Figure 4.79), and let V denote the volume of the resulting box.

FIGURE 4.79

The dimensions of the open box are $(16 - 2x)$ by $(10 - 2x)$ by x inches.

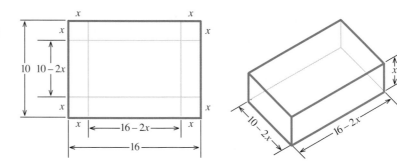

Step 2   The dimensions of the box are $(16 - 2x)$ inches long, $(10 - 2x)$ inches wide, and x inches high. Therefore, its volume,

$$\begin{aligned} V &= (16 - 2x)(10 - 2x)x \\ &= 4(x^3 - 13x^2 + 40x) \qquad \text{(Expanding the expression)} \end{aligned}$$

cubic inches, is the quantity to be maximized.

Step 3   Since each side of the box must be nonnegative, x must satisfy the inequalities $x \geq 0$, $16 - 2x \geq 0$, and $10 - 2x \geq 0$. This set of inequalities is satisfied if $0 \leq x \leq 5$. Thus the problem at hand is equivalent to that of finding the absolute maximum of

$$V = f(x) = 4(x^3 - 13x^2 + 40x)$$

on the closed interval $[0, 5]$.

Step 4 Observe that f is continuous on $[0, 5]$, so the absolute maximum value of f must be attained at the endpoint(s) or at the critical point(s) of f. Differentiating $f(x)$, we obtain

$$f'(x) = 4(3x^2 - 26x + 40)$$
$$= 4(3x - 20)(x - 2).$$

Upon setting $f'(x) = 0$ and solving the resulting equation for x, we obtain $x = \frac{20}{3}$ or $x = 2$. Since $\frac{20}{3}$ lies outside the interval $[0, 5]$, it is no longer considered and we are interested only in the critical point $x = 2$ of f. Next, evaluating $f(x)$ at $x = 0$, $x = 5$ (the endpoints of the interval $[0, 5]$), and $x = 2$, we obtain

$$f(0) = 0, \quad f(2) = 144, \quad \text{and} \quad f(5) = 0.$$

Thus the volume of the box is maximized by taking $x = 2$. The dimensions of the box are $12'' \times 6'' \times 2''$, and the volume is 144 cubic inches. ∎

EXAMPLE 3 A city's Metropolitan Transit Authority (MTA) operates a subway line for commuters from a certain suburb to the downtown metropolitan area. Currently, an average of 6000 passengers a day take the trains, paying a fare of $1.50 per ride. The Board of the MTA, contemplating raising the fare to $1.75 per ride in order to generate a larger revenue, engages the services of a consulting firm. The firm's study reveals that for each 25 cent increase in fare, the ridership will be reduced by an average of 1000 passengers a day. Thus the consulting firm recommends that MTA stick to the current fare of $1.50 per ride, which already yields a maximum revenue. Show that the consultants are correct.

Solution

Step 1 Let x denote the number of passengers per day, p denote the fare per ride, and R be MTA's revenue.

Step 2 In order to find a relationship between x and p, observe that the given data imply that when $x = 6000$, $p = 1.5$, and when $x = 5000$, $p = 1.75$. Therefore, the points $(6000, 1.5)$ and $(5000, 1.75)$ lie on a straight line. (Why?) To find the linear relationship between p and x, use the point-slope form of the equation of a straight line. Now, the slope of the line is

$$m = \frac{1.75 - 1.50}{5000 - 6000} = -0.00025.$$

Therefore, the required equation is

$$p - 1.5 = -0.00025(x - 6000)$$
$$= -0.00025x + 1.5$$
$$p = -0.00025x + 3.$$

Therefore, the revenue,

$$R = f(x) = xp = -0.00025x^2 + 3x,$$

(Number of riders
times unit fare)

is the quantity to be maximized.

Step 3 Since both p and x must be nonnegative, we see that $0 \leq x \leq 12{,}000$, and the problem is that of finding the absolute maximum of the function f on the closed interval $[0, 12{,}000]$.

Step 4 Observe that f is continuous on $[0, 12{,}000]$. To find the critical point of R, we compute

$$f'(x) = -0.0005x + 3$$

and set it equal to zero, giving $x = 6{,}000$. Evaluating the function f at $x = 6{,}000$, as well as at the endpoints $x = 0$ and $x = 12{,}000$, yields

$$f(0) = 0$$
$$f(6{,}000) = 9{,}000$$
$$f(12{,}000) = 0.$$

We conclude that a maximum revenue of \$9,000 per day is realized when the ridership is 6,000 per day. The optimum price of the fare per ride is therefore \$1.50, as recommended by the consultants. The graph of the revenue function R is shown in Figure 4.80. ■

FIGURE 4.80

f has an absolute maximum of 9000 when x = 6000.

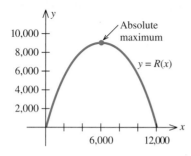

Minimization Problems

EXAMPLE 4   The Betty Moore Company requires that its corned beef hash containers have a capacity of 54 cubic inches, have the shape of right-circular cylinders, and be made of tin. Determine the radius and height of the container that requires the least amount of metal.

Solution

FIGURE 4.81

We want to minimize the amount of material used to construct the container.

Step 1 Let the radius and height of the container be r and h inches, respectively, and let S denote the surface area of the container (Figure 4.81).

Step 2 The amount of tin used to construct the container is given by the total surface area of the cylinder. Now, the area of the base and the top of the cylinder are each πr^2 square inches and the area of the side is $2\pi rh$ square inches. Therefore,

$$S = 2\pi r^2 + 2\pi rh, \tag{3}$$

the quantity to be minimized.

Step 3 The requirement that the volume of a container be 54 cubic inches implies that

$$\pi r^2 h = 54. \tag{4}$$

Solving (4) for h, we obtain

$$h = \frac{54}{\pi r^2}, \tag{5}$$

which, when substituted into (3), yields

$$S = 2\pi r^2 + 2\pi r\left(\frac{54}{\pi r^2}\right)$$

$$= 2\pi r^2 + \frac{108}{r}.$$

Clearly, the radius r of the container must satisfy the inequality $r > 0$. The problem now is reduced to finding the absolute minimum of the function $S = f(r)$ on the interval $(0, \infty)$.

Step 4 Using the curve-sketching techniques of Section 4.5, we obtain the graph of f in Figure 4.82.

To find the critical point of f, we compute

$$S' = 4\pi r - \frac{108}{r^2}$$

and solve the equation $S' = 0$ for r:

$$4\pi r - \frac{108}{r^2} = 0$$

$$4\pi r^3 - 108 = 0$$

$$r^3 = \frac{27}{\pi}$$

or

$$r = \frac{3}{\sqrt[3]{\pi}} \approx 2. \qquad (6)$$

FIGURE 4.82

The total surface area of the right cylindrical container is graphed as a function of r.

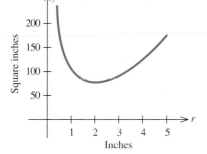

Square inches

Inches

Next, let us show that this value of r gives rise to the absolute minimum of f. To show this, we first compute

$$S'' = 4\pi + \frac{216}{r^2}.$$

Since $S'' > 0$ if $r = 3/\sqrt[3]{\pi}$, the Second Derivative Test implies that the value of r in (6) gives rise to a relative minimum of f. Finally, this relative minimum of f is also the absolute minimum of f, since f is always concave upward ($S'' > 0$ for all $r > 0$). To find the height of the given container, we substitute the value of r given in (6) into (5). Thus

$$h = \frac{54}{\pi r^2} = \frac{54}{\pi\left(\dfrac{3}{\pi^{1/3}}\right)^2}$$

$$= \frac{54\pi^{2/3}}{(\pi)9}$$

$$= \frac{6}{\pi^{1/3}} = \frac{6}{\sqrt[3]{\pi}}$$

$$= 2r.$$

We conclude that the required container has a radius of approximately 2 inches and a height of approximately 4 inches, or twice the size of the radius.

An Inventory Problem

One problem faced by many companies is that of controlling the inventory of goods carried. Ideally, the manager must ensure that the company has sufficient stock to meet customer demand at all times. At the same time she must make sure that this is accomplished without overstocking (incurring unnecessary storage costs) and also without having to place orders too frequently (incurring reordering costs).

EXAMPLE 5　The Dixie Import-Export Company is the sole agent for the Excalibur 250 cc motorcycle. Management estimates that the demand for these motorcycles is 10,000 per year and that they will sell at a uniform rate throughout the year. The cost incurred in ordering each shipment of motorcycles is $10,000 and the cost per year of storing each motorcycle is $200.

　　Dixie's management faces the following problem: Ordering too many motorcycles at one time ties up valuable storage space and increases the storage cost. On the other hand, placing orders too frequently increases the ordering costs. How large should each order be, and how often should orders be placed, to minimize ordering and storage costs?

Solution　Let x denote the number of motorcycles in each order (the lot size). Then, assuming that each shipment arrives just as the previous shipment has been sold, the average number of motorcycles in storage during the year is $x/2$. You can see that this is the case by examining Figure 4.83.

FIGURE 4.83

As each lot is depleted, the new lot arrives. The average inventory level is x/2 if x is the lot size.

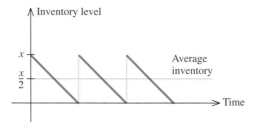

Thus, Dixie's storage cost for the year is given by $200(x/2)$, or $100x$ dollars.

　　Next, since the company requires 10,000 motorcycles for the year and since each order is for x motorcycles, the number of orders required is

$$\frac{10{,}000}{x}.$$

This gives an ordering cost of

$$10{,}000\left(\frac{10{,}000}{x}\right) = \frac{100{,}000{,}000}{x}$$

dollars for the year. Thus, the total yearly cost incurred by Dixie, which includes the ordering and storage costs attributed to the sale of these motorcycles, is given by

$$C(x) = 100x + \frac{100{,}000{,}000}{x}.$$

The problem is reduced to finding the absolute minimum of the function C in the interval $(0, 10{,}000]$. To accomplish this, we compute

$$C'(x) = 100 - \frac{100{,}000{,}000}{x^2}.$$

Setting $C'(x) = 0$ and solving the resulting equation, we obtain $x = \pm 1{,}000$. Since the number $-1{,}000$ is outside the domain of the function C, it is rejected, leaving $x = 1{,}000$ as the only critical point of C. Next, we find

$$C''(x) = \frac{200{,}000{,}000}{x^3}.$$

Since $C''(1{,}000) > 0$, the Second Derivative Test implies that the critical point $x = 1{,}000$ is a relative minimum of the function C (Figure 4.84). Also, since $C''(x) > 0$ for all x in $(0, 10{,}000]$, the function C is concave upward everywhere so that the point $x = 1{,}000$ also gives the absolute minimum of C. Thus, in order to minimize the ordering and storage costs, Dixie should place $10{,}000/1{,}000$, or 10, orders a year, each for a shipment of 1,000 motorcycles.

FIGURE 4.84
C has an absolute minimum at (1,000, 200,000).

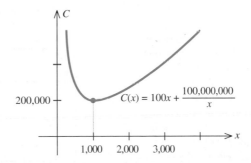

$$C(x) = 100x + \frac{100{,}000{,}000}{x}$$

SELF-CHECK EXERCISES 4.7

[c] 1. A man wishes to have an enclosed vegetable garden in his backyard. If the garden is to be a rectangular area of 300 square feet, find the dimensions of the garden that will minimize the amount of fencing material needed.

[c] 2. The demand for the Super Titan tires is 1,000,000 per year. The set-up cost for each production run is $4,000, and the manufacturing cost is $20 per tire. The cost of storing each tire over the year is $2. Assuming uniformity of demand throughout the year and instantaneous production, determine how many tires should be manufactured per production run in order to keep the production cost to a minimum.

Solutions to Self-Check Exercises 4.7 can be found on page 277.

4.7   EXERCISES

1. **Enclosing the Largest Area**   The owner of the Rancho Los Feliz has 3000 yards of fencing material with which to enclose a rectangular piece of grazing land along the straight portion of a river. If fencing is not required along the river, what are the dimensions of the largest area he can enclose? What is this area?

2. **Enclosing the Largest Area**   Refer to Exercise 1. As an alternative plan, the owner of the Rancho Los Feliz might use the 3000 yards of fencing material to enclose the rectangular piece of grazing land along the straight portion of the river and then subdivide it by means of a fence running parallel to the sides. Again, no fencing is required along the river. What are the dimensions of the largest area that can be enclosed? What is this area? (See the accompanying figure.)

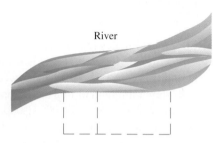

River

3. **Minimizing Construction Costs**   The management of the UNICO department store has decided to enclose an 800-square-foot area outside the building for displaying potted plants and flowers. One side will be formed by the external wall of the store, two sides will be constructed of pine boards, and the fourth side will be made of galvanized steel fencing material. If the pine board fencing costs $6 per running foot and the steel fencing costs $3 per running foot, determine the dimensions of the enclosure that can be erected at minimum cost.

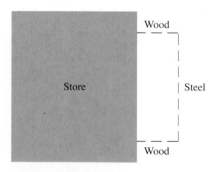

Store

Wood

Steel

Wood

4. **Packaging**   By cutting away identical squares from each corner of a rectangular piece of cardboard and folding up the resulting flaps, an open box may be made. If the cardboard is 15 inches long and 8 inches wide, find the dimensions of the box that will yield the maximum volume.

5. **Metal Fabrication**   If an open box is made from a tin sheet 8 inches square by cutting out identical squares from each corner and bending up the resulting flaps, determine the dimensions of the largest box that can be made.

6. **Minimizing Construction Costs**   If an open box has a square base and a volume of 108 cubic inches, and is constructed from a tin sheet, find the dimensions of such a box, assuming a minimum amount of material is used in its construction.

C 7. **Minimizing Construction Costs**   What are the dimensions of a closed rectangular box that has a square cross section, a capacity of 128 cubic inches, and is constructed using the least amount of material?

8. **Minimizing Construction Costs**   A rectangular box is to have a square base and a volume of 20 cubic feet. If the material for the base costs 30 cents per square foot, the material for the sides costs 10 cents per square foot, and the material for the top costs 20 cents per square foot, determine the dimensions of the box that can be constructed at minimum cost.

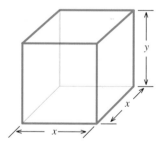

C 9. **Parcel Post Regulations**   Postal regulations specify that a parcel sent by parcel post may have a combined length and girth of no more than 108 inches. Find the dimensions of a rectangular package that has a square cross section and largest volume that may be sent through the mail. What is the volume of such a package? [*Hint:* The length plus the girth is $4x + h$ (see the accompanying figure).]

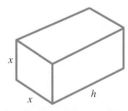

10. **Book Design**   A production editor at Saunders-Roe Publishing decided that the pages of a book should have one-inch margins at the top and bottom and half-inch margins on the sides. She further stipulated that each page should have an area of 50 square inches (see the

accompanying figure). Determine the page dimensions that will result in the maximum printed area on the page.

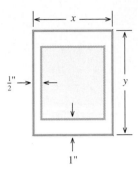

11. **Parcel Post Regulations** Postal regulations specify that a parcel sent by parcel post may have a combined length and girth of no more than 108 inches. Find the dimensions of the cylindrical package of greatest volume that may be sent through the mail. What is the volume of such a package? Compare with Exercise 9. [*Hint:* The length plus the girth is $2\pi r + l$.]

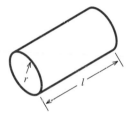

12. **Minimizing Costs** For its beef stew, the Betty Moore Company uses tin containers that have the form of right circular cylinders. Find the radius and height of a container if it has a capacity of 36 cubic inches and is constructed using the least amount of metal.

13. **Product Design** The cabinet that will enclose the Acrosonic Model D loudspeaker system will be rectangular and will have an internal volume of 2.4 cubic feet. For aesthetic reasons, it has been decided that the height of the cabinet is to be 1.5 times its width. If the top, bottom, and sides of the cabinet are constructed of veneer costing 40 cents per square foot, and the front (ignore the cutouts in the baffle) and rear are constructed of particle board costing 20 cents per square foot, what are the dimensions of the enclosure that can be constructed at a minimum cost?

14. **Designing a Norman Window** A norman window has the shape of a rectangle surmounted by a semicircle (see the accompanying figure). If a norman window is to have a perimeter of 28 ft, what should its dimensions be in order to allow the maximum amount of light through the window?

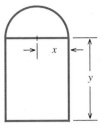

Figure for Exercise 14

15. **Optimal Charter Flight Fare** If exactly 200 people sign up for a charter flight, the Leisure World Travel Agency charges $300 per person. However, if more than 200 people sign up for the flight (assume this is the case), then each fare is reduced by $1 for each additional person. Determine how many passengers will result in a maximum revenue for the travel agency. What is the maximum revenue? What would the fare per passenger be in this case? [*Hint:* Let x denote the number of passengers beyond 200. Show that the revenue function R is given by $R(x) = (200 + x)(300 - x)$.]

16. **Maximizing Yield** An apple orchard has an average yield of 36 bushels of apples per tree if tree density is 22 trees per acre. For each unit increase in tree density, the yield decreases by 2 bushels. Find how many trees should be planted in order to maximize the yield.

17. **Strength of a Beam** A wooden beam has a rectangular cross section of height h inches and width w inches (see the accompanying figure). The strength S of the beam is directly proportional to its width and the square of its height. What are the dimensions of the cross section of the strongest beam that can be cut from a round log of diameter 24 inches? [*Hint:* $S = kh^2w$, where k is a constant of proportionality.]

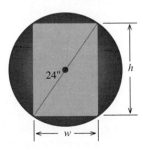

18. **Designing a Grain Silo** A grain silo has the shape of a right circular cylinder surmounted by a hemisphere (see the accompanying figure). If the silo is to have a capacity of 504π cubic feet, find the radius and height of the silo that requires the least amount of material to construct. [*Hint:* The volume of the silo is $\pi r^2 h + \frac{2}{3}\pi r^3$,

and the surface area (including the floor) is $\pi(3r^2 + 2rh)$.]

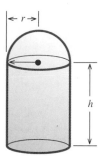

Figure for Exercise 18

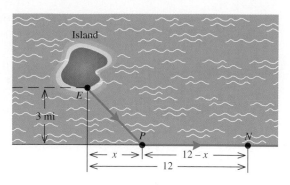

Figure for Exercise 20

19. **Minimizing Construction Costs**   In the following diagram, S represents the position of a power relay station located on a straight coast and E shows the location of a marine biology experimental station on an island. A cable is to be laid connecting the relay station with the experimental station. If the cost of running the cable on land is \$1 per running foot, and the cost of running the cable under water is \$3 per running foot, locate the point P that will result in a minimum cost (solve for x).

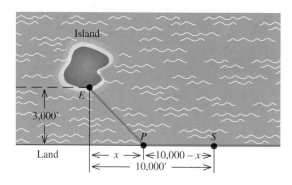

20. **Flights of Birds**   During daylight hours, some birds fly more slowly over water than over land because some of their energy is expended in overcoming the downdrafts of air over open bodies of water. Suppose a bird that flies at a constant speed of 4 mph over water and 6 mph over land starts its journey at the point E on an island and ends at its nest N on the shore of the mainland, as shown in the accompanying figure. Find the location of the point P that allows the bird to complete its journey in the minimum time (solve for x).

C 21. **Optimal Speed of a Truck**   A truck gets $400/x$ miles per gallon when driven at a constant speed of x miles per hour (between 50 and 70 miles per hour). If the price of fuel is \$1 per gallon and the driver is paid \$8 an hour, at what speed between 50 and 70 miles per hour is it most economical to drive?

22. **Inventory Control and Planning**   The demand for motorcycle tires imported by the Dixie Import-Export Company is 40,000 per year and may be assumed to be uniform throughout the year. The cost of ordering a shipment of tires is \$400, and the cost of storing each tire for a year is \$2. Determine how many tires should be in each shipment if the ordering and storage costs are to be minimized. (Assume that each shipment arrives just as the previous one has been sold.)

C 23. **Inventory Control and Planning**   The McDuff Preserves Company expects to bottle and sell 2,000,000 32-ounce jars of jam. The company orders its containers from the Consolidated Bottle Company. The cost of ordering a shipment of bottles is \$200, and the cost of storing each empty bottle for a year is 40 cents. Determine how many orders McDuff should place per year and how many bottles should be in each shipment if the ordering and storage costs are to be minimized. (Assume that each shipment of bottles is used up before the next shipment arrives.)

24. **Inventory Control and Planning**   The Neilsen Cookie Company sells its assorted butter cookies in containers that have a net content of 1 pound. The estimated demand for the cookies is 1,000,000 units. The set-up

cost for each production run is $500, and the manu-facturing cost is 50 cents for each container of cookies. The cost of storing each container of cookies over the year is 40 cents. Assuming uniformity of demand throughout the year and instantaneous production, determine how many containers of cookies Neilsen should produce per production run in order to minimize the production cost. [*Hint:* Following the method of Exam-

ple 5, show that the total production cost is given by the function

$$C(x) = \frac{500{,}000{,}000}{x} + 0.2x + 500{,}000.$$

Then minimize the function C on the interval $(0, 1{,}000{,}000)$.]

SOLUTIONS TO
SELF-CHECK
EXERCISES 4.7

1. Let x and y (measured in feet) denote the length and width of the rectangular garden. Since the area is to be 300 square feet, we have

$$xy = 300.$$

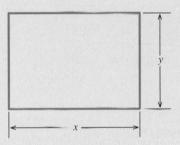

Next, the amount of fencing to be used is given by the perimeter, and this quantity is to be minimized. Thus, we want to minimize

$$2x + 2y,$$

or, since $y = \dfrac{300}{x}$ (obtained by solving for y in the first equation), we see that the expression to be minimized is

$$f(x) = 2x + 2\left(\frac{300}{x}\right)$$
$$= 2x + \frac{600}{x}$$

for positive values of x. Now,

$$f'(x) = 2 - \frac{600}{x^2}.$$

Setting $f'(x) = 0$ yields $x = -\sqrt{300}$ or $x = \sqrt{300}$. We consider only the critical point $x = \sqrt{300}$, since $-\sqrt{300}$ lies outside the interval $(0, \infty)$. We then compute

$$f''(x) = \frac{1200}{x^3}.$$

Since $$f''(\sqrt{300}) > 0,$$

the Second Derivative Test implies that a relative minimum of f occurs at $x = \sqrt{300}$. In fact, since $f''(x) > 0$ for all x in $(0, \infty)$, we conclude that $x = \sqrt{300}$ gives rise to the absolute minimum of f. The corresponding value of y, obtained by substituting this value of x into the equation $xy = 300$, is $y = \sqrt{300}$. Therefore, the required dimensions of the vegetable garden are approximately 17.3 ft $\times$ 17.3 ft.

2. Let x denote the number of tires in each production run. Then the average number of tires in storage is $x/2$, so the storage cost incurred by the company is $2(x/2)$, or x dollars. Next, since the company needs to manufacture 1,000,000 tires for the year in order to meet the demand, the number of production runs is $1,000,000/x$. This gives set-up costs amounting to

$$4,000\left(\frac{1,000,000}{x}\right) = \frac{4,000,000,000}{x}$$

dollars for the year. Thus, the total yearly cost incurred by the company is given by

$$C(x) = x + \frac{4,000,000,000}{x}.$$

Differentiating $C(x)$, we find

$$C'(x) = 1 - \frac{4,000,000,000}{x^2}.$$

Setting $C'(x) = 0$ gives $x = 63,246$ as the critical point in the interval $(0, 1,000,000)$. Next, we find

$$C''(x) = \frac{8,000,000,000}{x^3}.$$

Since $C''(x) > 0$ for all $x > 0$, we see that C is concave upward for all $x > 0$. Furthermore, $C''(63,246) > 0$ implies that $x = 63,246$ gives rise to a relative minimum of C (by the Second Derivative Test). Since C is always concave upward for $x > 0$, $x = 63,246$ gives the absolute minimum of C. Therefore, the company should manufacture 63,246 tires in each production run.

CHAPTER FOUR SUMMARY OF PRINCIPAL TERMS

Terms

Increasing function
Decreasing function
Relative maximum
Relative minimum
Relative extrema
Critical point
First Derivative Test
Concave upward
Concave downward

Inflection point
Second Derivative Test
Symmetry with respect to the y-axis
Symmetry with respect to the origin
Vertical asymptote
Horizontal asymptote
Absolute extrema
Absolute maximum value
Absolute minimum value

CHAPTER 4 REVIEW EXERCISES

In Exercises 1–10, (a) find the intervals where the given function f is increasing and where it is decreasing, (b) find the relative extrema of f, (c) find the intervals where f is concave upward and where it is concave downward, and (d) find the inflection points, if any, of f.

1. $f(x) = \frac{1}{3}x^3 - x^2 + x - 6$

2. $f(x) = (x - 2)^3$ 3. $f(x) = x^4 - 2x^2$

4. $f(x) = x + \frac{4}{x}$ 5. $f(x) = \frac{x^2}{x - 1}$

6. $f(x) = \sqrt{x - 1}$ 7. $f(x) = (1 - x)^{1/3}$

8. $f(x) = x\sqrt{x - 1}$ 9. $f(x) = \frac{2x}{x + 1}$

10. $f(x) = \frac{-1}{1 + x^2}$

In Exercises 11–18, obtain as much information as possible on each of the given functions. Then use this information to sketch the graph of the function.

11. $f(x) = x^2 - 5x + 5$ 12. $f(x) = -2x^2 - x + 1$

13. $g(x) = 2x^3 - 6x^2 + 6x + 1$

14. $g(x) = \frac{1}{3}x^3 - x^2 + x - 3$

15. $h(x) = x\sqrt{x - 2}$ 16. $h(x) = \frac{2x}{1 + x^2}$

17. $f(x) = \frac{x - 2}{x + 2}$ 18. $f(x) = x - \frac{1}{x}$

In Exercises 19–22, find the horizontal and vertical asymptotes of the graphs of the given functions. Do not sketch the graphs.

19. $f(x) = \frac{1}{2x + 3}$ 20. $f(x) = \frac{2x}{x + 1}$

21. $f(x) = \frac{5x}{x^2 - 2x - 8}$ 22. $f(x) = \frac{x^2 + x}{x(x - 1)}$

In Exercises 23–32, find the absolute maximum value and the absolute minimum value, if any, of the given function.

23. $f(x) = 2x^2 + 3x - 2$ 24. $g(x) = x^{2/3}$

25. $g(t) = \sqrt{25 - t^2}$

26. $f(x) = \frac{1}{3}x^3 - x^2 + x + 1$ on $[0, 2]$

27. $h(t) = t^3 - 6t^2$ on $[2, 5]$

28. $g(x) = \frac{x}{x^2 + 1}$ on $[0, 5]$

29. $f(x) = x - \frac{1}{x}$ on $[1, 3]$

30. $h(t) = 8t - \frac{1}{t^2}$ on $[1, 3]$

31. $f(s) = s\sqrt{1 - s^2}$ on $[-1, 1]$

32. $f(x) = \frac{x^2}{x - 1}$ on $[-1, 3]$

33. Odyssey Travel Agency's monthly profit (measured in thousands of dollars) depends on the amount of money x spent on advertising per month according to the rule

$$P(x) = -x^2 + 8x + 20,$$

where x is also measured in thousands of dollars. What should Odyssey's monthly advertising budget be in order to maximize its monthly profits?

34. The Department of the Interior of an African country began to record an index of environmental quality to measure progress or decline in the environmental quality of its wildlife. The index for the years 1984 through 1994 is approximated by the function

$$I(t) = \frac{50t^2 + 600}{t^2 + 10} \qquad (0 \le t \le 10).$$

   a. Compute $I'(t)$ and show that $I(t)$ is decreasing on the interval $(0, 10)$.
   b. Compute $I''(t)$. Study the concavity of the graph of I.
   c. Sketch the graph of I.
   d. Interpret your results.

35. The weekly demand for video discs manufactured by the Herald Record Company is given by

$$p = -0.0005x^2 + 60,$$

where p denotes the unit price in dollars and x denotes the quantity demanded. The weekly total cost function associated with producing these discs is given by

$$C(x) = -0.001x^2 + 18x + 4000,$$

where $C(x)$ denotes the total cost incurred in pressing x discs. Find the production level that will yield a maximum profit for the manufacturer. [*Hint:* Use the quadratic formula.]

36. The total monthly cost (in dollars) incurred by the Carlota Music Company in manufacturing x units of its Professional Series guitars is given by the function

$$C(x) = 0.001x^2 + 100x + 4000.$$

a. Find the average cost function $\overline{C}$.
b. Determine the production level that will result in the smallest average production cost.

37. The average worker at Wakefield Avionics, Inc., can assemble

$$N(t) = -2t^3 + 12t^2 + 2t \qquad (0 \leq t \leq 4)$$

ready-to-fly radio-controlled model airplanes t hours into the 8 A.M. to 12 noon morning shift. At what time during this shift is the average worker performing at peak efficiency?

38. You wish to construct a closed rectangular box that has a volume of 4 cubic feet. The length of the base of the box will be twice as long as its width. The material for the top and bottom of the box costs 30 cents per square foot. The material for the sides of the box costs 20 cents per square foot. Find the dimensions of the least expensive box that can be constructed.

39. The Lehen Vinters Company imports a certain brand of beer. The demand, which may be assumed to be uniform, is 800,000 cases per year. The cost of ordering a shipment of beer is $500, and the cost of storing each case of beer for a year is $2. Determine how many cases of beer should be in each shipment if the ordering and storage costs are to be kept to a minimum. (Assume that each shipment of beer arrives just as the previous one has been sold.)

How many bacteria will there be in a culture at the end of a certain period of time? How fast will the bacteria population be growing at the end of that time? Example 1, page 325, answers these questions.

Exponential and Logarithmic Functions

The exponential function is, without doubt, the most important function in mathematics and its applications. After a brief introduction to the exponential function and its *inverse*, the logarithmic function, we will learn how to differentiate such functions. This will lay the foundation for exploring the many applications involving exponential functions. For example, we will look at the role played by exponential functions in computing earned interest on a bank account, in studying the growth of a bacteria population in the laboratory, in studying the way radioactive matter decays, in studying the rate at which a factory worker learns a certain process, and in studying the rate at which a communicable disease is spread over time.

5.1   Exponential Functions

Exponential Functions and Their Graphs

Suppose that you deposit a sum of $1000 in an account earning interest at the rate of 10 percent per year *compounded continuously* (the way most financial institutions compute interest). Then the accumulated amount at the end of t years ($0 \leq t \leq 20$) is described by the function f, whose graph appears in Figure 5.1.[†] Such a function is called an *exponential function*. Observe that the graph of f rises rather slowly at first but very rapidly as time goes by. For purposes of comparison we have also shown the graph of the function $y = g(t) = 1000(1 + 0.10t)$ giving the accumulated amount for the same principal ($1000) but earning *simple* interest at the rate of 10 percent per year. The moral of the story: It is never too early to save.

Exponential functions play an important role in many real-world applications, as you will see throughout this chapter.

FIGURE 5.1
Under continuous compounding, a sum of money grows exponentially.

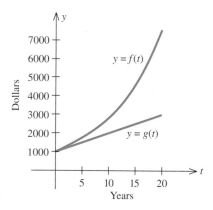

In order to define an exponential function, observe that whenever b is a positive number and n is any real number, the expression b^n is a real number. This enables us to define an *exponential function* as follows:

EXPONENTIAL FUNCTION

The function defined by

$$f(x) = b^x \qquad (b > 0, b \neq 1)$$

is called an **exponential function with base b and exponent x.** The domain of f is the set of all real numbers.

For example, the exponential function with base 2 is the function

$$f(x) = 2^x$$

with domain $(-\infty, \infty)$. The values of $f(x)$ for selected values of x follow.

[†] We will derive the rule for f in Section 5.3.

$$f(3) = 2^3 = 8, \qquad f\left(\frac{3}{2}\right) = 2^{3/2} = 2 \cdot 2^{1/2} = 2\sqrt{2}, \qquad f(0) = 2^0 = 1,$$

$$f(-1) = 2^{-1} = \frac{1}{2}, \quad \text{and} \quad f\left(-\frac{2}{3}\right) = 2^{-2/3} = \frac{1}{2^{2/3}} = \frac{1}{\sqrt[3]{4}}$$

Computations involving exponentials are facilitated by the laws of exponents. These laws were stated in Section 1.1, and you might want to review the material there. For convenience, however, we will restate these laws.

THE LAWS OF EXPONENTS

Let a and b be positive numbers and let x and y be real numbers. Then,

1. $b^x \cdot b^y = b^{x+y}$

2. $\dfrac{b^x}{b^y} = b^{x-y}$

3. $(b^x)^y = b^{xy}$

4. $(ab)^x = a^x b^x$

5. $\left(\dfrac{a}{b}\right)^x = \dfrac{a^x}{b^x}$

The use of the laws of exponents is illustrated in the next example.

EXAMPLE 1

a. $16^{7/4} \cdot 16^{-1/2} = 16^{7/4 - 1/2} = 16^{5/4} = 2^5 = 32$ (Law 1)

b. $\dfrac{8^{5/3}}{8^{-1/3}} = 8^{5/3 - (-1/3)} = 8^2 = 64$ (Law 2)

c. $(64^{4/3})^{-1/2} = 64^{(4/3)(-1/2)} = 64^{-2/3}$

$\qquad = \dfrac{1}{64^{2/3}} = \dfrac{1}{(64^{1/3})^2} = \dfrac{1}{4^2} = \dfrac{1}{16}$ (Law 3)

d. $(16 \cdot 81)^{-1/4} = 16^{-1/4} \cdot 81^{-1/4} = \dfrac{1}{16^{1/4}} \cdot \dfrac{1}{81^{1/4}} = \dfrac{1}{2} \cdot \dfrac{1}{3} = \dfrac{1}{6}$ (Law 4)

e. $\left(\dfrac{3^{1/2}}{2^{1/3}}\right)^4 = \dfrac{3^{4/2}}{2^{4/3}} = \dfrac{9}{2^{4/3}}$ (Law 5)   ∎

EXAMPLE 2   Let $f(x) = 2^{2x-1}$. Find the value of x for which $f(x) = 16$.

Solution   We want to solve the equation

$$2^{2x-1} = 16 = 2^4.$$

But this equation holds if and only if

$$2x - 1 = 4, \qquad (b^m = b^n \Rightarrow m = n)$$

giving $x = \frac{5}{2}$.   ∎

Exponential functions play an important role in mathematical analysis. Because of their special characteristics, they are some of the most useful functions and are found in virtually every field where mathematics is applied. Under ideal conditions the number of bacteria present at any time t in a culture may be described by an exponential function of t; radioactive substances decay over time in accordance with an "exponential" law of decay; money left on fixed deposit and earning compound interest grows exponentially; and some of the most important distribution functions encountered in statistics are exponential, just to mention a few examples.

Let us begin our investigation into the properties of exponential functions by studying their graphs.

EXAMPLE 3   Sketch the graph of the exponential function $y = 2^x$.

Solution   First, as discussed earlier, the domain of the exponential function $y = f(x) = 2^x$ is the set of real numbers. Next, putting $x = 0$ gives $y = 2^0 = 1$, the y-intercept of f. There is no x-intercept since there is no value of x for which $y = 0$. To find the range of f, consider the following table of values:

FIGURE 5.2

The graph of $y = 2^x$

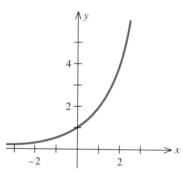

x	-5	-4	-3	-2	-1	0	1	2	3	4	5
y	1/32	1/16	1/8	1/4	1/2	1	2	4	8	16	32

We see from these computations that 2^x decreases and approaches zero as x decreases without bound and that 2^x increases without bound as x increases without bound. Thus, the range of f is the interval $(0, \infty)$, that is, the set of positive real numbers. Finally, we sketch the graph of $y = f(x) = 2^x$ in Figure 5.2.   ■

EXAMPLE 4   Sketch the graph of the exponential function $y = (1/2)^x$.

Solution   The domain of the exponential function $y = (1/2)^x$ is the set of all real numbers. The y-intercept is $(1/2)^0 = 1$; there is no x-intercept since there is no value of x for which $y = 0$. From the following table of values

FIGURE 5.3

The graph of $y = (1/2)^x$

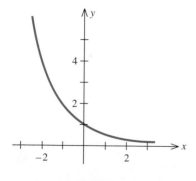

x	-5	-4	-3	-2	-1	0	1	2	3	4	5
y	32	16	8	4	2	1	1/2	1/4	1/8	1/16	1/32

we deduce that $(1/2)^x = 1/2^x$ increases without bound as x decreases without bound and that $(1/2)^x$ decreases and approaches zero as x increases without bound. Thus, the range of f is the interval $(0, \infty)$. The graph of $y = f(x) = (1/2)^x$ is sketched in Figure 5.3.   ■

The functions $y = 2^x$ and $y = (1/2)^x$, whose graphs you studied in Examples 1 and 2, are special cases of the exponential function $y = f(x) = b^x$, obtained by setting $b = 2$ and $b = 1/2$, respectively. In general, the exponential function $y = b^x$ with $b > 1$ has a graph similar to $y = 2^x$, while the graph of $y = b^x$ for $0 < b < 1$ is similar to that of $y = (1/2)^x$ (Exercises 27 and 28). When $b = 1$, the function $y = b^x$ reduces to the constant function $y = 1$. For comparison, the graphs of all three functions are sketched in Figure 5.4.

FIGURE 5.4
$y = b^x$ is an increasing function of x if $b > 1$, a constant function if $b = 1$, and a decreasing function if $0 < b < 1$.

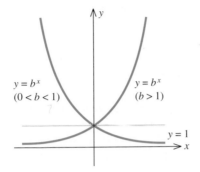

PROPERTIES OF THE EXPONENTIAL FUNCTION

The exponential function $y = b^x$ ($b > 0$, $b \neq 1$) has the following properties.

1. Its domain is $(-\infty, \infty)$.

2. Its range is $(0, \infty)$.

3. Its graph passes through the point $(0, 1)$.

4. It is continuous on $(-\infty, \infty)$.

5. It is increasing on $(-\infty, \infty)$ if $b > 1$ and decreasing on $(-\infty, \infty)$ if $b < 1$.

The Base *e*

Exponential functions to the base e, where e is an irrational number whose value is $2.7182818 \ldots$, play an important role in both theoretical and applied problems. It can be shown, although we will not do so here, that

$$e = \lim_{m \to \infty} \left(1 + \frac{1}{m}\right)^m. \tag{1}$$

However, you may convince yourself of the plausibility of this definition of the number e by examining Table 5.1, which may be constructed with the help of a calculator.

TABLE 5.1

m	10	100	1000	10,000	100,000	1,000,000
$\left(1 + \dfrac{1}{m}\right)^m$	2.59374	2.70481	2.71692	2.71815	2.71827	2.71828

FIGURE 5.5
The graph of $y = e^x$

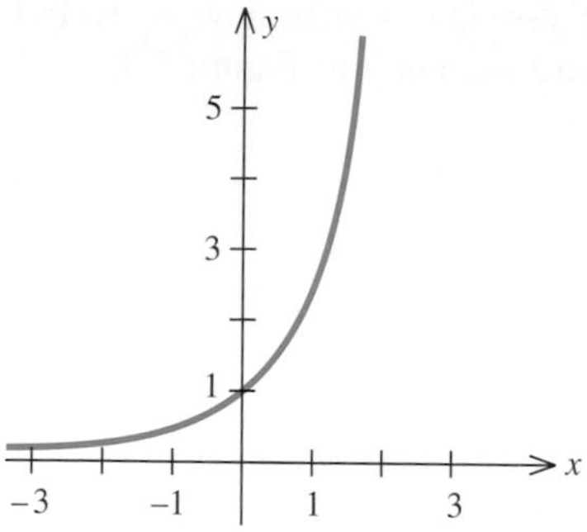

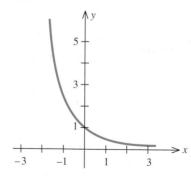

    EXAMPLE 5   Sketch the graph of the function $y = e^x$.

Solution   Since $e > 1$, it follows from our previous discussion that the graph of $y = e^x$ is similar to the graph of $y = 2^x$ (Figure 5.2). With the aid of a calculator, we obtain the following table.

x	−3	−2	−1	0	1	2	3
y	0.05	0.14	0.37	1	2.72	7.39	20.09

The graph of $y = e^x$ is sketched in Figure 5.5.   ∎

Next, we will consider another exponential function to the base e that is closely related to the previous function and is particularly useful in constructing models that describe "exponential decay."

    EXAMPLE 6   Using a calculator, sketch the graph of the function $y = e^{-x}$.

FIGURE 5.6
The graph of $y = e^{-x}$

Solution   Since $e > 1$, it follows that $0 < 1/e < 1$, so $f(x) = e^{-x} = 1/e^x = (1/e)^x$ is an exponential function with base less than 1. Therefore it has a graph similar to that of the exponential function $y = (1/2)^x$. As before, we construct the following table of values of $y = e^{-x}$ for selected values of x.

x	−3	−2	−1	0	1	2	3
y	20.09	7.39	2.72	1	0.37	0.14	0.05

Using this table, we sketch the graph of $y = e^{-x}$ in Figure 5.6.   ∎

**SELF-CHECK
EXERCISES 5.1**

1. Solve the equation $2^{2x+1} \cdot 2^{-3} = 2^{x-1}$.
2. Sketch the graph of $y = e^{0.4x}$.

Solutions to Self-Check Exercises 5.1 can be found on page 289.

5.1   EXERCISES

In Exercises 1–8, evaluate the given expressions.

1. a. $4^{-3} \cdot 4^5$ b. $3^{-3} \cdot 3^6$

2. a. $(2^{-1})^3$ b. $(3^{-2})^3$

3. a. $9(9)^{-1/2}$ b. $5(5)^{-1/2}$

4. a. $\left[\left(-\tfrac{1}{2}\right)^3\right]^{-2}$ b. $\left[\left(-\tfrac{1}{3}\right)^2\right]^{-3}$

5. a. $\dfrac{(-3)^4(-3)^5}{(-3)^8}$ b. $\dfrac{(2^{-4})(2^6)}{2^{-1}}$

6. a. $3^{1/4} \cdot (9)^{-5/8}$ b. $2^{3/4} \cdot (4)^{-3/2}$

7. a. $\dfrac{5^{3.3} \cdot 5^{-1.6}}{5^{-0.3}}$ b. $\dfrac{4^{2.7} \cdot 4^{-1.3}}{4^{-0.4}}$

8. a. $\left(\tfrac{1}{16}\right)^{-1/4}\left(\tfrac{27}{64}\right)^{-1/3}$ b. $\left(\tfrac{8}{27}\right)^{-1/3}\left(\tfrac{81}{256}\right)^{-1/4}$

In Exercises 9–16, simplify the given expressions.

9. a. $(64x^9)^{1/3}$ b. $(25x^3y^4)^{1/2}$

10. a. $(2x^3)(-4x^{-2})$ b. $(4x^{-2})(-3x^5)$

11. a. $\dfrac{6a^{-5}}{3a^{-3}}$ b. $\dfrac{4b^{-4}}{12b^{-6}}$

12. a. $y^{-3/2}y^{5/3}$ b. $x^{-3/5}x^{8/3}$

13. a. $(2x^3y^2)^3$ b. $(4x^2y^2z^3)^2$

14. a. $(x^{r/s})^{s/r}$ b. $(x^{-b/a})^{-a/b}$

15. a. $\dfrac{5^0}{(2^{-3}x^{-3}y^2)^2}$ b. $\dfrac{(x+y)(x-y)}{(x-y)^0}$

16. a. $\dfrac{(a^m \cdot a^{-n})^{-2}}{(a^{m+n})^2}$ b. $\left(\dfrac{x^{2n-2}y^{2n}}{x^{5n+1}y^{-n}}\right)^{1/3}$

In Exercises 17–26, solve the given equation for x.

17. $6^{2x} = 6^4$ 18. $5^{-x} = 5^3$

19. $3^{3x-4} = 3^5$ 20. $10^{2x-1} = 10^{x+3}$

21. $(2.1)^{x+2} = (2.1)^5$ 22. $(-1.3)^{x-2} = (-1.3)^{2x+1}$

23. $8^x = \left(\dfrac{1}{32}\right)^{x-2}$ 24. $3^{x-x^2} = \dfrac{1}{9^x}$

25. $3^{2x} - 12 \cdot 3^x + 27 = 0$ 26. $2^{2x} - 4 \cdot 2^x + 4 = 0$

C *In Exercises 27–35, sketch the graphs of the given functions on the same axes. A calculator is recommended for these exercises.*

27. $y = 2^x$, $y = 3^x$, and $y = 4^x$

28. $y = \left(\tfrac{1}{2}\right)^x$, $y = \left(\tfrac{1}{3}\right)^x$, and $y = \left(\tfrac{1}{4}\right)^x$

29. $y = 2^{-x}$, $y = 3^{-x}$, and $y = 4^{-x}$

30. $y = 4^{0.5x}$ and $y = 4^{0.5x}$

31. $y = 4^{0.5x}$, $y = 4^x$, and $y = 4^{2x}$

32. $y = e^x$, $y = 2e^x$, and $y = 3e^x$

33. $y = e^{0.5x}$, $y = e^x$, and $y = e^{1.5x}$

34. $y = e^{-0.5x}$, $y = e^{-x}$, and $y = e^{-1.5x}$

35. $y = 0.5e^{-x}$, $y = e^{-x}$, and $y = 2e^{-x}$

C *In Exercises 36 and 37, use a computer or graphing calculator to sketch the graphs of the given functions on the same axes.*

36. $y = 0.5e^{-x^2}$, $y = e^{-x^2}$, and $y = 2e^{-x^2}$

37. $y = e^{-x^2/2}$, $y = e^{-x^2}$, and $y = e^{-2x^2}$

SOLUTIONS TO SELF-CHECK EXERCISES 5.1

1. $2^{2x+1} \cdot 2^{-3} = 2^{x-1}$

$\dfrac{2^{2x+1}}{2^{x-1}} \cdot 2^{-3} = 1$ (Dividing both sides by 2^{x-1})

$2^{(2x+1)-(x-1)-3} = 1$

$2^{x-1} = 1$

This is true if and only if $x - 1 = 0$ or $x = 1$.

2. We first construct the following table of values.

x	-3	-2	-1	0	1	2	3	4
$y = e^{0.4x}$	0.3	0.5	0.7	1	1.5	2.2	3.3	5

Next, we plot these points and join them by a smooth curve to obtain the graph of f shown in the accompanying figure.

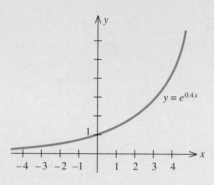

5.2   Logarithmic Functions

Logarithms

You are already familiar with exponential equations of the form

$$b^y = x \qquad (b > 0, \, b \neq 1),$$

where the variable x is expressed in terms of a real number b and a variable y. But what about solving this same equation for y? You may recall from your study of algebra that the number y is called the **logarithm of x to the base b** and is denoted by $\log_b x$. It is the exponent to which the base b must be raised in order to obtain the number x.

THE LOGARITHM OF x TO THE BASE b

$$y = \log_b x \quad \text{if and only if} \quad x = b^y \qquad (x > 0)$$

    Observe that the logarithm $\log_b x$ is defined only for positive values of x.

EXAMPLE 1

a. $\log_{10} 100 = 2$, since $100 = 10^2$
b. $\log_5 125 = 3$, since $125 = 5^3$
c. $\log_3 \dfrac{1}{27} = -3$, since $\dfrac{1}{27} = \dfrac{1}{3^3} = 3^{-3}$
d. $\log_{20} 20 = 1$, since $20 = 20^1$ ■

EXAMPLE 2   Solve each of the following equations for x.

a. $\log_3 x = 4$ **b.** $\log_{16} 4 = x$ **c.** $\log_x 8 = 3$

Solution

a. By definition, $\log_3 x = 4$ implies $x = 3^4 = 81$.

b. $\log_{16} 4 = x$ is equivalent to $4 = 16^x = (4^2)^x = 4^{2x}$, from which we deduce that $x = \frac{1}{2}$.

c. Referring once again to the definition, we see that the equation $\log_x 8 = 3$ is equivalent to the equation $8 = x^3$, so $x = 2$. ∎

The two widely used systems of logarithms are the system of **common logarithms,** which uses the number 10 as the base, and the system of **natural logarithms,** which uses the irrational number $e = 2.71828\ldots$ as the base. Also, it is standard practice to write **log** for $\log_{10}$ and **ln** for $\log_e$.

LOGARITHMIC NOTATION

$$\log x = \log_{10} x \qquad \text{(Common logarithm)}$$
$$\ln x = \log_e x \qquad \text{(Natural logarithm)}$$

The system of natural logarithms is widely used in theoretical work. Using natural logarithms rather than logarithms to other bases often leads to simpler expressions.

Laws of Logarithms

Computations involving logarithms are facilitated by the following **laws of logarithms.**

LAWS OF LOGARITHMS

If m and n are positive numbers, then

1. $\log_b mn = \log_b m + \log_b n$

2. $\log_b \dfrac{m}{n} = \log_b m - \log_b n$

3. $\log_b m^n = n \log_b m$

4. $\log_b 1 = 0$

5. $\log_b b = 1$

You will be asked to prove these laws in Exercises 49–51. Their derivations are based on the definition of a logarithm and the corresponding laws of exponents. The following examples illustrate the properties of logarithms.

EXAMPLE 3

a. $\log(2 \cdot 3) = \log 2 + \log 3$

b. $\ln \dfrac{5}{3} = \ln 5 - \ln 3$ **c.** $\log\sqrt{7} = \log 7^{1/2} = \dfrac{1}{2} \log 7$

d. $\log_5 1 = 0$ **e.** $\log_{45} 45 = 1$ ∎

EXAMPLE 4   Given that $\log 2 \approx 0.3010$, $\log 3 \approx 0.4771$, and $\log 5 \approx 0.6990$, use the laws of logarithms to find

a. $\log 15$ **b.** $\log 7.5$ **c.** $\log 81$ **d.** $\log 50$

Solution

a. Note that $15 = 3 \cdot 5$, so by Law 1 for logarithms,

$$
\begin{aligned}
\log 15 &= \log 3 \cdot 5 \\
&= \log 3 + \log 5 \\
&\approx 0.4771 + 0.6990 \\
&= 1.1761.
\end{aligned}
$$

b. Observing that $7.5 = 15/2 = (3 \cdot 5)/2$, we apply Laws 1 and 2, obtaining

$$
\begin{aligned}
\log 7.5 &= \log \frac{(3)(5)}{2} \\
&= \log 3 + \log 5 - \log 2 \\
&\approx 0.4771 + 0.6990 - 0.3010 \\
&= 0.8751.
\end{aligned}
$$

c. Since $81 = 3^4$, we apply Law 3 to obtain

$$
\begin{aligned}
\log 81 &= \log 3^4 \\
&= 4 \log 3 \\
&\approx 4(0.4771) \\
&= 1.9084.
\end{aligned}
$$

d. We write $50 = 5 \cdot 10$ and find

$$
\begin{aligned}
\log 50 &= \log(5)(10) \\
&= \log 5 + \log 10 \\
&\approx 0.6990 + 1 \qquad \text{(Using Law 5)} \\
&= 1.6990.
\end{aligned}
$$
■

EXAMPLE 5   Expand and simplify the following expressions.

a. $\log_3 x^2 y^3$ **b.** $\log_2 \dfrac{x^2 + 1}{2^x}$ **c.** $\ln \dfrac{x^2 \sqrt{x^2 - 1}}{e^x}$

Solution

a.
$$
\begin{aligned}
\log_3 x^2 y^3 &= \log_3 x^2 + \log_3 y^3 && \text{(Law 1)} \\
&= 2 \log_3 x + 3 \log_3 y && \text{(Law 3)}
\end{aligned}
$$

b.
$$
\begin{aligned}
\log_2 \frac{x^2 + 1}{2^x} &= \log_2(x^2 + 1) - \log_2 2^x && \text{(Law 2)} \\
&= \log_2(x^2 + 1) - x \log_2 2 && \text{(Law 3)} \\
&= \log_2(x^2 + 1) - x && \text{(Law 5)}
\end{aligned}
$$

c. $\ln \dfrac{x^2\sqrt{x^2-1}}{e^x} = \ln \dfrac{x^2(x^2-1)^{1/2}}{e^x}$ (Rewriting)

$\quad\quad\quad\quad\quad\quad = \ln x^2 + \ln (x^2-1)^{1/2} - \ln e^x$ (Laws 1 and 2)

$\quad\quad\quad\quad\quad\quad = 2 \ln x + \tfrac{1}{2} \ln (x^2-1) - x \ln e$ (Law 3)

$\quad\quad\quad\quad\quad\quad = 2 \ln x + \tfrac{1}{2} \ln (x^2-1) - x$ (Law 5)   ■

Logarithmic Functions and Their Graphs

The definition of the logarithm implies that if b and n are positive numbers and b is different from 1, then the expression $\log_b n$ is a real number. This enables us to define a *logarithmic function* as follows:

LOGARITHMIC FUNCTION

The function defined by

$$f(x) = \log_b x \quad\quad (b > 0, b \neq 1)$$

is called the **logarithmic function with base b.** The domain of f is the set of all positive numbers.

FIGURE 5.7

The points (u, v) and (v, u) are mirror reflections of each other.

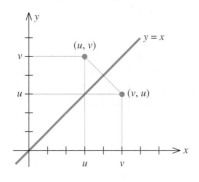

One easy way to obtain the graph of the logarithmic function $y = \log_b x$ is to construct a table of values of the logarithm (base b). However, another method—and a more instructive one—is based on exploiting the intimate relationship between logarithmic and exponential functions.

If a point (u, v) lies on the graph of $y = \log_b x$, then

$$v = \log_b u.$$

But we can also write this equation in exponential form as

$$u = b^v.$$

So the point (v, u) also lies on the graph of the function $y = b^x$. Let us look at the relationship between the points (u, v) and (v, u) and the line $y = x$ (Figure 5.7). If we think of the line $y = x$ as a mirror, then the point (v, u) is the mirror reflection of the point (u, v). Similarly, the point (u, v) is a mirror reflection of the point (v, u). We can take advantage of this relationship to help us draw the graph of logarithmic functions. For example, if we wish to draw the graph of $y = \log_b x$ where $b > 1$, then we need only draw the mirror reflection of the graph of $y = b^x$ with respect to the line $y = x$ (Figure 5.8).

FIGURE 5.8

The graphs of $y = b^x$ and $y = \log_b x$ are mirror reflections of each other.

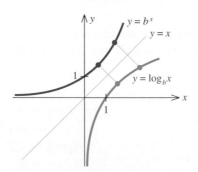

You may discover the following properties of the logarithmic function by taking the reflection of the graph of an appropriate exponential function (Exercises 31 and 32).

PROPERTIES OF THE LOGARITHMIC FUNCTION

The logarithmic function $y = \log_b x$ $(b > 0, b \neq 1)$ has the following properties.

1. Its domain is $(0, \infty)$.
2. Its range is $(-\infty, \infty)$.
3. Its graph passes through the point $(1, 0)$.
4. It is continuous on $(0, \infty)$.
5. It is increasing on $(0, \infty)$ if $b > 1$ and decreasing on $(0, \infty)$ if $b < 1$.

FIGURE 5.9
The graph of $y = \ln x$ is the mirror reflection of the graph of $y = e^x$.

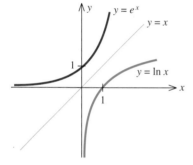

EXAMPLE 6   Sketch the graph of the function $y = \ln x$.

Solution   We first sketch the graph of $y = e^x$. Then the required graph is obtained by tracing the mirror reflection of the graph of $y = e^x$ with respect to the line $y = x$ (Figure 5.9). ∎

Properties Relating the Exponential and Logarithmic Functions

We made use of the relationship that exists between the exponential function $f(x) = e^x$ and the logarithmic function $g(x) = \ln x$ when we sketched the graph of g in Example 6. This relationship is further described by the following properties, which are an immediate consequence of the definition of the logarithm of a number.

PROPERTIES RELATING e^x and $\ln x$

$$e^{\ln x} = x \qquad x > 0 \qquad\qquad (2)$$
$$\ln e^x = x \qquad \text{for any real number } x \qquad\qquad (3)$$

(Try to verify these properties.)

From Properties (2) and (3), we conclude that the composite function

$$(f \circ g)(x) = f[g(x)]$$
$$= e^{\ln x} = x$$

and

$$(g \circ f)(x) = g[f(x)]$$
$$= \ln e^x = x.$$

Thus,

$$f[g(x)] = g[f(x)]$$
$$= x.$$

Any two functions f and g that satisfy this relationship are said to be **inverses** of each other. Note that the function f undoes what the function g does, and vice versa, so that the composition of the two functions in any order results in the identity function $F(x) = x$.

The relationship expressed in (2) and (3) are useful in solving equations that involve exponentials and logarithms.

EXAMPLE 7   Solve the equation $2e^{x+2} = 5$.

Solution   We first divide both sides of the equation by 2 to obtain

$$e^{x+2} = \frac{5}{2} = 2.5.$$

Next, taking the natural logarithm of each side of the equation and using (3), we have

$$\ln e^{x+2} = \ln 2.5$$
$$x + 2 = \ln 2.5$$
$$x = \quad 2 + \ln 2.5$$
$$\approx -1.08. \qquad \blacksquare$$

EXAMPLE 8   Solve the equation $5 \ln x + 3 = 0$.

Solution   Adding -3 to both sides of the equation leads to

$$5 \ln x = -3$$
$$\ln x - -\frac{3}{5} = \quad 0.6,$$

and so

$$e^{\ln x} = e^{-0.6}.$$

Using (2), we conclude that

$$x = e^{-0.6}$$
$$\approx 0.55. \qquad \blacksquare$$

SELF-CHECK EXERCISES 5.2	**1.** Sketch the graph of $y = 3^x$ and $y = \log_3 x$ on the same set of axes. **2.** Solve the equation $3e^{x+1} - 2 = 4$.

Solutions to Self-Check Exercises 5.2 can be found on page 297.

5.2 EXERCISES

In Exercises 1–10, express the given equation in logarithmic form.

1. $2^6 = 64$

2. $3^5 = 243$

3. $3^{-2} = \frac{1}{9}$

4. $5^{-3} = \frac{1}{125}$

5. $\left(\frac{1}{3}\right)^1 = \frac{1}{3}$

6. $\left(\frac{1}{2}\right)^{-4} = 16$

7. $32^{3/5} = 8$

8. $81^{3/4} = 27$

9. $10^{-3} = 0.001$

10. $16^{-1/4} = 0.5$

In Exercises 11–16, use the facts that $\log 3 = 0.4771$ *and* $\log 4 = 0.6021$ *to find the value of the given logarithm.*

11. $\log 12$

12. $\log \frac{3}{4}$

13. $\log 16$

14. $\log \sqrt{3}$

15. $\log 48$

16. $\log \frac{1}{300}$

In Exercises 17–26, use the laws of logarithms to simplify the given expression.

17. $\log x(x + 1)^4$

18. $\log x(x^2 + 1)^{-1/2}$

19. $\log \dfrac{\sqrt{x + 1}}{x^2 + 1}$

20. $\ln \dfrac{e^x}{1 + e^x}$

21. $\ln xe^{-x^2}$

22. $\ln x(x + 1)(x + 2)$

23. $\ln \dfrac{x^{1/2}}{x^2\sqrt{1 + x^2}}$

24. $\ln \dfrac{x^2}{\sqrt{x}(1 + x)^2}$

25. $\ln x^x$

26. $\ln x^{x^2 + 1}$

In Exercises 27–30, sketch the graph of the given equation.

27. $y = \log_3 x$

28. $y = \log_{1/3} x$

29. $y = \ln 2x$

30. $y = \ln \frac{1}{2} x$

In Exercises 31 and 32, sketch the graphs of the given equations on the same coordinate axes.

31. $y = 2^x$ and $y = \log_2 x$

32. $y = e^{3x}$ and $y = \ln 3x$

In Exercises 33–42, use logarithms to solve the given equation for t.

33. $e^{0.4t} = 8$

34. $\frac{1}{3} e^{-3t} = 0.9$

35. $5e^{-2t} = 6$

36. $4e^{t-1} = 4$

37. $2e^{-0.2t} - 4 = 6$

38. $12 - e^{0.4t} = 3$

39. $\dfrac{50}{1 + 4e^{0.2t}} = 20$

40. $\dfrac{200}{1 + 3e^{-0.3t}} = 100$

41. $A = Be^{-t/2}$

42. $\dfrac{A}{1 + Be^{t/2}} = C$

43. Blood Pressure A normal child's systolic blood pressure may be approximated by the function

$$p(x) = m(\ln x) + b,$$

where $p(x)$ is measured in millimeters of mercury, x is measured in pounds, and m and b are constants. Given that $m = 19.4$ and $b = 18$, determine the systolic blood pressure of a child who weighs 92 pounds.

44. Magnitude of Earthquakes On the Richter scale, the magnitude R of an earthquake is given by the formula

$$R = \log \frac{I}{I_0},$$

where I is the intensity of the earthquake being measured and I_0 is the standard reference intensity.
   a. Express the intensity I of an earthquake of magnitude $R = 5$ in terms of the standard intensity I_0.
   b. Express the intensity I of an earthquake of magnitude $R = 8$ in terms of the standard intensity I_0. How many times greater is the intensity of an earthquake of magnitude 8 than one of magnitude 5?
   c. In modern times, the greatest loss of life attributable to an earthquake occurred in eastern China in 1976. Known as the Tangshan earthquake, it registered 8.2 on the Richter scale. How does the intensity of this earthquake compare with the intensity of an earthquake of magnitude $R = 5$?

45. Sound Intensity The relative loudness of a sound D of intensity I is measured in decibels, where

$$D = 10 \log \frac{I}{I_0}$$

and I_0 is the standard threshold of audibility.
   a. Express the intensity I of a 30-decibel sound (the sound level of normal conversation) in terms of I_0.
   b. Determine how many times greater the intensity of an 80-decibel sound (rock music) is than that of a 30-decibel sound.
   c. Prolonged noise above 150 decibels causes immediate and permanent deafness. How does the intensity of a 150-decibel sound compare with the intensity of an 80-decibel sound?

46. Barometric Pressure Halley's Law states that the barometric pressure (in inches of mercury) at an altitude of x miles above sea level is approximately given by the equation

$$p(x) = 29.92e^{-0.2x} \qquad (x \geq 0).$$

If the barometric pressure as measured by a hot-air balloonist is 20 inches of mercury, what is the balloonist's altitude?

47. Forensic Science Forensic scientists use the following law to determine the time of death of accident or murder victims. If T denotes the temperature of a body t hours after death, then

$$T = T_0 + (T_1 - T_0)(0.97)^t,$$

where T_0 is the air temperature and T_1 is the body temperature at the time of death. John Doe was found murdered at midnight in his house, when the room temperature was 70°F and his body temperature was 80°F. When was he killed? Assume that the normal body temperature is 98.6°F.

48. a. Given that $2^x = e^{kx}$, find k.
    b. Show that, in general, if b is a nonnegative real number, then any equation of the form $y = b^x$ may be written in the form $y = e^{kx}$, for some real number k.

49. Use the definition of a logarithm to prove
    a. $\log_b mn = \log_b m + \log_b n$
    b. $\log_b \dfrac{m}{n} = \log_b m - \log_b n$

    [*Hint:* Let $\log_b m = p$ and $\log_b n = q$. Then $b^p = m$ and $b^q = n$.]

50. Use the definition of a logarithm to prove
    $$\log_b m^n = n \log_b m.$$

51. Use the definition of a logarithm to prove
    a. $\log_b 1 = 0$ b. $\log_b b = 1$

SOLUTIONS TO SELF-CHECK EXERCISES 5.2

1. First, sketch the graph of $y = 3^x$ with the help of the following table of values.

x	-3	-2	-1	0	1	2	3
$y = 3^x$	$1/27$	$1/9$	$1/3$	0	3	9	27

Next, take the mirror reflection of this graph with respect to the line $y = x$ to obtain the graph of $y = \log_3 x$.

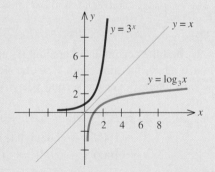

2. $3e^{x+1} - 2 = 4$
   $3e^{x+1} = 6$
   $e^{x+1} = 2$
   $\ln e^{x+1} = \ln 2$
   $(x + 1)\ln e = \ln 2$ (Law 3)
   $x + 1 = \ln 2$ (Law 5)
   $x = \ln 2 - 1$
   ≈ -0.3069

5.3   Compound Interest

Compound Interest

Compound interest is a natural application of the exponential function to the business world. We begin by recalling that simple interest is interest that is computed only on the original principal. Thus, if I denotes the interest on a principal P (in dollars) at an interest rate of r per year for t years, then we have

$$I = Prt.$$

The **accumulated amount** A, the sum of the principal and interest after t years, is given by

$$A = P + I = P + Prt$$
$$= P(1 + rt). \quad \text{(Simple interest formula)} \quad \textbf{(4)}$$

Frequently, interest earned is periodically added to the principal and thereafter earns interest itself at the same rate. This is called **compound interest.** In order to find a formula for the accumulated amount, let us consider a numerical example. Suppose that $1000 (the principal) is deposited in a bank for a **term** of three years earning interest at the rate of 8 percent per year (called the **nominal,** or **stated, rate**) compounded annually. Then, using (4) with $P = 1000$, $r = 0.08$, and $t = 1$, we see that the accumulated amount at the end of the first year is

$$A_1 = P(1 + rt)$$
$$= 1000[1 + 0.08(1)] = 1000(1.08) = 1080,$$

or $1080.

To find the accumulated amount A_2 at the end of the second year, we use (4) once again, this time with $P = A_1$. (Remember, the principal *and* interest now earn interest over the second year.) We obtain

$$A_2 = P(1 + rt) = A_1(1 + rt)$$
$$= 1000[1 + 0.08(1)][1 + 0.08(1)]$$
$$= 1000[1 + 0.08]^2 = 1000(1.08)^2 \approx 1166.40,$$

or approximately $1166.40.

Finally, the accumulated amount A_3 at the end of the third year is found using (4) with $P = A_2$, giving

$$A_3 = P(1 + rt) = A_2(1 + rt)$$
$$= 1000[1 + 0.08(1)]^2[1 + 0.08(1)]$$
$$= 1000[1 + 0.08]^3 = 1000(1.08)^3 \approx 1259.71,$$

or approximately $1259.71.

If you re-examine our calculations in this example, you will see that the accumulated amounts at the end of each year have the following form:

First year: $A_1 = 1000(1 + 0.08)$ or $A_1 = P(1 + r)$
Second year: $A_2 = 1000(1 + 0.08)^2$ or $A_2 = P(1 + r)^2$
Third year: $A_3 = 1000(1 + 0.08)^3$ or $A_3 = P(1 + r)^3$

These observations suggest the following general result: If P dollars are invested over a term of t years earning interest at the rate of r per year compounded annually, then the accumulated amount is

$$A = P(1 + r)^t. \tag{5}$$

Formula (5) was derived under the assumption that interest was compounded *annually*. In practice, however, interest is usually compounded more than once a year. The interval of time between successive interest calculations is called the **conversion period.**

If interest at a nominal rate of r per year is compounded m times a year on a principal of P dollars, then the simple interest rate per conversion period is

$$i = \frac{r}{m}. \qquad \frac{\text{(Annual interest rate)}}{\text{(Periods per year)}}$$

For example, if the nominal interest rate is 8 percent per year ($r = 0.08$) and interest is compounded quarterly ($m = 4$), then

$$i = \frac{r}{m} = \frac{0.08}{4} = 0.02,$$

or 2 percent per period.

In order to find a general formula for the accumulated amount when a principal of P dollars is deposited in a bank for a term of t years and earns interest at the (nominal) rate of r per year compounded m times per year, we proceed as before using (5) repeatedly with the interest rate $i = r/m$. We see that the accumulated amount at the end of each period is as follows:

First period: $A_1 = P(1 + i)$
Second period: $A_2 = A_1(1 + i)$ $= [P(1 + i)](1 + i) = P(1 + i)^2$
Third period: $A_3 = A_2(1 + i)$ $= [P(1 + i)^2](1 + i) = P(1 + i)^3$

$\qquad \vdots$ $\qquad\qquad\qquad\qquad\qquad \vdots$

nth period: $A_n = A_{n-1}(1 + i) = [P(1 + i)^{n-1}](1 + i) = P(1 + i)^n.$

But there are $n = mt$ periods in t years (number of conversion periods times the term). Therefore, the accumulated amount at the end of t years is given by

$$A = P(1 + i)^n.$$

COMPOUND INTEREST FORMULA	$$A = P(1 + i)^n, \qquad\qquad (6)$$

where $i = r/m$, $n = mt$, and

A = accumulated amount at the end of n conversion periods
P = principal
r = nominal interest rate per year
m = number of conversion periods per year
t = term (number of years)

EXAMPLE 1   Find the accumulated amount after three years if $1000 is invested at 8 percent per year compounded (a) annually, (b) semiannually, (c) quarterly, and (d) monthly.

Solution

a. Here $P = 1000$, $r = 0.08$, and $m = 1$. Thus, $i = r = 0.08$ and $n = 3$, so (6) gives

$$A = 1000(1.08)^3$$
$$= 1259.71,$$

or $1259.71.

b. Here $P = 1000$, $r = 0.08$, and $m = 2$. Thus, $i = 0.08/2 = 0.04$ and $n = (3)(2) = 6$, so that (6) gives

$$A = 1000(1.04)^6$$
$$= 1265.32,$$

or $1265.32.

c. In this case, $P = 1000$, $r = 0.08$, and $m = 4$. Thus, $i = 0.08/4 = 0.02$ and $n = (3)(4) = 12$, so (6) gives

$$A = 1000(1.02)^{12}$$
$$= 1268.24,$$

or $1268.24.

d. Here $P = 1000$, $r = 0.08$, and $m = 12$. Thus, $i = 0.08/12 = 0.0067$ and $n = (3)(12) = 36$, so (6) gives

$$A = 1000(1.0067)^{36}$$
$$= 1271.75,$$

or $1271.75. These results are summarized in Table 5.2.

TABLE 5.2

Nominal rate (r)	Conversion period	Interest rate/ conversion period	Initial investment	Accumulated amount
8%	Annual ($m = 1$)	8%	$1000	$1259.71
8%	Semiannual ($m = 2$)	4%	$1000	$1265.32
8%	Quarterly ($m = 4$)	2%	$1000	$1268.24
8%	Monthly ($m = 12$)	2/3%	$1000	$1271.75

Effective Rate of Interest

In the last example we saw that the interest actually earned on an investment depends on the frequency with which the interest is compounded. Thus, the stated, or nominal, rate of 8 percent per year does not reflect the actual rate at which interest is earned. This suggests that we need to find a common basis for comparing interest rates. One such way of comparing interest rates is provided by using the *effective rate*. The **effective rate** is the *simple* interest rate that would produce the same accumulated amount in one year as the nominal rate compounded m times a year. The effective rate is also called the **true rate.**

To derive a relation between the nominal interest rate, r per year compounded m times, and its corresponding effective rate, r_{eff} per year, let us assume an initial investment of P dollars. Then the accumulated amount after one year at a simple interest rate of r_{eff} per year is

$$A = P(1 + r_{eff}).$$

Also, the accumulated amount after one year at an interest rate of r per year compounded m times a year is

$$A = P(1 + i)^n = P\left(1 + \frac{r}{m}\right)^m. \qquad \text{(Since } i = r/m\text{)}$$

Equating the two expressions gives

$$P(1 + r_{eff}) = P\left(1 + \frac{r}{m}\right)^m$$

$$1 + r_{eff} = \left(1 + \frac{r}{m}\right)^m, \qquad \text{(Dividing both sides by } P\text{)}$$

or, upon solving for r_{eff}, we obtain the formula for computing the effective rate of interest:

EFFECTIVE RATE OF INTEREST FORMULA

$$r_{eff} = \left(1 + \frac{r}{m}\right)^m - 1 \qquad (7)$$

where
r_{eff} = effective rate of interest
r = nominal interest rate per year
m = number of conversion periods per year

EXAMPLE 2   Find the effective rate of interest corresponding to a nominal rate of 8 percent per year compounded (a) annually, (b) semiannually, (c) quarterly, and (d) monthly.

Solution

a. The effective rate of interest corresponding to a nominal rate of 8 percent per year compounded annually is, of course, given by 8 percent per year. This result is also confirmed by using (7) with $r = 0.08$ and $m = 1$.

Thus, $$r_{\text{eff}} = (1 + 0.08) - 1 = 0.08.$$

b. Let $r = 0.08$ and $m = 2$. Then (7) yields

$$r_{\text{eff}} = \left(1 + \frac{0.08}{2}\right)^2 - 1$$
$$= (1.04)^2 - 1$$
$$= 0.0816,$$

so the required effective rate is 8.16 percent per year.

c. Let $r = 0.08$ and $m = 4$. Then (7) yields

$$r_{\text{eff}} = \left(1 + \frac{0.08}{4}\right)^4 - 1$$
$$= (1.02)^4 - 1$$
$$= 0.08243,$$

so the corresponding effective rate in this case is 8.243 percent per year.

d. Let $r = 0.08$ and $m = 12$. Then (7) yields

$$r_{\text{eff}} = \left(1 + \frac{0.08}{12}\right)^{12} - 1$$
$$= (1.0067)^{12} - 1$$
$$= 0.08343,$$

so the corresponding effective rate in this case is 8.343 percent per year.

◼

Now, if the effective rate of interest r_{eff} is known, then the accumulated amount after t years on an investment of P dollars may be more readily computed by using the formula

$$A = P(1 + r_{\text{eff}})^t.$$

The Truth in Lending Act passed by Congress in 1968 requires that the effective rate of interest be disclosed in all contracts involving interest charges. The passage of this act has benefited consumers because they now have a common basis for comparing the various nominal rates quoted by different financial institutions. Furthermore, knowing the effective rate enables consumers to compute the actual charges involved in a transaction. Thus, if the effective rates of interest found in Example 2 were known, the accumulated values of Example 1, shown in Table 5.3, could have been readily found.

TABLE 5.3

Nominal rate	Frequency of interest payment	Effective rate	Initial investment	Accumulated amount after 3 years	
8%	Annually	8%	$1000	$1000(1 + 0.08)^3$	$= \$1259.71$
8%	Semiannually	8.16%	$1000	$1000(1 + 0.0816)^3$	$= \$1265.32$
8%	Quarterly	8.243%	$1000	$1000(1 + 0.08243)^3$	$= \$1268.23$
8%	Monthly	8.343%	$1000	$1000(1 + 0.08343)^3$	$= \$1271.75$

Present Value

Let us return to the compound interest formula (6), which expresses the accumulated amount at the end of n periods when interest at the rate of r is compounded m times a year. The principal P in (6) is often referred to as the **present value,** and the accumulated value A is called the **future value,** since it is realized at a future date. In certain instances, an investor may wish to determine how much money he should invest now, at a fixed rate of interest, so that he will realize a certain sum at some future date. This problem may be solved by expressing P in terms of A. Thus, from (6) we find

$$P = A(1 + i)^{-n}.$$

Here, as before, $i = r/m$, where m is the number of conversion periods per year.

PRESENT VALUE FORMULA FOR COMPOUND INTEREST

$$P = A (1 + i)^{-n} \qquad\qquad (8)$$

EXAMPLE 3   Find how much money should be deposited in a bank paying interest at the rate of 6 percent per year compounded monthly so that at the end of three years the accumulated amount will be $20,000.

Solution   Here, $r = 0.06$ and $m = 12$, so $i = 0.06/12 = 0.005$ and $n = (3)(12) = 36$. Thus, the problem is to determine P given that $A = 20,000$. Using (8), we obtain

$$P = 20{,}000(1.005)^{-36}$$
$$= 16{,}713,$$

or $16,713.

EXAMPLE 4   Find the present value of $49,158.60 due in five years at an interest rate of 10 percent per year compounded quarterly.

Solution   Using (8) with $r = 0.1$ and $m = 4$, so that $i = 0.1/4 = 0.025$, $n = (4)(5) = 20$, and $A = 49{,}158.6$, we obtain

$$P = (49{,}158.6)(1.025)^{-20} = 30{,}000,$$

or $30,000.   ■

Continuous Compounding of Interest

One question that arises naturally in the study of compound interest is: What happens to the accumulated amount over a fixed period of time if the interest is computed more and more frequently?

Intuition suggests that the more often interest is compounded, the larger the accumulated amount will be. This is confirmed by the results of Example 1, where we found that the accumulated amounts did, in fact, increase when we increased the number of conversion periods per year.

This leads us to another question: Does the accumulated amount approach a limit when the interest is computed more and more frequently over a fixed period of time?

To answer this question, let us look again at the compound interest formula:

$$A = P\left(1 + \frac{r}{m}\right)^{mt}. \tag{9}$$

Recall that m is the number of conversion periods per year. So to find an answer to our problem, we should let m approach infinity (get larger and larger) in (9). But first we will rewrite this equation in the form

$$A = P\left[\left(1 + \frac{r}{m}\right)^{m}\right]^{t}. \qquad \text{[Since } b^{xy} = (b^x)^y]$$

Now, letting $m \to \infty$, we find that

$$\lim_{m \to \infty}\left[P\left(1 + \frac{r}{m}\right)^{m}\right]^{t} = P\left[\lim_{m \to \infty}\left(1 + \frac{r}{m}\right)^{m}\right]^{t} \qquad \text{(Why?)}$$

Next, upon making the substitution $u = m/r$ and observing that $u \to \infty$ as $m \to \infty$, the foregoing expression reduces to

$$P\left[\lim_{u \to \infty}\left(1 + \frac{1}{u}\right)^{ur}\right]^{t} = P\left[\lim_{u \to \infty}\left(1 + \frac{1}{u}\right)^{u}\right]^{rt}.$$

But

$$\lim_{u \to \infty}\left(1 + \frac{1}{u}\right)^{u} = e, \qquad \text{[Using (1)]}$$

so

$$\lim_{m \to \infty} P\left[\left(1 + \frac{r}{m}\right)^{m}\right]^{t} = Pe^{rt}.$$

Our computations tell us that as the frequency with which interest is compounded increases without bound, the accumulated amount approaches Pe^{rt}. In this situation, we say that interest is *compounded continuously*. Let us summarize this important result.

CONTINUOUS
COMPOUND
INTEREST FORMULA

$$A = Pe^{rt} \qquad\qquad\qquad (10)$$

where

P = principal
r = annual interest rate compounded continuously
t = time in years
A = accumulated amount at the end of t years

EXAMPLE 5   Find the accumulated amount after three years if $1000 is invested at 8 percent per year compounded (a) daily (take the number of days in a year to be 365) and (b) continuously.

Solution
a. Using (6) with $P = 1000$, $r = 0.08$, $m = 365$, and $n = (365)(3) = 1095$, we find

$$A = 1000\left(1 + \frac{0.08}{365}\right)^{1095} \approx 1271.22,$$

or $1271.22.
b. Here we use (10) with $P = 1000$, $r = 0.08$, and $t = 3$, obtaining

$$A = 1000e^{(0.08)(3)}$$
$$\approx 1271.25, \qquad \text{(Using the "e^x" key)}$$

or $1271.25.   ■

  Observe that the accumulated amounts corresponding to interest compounded daily and interest compounded continuously differ by very little. The continuous compound interest formula is a very important tool in theoretical work in financial analysis.
  If we solve (10) for P, we obtain

$$P = Ae^{-rt}, \qquad\qquad\qquad (11)$$

which gives the present value in terms of the future (accumulated) value for the case of continuous compounding.

EXAMPLE 6   The Blakely Investment Company owns an office building located in the commercial district of a city. As a result of the continued success of an urban renewal program, local business is enjoying a miniboom. The market value of Blakely's property is

$$V(t) = 300{,}000e^{\sqrt{t}/2},$$

where $V(t)$ is measured in dollars and t is the time in years from the present. If the expected rate of inflation is 9 percent compounded continuously for the next ten years, find an expression for the present value $P(t)$ of the market price of the property valid for the next ten years. Compute $P(7)$, $P(8)$, and $P(9)$ and interpret your results.

Solution Using (11) with $A = V(t)$ and $r = 0.09$, we find that the present value of the market price of the property t years from now is

$$P(t) = V(t)e^{-0.09t}$$
$$= 300,000e^{-0.09t + \sqrt{t}/2} \qquad (0 \le t \le 10).$$

Letting $t = 7$, 8, and 9, respectively, we find that

$$P(7) = 300,000e^{-0.09(7) + \sqrt{7}/2} = 599,837 \quad \text{or} \quad \$599,837$$
$$P(8) = 300,000e^{-0.09(8) + \sqrt{8}/2} = 600,640 \quad \text{or} \quad \$600,640$$
$$P(9) = 300,000e^{-0.09(9) + \sqrt{9}/2} = 598,115 \quad \text{or} \quad \$598,115.$$

From the results of these computations, we see that the present value of the property's market price seems to decrease after a certain period of growth. This suggests that there is an optimal time for the owners to sell. Later we will show that the highest present value of the property's market price is $600,779, which occurs at time $t = 7.72$ years. ∎

SELF-CHECK EXERCISES 5.3

c **1.** Find the present value of $20,000 due in three years at an interest rate of 12 percent per year compounded monthly.

c **2.** Mr. Baker is a retiree living on social security and the income from his investment. Currently, his $100,000 investment in a one-year CD is yielding 11.6 percent interest compounded daily. If he reinvests the principal ($100,000) on the due date of the CD in another one-year CD paying 9.2 percent interest compounded daily, find the net decrease in his yearly income from his investment.

c **3. a.** What is the accumulated amount after five years if $10,000 is invested at 10 percent per year compounded continuously?
  b. Find the present value of $10,000 due in five years at an interest rate of 10 percent per year compounded continuously.

Solutions to Self-Check Exercises 5.3 can be found on page 309.

5.3 EXERCISES

c *A calculator is recommended for these exercises.*

In Exercises 1–4, find the accumulated amount A if the principal P is invested at an interest rate of r per year for t years.

1. $P = \$2500$, $r = 7$ percent, $t = 10$, compounded semiannually

2. $P = \$12,000$, $r = 8$ percent, $t = 10$, compounded quarterly

3. $P = \$150,000$, $r = 10$ percent, $t = 4$, compounded monthly

4. $P = \$150,000$, $r = 9$ percent, $t = 3$, compounded daily

In Exercises 5 and 6, find the effective rate corresponding to the given nominal rate.

5. a. 10 percent per year compounded semiannually
  b. 9 percent per year compounded quarterly

6. a. 8 percent per year compounded monthly
  b. 8 percent per year compounded daily

In Exercises 7 and 8, find the present value of $40,000 due in four years at the given rate of interest.

7. a. 8 percent per year compounded semiannually
  b. 8 percent per year compounded quarterly

8. a. 7 percent per year compounded monthly
    b. 9 percent per year compounded daily

9. Find the accumulated amount after four years if $5000 is invested at 8 percent per year compounded continuously.

10. An amount of $25,000 is deposited in a bank that pays interest at the rate of 7 percent per year, compounded annually. What is the total amount on deposit at the end of six years, assuming there are no deposits or withdrawals during those six years? What is the interest earned in that period of time?

11. **Housing Prices** The Brennans are planning to buy a house four years from now. Housing experts in their area have estimated that the cost of a home will increase at a rate of 9 percent per year during that four-year period. If this economic prediction holds true, how much can they expect to pay for a house that currently costs $80,000?

12. **Energy Consumption** A metropolitan utility company in a western city of the United States expects the consumption of electricity to increase by 8 percent per year during the next decade, due mainly to the expected population increase. If consumption does increase at this rate, find the amount by which the utility company will have to increase its generating capacity in order to meet the area's needs at the end of the decade.

13. **Pension Funds** The managers of a pension fund have invested $1.5 million in U.S. government certificates of deposit (CDs) that pay interest at the rate of 9.5 percent per year compounded semiannually over a period of ten years. At the end of this period, how much will the investment be worth?

14. **Savings Accounts** Mr. Kaplan invested a sum of money five years ago in a savings account, which has since paid interest at the rate of 8 percent compounded quarterly. His investment is now worth $22,289.22. How much did he originally invest?

15. **Loan Consolidation** The proprietors of the Coachmen Inn secured two loans from the Union Bank: one for $8,000 due in three years and one for $15,000 due in six years, both at an interest rate of 10 percent compounded semiannually. The bank agreed to allow the two loans to be consolidated into one loan payable in five years at the same interest rate. How much will the proprietors have to pay the bank at the end of five years?

16. **Tax-Deferred Annuities** Mrs. Bennett is in the 28 percent tax bracket and has $25,000 available for investment during her current tax year. Assume that she remains in the same tax bracket over the next ten years

and determine the accumulated amount of her investment if

a. she puts the $25,000 into a tax-deferred annuity that pays 12 percent per year, tax deferred for ten years.

b. she puts the $25,000 into a taxable instrument that pays 12 percent per year for ten years. [*Hint:* In this case the yield after taxes is 8.64 percent per year.]

17. **Consumer Price Index** At an annual inflation rate of 7.5 percent, how long will it take the Consumer Price Index (CPI) to double?

18. **Investment Returns** Ms. Collins purchased a house in 1987 for $80,000. In 1993 she sold the house and made a net profit of $28,000. Find the effective annual rate of return on her investment over the six-year period.

19. **Investment Returns** Mr. Stevens purchased 1,000 shares of a certain stock for $25,250 (including commissions). He sold the shares two years later and received $32,100 after deducting commissions. Find the effective annual rate of return on his investment over the two-year period.

20. **Investment Options** Investment A offers a 10 percent return compounded semiannually, and investment B offers a 9.75 percent return compounded continuously. Which investment has a higher rate of return over a four-year period?

21. **Present Value** Find the present value of $59,673 due in five years at an interest rate of 8 percent per year compounded continuously.

22. **Real Estate Investments** A condominium complex was purchased by a group of private investors for $1.4 million and sold six years later for $3.6 million. Find the annual rate of return (compounded continuously) on their investment.

23. **Saving for College** Having received a large inheritance, a child's parents wish to establish a trust for the child's college education. If they need an estimated $70,000 seven years from now, how much should they set aside in trust now, if they invest the money at 10.5 percent compounded (a) quarterly? (b) continuously?

24. **Effect of Inflation on Salaries** Mr. Lyons's current annual salary is $25,000. Ten years from now, how much will he need to earn in order to retain his present purchasing power if the rate of inflation over that period is 6 percent per year? Assume that inflation is continuously compounded.

25. **Pensions** Ms. Lindstrom, who is now 50 years old, is employed by a firm that guarantees her a pension of $40,000 per year at age 65. What is the present value of her first year's pension if inflation over the next 15

Misato Nakazaki

Title: Assistant Vice President
Institution: A large investment corporation.

In the securities industry, buying and selling stocks and bonds has always required a mastery of concepts and formulas that outsiders find confusing. As a bond seller, Nakazaki routinely uses terms such as *issue, maturity, current yield,* and *callable* and *convertible bonds* and so on.

These terms, however, are easily defined. When corporations issue bonds, they are borrowing money at a fixed rate of interest. The bonds are scheduled to mature—to be paid back—on a specific date as much as thirty years into the future. Callable bonds allow the issuer to pay off the loans prior to their expected maturity, reducing overall interest payments. In its simplest terms, current yield is the price of a bond multiplied by the interest rate at which the bond is issued. For example, a bond with a face value of $1000 and an interest rate of 10 percent yields $100 per year in interest payments. When that same bond is resold at a premium on the secondary market for $1200, its current yield nets only an 8.3 percent rate of return based on the higher purchase price.

Bonds attract investors for many reasons. A key variable is the sensitivity of the bond's price to future changes in interest rates. If investors get locked into a low-paying bond when future bonds pay higher yields, they lose money. Nakazaki stresses that "no one knows for sure what rates will be over time." Employing differentials allows her to calculate interest-rate sensitivity for clients as they ponder purchase decisions.

Computerized formulas, "whose basis is calculus," says Nakazaki, help her factor the endless stream of numbers flowing across her desk.

On a typical day, Nakazaki might be given a bid on "10 million, GMAC, 8.5 percent, January 2003." Translation: Her customer wants her to buy General Motors Acceptance Corporation bonds with a face value of $10 million and an interest rate of 8.5 percent, maturing in January 2003.

After she calls her firm's trader to find out the yield on the bond in question, Nakazaki enters the price and other variables, such as the interest rate and date of maturity, and the computer prints out the answers. Nakazaki can then relay to her client the bond's current yield, accrued interest, and so on. In Nakazaki's rapid-fire work environment, such speed is essential. Nakazaki cautions that "computer users have to understand what's behind the formulas." The software "relies on the basics of calculus. If people don't understand the formula, it's useless for them to use the calculations."

With an MBA from New York University, Nakazaki typifies the younger generation of Japanese women who have chosen to succeed in the business world. Since earning her degree, she has sold bonds for a global securities firm in New York City.

Nakazaki's client list reads like a who's who of the leading Japanese banks, insurance companies, mutual funds, and corporations. As institutional buyers, her clients purchase large blocks of American corporate bonds and mortgage-backed securities such as Ginnie Maes.

years is (a) 6 percent? (b) 8 percent? (c) 12 percent? Assume that inflation is continuously compounded.

26. **Real Estate Investments** An investor purchased a piece of waterfront property. Because of the development of a marina in the vicinity, the market value of the property is expected to increase according to the rule

$$V(t) = 80,000e^{\sqrt{t}/2},$$

where $V(t)$ is measured in dollars and t is the time in years from the present. If the rate of inflation is expected to be 9 percent compounded continuously for the next eight years, find an expression for the present value $P(t)$ of the property's market price valid for the next eight years. What is $P(t)$ expected to be in four years?

27. Show that the effective rate of interest r_{eff} that corresponds to a nominal interest rate r per year compounded continuously is given by

$$r_{\text{eff}} = e^r - 1.$$

[*Hint:* From (7) we see that the effective rate r_{eff} corresponding to a nominal interest rate r per year compounded m times a year is given by

$$r_{\text{eff}} = \left(1 + \frac{r}{m}\right)^m - 1.$$

Let m tend to infinity in this expression.]

28. Refer to Exercise 27. Find the effective rate of interest that corresponds to a nominal rate of 10 percent per year compounded (a) quarterly, (b) monthly, and (c) continuously.

29. **Investment Analysis** Refer to Exercise 27. Bank A pays interest on deposits at a 7 percent annual rate compounded quarterly, and Bank B pays interest on deposits at a $7\frac{1}{8}$ percent annual rate compounded continuously. Which bank has the higher effective rate of interest?

30. **Investment Analysis** Find the nominal rate of interest that, when compounded monthly, yields an effective rate of interest of 10 percent per year. [*Hint:* Use Equation (7).]

31. **Investment Analysis** Find the nominal rate of interest that, when compounded continuously, yields an effective rate of interest of 10 percent per year. [*Hint:* See Exercise 27.]

SOLUTIONS TO SELF-CHECK EXERCISES 5.3

1. Using (8) with $r = 0.12$ and $m = 12$, so that

$$i = \frac{0.12}{12} = 0.01, \quad n = (12)(3) = 36, \quad \text{and} \quad A = 20,000,$$

we find the required present value to be

$$P = 20,000(1.01)^{-36}$$
$$= 13,978.50,$$

or $13,978.50.

2. The accumulated amount of Mr. Baker's current investment is found by using (6) with $P = 100,000$, $r = 0.116$, and $m = 360$. Thus,

$$i = \frac{0.116}{360} = 0.0003222 \quad \text{and} \quad n = 360,$$

so the required accumulated amount is

$$A = 100,000(1.0003222)^{360}$$
$$= 112,296.59,$$

or $112,296.59. Next, we compute the accumulated amount of Mr. Baker's reinvestment. Once again, using (6) with $P = 100,000$, $r = 0.092$, and $m = 360$ so that

$$i = \frac{0.092}{360} = 0.0002556 \quad \text{and} \quad n = 360,$$

we find the required accumulated amount in this case to be

$$\bar{A} = 100,000(1.0002556)^{360},$$

or $109,636.95. Therefore, Mr. Baker can expect to experience a net decrease in yearly income of

$$112,296.59 - 109,636.95,$$

or $2,659.64.

3. a. Using (10) with $P = 10{,}000$, $r = 0.1$, and $t = 5$, we find that the required accumulated amount is given by

$$A = 10{,}000e^{(0.1)(5)}$$
$$= 16{,}487.21,$$

or $16,487.21.

b. Using (11) with $A = 10{,}000$, $r = 0.1$, and $t = 5$, we see that the required present value is given by

$$P = 10{,}000e^{-(0.1)(5)}$$
$$= 6{,}065.31,$$

or $6,065.31.

| 5.4 | **Differentiation of Exponential Functions** |

The Derivative of the Exponential Function

In order to study mathematical models involving exponential and logarithmic functions, we need to develop rules for computing the derivatives of these functions. We begin by looking at the rule for computing the derivative of the exponential function.

| **RULE 1   DERIVATIVE OF THE EXPONENTIAL FUNCTION** | $$\frac{d}{dx}e^x = e^x$$ |

Thus, the derivative of the exponential function with base e is equal to the function itself. To demonstrate the validity of this rule, we compute

$$f'(x) = \lim_{h \to 0} \frac{f(x + h) - f(x)}{h}$$

$$= \lim_{h \to 0} \frac{e^{x+h} - e^x}{h}$$

$$- \lim_{h \to 0} \frac{e^x(e^h - 1)}{h} \qquad \text{(Writing } e^{x+h} = e^x e^h \text{ and factoring)}$$

$$= e^x \lim_{h \to 0} \frac{e^h - 1}{h}. \qquad \text{(Why?)}$$

To evaluate

$$\lim_{h \to 0} \frac{e^h - 1}{h},$$

let us refer to Table 5.4, which is constructed with the aid of a calculator. From the table, we see that

$$\lim_{h \to 0} \frac{e^h - 1}{h} = 1.$$

TABLE 5.4

h	0.1	0.01	0.001	−0.1	−0.01	−0.001
$\dfrac{e^h - 1}{h}$	1.0517	1.0050	1.0005	0.9516	0.9950	0.9995

(Although a rigorous proof of this fact is possible, it is beyond the scope of this book.) Using this result, we conclude that

$$f'(x) = e^x \cdot 1 = e^x,$$

as we set out to show.

EXAMPLE 1   Compute the derivative of each of the following functions:

a. $f(x) = x^2 e^x$ **b.** $g(t) = (e^t + 2)^{3/2}$

Solution
a. The Product Rule gives

$$f'(x) = \frac{d}{dx}(x^2 e^x)$$

$$= x^2 \frac{d}{dx}(e^x) + e^x \frac{d}{dx}(x^2)$$

$$= x^2 e^x + e^x(2x)$$

$$= xe^x(x + 2).$$

b. Using the General Power Rule, we find

$$g'(t) = \frac{3}{2}(e^t + 2)^{1/2} \frac{d}{dt}(e^t + 2)$$

$$= \frac{3}{2}(e^t + 2)^{1/2} e^t$$

$$= \frac{3}{2} e^t (e^t + 2)^{1/2}.$$ ∎

Applying the Chain Rule to Exponential Functions

To enlarge the class of exponential functions to be differentiated, we appeal to the Chain Rule to obtain the following rule for differentiating composite functions of the form $h(x) = e^{f(x)}$. An example of such a function is $h(x) = e^{x^2 - 2x}$. Here $f(x) = x^2 - 2x$.

RULE 2   THE CHAIN RULE FOR EXPONENTIAL FUNCTIONS

If $f(x)$ is a differentiable function, then

$$\frac{d}{dx}(e^{f(x)}) = e^{f(x)}f'(x).$$

To see this, observe that if $h(x) = g[f(x)]$, where $g(x) = e^x$, then by virtue of the Chain Rule,

$$h'(x) = g'(f(x))f'(x) = e^{f(x)}f'(x)$$

since $g'(x) = e^x$.

As an aid to remembering the Chain Rule for exponential functions, observe that it has the following form:

$$\frac{d}{dx}(e^{f(x)}) = e^{f(x)} \cdot \text{derivative of exponent}$$
$$\underset{\text{same}}{\llcorner \qquad \lrcorner}$$

EXAMPLE 2   Find the derivative of each of the following functions.

a. $f(x) = e^{2x}$ **b.** $y = e^{-3x}$ **c.** $g(t) = e^{2t^2+t}$

Solution

a. $f'(x) = e^{2x}\dfrac{d}{dx}(2x) = e^{2x} \cdot 2 = 2e^{2x}$

b. $\dfrac{dy}{dx} = e^{-3x}\dfrac{d}{dx}(-3x) = -3e^{-3x}$

c. $g'(t) = e^{2t^2+t} \cdot \dfrac{d}{dt}(2t^2 + t) = (4t + 1)e^{2t^2+t}$ ■

EXAMPLE 3   Differentiate the function $y = xe^{-2x}$.

Solution   Using the Product Rule, followed by the Chain Rule, we find

$$\frac{dy}{dx} = x\frac{d}{dx}e^{-2x} + e^{-2x}\frac{d}{dx}(x)$$

$$= xe^{-2x}\frac{d}{dx}(-2x) + e^{-2x} \qquad \text{(Using the Chain Rule on the first term)}$$

$$= -2xe^{-2x} + e^{-2x}$$

$$= e^{-2x}(1 - 2x).$$ ■

EXAMPLE 4   Differentiate the function $g(t) = \dfrac{e^t}{e^t + e^{-t}}$.

Solution   Using the Quotient Rule, followed by the Chain Rule, we find

$$g'(t) = \frac{(e^t + e^{-t})\frac{d}{dt}(e^t) - e^t\frac{d}{dt}(e^t + e^{-t})}{(e^t + e^{-t})^2}$$

$$= \frac{(e^t + e^{-t})e^t - e^t(e^t - e^{-t})}{(e^t + e^{-t})^2}$$

$$= \frac{e^{2t} + 1 - e^{2t} + 1}{(e^t + e^{-t})^2} \qquad (e^0 = 1)$$

$$= \frac{2}{(e^t + e^{-t})^2}. \qquad\blacksquare$$

EXAMPLE 5   In Section 5.6 we will discuss some practical applications of the exponential function

$$Q(t) = Q_0 e^{kt},$$

where Q_0 and k are positive constants and $t \in [0, \infty)$. A quantity $Q(t)$ growing according to this law experiences exponential growth. Show that for a quantity $Q(t)$ experiencing exponential growth, the rate of growth of the quantity, $Q'(t)$, at any time t is directly proportional to the amount of the quantity present.

Solution   Using the Chain Rule for exponential functions, we compute the derivative Q' of the function Q. Thus,

$$Q'(t) = Q_0 e^{kt}\frac{d}{dt}(kt)$$

$$= Q_0 e^{kt}(k)$$

$$= kQ_0 e^{kt}$$

$$= kQ(t), \qquad (Q(t) = Q_0 e^{kt})$$

which is the desired conclusion. $\blacksquare$

EXAMPLE 6   Find the points of inflection of the function $f(x) = e^{-x^2}$.

Solution   The first derivative of f is

$$f'(x) = -2xe^{-x^2}.$$

Differentiating $f'(x)$ with respect to x yields

$$f''(x) = (-2x)(-2xe^{-x^2}) - 2e^{-x^2}$$
$$= 2e^{-x^2}(2x^2 - 1).$$

Setting $f''(x) = 0$ gives

$$2e^{-x^2}(2x^2 - 1) = 0.$$

FIGURE 5.10
Sign diagram for f''

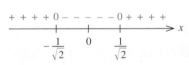

Since e^{-x^2} never equals zero for any real value of x, we see that $x = \pm 1/\sqrt{2}$ are the only candidates for inflection points of f. The sign diagram of f'' shown in Figure 5.10 tells us that both $x = -1/\sqrt{2}$ and $x = 1/\sqrt{2}$ give rise to inflection points of f.

Next,

$$f\left(-\frac{1}{\sqrt{2}}\right) = f\left(\frac{1}{\sqrt{2}}\right) = e^{-1/2},$$

and the inflection points of f are $(-1/\sqrt{2}, e^{-1/2})$ and $(1/\sqrt{2}, e^{-1/2})$. The graph of f appears in Figure 5.11.

FIGURE 5.11
The graph of $y = e^{-x^2}$ has two inflection points.

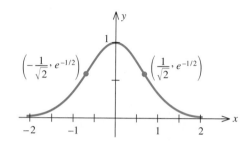

Application

Our final example involves finding the absolute maximum of an exponential function.

EXAMPLE 7   Refer to Example 6, Section 5.3. The present value of the market price of the Blakely Office Building is given by

$$P(t) = 300{,}000e^{-0.09t+\sqrt{t}/2} \qquad (0 \le t \le 10).$$

Find the optimal present value of the building's market price.

Solution   To find the maximum value of P over $[0, 10]$, we compute

$$P'(t) = 300{,}000e^{-0.09t+\sqrt{t}/2}\frac{d}{dt}\left(-0.09t + \frac{1}{2}t^{1/2}\right)$$

$$= 300{,}000e^{-0.09t+\sqrt{t}/2}\left(-0.09 + \frac{1}{4}t^{-1/2}\right)$$

Setting $P'(t) = 0$ gives

$$-0.09 + \frac{1}{4t^{1/2}} = 0,$$

since $e^{-0.09t+\sqrt{t}/2}$ is never zero for any value of t. Solving this equation, we find

$$\frac{1}{4t^{1/2}} = 0.09$$

$$t^{1/2} = \frac{1}{4(0.09)}$$

$$= \frac{1}{0.36}$$

or $t \approx 7.72,$

the sole critical point of the function P. Finally, evaluating $P(t)$ at the critical point as well as at the endpoints of $[0, 10]$, we have

t	$P(t)$
0	300,000
7.72	600,779
10	592,838

We conclude, accordingly, that the optimal present value of the property's market price is $600,779 and that this will occur 7.72 years from now. ■

SELF-CHECK EXERCISES 5.4

1. Let $f(x) = xe^{-x}$.
   a. Find the first and second derivatives of f.
   b. Find the relative extrema of f.
   c. Find the inflection points of f.

2. An industrial asset is being depreciated at a rate so that its book value t years from now will be

$$V(t) = 50{,}000e^{-0.4t}$$

dollars. How fast will the book value of the asset be changing three years from now?

Solutions to Self-Check Exercises 5.4 can be found on page 317.

5.4 EXERCISES

In Exercises 1–28, find the derivative of the given function.

1. $f(x) = e^{3x}$

2. $f(x) = 3e^x$

3. $g(t) = e^{-t}$

4. $f(x) = e^{-2x}$

5. $f(x) = e^x + x$

6. $f(x) = 2e^x - x^2$

7. $f(x) = x^3 e^x$

8. $f(u) = u^2 e^{-u}$

9. $f(x) = \dfrac{2e^x}{x}$

10. $f(x) = \dfrac{x}{e^x}$

11. $f(x) = 3(e^x + e^{-x})$

12. $f(x) = \dfrac{e^x + e^{-x}}{2}$

13. $f(w) = \dfrac{e^w + 1}{e^w}$

14. $f(x) = \dfrac{e^x}{e^x + 1}$

15. $f(x) = 2e^{3x-1}$

16. $f(t) = 4e^{3t+2}$

17. $h(x) = e^{-x^2}$

18. $f(x) = e^{x^2-1}$

19. $f(x) = 3e^{-1/x}$

20. $f(x) = e^{1/(2x)}$

21. $f(x) = (e^x + 1)^{25}$

22. $f(x) = (4 - e^{-3x})^3$

23. $f(x) = e^{\sqrt{x}}$

24. $f(t) = -e^{-\sqrt{2t}}$

25. $f(x) = (x - 1)e^{3x+2}$

26. $f(s) = (s^2 + 1)e^{-s^2}$

27. $f(x) = \dfrac{e^x - 1}{e^x + 1}$

28. $g(t) = \dfrac{e^{-t}}{1 + t^2}$

In Exercises 29–32, find the second derivative of the given function.

29. $f(x) = e^{-4x} + 2e^{3x}$

30. $f(t) = 3e^{-2t} - 5e^{-t}$

31. $f(x) = 2xe^{3x}$

32. $f(t) = t^2 e^{-2t}$

33. Find an equation of the tangent line to the graph of $y = e^{2x-3}$ at the point $\left(\frac{3}{2}, 1\right)$.

34. Find an equation of the tangent line to the graph of $y = e^{-x^2}$ at the point $(1, 1/e)$.

35. Determine the intervals where the function $f(x) = e^{-x^2/2}$ is increasing and where it is decreasing.

36. Determine the intervals where the function $f(x) = x^2 e^{-x}$ is increasing and where it is decreasing.

37. Determine the intervals of concavity for the function $f(x) = \dfrac{e^x - e^{-x}}{2}$.

38. Determine the intervals of concavity for the function $f(x) = xe^x$.

39. Find the inflection point of the function $f(x) = xe^{-2x}$.

40. Find the inflection point(s) of the function $f(x) = 2e^{-x^2}$.

In Exercises 41–44, find the absolute extrema of the given function.

41. $f(x) = e^{-x^2}$ on $[-1, 1]$

42. $h(x) = e^{x^2-4}$ on $[-2, 2]$

43. $g(x) = (2x - 1)e^{-x}$ on $[0, \infty)$

44. $f(x) = xe^{-x^2}$ on $[0, 2]$

In Exercises 45–48, use the curve-sketching guidelines of Chapter 4, page 246, to sketch the graph of the given function.

45. $f(t) = e^t - t$

46. $h(x) = \dfrac{e^x + e^{-x}}{2}$

47. $f(x) = 2 - e^{-x}$

48. $f(x) = \dfrac{3}{1 + e^{-x}}$

C A calculator is recommended for the remainder of these exercises.

49. **Sales Promotion**   The Lady Bug, a women's clothing chain store, found that t days after the end of a sales promotion the volume of sales was given by

$$S(t) = 20{,}000(1 + e^{-0.5t}) \qquad (0 \le t \le 5)$$

dollars. Find the rate of change of The Lady Bug's sales volume when $t = 1$, $t = 2$, $t = 3$, and $t = 4$.

50. **Energy Consumption of Appliances**   The average energy consumption of the typical refrigerator/freezer manufactured by York Industries is approximately

$$C(t) = 1486e^{-0.073t} + 500 \qquad (0 \le t \le 20)$$

kilowatt hours per year, where t is measured in years, with $t = 0$ corresponding to 1972.
   a. What was the average energy consumption of the York refrigerator/freezer at the beginning of 1972?

   b. Prove that the average energy consumption of the York refrigerator/freezer is decreasing over the years in question.
   c. All refrigerator/freezers manufactured as of January 1, 1990 must meet the 950-kilowatt-hours-per-year maximum energy-consumption standard set by the National Appliance Conservation Act. Does the York refrigerator/freezer meet this requirement?

51. **Polio Immunization**   Polio, a once-feared killer, declined markedly in the United States in the 1950s after Jonas Salk developed the inactivated polio vaccine and mass immunization of children took place. The number of polio cases in the United States from the beginning of 1959 to the beginning of 1963 is approximated by the function

$$N(t) = 5.3e^{0.095t^2 - 0.85t} \qquad (0 \le t \le 4),$$

where $N(t)$ gives the number of polio cases (in thousands) and t is measured in years, with $t = 0$ corresponding to the beginning of 1959.
   a. Show that the function N is decreasing over the time interval under consideration.
   b. How fast was the number of polio cases decreasing at the beginning of 1959? At the beginning of 1962? [*Comment:* Following the introduction of the oral vaccine developed by Dr. Albert B. Sabin in 1963, polio in the United States has, for all practical purposes been eliminated.]

52. **Price of Perfume**   The monthly demand for a certain brand of perfume is given by the demand equation

$$p = 100e^{-0.0002x} + 150,$$

where p denotes the retail unit price in dollars and x denotes the quantity (in 1-ounce bottles) demanded.
   a. Find the rate of change of the price per bottle when $x = 1000$; when $x = 2000$.
   b. What is the price per bottle when $x = 1000$? when $x = 2000$?

53. **Price of Wine**   The monthly demand for a certain brand of table wine is given by the demand equation

$$p = 240\left(1 - \frac{3}{3 + e^{-0.0005x}}\right),$$

where p denotes the wholesale price per case (in dollars) and x denotes the number of cases demanded.
   a. Find the rate of change of the price per case when $x = 1000$.
   b. What is the price per case when $x = 1000$?

54. **Spread of an Epidemic**   During a flu epidemic, the total number of students on a state university campus

who had contracted influenza by the xth day was given by

$$N(x) = \frac{3000}{1 + 99e^{-x}} \qquad (x \geq 0).$$

a. How many students had influenza initially?
b. Derive an expression for the rate at which the disease was being spread and prove that the function N is increasing on the interval $(0, \infty)$.
c. Sketch the graph of N. What was the total number of students who contracted influenza during that particular epidemic?

55. **Maximum Oil Production**   It has been estimated that the total production of oil from a certain oil well is given by

$$T(t) = -1,000(t + 10)e^{-0.1t} + 10,000$$

thousand barrels t years after production has begun. Determine the year when the oil well will be producing at maximum capacity.

56. **Optimal Selling Time**   Refer to Exercise 26, page 309. The present value of a piece of waterfront property purchased by an investor is given by the function

$$P(t) = 80,000e^{\sqrt{t/2}\,0.09t} \qquad (0 \leq t \leq 8).$$

Determine the optimal time (based on present value) for the investor to sell the property. What is the property's optimal present value?

57. **Oil Used to Fuel Productivity**   A study on worldwide oil use was prepared for a major oil company. The study predicted that the amount of oil used to fuel productivity in a certain country is given by

$$f(t) = 1.5 + 1.8te^{-1.2t} \qquad (0 \leq t \leq 4),$$

where $f(t)$ denotes the number of barrels per \$1000 of economic output and t is measured in decades ($t = 0$ corresponds to 1965). Compute $f'(0)$, $f'(1)$, $f'(2)$, and $f'(3)$, and interpret your results.

58. **Percentage of Population Relocating**   Based on data obtained from the Census Bureau, the manager of Plymouth Van Lines estimates that the percentage of the total population relocating in year t ($t = 0$ corresponds to the year 1960) may be approximated by the formula

$$P(t) = 20.6e^{-0.009t} \qquad (0 \leq t \leq 35).$$

Compute $P'(10)$, $P'(20)$, and $P'(30)$ and interpret your results.

C   *In Exercises 59 and 60, use a computer or graphing calculator to plot the graph of the given function.*

59. $f(x) = xe^{-2x}$ 60. $g(t) = t^2 e^{-t}$

61. **Percentage of Females in the Labor Force**   Based on data from the United States Census Bureau, the chief economist of Manpower, Inc., constructed the following formula giving the percentage of the total female population in the civilian labor force, $P(t)$, at the beginning of the tth decade ($t = 0$ corresponds to the year 1900):

$$P(t) = \frac{74}{1 + 2.6e^{-0.166t + 0.04536t^2 - 0.0066t^3}} \qquad (0 \leq t \leq 10).$$

If this trend continues for the rest of the twentieth century,

a. what will the percentage of the total female population in the civilian labor force be at the beginning of the year 2000?
b. what will the growth rate of the percentage of the total female population in the civilian labor force be at the beginning of the year 2000?

SOLUTIONS TO SELF-CHECK EXERCISES 5.4

1. a. Using the Product Rule, we obtain

$$f'(x) = x\frac{d}{dx}e^{-x} + e^{-x}\frac{d}{dx}x$$
$$= -xe^{-x} + e^{-x} = (1 - x)e^{-x}.$$

Using the Product Rule once again, we obtain

$$f''(x) = (1 - x)\frac{d}{dx}e^{-x} + e^{-x}\frac{d}{dx}(1 - x)$$
$$= (1 - x)(-e^{-x}) + e^{-x}(-1)$$
$$= -e^{-x} + xe^{-x} - e^{-x} = (x - 2)e^{-x}.$$

b. Setting $f'(x) = 0$ gives

$$(1 - x)e^{-x} = 0.$$

Since $e^{-x} \neq 0$, we see that $1 - x = 0$, and this gives $x = 1$ as the only critical point of f. The sign diagram of f' shown in the accompanying figure tells us that the point $(1, e^{-1})$ is a relative maximum of f.

$$+ + + + + + + + + + 0 - - - - -$$
$$\xrightarrow{\qquad\quad|\qquad\quad|\qquad\qquad} x$$
$$01$$

Sign diagram of f'

c. Setting $f''(x) = 0$ gives $x - 2 = 0$, so $x = 2$ is a candidate for an inflection point of f. The sign diagram of f'' (accompanying figure) shows that $(2, 2e^{-2})$ is an inflection point of f.

$$- - - - - - - - - - 0 + + + + +$$
$$\xrightarrow{\qquad\quad|\qquad\quad|\qquad\qquad} x$$
$$02$$

Sign diagram of f''

2. The rate of change of the book value of the asset t years from now is

$$V'(t) = 50{,}000 \,\frac{d}{dt}\, e^{-0.4t}$$
$$= 50{,}000(-0.4)e^{-0.4t} = -20{,}000e^{-0.4t}.$$

Therefore, three years from now the book value of the asset will be changing at the rate of

$$V'(3) = -20{,}000\,e^{-0.4(3)} = -20{,}000\,e^{-1.2} \approx -6{,}023.88;$$

that is, decreasing at the rate of approximately \$6,024 per year.

5.5 Differentiation of Logarithmic Functions

The Derivative of ln x

Let us now turn our attention to the differentiation of logarithmic functions.

RULE 3 DERIVATIVE OF ln x	$\dfrac{d}{dx}\ln|x| = \dfrac{1}{x} \quad (x \neq 0)$

To derive Rule 3, suppose that $x > 0$ and write $f(x) = \ln x$ in the equivalent form

$$x = e^{f(x)}.$$

Differentiating both sides of the equation with respect to x, we find, using the Chain Rule,

$$1 = e^{f(x)} \cdot f'(x),$$

from which we see that

$$f'(x) = \frac{1}{e^{f(x)}}$$

or, since $e^{f(x)} = x$,

$$f'(x) = \frac{1}{x},$$

as we set out to show. You are asked to prove the rule for the case $x < 0$ in Exercise 59.

EXAMPLE 1   Compute the derivative of each of the following functions:

a. $f(x) = x \ln x$ **b.** $g(x) = \dfrac{\ln x}{x}$

Solution

a. Using the Product Rule, we obtain

$$f'(x) = \frac{d}{dx}(x \ln x) = x \frac{d}{dx}(\ln x) + (\ln x) \frac{d}{dx}(x)$$

$$= x\left(\frac{1}{x}\right) + \ln x$$

$$= 1 + \ln x.$$

b. Using the Quotient Rule, we obtain

$$g'(x) = \frac{x \dfrac{d}{dx}(\ln x) - (\ln x) \dfrac{d}{dx}(x)}{x^2}$$

$$= \frac{x\left(\dfrac{1}{x}\right) - \ln x}{x^2} = \frac{1 - \ln x}{x^2}.$$ ∎

The Chain Rule for Logarithmic Functions

To enlarge the class of logarithmic functions to be differentiated, we appeal once more to the Chain Rule to obtain the following rule for differentiating composite functions of the form $h(x) = \ln f(x)$, where $f(x)$ is assumed to be a positive differentiable function.

RULE 4 THE CHAIN RULE FOR LOGARITHMIC FUNCTIONS	If $f(x)$ is a differentiable function, then $$\frac{d}{dx}[\ln f(x)] = \frac{f'(x)}{f(x)} \qquad (f(x) > 0).$$

To see this, observe that $h(x) = g[f(x)]$ where $g(x) = \ln x \ (x > 0)$. Since $g'(x) = 1/x$, we have, using the Chain Rule,

$$h'(x) = g'(f(x))f'(x)$$

$$= \frac{1}{f(x)} f'(x)$$

$$= \frac{f'(x)}{f(x)}.$$

Observe that in the special case $f(x) = x$, $h(x) = \ln x$, so the derivative of h is, by Rule 3, given by $h'(x) = 1/x$.

EXAMPLE 2   Find the derivative of the function $f(x) = \ln(x^2 + 1)$.

Solution   Using Rule 4, we see immediately that

$$f'(x) = \frac{\dfrac{d}{dx}(x^2 + 1)}{x^2 + 1}$$

$$= \frac{2x}{x^2 + 1}.$$ ∎

When differentiating functions involving logarithms, the rules of logarithms may be used to advantage, as shown in Examples 3 and 4.

EXAMPLE 3   Differentiate the function $y = \ln[(x^2 + 1)(x^3 + 2)^6]$.

Solution   We first rewrite the given function using the properties of logarithms:

$$\begin{aligned}
y &= \ln[(x^2 + 1)(x^3 + 2)^6] \\
&= \ln(x^2 + 1) + \ln(x^3 + 2)^6 && (\ln mn = \ln m + \ln n) \\
&= \ln(x^2 + 1) + 6 \ln(x^3 + 2). && (\ln m^n = n \ln m)
\end{aligned}$$

Differentiating and using Rule 4, we obtain

$$\begin{aligned}
y' &= \frac{\dfrac{d}{dx}(x^2 + 1)}{x^2 + 1} + \frac{6\dfrac{d}{dx}(x^3 + 2)}{x^3 + 2} \\
&= \frac{2x}{x^2 + 1} + \frac{6(3x^2)}{x^3 + 2} \\
&= \frac{2x}{x^2 + 1} + \frac{18x^2}{x^3 + 2}.
\end{aligned}$$ ∎

EXAMPLE 4   Find the derivative of the function $g(t) = \ln(t^2 e^{-t^2})$.

Solution   Here again, to save a lot of work, we first simplify the given expression using the properties of logarithms. We have

$$\begin{aligned}
g(t) &= \ln(t^2 e^{-t^2}) \\
&= \ln t^2 + \ln e^{-t^2} && (\ln mn = \ln m + \ln n) \\
&= 2 \ln t - t^2. && (\ln m^n = n \ln m \text{ and } \ln e = 1)
\end{aligned}$$

Therefore, $$g'(t) = \frac{2}{t} - 2t = \frac{2(1 - t^2)}{t}.$$ ∎

Logarithmic Differentiation

As we saw in the last two examples, the task of finding the derivative of a given function can be made easier by first applying the laws of logarithms to simplify the function. We will now illustrate a process called **logarithmic differentiation**, which not only simplifies the calculation of the derivatives of certain functions but also enables us to compute the derivatives of functions we could not otherwise differentiate using the techniques developed thus far.

EXAMPLE 5   Differentiate $y = x(x + 1)(x^2 + 1)$ using logarithmic differentiation.

Solution   First we take the natural logarithm on both sides of the given equation, obtaining

$$\ln y = \ln x(x + 1)(x^2 + 1).$$

Next, we use the properties of logarithms to rewrite the right-hand side of this equation, obtaining

$$\ln y = \ln x + \ln(x + 1) + \ln(x^2 + 1).$$

If we differentiate both sides of this equation, we have

$$\frac{d}{dx} \ln y = \frac{d}{dx} [\ln x + \ln(x + 1) + \ln(x^2 + 1)]$$

$$= \frac{1}{x} + \frac{1}{x + 1} + \frac{2x}{x^2 + 1}. \qquad \text{(Using Rule 4)}$$

In order to evaluate the expression on the left-hand side, note that y is a function of x. Therefore, writing $y = f(x)$ to remind us of this fact, we have

$$\frac{d}{dx} \ln y = \frac{d}{dx} \ln[f(x)] \qquad \text{[Writing } y = f(x)]$$

$$= \frac{f'(x)}{f(x)} \qquad \text{(Using Rule 4)}$$

$$= \frac{y'}{y}. \qquad \text{[Returning to using } y \text{ instead of } f(x)]$$

Therefore, we have

$$\frac{y'}{y} = \frac{1}{x} + \frac{1}{x + 1} + \frac{2x}{x^2 + 1}.$$

Finally, solving for y', we have

$$y' = y\left(\frac{1}{x} + \frac{1}{x + 1} + \frac{2x}{x^2 + 1}\right)$$

$$= x(x + 1)(x^2 + 1)\left(\frac{1}{x} + \frac{1}{x + 1} + \frac{2x}{x^2 + 1}\right). \qquad \blacksquare$$

Before considering other examples, let us summarize the important steps involved in logarithmic differentiation.

FINDING $\dfrac{dy}{dx}$ BY LOGARITHMIC DIFFERENTIATION

1. Take the natural logarithm on both sides of the equation and use the properties of logarithms to write any "complicated expression" as a sum of simpler terms.

2. Differentiate both sides of the equation with respect to x.

3. Solve the resulting equation for $\dfrac{dy}{dx}$.

EXAMPLE 6   Differentiate $y = x^2(x - 1)(x^2 + 4)^3$.

Solution   Taking the natural logarithm on both sides of the given equation and using the laws of logarithms, we obtain

$$\begin{aligned}
\ln y &= \ln x^2(x - 1)(x^2 + 4)^3 \\
&= \ln x^2 + \ln(x - 1) + \ln(x^2 + 4)^3 \\
&= 2 \ln x + \ln(x - 1) + 3 \ln(x^2 + 4).
\end{aligned}$$

Differentiating both sides of the equation with respect to x, we have

$$\frac{d}{dx} \ln y = \frac{y'}{y} = \frac{2}{x} + \frac{1}{x - 1} + 3 \cdot \frac{2x}{x^2 + 4}.$$

Finally, solving for y', we have

$$\begin{aligned}
y' &= y\left(\frac{2}{x} + \frac{1}{x - 1} + \frac{6x}{x^2 + 4}\right) \\
&= x^2(x - 1)(x^2 + 4)^3\left(\frac{2}{x} + \frac{1}{x - 1} + \frac{6x}{x^2 + 4}\right). \quad \blacksquare
\end{aligned}$$

EXAMPLE 7   Find the derivative of $f(x) = x^x$, $x > 0$.

Solution   A word of caution! This function is neither a power function nor an exponential function. Taking the natural logarithm on both sides of the equation gives

$$\ln f(x) = \ln x^x = x \ln x.$$

Differentiating both sides of the equation with respect to x, we obtain

$$\begin{aligned}
\frac{f'(x)}{f(x)} &= x \frac{d}{dx} \ln x + (\ln x) \frac{d}{dx} x \\
&= x\left(\frac{1}{x}\right) + \ln x \\
&= 1 + \ln x.
\end{aligned}$$

Therefore,

$$f'(x) = f(x)(1 + \ln x) = x^x(1 + \ln x). \quad \blacksquare$$

SELF-CHECK EXERCISES 5.5

1. Find an equation of the tangent line to the graph of $f(x) = x \ln(2x + 3)$ at the point $(-1, 0)$.

2. Use logarithmic differentiation to compute y' given $y = (2x + 1)^3(3x + 4)^5$.

Solutions to Self-Check Exercises 5.5 can be found on page 324.

5.5 EXERCISES

In Exercises 1–32, find the derivative of the given function.

1. $f(x) = 5 \ln x$

2. $f(x) = \ln 5x$

3. $f(x) = \ln(x + 1)$

4. $g(x) = \ln(2x + 1)$

5. $f(x) = \ln x^8$

6. $h(t) = 2 \ln t^5$

7. $f(x) = \ln \sqrt{x}$

8. $f(x) = \ln(\sqrt{x} + 1)$

9. $f(x) = \ln \dfrac{1}{x^2}$

10. $f(x) = \ln \dfrac{1}{2x^3}$

11. $f(x) = \ln(4x^2 - 6x + 3)$

12. $f(x) = \ln(3x^2 - 2x + 1)$

13. $f(x) = \ln \dfrac{2x}{x + 1}$

14. $f(x) = \ln \dfrac{x + 1}{x - 1}$

15. $f(x) = x^2 \ln x$

16. $f(x) = 3x^2 \ln 2x$

17. $f(x) = \dfrac{2 \ln x}{x}$

18. $f(x) = \dfrac{3 \ln x}{x^2}$

19. $f(u) = \ln(u - 2)^3$

20. $f(x) = \ln(x^3 - 3)^4$

21. $f(x) = \sqrt{\ln x}$

22. $f(x) = \sqrt{\ln x + x}$

23. $f(x) = (\ln x)^3$

24. $f(x) = 2(\ln x)^{3/2}$

25. $f(x) = \ln(x^3 + 1)$

26. $f(x) = \ln \sqrt{x^2 - 4}$

27. $f(x) = e^x \ln x$

28. $f(x) = e^x \ln \sqrt{x + 3}$

29. $f(t) = e^{2t} \ln(t + 1)$

30. $g(t) = t^2 \ln(e^{2t} + 1)$

31. $f(x) = \dfrac{\ln x}{x}$

32. $g(t) = \dfrac{t}{\ln t}$

In Exercises 33–36, find the second derivative of the given function.

33. $f(x) = \ln 2x$

34. $f(x) = \ln(x + 5)$

35. $f(x) = \ln(x^2 + 2)$

36. $f(x) = (\ln x)^2$

In Exercises 37–46, use logarithmic differentiation to find the derivative of the given function.

37. $y = (x + 1)^2(x + 2)^3$

38. $y = (3x + 2)^4(5x - 1)^2$

39. $y = (x - 1)^2(x + 1)^3(x + 3)^4$

40. $y = \sqrt{3x + 5}\,(2x - 3)^4$

41. $y = \dfrac{(2x^2 - 1)^5}{\sqrt{x + 1}}$

42. $y = \dfrac{\sqrt{4 + 3x^2}}{\sqrt[3]{x^2 + 1}}$

43. $y = 3^x$

44. $y = x^{x+2}$

45. $y = (x^2 + 1)^x$

46. $y = x^{\ln x}$

47. Find an equation of the tangent line to the graph of $y = x \ln x$ at the point $(1, 0)$.

48. Find an equation of the tangent line to the graph of $y = \ln x^2$ at the point $(2, \ln 4)$.

49. Determine the intervals where the function $f(x) = \ln x^2$ is increasing and where it is decreasing.

50. Determine the intervals where the function $f(x) = \dfrac{\ln x}{x}$ is increasing and where it is decreasing.

51. Determine the intervals of concavity for the function $f(x) = x^2 + \ln x^2$.

52. Determine the intervals of concavity for the function $f(x) = \dfrac{\ln x}{x}$.

53. Find the inflection points of the function $f(x) = \ln(x^2 + 1)$.

54. Find the inflection points of the function $f(x) = x^2 \ln x$.

55. Find the absolute extrema of the function $f(x) = x - \ln x$ on $\left[\frac{1}{2}, 3\right]$.

324 CHAPTER 5 Exponential and Logarithmic Functions

56. Find the absolute extrema of the function $g(x) = x/\ln x$ on $[2, \infty)$.

In Exercises 57 and 58, use the guidelines on page 246 to sketch the graph of the given function.

57. $f(x) = \ln(x - 1)$ **58.** $f(x) = 2x - \ln x$

59. Prove that $\dfrac{d}{dx} \ln |x| = \dfrac{1}{x}$ $(x \neq 0)$ for the case $x < 0$.

C *In Exercises 60 and 61, use a computer or graphing calculator to plot the graph of the given function.*

60. $f(x) = \ln(x^2 + 2x + 5)$ **61.** $f(x) = x \ln x$

SOLUTIONS TO SELF-CHECK EXERCISES 5.5

1. The slope of the tangent line to the graph of f at any point $(x, f(x))$ lying on the graph of f is given by $f'(x)$. Using the Product Rule, we find

$$f'(x) = \frac{d}{dx}[x \ln(2x + 3)]$$

$$= x \frac{d}{dx} \ln(2x + 3) + \ln(2x + 3) \cdot \frac{d}{dx}(x)$$

$$= x\left(\frac{2}{2x + 3}\right) + \ln(2x + 3) \cdot 1$$

$$= \frac{2x}{2x + 3} + \ln(2x + 3).$$

In particular, the slope of the tangent line to the graph of f at the point $(-1, 0)$ is

$$f'(-1) = \frac{-2}{-2 + 3} + \ln 1 = -2.$$

Therefore, using the point-slope form of the equation of a line, we see that a required equation is

$$y - 0 = -2(x + 1),$$

or

$$y = -2x - 2.$$

2. Taking the logarithm on both sides of the equation gives

$$\ln y = \ln(2x + 1)^3(3x + 4)^5$$

$$= \ln(2x + 1)^3 + \ln(3x + 4)^5$$

$$= 3 \ln(2x + 1) + 5 \ln(3x + 4).$$

Differentiating both sides of the equation with respect to x, keeping in mind that y is a function of x, we obtain

$$\frac{d}{dx}(\ln y) = \frac{y'}{y} = 3 \cdot \frac{2}{2x + 1} + 5 \cdot \frac{3}{3x + 4}$$

$$= 3\left[\frac{2}{2x + 1} + \frac{5}{3x + 4}\right]$$

$$= \left(\frac{6}{2x + 1} + \frac{15}{3x + 4}\right),$$

and

$$y' = (2x + 1)^3(3x + 4)^5 \cdot \left(\frac{6}{2x + 1} + \frac{15}{3x + 4}\right).$$

5.6 Exponential Functions as Mathematical Models

Exponential Growth

FIGURE 5.12

Exponential growth

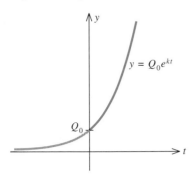

Many problems arising from practical situations can be described mathematically in terms of exponential functions or functions closely related to the exponential function. In this section, we will look at some applications involving exponential functions from the fields of the life and social sciences.

In Section 5.1 we saw that the exponential function $f(x) = b^x$ is an increasing function when $b > 1$. In particular, the function $f(x) = e^x$ shares this property. From this result one may deduce that the function $Q(t) = Q_0 e^{kt}$, where Q_0 and k are positive constants, has the following properties:

1. $Q(0) - Q_0$
2. $Q(t)$ increases "rapidly" without bound as t increases without bound (Figure 5.12).

Property 1 follows from the computation

$$Q(0) = Q_0 e^0 = Q_0.$$

Next, to study the rate of change of the function $Q(t)$, we differentiate it with respect to t, obtaining

$$Q'(t) = \frac{d}{dt}(Q_0 e^{kt}) \qquad (12)$$

$$= Q_0 \frac{d}{dt}(e^{kt})$$

$$- kQ_0 e^{kt}$$

$$= kQ(t).$$

Since $Q(t) > 0$ (because Q_0 is assumed to be positive) and $k > 0$, we see that $Q'(t) > 0$ and so $Q(t)$ is an increasing function of t. Our computation has, in fact, shed more light on an important property of the function $Q(t)$. Equation (12) says that the rate of increase of the function $Q(t)$ is proportional to the amount $Q(t)$ of the quantity present at time t. The implication is that as $Q(t)$ increases, so does the *rate of increase* of $Q(t)$, resulting in a very rapid increase in $Q(t)$ as t increases without bound.

Thus, the exponential function

$$Q(t) = Q_0 e^{kt} \qquad (0 \leq t < \infty) \qquad (13)$$

provides us with a mathematical model of a quantity $Q(t)$ that is initially present in the amount of $Q(0) = Q_0$ and whose rate of growth at any time t is directly proportional to the amount of the quantity present at time t. Such a quantity is said to exhibit **exponential growth,** and the constant k is called the **growth constant.** Interest earned on a fixed deposit when compounded continuously exhibits exponential growth. Other examples of exponential growth follow.

EXAMPLE 1   Under ideal laboratory conditions, the number of bacteria in a culture grows in accordance with the law $Q(t) = Q_0 e^{kt}$, where Q_0 denotes

the number of bacteria initially present in the culture, k is some constant determined by the strain of bacteria under consideration, and t is the elapsed time measured in hours. Suppose that there are 10,000 bacteria present initially in the culture and 60,000 present 2 hours later.

a. How many bacteria will there be in the culture at the end of 4 hours?
b. What is the rate of growth of the population after 4 hours?

Solution

a. We are given that $Q(0) = Q_0 = 10,000$, so $Q(t) = 10,000e^{kt}$. Next, the fact that 60,000 bacteria are present 2 hours later translates into $Q(2) = 60,000$. Thus

$$60,000 = 10,000e^{2k}$$
$$e^{2k} = 6.$$

Taking the natural logarithm on both sides of the equation, we obtain

$$\ln e^{2k} = \ln 6$$
$$2k = \ln 6, \qquad \text{(Since ln } e = 1\text{)}$$

or
$$k \approx 0.8959.$$

Thus, the number of bacteria present at any time t is given by

$$Q(t) = 10,000e^{0.8959t}.$$

In particular, the number of bacteria present in the culture at the end of 4 hours is given by

$$Q(4) = 10,000e^{0.8959(4)}$$
$$= 360,029.$$

b. The rate of growth of the bacteria population at any time t is given by

$$Q'(t) = kQ(t).$$

Thus, using the result from (a), we find that the rate at which the population is growing at the end of 4 hours is

$$Q'(4) = kQ(4)$$
$$\approx (0.8959)(360,029)$$
$$\approx 322,550,$$

or approximately 322,550 bacteria per hour.   ■

FIGURE 5.13
Exponential decay

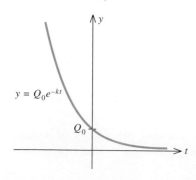

Exponential Decay

In contrast to exponential growth, a quantity exhibits **exponential decay** if it decreases at a rate that is directly proportional to its size. Such a quantity may be described by the exponential function

$$Q(t) = Q_0 e^{-kt} \qquad (t \in [0, \infty)) \qquad (14)$$

where the positive constant Q_0 measures the amount present initially ($t = 0$) and k is some suitable positive number, called the **decay constant**. The choice of this number is determined by the nature of the substance under consideration. The graph of this function is sketched in Figure 5.13.

To verify the properties ascribed to the function $Q(t)$, we simply compute

$$Q(0) = Q_0 e^0 = Q_0$$

$$Q'(t) = \frac{d}{dt}(Q_0 e^{-kt})$$

and

$$= Q_0 \frac{d}{dt}(e^{-kt})$$

$$= -k Q_0 e^{-kt} = -k Q(t).$$

 EXAMPLE 2   Radioactive substances decay exponentially. For example, the amount of radium present at any time t obeys the law $Q(t) = Q_0 e^{-kt}$, where Q_0 is the initial amount present and k is a suitable positive constant. The **half-life** of a radioactive substance is the time required for a given amount to be reduced by one-half. Now, it is known that the half-life of radium is approximately 1600 years. Suppose that initially there are 200 milligrams of pure radium. Find the amount left after t years. What is the amount left after 800 years?

Solution   The initial amount of radium present is 200 milligrams, so $Q(0) = Q_0 = 200$. Thus $Q(t) = 200 e^{-kt}$. Next, the datum concerning the half-life of radium implies that $Q(1600) = 100$, and this gives

$$100 = 200 e^{-1600k}$$

$$e^{-1600k} = \frac{1}{2}.$$

Taking the natural logarithm on both sides of this equation yields

$$-1600k \ln e = \ln \frac{1}{2}$$

$$-1600k = \ln \frac{1}{2} \qquad (\ln e = 1)$$

$$k = -\frac{1}{1600} \ln \left(\frac{1}{2}\right) = 0.0004332.$$

Therefore, the amount of radium left after t years is

$$Q(t) = 200 e^{-0.0004332t}.$$

In particular, the amount of radium left after 800 years is

$$Q(800) = 200 e^{-0.0004332(800)} \approx 141.42,$$

or approximately 141 milligrams.   ∎

 EXAMPLE 3   Carbon 14, a radioactive isotope of carbon, has a half-life of 5770 years. What is its decay constant?

Solution   We have $Q(t) = Q_0 e^{-kt}$. Since the half-life of the element is 5770 years, half of the substance is left at the end of that period. That is,

$$Q(5770) = Q_0 e^{-5770k} = \frac{1}{2} Q_0,$$

or

$$e^{-5770k} = \frac{1}{2}.$$

Taking the natural logarithm on both sides of this equation, we have

$$\ln e^{-5770k} = \ln \frac{1}{2}$$

$$-5770k = -0.693147$$

$$\approx 0.00012. \qquad \blacksquare$$

Carbon-14 dating is a well-known method used by anthropologists to establish the age of animal and plant fossils. This method assumes that the proportion of carbon 14 (C-14) present in the atmosphere has remained constant over the past 50,000 years. Professor Willard Libby, recipient of the Nobel Prize in chemistry in 1960, proposed this theory.

The amount of C-14 in the tissues of a living plant or animal is constant. However, when an organism dies it stops absorbing new quantities of C-14 and the amount of C-14 in the remains diminishes because of the natural decay of the radioactive substance. Thus, the approximate age of a plant or animal fossil can be determined by measuring the amount of C-14 present in the remains.

EXAMPLE 4   A skull from an archeological site has one-tenth the amount of C-14 that it originally contained. Determine the approximate age of the skull.

Solution   Here

$$Q(t) = Q_0 e^{-kt}$$

$$= Q_0 e^{-0.00012t},$$

where Q_0 is the amount of C-14 present originally and k, the decay constant, is equal to 0.00012 (see Example 3). Since $Q(t) = (1/10)Q_0$, we have

$$\frac{1}{10} Q_0 = Q_0 e^{-0.00012t}$$

$$\ln \frac{1}{10} = -0.00012t \qquad \text{(Taking the natural logarithm on both sides)}$$

$$t = \frac{\ln \dfrac{1}{10}}{-0.00012}$$

$$\approx 19,200,$$

or approximately 19,200 years. $\qquad \blacksquare$

Learning Curves

The next example shows how the exponential function may be applied to describe certain types of learning processes. Consider the function

$$Q(t) = C - Ae^{-kt}$$

where C, A, and k are positive constants. To sketch the graph of the function Q, observe that its y-intercept is given by $Q(0) = C - A$. Next, we compute

$$Q'(t) = kAe^{-kt}.$$

FIGURE 5.14

A learning curve

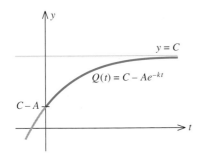

Since both k and A are positive, we see that $Q'(t) > 0$ for all values of t. Thus $Q(t)$ is an increasing function of t. Also,

$$\lim_{t \to \infty} Q(t) = \lim_{t \to \infty} (C - Ae^{-kt})$$
$$= \lim_{t \to \infty} C - \lim_{t \to \infty} Ae^{-kt}$$
$$= C,$$

so $y = C$ is a horizontal asymptote of Q. Thus, $Q(t)$ increases and approaches the number C as t increases without bound. The graph of the function Q is shown in Figure 5.14, where that part of the graph corresponding to the negative values of t is drawn with a gray line since, in practice, one normally restricts the domain of the function to the interval $[0, \infty)$.

Observe that $Q(t)$ $(t > 0)$ increases rather rapidly initially but that the rate of increase slows down considerably after a while. To see this, we compute

$$\lim_{t \to \infty} Q'(t) = \lim_{t \to \infty} kAe^{-kt} = 0.$$

This behavior of the graph of the function Q closely resembles the learning pattern experienced by workers engaged in highly repetitive work. For example, the productivity of an assembly-line worker increases very rapidly in the early stages of the training period. This productivity increase is a direct result of the worker's training and accumulated experience. But the rate of increase of productivity slows down as time goes by and the worker's productivity level approaches some fixed level due to the limitations of the worker and the machine. Because of this characteristic, the graph of the function $Q(t) = C - Ae^{-kt}$ is often called a **learning curve.**

EXAMPLE 5   The Camera Division of the Eastman Optical Company produces a 35-mm single-lens reflex camera. Eastman's training department determines that after completing the basic training program, a new, previously inexperienced employee will be able to assemble

$$Q(t) = 50 - 30e^{-0.5t}$$

model F cameras per day, t months after the employee starts work on the assembly line.

a. How many model F cameras can a new employee assemble per day after basic training?
b. How many model F cameras can an employee with one month of experience assemble per day? An employee with two months of experience? An employee with six months of experience?
c. How many model F cameras can the average experienced employee assemble per day?

Solution
a. The number of model F cameras a new employee can assemble is given by

$$Q(0) = 50 - 30 = 20.$$

b. The number of model F cameras that an employee with one month of experience, two months of experience, and six months of experience can

assemble per day is given by

$$Q(1) = 50 - 30e^{-0.5} \approx 31.80$$
$$Q(2) = 50 - 30e^{-1} \approx 38.96$$
$$Q(6) = 50 - 30e^{-3} \approx 48.51,$$

or approximately 32, 39, and 49, respectively.

c. As t increases without bound, $Q(t)$ approaches 50. Hence, the average experienced employee can ultimately be expected to assemble 50 model F cameras per day.   ■

Other applications of the learning curve are found in models that describe the dissemination of information about a product or the velocity of an object dropped into a viscous medium.

FIGURE 5.15

A logistic curve

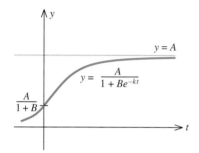

Logistic Growth Functions

Our last example of an application of exponential functions to the description of natural phenomena involves the **logistic** (also called the **S-shaped,** or **sigmoidal**) **curve,** which is the graph of the function

$$Q(t) = \frac{A}{1 + Be^{-kt}},$$

where A, B, and k are positive constants. The graph of the function Q is sketched in Figure 5.15.

Observe that $Q(t)$ increases rather rapidly for small values of t. In fact, for small values of t, the logistic curve resembles an exponential growth curve. However, the *rate of growth* of $Q(t)$ decreases quite rapidly as t increases and $Q(t)$ approaches the number A as x increases without bound.

Thus, the logistic curve exhibits both the property of rapid growth of the exponential growth curve as well as the "saturation" property of the learning curve. Because of these characteristics, the logistic curve serves as a suitable mathematical model for describing many natural phenomena. For example, if a small number of rabbits were introduced to a tiny island in the South Pacific, the rabbit population might be expected to grow very rapidly at first, but the growth rate would decrease quickly as overcrowding, scarcity of food, and other environmental factors affected it. The population would eventually stabilize at a level compatible with the life-support capacity of the environment. Models describing the spread of rumors and epidemics are other examples of the application of the logistic curve.

EXAMPLE 6   The number of men at Fort MacArthur who contracted influenza after t days during a flu epidemic is approximated by the exponential model

$$Q(t) = \frac{5000}{1 + 1249e^{-kt}}.$$

If 40 men contracted the flu by the seventh day, find how many men contracted the flu by the fifteenth day.

Solution   The given information implies that

$$Q(7) = 40.$$

$$Q(7) = \frac{5000}{1 + 1249e^{-7k}}$$

Thus,
$$= 40$$

$$40(1 + 1249e^{-7k}) = 5000$$

$$1 + 1249e^{-7k} = \frac{5000}{40} = 125$$

or
$$e^{-7k} = \frac{124}{1249}$$

$$-7k = \ln \frac{124}{1249}$$

$$k = -\frac{\ln \dfrac{124}{1249}}{7} \approx 0.33.$$

Therefore, the number of men who contracted the flu after t days is given by

$$Q(t) = \frac{5000}{1 + 1249e^{-0.33t}}.$$

In particular, the number of men who contracted the flu by the fifteenth day is given by

$$Q(15) = \frac{5000}{1 + 1249e^{-15(0.33)}}$$

$$\approx 508,$$

or approximately 508 men.   ■

SELF-CHECK EXERCISES 5.6

1. Suppose that the population of a country (in millions) at any time t grows in accordance with the rule

$$P = \left(P_0 + \frac{I}{k}\right)e^{kt} - \frac{I}{k},$$

where P denotes the population at any time t, k is a constant reflecting the natural growth rate of the population, I is a constant giving the (constant) rate of immigration into the country, and P_0 is the total population of the country at time $t = 0$. The population of the United States in the year 1980 ($t = 0$) was 226.5 million. If the natural growth rate is 0.8 percent annually ($k = 0.008$) and net immigration is allowed at the rate of half a million people per year ($I = 0.5$) until the end of the century, what will the population of the United States be in the year 2000?

Solutions to Self-Check Exercises 5.6 can be found on page 334.

5.6   EXERCISES

A calculator is recommended for this exercise set.

1. **Exponential Growth**   Given that a quantity $Q(t)$ is described by the exponential growth function

$$Q(t) = 400\,e^{0.05t},$$

where t is measured in minutes, answer the following questions.
   a. What is the growth constant?
   b. What quantity is present initially?
   c. Using Table 1 in the appendix or a calculator, complete the following table of values:

t	0	10	20	100	1000
Q					

2. **Exponential Decay**   Given that a quantity $Q(t)$ exhibiting exponential decay is described by the function

$$Q(t) = 2000\,e^{-0.06t},$$

where t is measured in years, answer the following questions.
   a. What is the decay constant?
   b. What quantity is present initially?
   c. Using Table 1 in the appendix or a calculator, complete the following table of values:

t	0	5	10	20	100
Q					

3. **Growth of Bacteria**   The growth rate of the bacterium *Escherichia coli,* a common bacterium found in the human intestine, is proportional to its size. Under ideal laboratory conditions, when this bacterium is grown in a nutrient broth medium, the number of cells in a culture doubles approximately every 20 minutes.
   a. If the initial cell population is 100, determine the function $Q(t)$ that expresses the exponential growth of the number of cells of this bacterium as a function of time t (in minutes).
   b. How long will it take for a colony of 100 cells to increase to a population of one million?
   c. If the initial cell population were 1000, how would this alter our model?

4. **World Population**   The world population at the beginning of 1990 was 5.3 billion. Assuming that the population continues to grow at its present rate of approximately 2 percent per year, find the function $Q(t)$ that expresses the world population (in billions) as a function of time t (in years).
   a. Using this function, complete the following table of values and sketch the graph of the function Q.

Year	1990	1995	2000	2005
World Population				

Year	2010	2015	2020	2025
World Population				

   b. Find the estimated rate of growth in the year 2000.

5. **World Population**   Refer to Exercise 4.
   a. If the world population continues to grow at its present rate of approximately 2 percent per year, find the length of time t_0 required for the world population to triple in size.
   b. Using the time t_0 found in (a), what would the world population be if the growth rate were reduced to 1.8 percent?

6. **Resale Value**   A certain piece of machinery was purchased three years ago by the Garland Mills Company for $500,000. Its present resale value is $320,000. Assuming that the machine's resale value decreases exponentially, what will it be four years from now?

7. **Radioactive Decay**   If the temperature is constant, then the atmospheric pressure P (in pounds per square inch) varies with the altitude above sea level h in accordance with the law

$$P = p_0 e^{-kh}$$

where p_0 is the atmospheric pressure at sea level and k is a constant. If the atmospheric pressure is 15 pounds per square inch at sea level and 12.5 pounds per square inch at 4,000 feet, find the atmospheric pressure at an altitude of 12,000 feet.

8. **Radioactive Decay** The radioactive element polonium decays according to the law

$$Q(t) = Q_0 \cdot 2^{-(t/140)},$$

where Q_0 is the initial amount and the time t is measured in days. If the amount of polonium left after 280 days is 20 milligrams, what was the initial amount present?

9. **Radioactive Decay** Phosphorus 32 has a half-life of 14.2 days. If 100 grams of this substance are present initially, find the amount present after t days. What amount will be left after 7.1 days?

10. **Nuclear Fallout** Strontium 90, a radioactive isotope of strontium, is present in the fallout resulting from nuclear explosions. It is especially hazardous to animal life, including humans, because upon ingestion of contaminated food it is absorbed into the bone structure. Its half-life is 27 years. If the amount of strontium 90 in a certain area is found to be four times the "safe" level, find how much time must elapse before an "acceptable level" is reached.

11. **Carbon-14 Dating** Wood deposits recovered from an archeological site contain 20 percent of the carbon 14 they originally contained. How long ago did the tree from which the wood was obtained die?

12. **Carbon-14 Dating** Skeletal remains of the so-called "Pittsburgh Man," unearthed in Pennsylvania, had lost 82 percent of the carbon 14 they originally contained. Determine the approximate age of the bones.

13. **Learning Curves** The American Stenographic Institute finds that the average student taking Advanced Shorthand, an intensive 20-week course, progresses according to the function

$$Q(t) = 120(1 - e^{-0.05t}) + 60 \qquad (0 \le t \le 20),$$

where $Q(t)$ measures the number of words (per minute) of dictation that the student can take in shorthand after t weeks in the course. Sketch the graph of the function Q and answer the following questions.
a. What is the beginning shorthand speed for the average student in this course?
b. What shorthand speed does the average student attain halfway through the course?
c. How many words per minute can the average student take after completing this course?

14. **Effect of Advertising on Sales** The Metro Department Store found that t weeks after the end of a sales promotion the volume of sales was given by a function of the form

$$S(t) = B + Ae^{-kt} \qquad (0 \le t \le 4),$$

where $B = 50,000$ and is equal to the average weekly volume of sales before the promotion. The sales volumes at the end of the first and third weeks were \$83,515 and \$65,055, respectively. Assume that the sales volume is decreasing exponentially and find
a. the decay constant k.
b. the sales volume at the end of the fourth week.

15. **Demand for Computers** The Universal Instruments Company found that the monthly demand for its new line of Galaxy Home Computers t months after placing the line on the market was given by

$$D(t) = 2000 - 1500e^{-0.05t} \qquad (t > 0).$$

Graph this function and answer the following questions.
a. What is the demand after one month? One year? Two years? Five years?
b. At what level is the demand expected to stabilize?
c. Find the rate of growth of the demand after the tenth month.

16. **Newton's Law of Cooling** Newton's law of cooling states that the rate at which the temperature of an object changes is proportional to the difference in temperature between the object and that of the surrounding medium. Thus, the temperature $F(t)$ of an object that is greater than the temperature of its surrounding medium is given by

$$F(t) = T + Ae^{-kt},$$

where t is the time expressed in minutes, T is the temperature of the surrounding medium, and A and k are constants. Suppose that a cup of instant coffee is prepared with boiling water (212°F) and left to cool on the counter in a room where the temperature is 72°F. If $k = 0.1865$, determine when the coffee will be cool enough to drink (say, 110°F).

17. **Spread of an Epidemic** During a flu epidemic, the number of children in the Woodbridge Community School System who contracted influenza after t days was given by

$$Q(t) = \frac{1000}{1 + 199e^{-0.8t}}.$$

a. How many children were stricken by the flu after the first day?
b. How many children had the flu after ten days?
c. How many children eventually contracted the disease?

18. **Growth of a Fruit-Fly Population** On the basis of data collected during an experiment, a biologist found

that the growth of the fruit fly (*Drosophila*) with a limited food supply could be approximated by the exponential model

$$N(t) = \frac{400}{1 + 39e^{-0.16t}},$$

where t denotes the number of days since the beginning of the experiment.
a. What was the initial fruit-fly population in the experiment?
b. What was the maximum fruit-fly population that could be expected under this laboratory condition?
c. What was the population of the fruit-fly colony on the twentieth day?

19. **Percentage of Households with VCRs** According to estimates by Paul Kroger Associates, the percentage of households that own videocassette recorders (VCRs) is given by

$$P(t) = \frac{68}{1 + 21.67e^{-0.62t}} \qquad (0 \le t \le 12),$$

where t is measured in years, with $t = 0$ corresponding to the beginning of 1985. What percentage of households owned VCRs at the beginning of 1985? At the beginning of 1995?

20. **Spread of a Rumor** Three hundred students attended the dedication ceremony of a new building on a college campus. The president of the traditionally female college announced a new expansion program, which included plans to make the college coeducational. The number of students who learned of the new program t hours later is given by the function

$$f(t) = \frac{3000}{1 + Be^{-kt}}.$$

If 600 students on campus had heard about the new program 2 hours after the ceremony, how many students had heard about the policy after 4 hours?

21. **Concentration of Glucose in the Bloodstream** A glucose solution is administered intravenously into the bloodstream at a constant rate of r mg/hr. As the glucose is being administered, it is converted into other substances and removed from the bloodstream. Suppose the concentration of the glucose solution at time t is given by

$$C(t) = \frac{r}{k} - \left[\left(\frac{r}{k} \right) - C_0 \right] e^{-kt},$$

where C_0 is the concentration at time $t = 0$ and k is a constant.
a. Assuming that $C_0 < r/k$, evaluate

$$\lim_{t \to \infty} C(t)$$

and interpret your result.
b. Sketch the graph of the function C.

22. **Gompertz Growth Curve** Consider the function

$$Q(t) = Ce^{-Ae^{-kt}},$$

where $Q(t)$ is the size of a quantity at time t and A, C, and k are positive constants. The graph of this function, called the **Gompertz growth curve**, is used by biologists to describe restricted population growth.
a. Show that the function Q is always increasing.
b. Find the time t at which the growth rate $Q'(t)$ is increasing most rapidly. [*Hint:* Find the inflection point of Q.]
c. Show that $\lim_{t \to \infty} Q(t) = C$ and interpret your result.

SOLUTIONS TO SELF-CHECK EXERCISES 5.6

1. We are given that $P_0 = 226.5$, $k = 0.008$, and $I = 0.5$. So

$$P = \left(226.5 + \frac{0.5}{0.008} \right) e^{0.008t} - \frac{0.5}{0.008}$$
$$= 289e^{0.008t} - 62.5.$$

Therefore, the population in the year 2000 will be given by

$$P(20) = 289e^{0.16} - 62.5$$
$$\approx 276.7,$$

or approximately 276.7 million.

CHAPTER FIVE SUMMARY OF PRINCIPAL FORMULAS AND TERMS

Formulas

1. Exponential function with base b

$$y = b^x$$

2. The number e

$$e = \lim_{m \to \infty} \left(1 + \frac{1}{m}\right)^m = 2.71828$$

3. Exponential function with base e

$$y = e^x$$

4. Logarithmic function with base b

$$y = \log_b x$$

5. Logarithmic function with base e

$$y = \ln x$$

6. Inverse properties of $\ln x$ and e

$$\ln e^x = x \quad \text{and} \quad e^{\ln x} = x$$

7. Compound interest (accumulated amount)

$$A = P(1 + i)^n, \text{ where } i = r/m, \text{ and } n = mt$$

8. Effective rate of interest

$$r_{\text{eff}} = \left(1 + \frac{r}{m}\right)^m - 1$$

9. Compound interest (present value)

$$P = A(1 + i)^{-n}, \text{ where } i = r/m, \text{ and } n = mt$$

10. Continuous compound interest

$$A = Pe^{rt}$$

11. Derivative of the exponential function

$$\frac{d}{dx}(e^x) = e^x$$

12. Chain rule for exponential functions

$$\frac{d}{dx}(e^u) = e^u \frac{du}{dx}$$

13. Derivative of the logarithmic function

$$\frac{d}{dx}(\ln x) = \frac{1}{x}$$

14. Chain rule for logarithmic functions

$$\frac{d}{dx}(\ln u) = \frac{1}{u}\frac{du}{dx}$$

Terms

Common logarithm   Exponential decay
Natural logarithm   Decay constant
Logarithmic differentiation   Half-life of a radioactive element
Exponential growth   Logistic growth function
Growth constant

CHAPTER 5 REVIEW EXERCISES

1. Sketch the graphs of the exponential functions defined by the following equations on the same set of coordinate axes.

   a. $y = 2^{-x}$ **b.** $y = \left(\frac{1}{2}\right)^x$

In Exercises 2 and 3, express each equation in logarithmic form.

2. $\left(\frac{2}{3}\right)^{-3} = \frac{27}{8}$ 3. $16^{-3/4} = 0.125$

In Exercises 4 and 5, solve each equation for x.

4. $\log_4(2x + 1) = 2$

5. $\ln(x - 1) + \ln 4 = \ln(2x + 4) - \ln 2$

In Exercises 6–8, given that $\ln 2 = x$, $\ln 3 = y$, and $\ln 5 = z$, express each of the given logarithmic values in terms of x, y, and z.

6. $\ln 30$ 7. $\ln 3.6$ 8. $\ln 75$

9. Sketch the graph of the function $y = \log_2(x + 3)$.

10. Sketch the graph of the function $y = \log_3(x + 1)$.

In Exercises 11–28, find the derivative of the given function.

11. $f(x) = xe^{2x}$ 12. $f(t) = \sqrt{t}e^t + t$

13. $g(t) = \sqrt{t}e^{-2t}$ 14. $g(x) = e^x\sqrt{1 + x^2}$

15. $y = \dfrac{e^{2x}}{1 + e^{-2x}}$ 16. $f(x) = e^{2x^2 - 1}$

17. $f(x) = xe^{-x^2}$ 18. $g(x) = (1 + e^{2x})^{3/2}$

19. $f(x) = x^2e^x + e^x$ 20. $g(t) = t \ln t$

21. $f(x) = \ln(e^{x^2} + 1)$ 22. $f(x) = \dfrac{x}{\ln x}$

23. $f(x) = \dfrac{\ln x}{x + 1}$ 24. $y = (x + 1)e^x$

25. $y = \ln(e^{4x} + 3)$ 26. $f(r) = \dfrac{re^r}{1 + r^2}$

27. $f(x) = \dfrac{\ln x}{1 + e^x}$ 28. $g(x) = \dfrac{e^{x^2}}{1 + \ln x}$

29. Find the second derivative of the function $y = \ln(3x + 1)$.

30. Find the second derivative of the function $y = x \ln x$.

31. Find $h'(0)$ if $h(x) = g(f(x))$, $g(x) = x + \dfrac{1}{x}$, and $f(x) = e^x$.

32. Find $h'(1)$ if $h(x) = g(f(x))$, $g(x) = \dfrac{x + 1}{x - 1}$, and $f(x) = \ln x$.

33. Use logarithmic differentiation to find the derivative of $f(x) = (2x^3 + 1)(x^2 + 2)^3$.

34. Use logarithmic differentiation to find the derivative of $f(x) = \dfrac{x(x^2 - 2)^2}{(x - 1)}$.

35. Find an equation of the tangent line to the graph of $y = e^{-2x}$ at the point $(1, e^{-2})$.

36. Find an equation of the tangent line to the graph of $y = xe^{-x}$ at the point $(1, e^{-1})$.

37. Sketch the graph of the function $f(x) = xe^{-2x}$.

38. Sketch the graph of the function $f(x) = x^2 - \ln x$.

39. Find the absolute extrema of the function $f(t) = te^{-t}$.

40. Find the absolute extrema of the function $g(t) = \dfrac{\ln t}{t}$ on $[1, 2]$.

41. A hotel was purchased by a conglomerate for \$4.5 million and sold five years later for \$8.2 million. Find the annual rate of return (compounded continuously).

42. Find the present value of \$119,346 due in four years at an interest rate of 10 percent per year compounded continuously.

43. A culture of bacteria that initially contained 2,000 bacteria has a count of 18,000 bacteria after two hours.
    a. Determine the function $Q(t)$ that expresses the exponential growth of the number of cells of this bacterium as a function of time t (in minutes).
    b. Find the number of bacteria present after four hours.

44. The radioactive element radium has a half-life of 1600 years. What is its decay constant?

45. The V.C.A. Television Company found that the monthly demand for its new line of video-disc players t months after placing the players on the market is given by

$$D(t) = 4000 - 3000e^{-0.06t} \qquad (t \geq 0).$$

Graph this function and answer the following questions.
   a. What was the demand after one month? After one year? After two years?
   b. At what level is the demand expected to stabilize?

46. During a flu epidemic, it was found that the number of students at a certain university who contracted influenza after t days could be approximated by the exponential model

$$Q(t) = \frac{3000}{1 + 499e^{-kt}}.$$

If 90 students contracted the flu by the tenth day, find how many students contracted the flu by the twentieth day.

How much will the solar cell panels cost? The head of Soloron Corporation's research and development department has projected that the cost of producing solar cell panels will drop at a certain rate in the next several years. In Example 7, page 358, you will see how this information can be used to predict the cost of solar cell panels in the coming year.

6

Integration

Differential calculus is concerned with the problem of finding the rate of change of one quantity with respect to another. In this chapter we begin the study of the other branch of calculus known as integral calculus. Here we are interested in precisely the opposite problem: If we know the rate of change of one quantity with respect to another, can we find the relationship between these two quantities? The principal tool used in the study of integral calculus is the *antiderivative* of a function, and we develop rules for antidifferentiation, or *integration,* as the process of finding the antiderivative is called. We will also show that a link is established between differential and integral calculus—via the Fundamental Theorem of Calculus.

<table>
<tr><td>6.1</td><td></td></tr>
</table>

6.1   Antiderivatives and the Rules of Integration

Antiderivatives

In Chapter 2 we considered this problem:

Given the position of a car at any time t, can we find its velocity at time t?

As we saw, the velocity of the car is given by $f'(t)$, where f is the function giving the position of the car at any time t (Figure 6.1). The function f' is just the derivative of f.

FIGURE 6.1

The position of a car moving along a straight road at any time t is given by s = f(t).

$f(0)$ $f(t)$

In Chapters 6 and 7, we will consider precisely the opposite problem:

Given the velocity of a car at any time t, can we find its position at time t?

Put another way, if we know the car's velocity function f', can we find its position function f?

In order to solve this problem, we need the concept of an *antiderivative* of a function.

ANTIDERIVATIVE   A function F is an **antiderivative** of f on an interval I if $F'(x) = f(x)$ for all x in I.

Thus, an antiderivative of a function f is a function F whose derivative is f. For example, $F(x) = x^2$ is an antiderivative of $f(x) = 2x$ because

$$F'(x) = \frac{d}{dx}(x^2) = 2x = f(x),$$

and $F(x) = x^3 + 2x + 1$ is an antiderivative of $f(x) = 3x^2 + 2$ because

$$F'(x) = \frac{d}{dx}(x^3 + 2x + 1) = 3x^2 + 2 = f(x).$$

EXAMPLE 1   Let $F(x) = \frac{1}{3}x^3 - 2x^2 + x - 1$. Show that F is an antiderivative of $f(x) = x^2 - 4x + 1$.

Solution   Differentiating the function F, we obtain

$$F'(x) = x^2 - 4x + 1 = f(x),$$

and the desired result follows.   ∎

EXAMPLE 2   Let $F(x) = x$, $G(x) = x + 2$, and $H(x) = x + C$, where C is a constant. Show that F, G, and H are all antiderivatives of the function f defined by $f(x) = 1$.

Solution   Since

$$F'(x) = \frac{d}{dx}(x) = 1 = f(x),$$

$$G'(x) = \frac{d}{dx}(x + 2) = 1 = f(x),$$

and $$H'(x) = \frac{d}{dx}(x + C) = 1 = f(x),$$

we see that F, G, and H are indeed antiderivatives of f. ∎

Example 2 shows that once an antiderivative G of a function f is known, then another antiderivative of f may be found by adding an arbitrary constant to the function G. The following theorem states that no function other than one obtained in this manner can be an antiderivative of f. (We omit the proof.)

THEOREM 1

Let G be an antiderivative of a function f. Then every antiderivative F of f must be of the form $F(x) = G(x) + C$, where C is a constant.

Returning to Example 2, we see that there are infinitely many antiderivatives of the function $f(x) = 1$. We obtain each one by specifying the constant C in the function $F(x) = x + C$. Figure 6.2 shows the graphs of some of these antiderivatives for selected values of C. These graphs constitute part of a family of infinitely many parallel straight lines, each having a slope equal to 1. This result is expected, since there are infinitely many curves (straight lines) with a given slope equal to 1. The antiderivatives $F(x) = x + C$ (C, a constant) are precisely the functions representing this family of straight lines.

FIGURE 6.2

The graphs of some antiderivatives of $f(x) = 1$

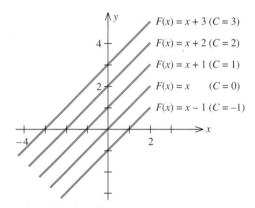

$F(x) = x + 3$ ($C = 3$)
$F(x) = x + 2$ ($C = 2$)
$F(x) = x + 1$ ($C = 1$)
$F(x) = x$ ($C = 0$)
$F(x) = x - 1$ ($C = -1$)

EXAMPLE 3   Prove that the function $G(x) = x^2$ is an antiderivative of the function $f(x) = 2x$. Write a general expression for the antiderivatives of f.

Solution   Since $G'(x) = 2x = f(x)$, we have shown that $G(x) = x^2$ is an antiderivative of $f(x) = 2x$. By Theorem 1, every antiderivative of the function $f(x) = 2x$ has the form $F(x) = x^2 + C$, where C is some constant. The graphs of a few of the antiderivatives of f are shown in Figure 6.3.

FIGURE 6.3
The graphs of some antiderivatives of $f(x) = 2x$

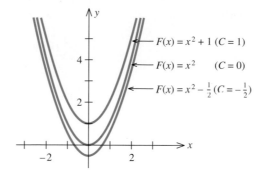

$F(x) = x^2 + 1 \ (C = 1)$

$F(x) = x^2 \quad (C = 0)$

$F(x) = x^2 - \frac{1}{2} (C = -\frac{1}{2})$

Application

The next example involves a familiar interpretation of the antiderivative.

EXAMPLE 4   The velocity of a car (in ft/sec) t seconds after starting from rest and moving in a straight line is given by

$$f(t) = 2t \qquad (0 \le t \le 20).$$

Find the position of the car at any time t.

Solution   Let $s = F(t)$ be the function that describes the position of the car at any time t. Then the velocity of the car at any time t is given by

$$v = \frac{ds}{dt} = F'(t).$$

But the car's velocity at time t is given by $f(t) = 2t$, so we have

$$F'(t) = f(t),$$

and this problem is equivalent to finding the antiderivative F of the function $f(t) = 2t$. Using the result of Example 3, we have

$$s = F(t) = t^2 + C \qquad (C, \text{ a constant}).$$

Next, observe that when $t = 0$, the position $s = 0$, so that we have

$$0 = 0^2 + C,$$

or $C = 0$. Thus the required expression for the position of the car at any time t is given by

$$s = F(t) = t^2.$$

Notice that the "initial condition" helped us select the required function from the infinitely many antiderivatives of the function f. This situation is typical of problems involving antidifferentiation.

The Indefinite Integral

The process of finding all antiderivatives of a function is called **antidifferentia-tion,** or **integration.** We use the symbol $\int$, called an **integral sign,** to indicate that the operation of integration is to be performed on some function f. Thus,

$$\int f(x)\ dx = F(x) + C$$

(read "the indefinite integral of $f(x)$ with respect to x equals $F(x)$ plus C") tells us that the **indefinite integral** of f is the family of functions given by $F(x) + C$, where $F'(x) = f(x)$. The function f to be integrated is called the **integrand** and the constant C is called a **constant of integration.** The expression dx following the integrand $f(x)$ reminds us that the operation is performed with respect to x. If the independent variable is t, we write $\int f(t)\ dt$ instead. In this sense both t and x are "dummy variables."

Using this notation, the results of Examples 2 and 3 may be written as

$$\int 1\ dx = x + C \quad \text{and} \quad \int 2x\ dx = x^2 + K,$$

where C and K are arbitrary constants.

Basic Integration Rules

Our next task is to develop some rules for finding the indefinite integral of a given function f. Because integration and differentiation are reverse operations, we discover many of the rules of integration by first making an "educated guess" at the antiderivative F of the function f to be integrated. Then this result is verified by demonstrating that $F' = f$.

RULE 1   INDEFINITE INTEGRAL OF A CONSTANT

$$\int k\ dx = kx + C \quad (k, \text{ a constant})$$

To prove this result, observe that

$$F'(x) = \frac{d}{dx}(kx + C) = k.$$

EXAMPLE 5   Evaluate each of the following indefinite integrals.

a. $\int 2\ dx$ **b.** $\int \pi^2\ dx$

Solution   Each of the integrands has the form $f(x) = k$, where k is a constant. Applying Rule 1 in each case yields

a. $\int 2\ dx = 2x + C$ **b.** $\int \pi^2\ dx = \pi^2 x + C.$

Next, from the rule of differentiation,

$$\frac{d}{dx}x^n = nx^{n-1},$$

we obtain the following rule of integration.

RULE 2   THE POWER RULE

$$\int x^n \, dx = \frac{1}{n+1} x^{n+1} + C \quad (n \neq -1)$$

An antiderivative of a power function is another power function obtained from the integrand by increasing its power by 1 and dividing the resulting expression by the new power.

To prove this result, observe that

$$F'(x) = \frac{d}{dx}\left[\frac{1}{n+1} x^{n+1} + C\right]$$

$$= \frac{n+1}{n+1} x^n$$

$$= x^n$$

$$= f(x).$$

EXAMPLE 6   Evaluate each of the following indefinite integrals:

a. $\displaystyle\int x^3 \, dx$ **b.** $\displaystyle\int x^{3/2} \, dx$ **c.** $\displaystyle\int \frac{1}{x^{3/2}} \, dx$

Solution   Each integrand is a power function with exponent $n \neq -1$. Applying Rule 2 in each case yields the following results:

a. $\displaystyle\int x^3 \, dx = \frac{1}{4}x^4 + C$

b. $\displaystyle\int x^{3/2} \, dx = \frac{1}{5/2}x^{5/2} + C = \frac{2}{5}x^{5/2} + C$

c. $\displaystyle\int \frac{1}{x^{3/2}} \, dx = \int x^{-3/2} \, dx = \frac{1}{-1/2}x^{-1/2} + C = -2x^{-1/2} + C$

These results may be verified by differentiating each of the antiderivatives and showing that the result is equal to the corresponding integrand.   ■

The next rule tells us that a constant factor may be moved through an integral sign.

RULE 3   INDEFINITE INTEGRAL OF A CONSTANT MULTIPLE OF A FUNCTION

$$\int cf(x) \, dx = c \int f(x) \, dx \quad (c, \text{ a constant})$$

The indefinite integral of a constant multiple of a function is equal to the constant multiple of the indefinite integral of the function.

This result follows from the corresponding rule of differentiation (see Rule 3, Section 3.1).

 Only a constant can be "moved out" of an integral sign. For example, it would be incorrect to write

$$\int x^2 \, dx = x^2 \int 1 \, dx.$$

In fact, $\int x^2 \, dx = \frac{1}{3}x^3 + C$, whereas $x^2 \int 1 \, dx = x^2(x + C) = x^3 + Cx^2$.

EXAMPLE 7

a. $\displaystyle\int 2t^3 \, dt$ **b.** $\displaystyle\int -3x^{-2} \, dx$

Solution   Each integrand has the form $cf(x)$, where c is a constant. Applying Rule 3, we obtain

a. $\displaystyle\int 2t^3 \, dt = 2\int t^3 \, dt = 2\left[\frac{1}{4}t^4 + K\right] = \frac{1}{2}t^4 + 2K = \frac{1}{2}t^4 + C,$

where $C = 2K$. From now on we will write the constant of integration as C, since any nonzero multiple of an arbitrary constant is an arbitrary constant.

b. $\displaystyle\int -3x^{-2} \, dx = -3\int x^{-2} \, dx = (-3)(-1)x^{-1} + C = \frac{3}{x} + C$ ∎

RULE 4   THE SUM RULE

$$\int [f(x) + g(x)] \, dx = \int f(x) \, dx + \int g(x) \, dx$$

and

$$\int [f(x) - g(x)] \, dx = \int f(x) \, dx - \int g(x) \, dx$$

The indefinite integral of a sum (difference) of two integrable functions is equal to the sum (difference) of their indefinite integrals.

This result is easily extended to the case involving the sum and difference of any finite number of functions. As in Rule 3, the proof of Rule 4 follows from the corresponding rule of differentiation (see Rule 4, Section 3.1).

EXAMPLE 8   Evaluate the indefinite integral

$$\int (3x^5 + 4x^{3/2} - 2x^{-1/2}) \, dx.$$

Solution Applying the extended version of Rule 4, we find that

$$\int (3x^5 + 4x^{3/2} - 2x^{-1/2})\, dx$$

$$= \int 3x^5\, dx + \int 4x^{3/2}\, dx - \int 2x^{-1/2}\, dx$$

$$= 3\int x^5\, dx + 4\int x^{3/2}\, dx - 2\int x^{-1/2}\, dx \qquad \text{(Rule 3)}$$

$$= (3)\left(\frac{1}{6}\right)x^6 + (4)\left(\frac{2}{5}\right)x^{5/2} - (2)(2)x^{1/2} + C \qquad \text{(Rule 2)}$$

$$= \frac{1}{2}x^6 + \frac{8}{5}x^{5/2} - 4x^{1/2} + C. \qquad \blacksquare$$

Observe that we have combined the three constants of integration, which arise from evaluating the three indefinite integrals, to obtain one constant C. After all, the sum of three arbitrary constants is also an arbitrary constant.

RULE 5   THE INDEFINITE INTEGRAL OF THE EXPONENTIAL FUNCTION

$$\int e^x\, dx = e^x + C$$

The indefinite integral of the exponential function with base e is equal to the function itself (except, of course, for the constant of integration).

EXAMPLE 9   Evaluate the indefinite integral

$$\int (2e^x - x^3)\, dx.$$

Solution We have

$$\int (2e^x - x^3)\, dx = \int 2e^x\, dx - \int x^3\, dx$$

$$= 2\int e^x\, dx - \int x^3\, dx$$

$$= 2e^x - \frac{1}{4}x^4 + C. \qquad \blacksquare$$

The last rule of integration in this section covers the integration of the function $f(x) = x^{-1}$. Remember that this function constituted the only exceptional case in the integration of the power function $f(x) = x^n$ (see Rule 2).

RULE 6   THE INDEFINITE INTEGRAL OF THE FUNCTION $f(x) = x^{-1}$

$$\int x^{-1}\, dx = \int \frac{1}{x}\, dx = \ln |x| + C \qquad (x \neq 0)$$

To prove Rule 6, observe that

$$\frac{d}{dx} \ln |x| = \frac{1}{x}. \qquad \text{(See Rule 3, Section 5.5.)}$$

EXAMPLE 10   Evaluate the indefinite integral

$$\int \left(2x + \frac{3}{x} + \frac{4}{x^2} \right) dx.$$

Solution

$$\int \left(2x + \frac{3}{x} + \frac{4}{x^2} \right) dx = \int 2x\, dx + \int \frac{3}{x}\, dx + \int \frac{4}{x^2}\, dx$$

$$= 2 \int x\, dx + 3 \int \frac{1}{x}\, dx + 4 \int x^{-2}\, dx$$

$$= 2 \left(\frac{1}{2} \right) x^2 + 3 \ln |x| + 4(-1)x^{-1} + C$$

$$= x^2 + 3 \ln |x| - \frac{4}{x} + C \qquad\blacksquare$$

In many applications we are given the derivative f' of a function f and we are required to find f. In such instances, a condition on f must be specified. This is illustrated in Examples 11–13.

EXAMPLE 11   Find the function f if it is known that

$$f'(x) = 3x^2 - 4x + 8 \quad \text{and} \quad f(1) = 9.$$

Solution   Integrating the function f', we find

$$f(x) = \int f'(x)\, dx$$

$$= \int (3x^2 - 4x + 8)\, dx$$

$$= x^3 - 2x^2 + 8x + C.$$

Using the condition $f(1) = 9$, we have

$$9 = f(1) = 1^3 - 2(1)^2 + 8(1) + C = 7 + C,$$

or $\qquad\qquad C = 2.$

Therefore, the required function f is given by $f(x) = x^3 - 2x^2 + 8x + 2.$ $\blacksquare$

Additional Applications

EXAMPLE 12   The management of the Staedtler Office Equipment Company determined that the daily marginal revenue function associated with producing and selling their battery-operated pencil sharpeners is given by

$$R'(x) = -0.0006x + 6,$$

where x denotes the number of units produced and sold and $R'(x)$ is measured in dollars per unit.

a. Determine the revenue function $R(x)$ associated with producing and selling these pencil sharpeners.

b. What is the demand equation that relates the wholesale unit price of the pencil sharpeners to the quantity demanded?

Solution

a. The revenue function R is found by integrating the marginal revenue function $R'(x)$. Thus,

$$R(x) = \int R'(x)\, dx$$

$$= \int (-0.0006x + 6)\, dx$$

$$= -0.0003x^2 + 6x + C,$$

where C is a constant. To determine the value of C, observe that the total revenue realized by Staedtler is nil when the production and sales level is zero; that is, $R(0) = 0$. This condition implies that

$$R(0) = -0.0003(0)^2 + 6(0) + C = 0,$$

or $C = 0$. Thus, the required revenue function is given by

$$R(x) = -0.0003x^2 + 6x.$$

b. Let p denote the unit wholesale price of the pencil sharpeners. Then

$$R(x) = px.$$

Thus, $$p = \frac{R(x)}{x} = -0.0003x + 6,$$

so the required demand equation is

$$p = -0.0003x + 6. \qquad \blacksquare$$

EXAMPLE 13   The current circulation of *Investor's Digest* is 2000 copies per week. Circulation is expected to grow at the rate of

$$5 + 2t^{2/3}$$

copies per week, t weeks from now, for the next three years. Determine what the digest's circulation will be 125 weeks from now.

Solution   Let $S(t)$ denote the digest's circulation t weeks from now. Then $S'(t)$, the rate of change in the circulation per week, is given by

$$S'(t) = 5 + 2t^{2/3}.$$

Therefore,
$$S(t) = \int (5 + 2t^{2/3})\, dt$$
$$= 5t + (2)\left(\frac{3}{5}\right)t^{5/3} + C$$
$$= 5t + \frac{6}{5}t^{5/3} + C.$$

To determine the value of C, observe that the circulation at $t = 0$ is 2000. Using this condition, we have

$$S(0) = 2000 = 5(0) + \frac{6}{5}(0) + C,$$

and $C = 2000$, so

$$S(t) = 5t + \frac{6}{5}t^{5/3} + 2000.$$

Therefore, 125 weeks from now the circulation will be

$$S(125) = 5(125) + \frac{6}{5}(125)^{5/3} + 2000 = 6375,$$

or 6375 copies per week.   ∎

SELF-CHECK EXERCISES 6.1

1. Evaluate $\int \left(\frac{1}{\sqrt{x}} - \frac{2}{x} + 3e^x\right) dx.$

2. Find the rule for the function f given that (1) the slope of the tangent line to the graph of f at any point $P(x, f(x))$ is given by the expression $3x^2 - 6x + 3$ and (2) the graph of f passes through the point $(2, 9)$.

c 3. Suppose that United Motors' share of the new cars sold in a certain country is changing at the rate of

$$f(t) = -0.01875t^2 + 0.15t - 1.2 \qquad (0 \le t \le 12)$$

percent at year t ($t = 0$ corresponds to the beginning of 1982). The company's market share at the beginning of 1982 was 48.4 percent. What was United Motors' market share at the beginning of 1994?

Solutions to Self-Check Exercises 6.1 can be found on page 352.

6.1   EXERCISES

In Exercises 1–4, verify directly that F is an antiderivative of f.

1. $F(x) = \frac{1}{3}x^3 + 2x^2 - x + 2$; $f(x) = x^2 + 4x - 1$

2. $F(x) = xe^x + \pi$; $f(x) = e^x(1 + x)$

3. $F(x) = \sqrt{2x^2 - 1}$; $f(x) = \dfrac{2x}{\sqrt{2x^2 - 1}}$

4. $F(x) = x \ln x - x$; $f(x) = \ln x$

In Exercises 5–8, (a) verify that G is an antiderivative of f, (b) find all antiderivatives of f, and (c) sketch the graphs of a few of the family of antiderivatives found in (b).

5. $G(x) = 2x$; $f(x) = 2$

6. $G(x) = 2x^2$; $f(x) = 4x$

7. $G(x) = \frac{1}{3}x^3$; $f(x^2) = x^2$

8. $G(x) = e^x$; $f(x) = e^x$

In Exercises 9–50, find the indefinite integral.

9. $\displaystyle\int 6\, dx$ 10. $\displaystyle\int \sqrt{2}\, dx$

11. $\displaystyle\int x^3\, dx$ 12. $\displaystyle\int 2x^5\, dx$

13. $\displaystyle\int x^{-4}\, dx$ 14. $\displaystyle\int 3t^{-7}\, dt$

15. $\displaystyle\int x^{2/3}\, dx$ 16. $\displaystyle\int 2u^{3/4}\, du$

17. $\displaystyle\int x^{-5/4}\, dx$ 18. $\displaystyle\int 3x^{-2/3}\, dx$

19. $\displaystyle\int \frac{2}{x^2}\, dx$ 20. $\displaystyle\int \frac{1}{3x^5}\, dx$

21. $\displaystyle\int \pi\sqrt{t}\, dt$ 22. $\displaystyle\int \frac{3}{\sqrt{t}}\, dt$

23. $\displaystyle\int (3 - 2x)\, dx$ 24. $\displaystyle\int (1 + u + u^2)\, du$

25. $\displaystyle\int (x^2 + x + x^{-3})\, dx$

26. $\displaystyle\int (0.3t^2 + 0.02t + 2)\, dt$

27. $\displaystyle\int 4e^x\, dx$ 28. $\displaystyle\int (1 + e^x)\, dx$

29. $\displaystyle\int (1 + x + e^x)\, dx$

30. $\displaystyle\int (2 + x + 2x^2 + e^x)\, dx$

31. $\displaystyle\int \left(4x^3 - \frac{2}{x^2} - 1\right) dx$ 32. $\displaystyle\int \left(6x^3 + \frac{3}{x^2} - x\right) dx$

33. $\displaystyle\int (x^{5/2} + 2x^{3/2} - x)\, dx$

34. $\displaystyle\int (t^{3/2} + 2t^{1/2} - 4t^{-1/2})\, dt$

35. $\displaystyle\int \left(\sqrt{x} + \frac{3}{\sqrt{x}}\right) dx$ 36. $\displaystyle\int \left(\sqrt[3]{x^2} - \frac{1}{x^2}\right) dx$

37. $\displaystyle\int \left(\frac{u^3 + 2u^2 - u}{3u}\right) du$

$\left[\text{Hint: } \dfrac{u^3 + 2u^2 - u}{3u} = \dfrac{1}{3}u^2 + \dfrac{2}{3}u - \dfrac{1}{3}\right]$

38. $\displaystyle\int \frac{x^4 - 1}{x^2}\, dx$ $\left[\text{Hint: } \dfrac{x^4 - 1}{x^2} = x^2 - x^{-2}\right]$

39. $\displaystyle\int (2t + 1)(t - 2)\, dt$

40. $\displaystyle\int u^{-2}(1 - u^2 + u^4)\, du$

41. $\displaystyle\int \frac{1}{x^2}(x^4 - 2x^2 + 1)\, dx$

42. $\displaystyle\int \sqrt{t}(t^2 + t - 1)\, dt$ 43. $\displaystyle\int \frac{ds}{(s + 1)^{-2}}$

44. $\displaystyle\int \left(\sqrt{x} + \frac{3}{x} - 2e^x\right) dx$ 45. $\displaystyle\int (e^t + t^e)\, dt$

46. $\displaystyle\int \left(\frac{1}{x^2} - \frac{1}{\sqrt[3]{x^2}} + \frac{1}{\sqrt{x}}\right) dx$

47. $\displaystyle\int \left(\frac{x^3 + x^2 - x + 1}{x^2}\right) dx$ [*Hint:* Simplify the integrand first.]

48. $\displaystyle\int \frac{t^3 + \sqrt[3]{t}}{t^2}\, dt$ [*Hint:* Simplify the integrand first.]

49. $\displaystyle\int \frac{(\sqrt{x} - 1)^2}{x^2}\, dx$ [*Hint:* Simplify the integrand first.]

50. $\displaystyle\int (x + 1)^2\left(1 - \frac{1}{x}\right) dx$ [*Hint:* Simplify the integrand first.]

From the given information in Exercises 51–54, find the function f.

51. $f'(x) = 3x^2 + 4x - 1$; $f(2) = 9$

52. $f'(x) = e^x - 2x$; $f(0) = 2$

53. $f'(x) = \dfrac{x+1}{x}; \; f(1) = 1$

54. $f'(x) = 1 + e^x + \dfrac{1}{x}; \; f(1) = 3 + e$

In Exercises 55–58, find the function f given that the slope of the tangent line to the graph of f at any point $(x, f(x))$ is $f'(x)$ and that the graph of f passes through the given point.

55. $f'(x) = \tfrac{1}{2}x^{-1/2}; \; (2, \sqrt{2})$

56. $f'(t) = t^2 - 2t + 3; \; (1, 2)$

57. $f'(x) = e^x + x; \; (0, 3)$

58. $f'(x) = \dfrac{2}{x} + 1; \; (1, 2)$

59. **Velocity of a Car**   The velocity of a car (in ft/sec) t seconds after starting from rest is given by the function

$$f(t) = 2\sqrt{t} \qquad (0 \le t \le 30).$$

Find the car's position at any time t.

60. **Cost of Producing Clocks**   The Lorimar Watch Company manufactures travel clocks. The daily marginal cost function associated with producing these clocks is

$$C'(x) = 0.000009x^2 - 0.009x + 8,$$

where $C'(x)$ is measured in dollars per unit and x denotes the number of units produced. Management has determined that the daily fixed cost incurred in producing these clocks is $120. Find the total cost incurred by Lorimar in producing the first 500 travel clocks per day.

61. **Revenue Functions**   The management of the Lorimar Watch Company has determined that the daily marginal revenue function associated with producing and selling their travel clocks is given by

$$R'(x) = -0.009x + 12,$$

where x denotes the number of units produced and sold and $R'(x)$ is measured in dollars per unit.
a. Determine the revenue function $R(x)$ associated with producing and selling these clocks.
b. What is the demand equation that relates the wholesale unit price with the quantity of travel clocks demanded?

62. **Profit Functions**   The Cannon Precision Instruments Corporation makes an automatic electronic flash with Thyrister circuitry. The estimated marginal profit associated with producing and selling these electronic flashes is

$$(-0.004x + 20)$$

dollars per unit per month when the production level is x units per month. Cannon's fixed cost for producing and selling these electronic flashes is $16,000 per month. At what level of production does Cannon realize a maximum profit? What is the maximum monthly profit?

63. **Cost of Producing Guitars**   The Carlota Music Company estimates that the marginal cost of manufacturing its Professional Series guitars is

$$C'(x) = 0.002x + 100$$

dollars per month when the level of production is x guitars per month. The fixed costs incurred by Carlota are $4000 per month. Find the total monthly cost incurred by Carlota in manufacturing x guitars per month.

64. **Quality Control**   As part of a quality-control program, the chess sets manufactured by the Jones Brothers Company are subjected to a final inspection before packing. The rate of increase in the number of sets checked per hour by an inspector t hours into the 8 A.M. to 12 noon morning shift is approximately

$$N'(t) = -3t^2 + 12t + 45 \qquad (0 \le t \le 4).$$

Find an expression $N(t)$ that approximates the number of sets inspected at the end of t hours. [*Hint:* $N(0) = 0$.] How many sets does the average inspector check during a morning shift?

65. **Ballast Dropped from a Balloon**   A ballast is dropped from a stationary hot-air balloon that is hovering at an altitude of 400 ft. Its velocity after t seconds is $-32t$ ft/sec.
a. Find the height $h(t)$ of the ballast from the ground at time t. [*Hint:* $h'(t) = -32t$ and $h(0) = 400$.]
b. When will the ballast strike the ground?
c. Find the velocity of the ballast when it hits the ground.

Ballast

66. **Cable TV Subscribers** A study conducted by Tele-Cable, Inc., estimates that the number of cable television subscribers will grow at the rate of

$$100 + 210t^{3/4}$$

new subscribers per month t months from the start date of the service. If 5000 subscribers signed up for the service before the starting date, how many subscribers will there be 16 months from that date?

67. **Air Pollution** On an average summer day, the level of carbon monoxide in a city's air is 2 parts per million. An environmental protection agency's study predicts that, unless more stringent measures are taken to protect the city's atmosphere, the concentration of carbon monoxide present in the air will increase at the rate of

$$0.003t^2 + 0.06t + 0.1$$

parts per million per year t years from now. If no further pollution-control efforts are made, what will the concentration of carbon monoxide be on an average summer day five years from now?

68. **Population Growth** The development of Astro World ("The Amusement Park of the Future") on the outskirts of a city will increase the city's population at the rate of

$$4500\sqrt{t} + 1000$$

people per year t years from the start of construction. The population before construction is 30,000. Determine the projected population nine years after construction of the park has begun.

69. **Ozone Pollution** The rate of change of the level of ozone, an invisible gas that is an irritant and impairs breathing, present in the atmosphere on a certain May day in the city of Riverside is given by

$$R(t) = 3.2922t^2 - 0.366t^3 \qquad (0 < t < 11)$$

(measured in pollutant standard index per hour). Here t is measured in hours, with $t = 0$ corresponding to 7 A.M. Find the ozone level $A(t)$ at any time t assuming that at 7 A.M. it is zero. [*Hint:* $A'(t) = R(t)$ and $A(0) = 0$.]

70. **Surface Area of a Human** Empirical data suggest that the surface area of a 180-cm-tall human body changes at the rate of

$$S'(W) = 0.131773W^{-0.575}$$

square meters per kilogram, where W is the weight of the body in kg. If the surface area of a 180-cm-tall human body weighing 70 kg is 1.886277 square meters, what is the surface area of a human body of the same height weighing 75 kg?

71. **Flight of a Rocket** The velocity, in feet per second, of a rocket t seconds into vertical flight is given by

$$v(t) = -3t^2 + 192t + 120.$$

Find an expression $h(t)$ that gives the rocket's altitude, in feet, t seconds after lift-off. What is the altitude of the rocket 30 seconds after lift-off? [*Hint:* $h'(t) = v(t)$; $h(0) = 0$.]

72. **Flow of Blood in an Artery** Nineteenth-century physician Jean Louis Marie Poiseuille discovered that the rate of change of the velocity of blood r cm from the central axis of an artery (in cm/sec/cm) is given by

$$a(r) = -kr,$$

where k is a constant. If the radius of an artery is R cm, find an expression for the velocity of blood as a function of r. [*Hint:* $v'(r) = a(r)$ and $v(R) = 0$. (Why?)]

Blood vessel

SOLUTIONS TO SELF-CHECK EXERCISES 6.1

1.
$$\int \left(\frac{1}{\sqrt{x}} - \frac{2}{x} + 3e^x \right) dx = \int \left(x^{-1/2} - \frac{2}{x} + 3e^x \right) dx$$

$$= \int x^{-1/2} \, dx - 2 \int \frac{1}{x} \, dx + 3 \int e^x \, dx$$

$$= 2x^{1/2} - 2 \ln |x| + 3e^x + C$$

$$= 2\sqrt{x} - 2 \ln |x| + 3e^x + C$$

2. The slope of the tangent line to the graph of the function f at any point $P(x, f(x))$ is given by the derivative f' of f. Thus, the first condition implies that

$$f'(x) = 3x^2 - 6x + 3,$$

which, upon integration, yields

$$f(x) = \int (3x^2 - 6x + 3)\, dx$$

$$= x^3 - 3x^2 + 3x + k,$$

where k is the constant of integration.

To evaluate k, we use condition (2), which implies that $f(2) = 9$, or

$$9 = f(2) = 2^3 - 3(2)^2 + 3(2) + k,$$

or $k = 7$. Hence, the required rule of definition of the function f is

$$f(x) = x^3 - 3x^2 + 3x + 7.$$

3. Let $M(t)$ denote United Motors' market share at year t. Then

$$M(t) = \int f(t)\, dt$$

$$= \int (-0.01875t^2 + 0.15t - 1.2)\, dt$$

$$= -0.00625t^3 + 0.075t^2 - 1.2t + C.$$

To determine the value of C, we use the condition $M(0) = 48.4$, obtaining $C = 48.4$. Therefore,

$$M(t) = -0.00625t^3 + 0.075t^2 - 1.2t + 48.4.$$

In particular, United Motors' market share of new cars at the beginning of 1994 is given by

$$M(12) = -0.00625(12)^3 + 0.075(12)^2$$
$$-1.2(12) + 48.4 = 34,$$

or 34 percent.

6.2 Integration by Substitution

In Section 6.1 we developed certain rules of integration that are closely related to the corresponding rules of differentiation in Chapters 3 and 5. In this section, we introduce a method of integration called the **method of substitution,** which is related to the Chain Rule for differentiating functions. When used in conjunction with the rules of integration developed earlier, the method of substitution is a powerful tool for integrating a large class of functions.

How the Method of Substitution Works

Consider the indefinite integral

$$\int 2(2x + 4)^5\, dx. \qquad\qquad (1)$$

One way of evaluating this integral is to expand the expression $(2x + 4)^5$ and then integrate the resulting integrand term by term. As an alternative approach, let us see if we can simplify the integral by making a change of variable. Write

$$u = 2x + 4$$

with differential

$$du = 2 \, dx.$$

If we substitute these quantities into (1), we obtain

$$\int 2(2x + 4)^5 \, dx = \int (2x + 4)^5 (2 \, dx) = \int u^5 \, du$$

$$\uparrow \qquad\qquad\qquad\qquad\qquad \uparrow \quad \begin{cases} u = 2x + 4 \\ du = 2 \, dx \end{cases}$$

rewriting

Now, the last integral involves a power function and is easily evaluated using Rule 2 of Section 6.1. Thus,

$$\int u^5 \, du = \frac{1}{6} u^6 + C.$$

Therefore, using this result and replacing u by $u = 2x + 4$, we obtain

$$\int 2(2x + 4)^5 \, dx = \frac{1}{6}(2x + 4)^6 + C.$$

We can verify that the foregoing result is indeed correct by computing

$$\frac{d}{dx}\left[\frac{1}{6}\left(2x + 4 \right)^6 \right] = \frac{1}{6} \cdot 6(2x + 4)^5(2) \qquad \text{(Using the Chain Rule)}$$

$$= 2(2x + 4)^5$$

and observing that the last expression is just the integrand of (1).

The Method of Integration by Substitution

In order to see why the approach used in evaluating the integral in (1) is successful, write

$$f(x) = x^5 \quad \text{and} \quad g(x) = 2x + 4.$$

Then $g'(x) = 2 \, dx$. Furthermore, the integrand of (1) is just the composition of f and g. Thus,

$$(f \circ g)(x) = f(g(x))$$
$$= [g(x)]^5 = (2x + 4)^5.$$

Therefore, (1) can be written as

$$\int f(g(x))g'(x) \, dx. \qquad\qquad\qquad (2)$$

Next, let us show that an integral having the form (2) can always be written as

$$\int f(u) \, du. \qquad\qquad\qquad (3)$$

Suppose F is an antiderivative of f. By the Chain Rule, we have

$$\frac{d}{dx}[F(g(x))] = F'(g(x))g'(x)$$

and therefore,

$$\int F'(g(x))g'(x)\,dx = F(g(x)) + C.$$

Letting $F' = f$ and making the substitution $u = g(x)$, we have

$$\int f(g(x))g'(x)\,dx = F(u) + C = \int F'(u)\,du = \int f(u)\,du,$$

as we wished to show. Thus, if the transformed integral is readily evaluated, as is the case with the integral (1), then the method of substitution will prove successful.

Before we look at more examples, let us summarize the steps involved in integration by substitution.

INTEGRATION BY SUBSTITUTION

Step 1   Let $u = g(x)$, where $g(x)$ is part of the integrand, usually the "inside function" of the composite function $f(g(x))$.

Step 2   Compute $du = g'(x)\,dx$.

Step 3   Use the substitution $u = g(x)$ and $du = g'(x)\,dx$ to convert the *entire* integral into one involving *only u*.

Step 4   Evaluate the resulting integral.

Step 5   Replace u by $g(x)$ to obtain the final solution as a function of x.

REMARK   Sometimes we need to consider different choices of g for the substitution $u = g(x)$ in order to carry out Step 3 and/or Step 4.   ☐

EXAMPLE 1   Evaluate $\int 2x(x^2 + 3)^4\,dx$.

Solution

Step 1   Observe that the integrand involves the composite function $(x^2 + 3)^4$ with "inside function" $g(x) = x^2 + 3$. So, we choose $u = x^2 + 3$.

Step 2   Compute $du = 2x\,dx$.

Step 3   Making the substitution $u = x^2 + 3$ and $du = 2x\,dx$, we obtain

$$\int 2x(x^2 + 3)^4\,dx = \int (x^2 + 3)^4(2x\,dx) = \int u^4\,du,$$

$$\underset{\text{rewriting}}{\uparrow}$$

an integral involving only the variable u.

Step 4   Evaluate

$$\int u^4 \, du = \frac{1}{5}u^5 + C.$$

Step 5   Replacing u by $x^2 + 3$, we obtain

$$\int 2x(x^2 + 3)^4 \, dx = \frac{1}{5}(x^2 + 3)^5 + C.$$ ∎

EXAMPLE 2   Evaluate $\int 3\sqrt{3x + 1} \, dx$.

Solution

Step 1   The integrand involves the composite function $\sqrt{3x + 1}$ with "inside function" $g(x) = 3x + 1$. So, let $u = 3x + 1$.

Step 2   Compute $du = 3 \, dx$.

Step 3   Making the substitution $u = 3x + 1$ and $du = 3 \, dx$, we obtain

$$\int 3\sqrt{3x + 1} \, dx = \int \sqrt{3x + 1}(3 \, dx) = \int \sqrt{u} \, du,$$

an integral involving only the variable u.

Step 4   Evaluate

$$\int \sqrt{u} \, du = \int u^{1/2} \, du = \frac{2}{3}u^{3/2} + C.$$

Step 5   Replacing u by $u = 3x + 1$, we obtain

$$\int 3\sqrt{3x + 1} \, dx = \frac{2}{3}(3x + 1)^{3/2} + C.$$ ∎

EXAMPLE 3   Evaluate $\int x^2(x^3 + 1)^{3/2} \, dx$.

Solution

Step 1   The integrand contains the composite function $(x^3 + 1)^{3/2}$ with "inside function" $g(x) = x^3 + 1$. So, let $u = x^3 + 1$.

Step 2   Compute $du = 3x^2 \, dx$.

Step 3   Making the substitution $u = x^3 + 1$ and $du = 3x^2 \, dx$, or $x^2 \, dx = \frac{1}{3} \, du$, we obtain

$$\int x^2(x^3 + 1)^{3/2} \, dx = \int (x^3 + 1)^{3/2}(x^2 \, dx)$$

$$= \int u^{3/2}\left(\frac{1}{3} \, du\right) = \frac{1}{3}\int u^{3/2} \, du,$$

an integral involving only the variable u.

Step 4   We evaluate

$$\frac{1}{3}\int u^{3/2} \, du = \frac{1}{3} \cdot \frac{2}{5}u^{5/2} + C = \frac{2}{15}u^{5/2} + C.$$

Step 5   Replacing u by $x^3 + 1$, we obtain

$$\int x^2(x^3 + 1)^{3/2} \, dx = \frac{2}{15}(x^3 + 1)^{5/2} + C.$$ ■

In the remaining examples, we will drop the practice of labeling the steps involved in evaluating each integral.

EXAMPLE 4   Evaluate $\int e^{-3x} \, dx$.

Solution   Let $u = -3x$ so that $du = -3 \, dx$, or $dx = -\frac{1}{3} \, du$. Then

$$\int e^{-3x} \, dx = \int e^u \left(-\frac{1}{3} \, du \right) = -\frac{1}{3} \int e^u \, du$$

$$= -\frac{1}{3} e^u + C = -\frac{1}{3} e^{-3x} + C.$$ ■

EXAMPLE 5   Evaluate $\int \dfrac{x}{3x^2 + 1} \, dx$.

Solution   Let $u = 3x^2 + 1$. Then $du = 6x \, dx$, or $x \, dx = \frac{1}{6} \, du$. Making the appropriate substitutions, we have

$$\int \frac{x}{3x^2 + 1} \, dx = \int \frac{1/6}{u} \, du$$

$$= \frac{1}{6} \int \frac{1}{u} \, du$$

$$= \frac{1}{6} \ln |u| + C$$

$$= \frac{1}{6} \ln (3x^2 + 1) + C \quad \text{(since } 3x^2 + 1 > 0\text{).}$$ ■

EXAMPLE 6   Evaluate $\int \dfrac{(\ln x)^2}{2x} \, dx$.

Solution   Let $u = \ln x$.

Then

$$du = \frac{d}{dx} (\ln x) \, dx = \frac{1}{x} \, dx.$$

$$\int \frac{(\ln x)^2}{2x} \, dx = \frac{1}{2} \int \frac{(\ln x)^2}{x} \, dx$$

Thus,

$$= \frac{1}{2} \int u^2 \, du$$

$$= \frac{1}{6} u^3 + C$$

$$= \frac{1}{6} (\ln x)^3 + C.$$ ■

Applications

Examples 7 and 8 show how the method of substitution can be used in practical situations.

EXAMPLE 7   In 1988, the head of the research and development department of the Soloron Corporation claimed that the cost of producing solar cell panels would drop at the rate of

$$\frac{58}{(3t + 2)^2} \qquad (0 \le t \le 10)$$

dollars per peak watt for the next t years, with $t = 0$ corresponding to the beginning of the year 1988. (A peak watt is the power produced at noon on a sunny day.) In 1988, the panels, which are used for photovoltaic power systems, cost \$10 per peak watt. Find an expression giving the cost per peak watt of producing solar cell panels at the beginning of year t. What will the cost be at the beginning of 1998?

Solution   Let $C(t)$ denote the cost per peak watt for producing solar cell panels at the beginning of year t. Then

$$C'(t) = -\frac{58}{(3t + 2)^2}.$$

Integrating, we find that

$$C(t) = \int \frac{-58}{(3t + 2)^2}\, dt$$

$$= -58 \int (3t + 2)^{-2}\, dt.$$

Let $u = 3t + 2$ so that

$$du = 3\, dt, \quad \text{or} \quad dt = \frac{1}{3}\, du.$$

Then
$$C(t) = -58 \left(\frac{1}{3}\right) \int u^{-2}\, du$$

$$= -\frac{58}{3}(-1)u^{-1} + k$$

$$= \frac{58}{3(3t + 2)} + k,$$

where k is an arbitrary constant. To determine the value of k, note that the cost per peak watt of producing solar cell panels at the beginning of 1988 ($t = 0$) was 10, or $C(0) = 10$. This gives

$$C(0) = \frac{58}{3(2)} + k = 10,$$

or $k = \frac{1}{3}$. Therefore, the required expression is given by

$$C(t) = \frac{58}{3(3t + 2)} + \frac{1}{3}$$

$$= \frac{58 + (3t + 2)}{3(3t + 2)}$$

$$= \frac{t + 20}{3t + 2}.$$

The cost per peak watt for producing solar cell panels at the beginning of 1998 is given by

$$C(10) = \frac{10 + 20}{3(10) + 2} \approx 0.94,$$

or approximately \$.94 per peak watt.   ■

EXAMPLE 8   A study prepared by the marketing department of the Universal Instruments Company forecasts that, after its new line of Galaxy Home Computers is introduced into the market, sales will grow at the rate of

$$2000 - 1500e^{-0.05t} \qquad (0 \le t \le 60)$$

units per month. Find an expression that gives the total number of computers that will sell t months after they become available on the market. How many computers will Universal sell in the first year they are on the market?

Solution   Let $N(t)$ denote the total number of computers that may be expected to be sold t months after their introduction in the market. Then, the rate of growth of sales is given by $N'(t)$ units per month. Thus,

$$N'(t) = 2000 - 1500e^{-0.05t}$$

so that

$$N(t) = \int (2000 - 1500e^{-0.05t})\, dt$$

$$= \int 2000\, dt - 1500 \int e^{-0.05t}\, dt.$$

Upon integrating the second integral by the method of substitution, we obtain

$$N(t) = 2,000t + \frac{1,500}{0.05} e^{-0.05t} + C \qquad \text{(Let } u = -0.05t,$$
$$\text{then } du = -0.05\, dt.)$$

$$= 2,000t + 30,000e^{-0.05t} + C.$$

To determine the value of C, note that the number of computers sold at the end of month 0 is nil, so $N(0) = 0$. This gives

$$N(0) = 30,000 + C = 0, \qquad \text{(Since } e^0 = 1)$$

or $C = -30,000$. Therefore, the required expression is given by

$$N(t) = 2,000t + 30,000e^{-0.05t} - 30,000$$
$$= 2,000t + 30,000(e^{-0.05t} - 1).$$

The number of computers that Universal can expect to sell in the first year is given by

$$N(12) = 2,000(12) + 30,000(e^{-0.05(12)} - 1)$$
$$= 10,464,$$

or 10,464 units. ∎

SELF-CHECK
EXERCISES 6.2

1. Evaluate $\int \sqrt{2x + 5}\, dx$.

2. Evaluate $\int \dfrac{x^2}{(2x^3 + 1)^{3/2}}\, dx$.

3. Evaluate $\int x e^{2x^2 - 1}\, dx$.

C 4. According to a joint study conducted by Oxnard's Environmental Management Department and a state government agency, the concentration of carbon monoxide in the air due to automobile exhaust is increasing at the rate given by

$$f(t) = \dfrac{8(0.1t + 1)}{300(0.2t^2 + 4t + 64)^{1/3}}$$

parts per million per year t. Currently, the concentration of carbon monoxide due to automobile exhaust is 0.16 parts per million. Find an expression giving the concentration of carbon monoxide t years from now.

Solutions to Self-Check Exercises 6.2 can be found on page 362.

6.2 EXERCISES

In Exercises 1–50, evaluate the given indefinite integral.

1. $\int 4(4x + 3)^4\, dx$

2. $\int 4x(2x^2 + 1)^7\, dx$

3. $\int (x^3 - 2x)^2(3x^2 - 2)\, dx$

4. $\int (3x^2 - 2x + 1)(x^3 - x^2 + x)^4\, dx$

5. $\int \dfrac{4x}{(2x^2 + 3)^3}\, dx$

6. $\int \dfrac{3x^2 + 2}{(x^3 + 2x)^2}\, dx$

7. $\int 3t^2 \sqrt{t^3 + 2}\, dt$

8. $\int 3t^2(t^3 + 2)^{3/2}\, dt$

9. $\int (x^2 - 1)^9 x\, dx$

10. $\int x^2(2x^3 + 3)^4\, dx$

11. $\int \dfrac{x^4}{1 - x^5}\, dx$

12. $\int \dfrac{x^2}{\sqrt{x^3 - 1}}\, dx$

13. $\int \dfrac{2}{x - 2}\, dx$

14. $\int \dfrac{x^2}{x^3 - 3}\, dx$

15. $\int \dfrac{0.3x - 0.2}{0.3x^2 - 0.4x + 2}\, dx$

16. $\int \dfrac{2x^2 + 1}{0.2x^3 + 0.3x}\, dx$

17. $\int \dfrac{x}{3x^2 - 1}\, dx$

18. $\int \dfrac{x^2 - 1}{x^3 - 3x + 1}\, dx$

19. $\int e^{-2x}\, dx$

20. $\int e^{-0.02x}\, dx$

21. $\int e^{2-x}\, dx$

22. $\int e^{2t+3}\, dt$

23. $\int x e^{-x^2}\, dx$

24. $\int x^2 e^{x^3 - 1}\, dx$

25. $\int (e^x - e^{-x})\, dx$

26. $\int (e^{2x} + e^{-3x})\, dx$

27. $\int \dfrac{e^x}{1 + e^x}\, dx$

28. $\int \dfrac{e^{2x}}{1 + e^{2x}}\, dx$

29. $\displaystyle\int \frac{e^{\sqrt{x}}}{\sqrt{x}}\, dx$

30. $\displaystyle\int \frac{e^{-1/x}}{x^2}\, dx$

31. $\displaystyle\int \frac{e^{3x} + x^2}{(e^{3x} + x^3)^3}\, dx$

32. $\displaystyle\int \frac{e^x - e^{-x}}{(e^x + e^{-x})^{3/2}}\, dx$

33. $\displaystyle\int e^{2x}(e^{2x} + 1)^3\, dx$

34. $\displaystyle\int e^{-x}(1 + e^{-x})\, dx$

35. $\displaystyle\int \frac{\ln 5x}{x}\, dx$

36. $\displaystyle\int \frac{(\ln u)^3}{u}\, du$

37. $\displaystyle\int \frac{1}{x \ln x}\, dx$

38. $\displaystyle\int \frac{1}{x(\ln x)^2}\, dx$

39. $\displaystyle\int \frac{\sqrt{\ln x}}{x}\, dx$

40. $\displaystyle\int \frac{(\ln x)^{7/2}}{x}\, dx$

41. $\displaystyle\int \left(xe^{x^2} - \frac{x}{x^2 + 2} \right) dx$

42. $\displaystyle\int \left(xe^{-x^2} + \frac{e^x}{e^x + 3} \right) dx$

43. $\displaystyle\int \frac{x + 1}{\sqrt{x} - 1}\, dx$ [*Hint: Let* $u = \sqrt{x} - 1$.]

44. $\displaystyle\int \frac{e^{-u} - 1}{e^{-u} + u}\, du$ [*Hint: Let* $v = e^{-u} + u$.]

45. $\displaystyle\int x(x - 1)^5\, dx$ [*Hint:* $u = x - 1$ *implies* $x = u + 1$.]

46. $\displaystyle\int \frac{t}{t + 1}\, dt$ $\left[Hint: \dfrac{t}{t + 1} = 1 - \dfrac{1}{t + 1}. \right]$

47. $\displaystyle\int \frac{1 - \sqrt{x}}{1 + \sqrt{x}}\, dx$ [*Hint: Let* $u = 1 + \sqrt{x}$.]

48. $\displaystyle\int \frac{1 + \sqrt{x}}{1 - \sqrt{x}}\, dx$ [*Hint: Let* $u = 1 - \sqrt{x}$.]

49. $\displaystyle\int v^2(1 - v)^6\, dv$ [*Hint: Let* $u = 1 - v$.]

50. $\displaystyle\int x^3(x^2 + 1)^{3/2}\, dx$ [*Hint: Let* $u = x^2 + 1$.]

In Exercises 51–54, find the function f given that the slope of the tangent line to the graph of f at any point $(x, f(x))$ *is* $f'(x)$ *and that the graph of f passes through the given point.*

51. $f'(x) = 5(2x - 1)^4;\ (1, 3)$

52. $f'(x) = \dfrac{3x^2}{2\sqrt{x^3 - 1}};\ (1, 1)$

53. $f'(x) = -2xe^{-x^2+1};\ (1, 0)$

54. $f'(x) = 1 - \dfrac{2x}{x^2 + 1};\ (0, 2)$

[C] 55. **Student Enrollment**   The registrar of Kellogg University estimates that the total student enrollment in the Continuing Education division will grow at the rate of

$$N'(t) = 2000(1 + 0.2t)^{-3/2}$$

students per year t years from now. If the current student enrollment is 1000, find an expression giving the total student enrollment t years from now. What will the student enrollment be five years from now?

[C] 56. **TV Viewers: Newsmagazine Shows**   The number of viewers of a weekly TV newsmagazine show, introduced in the 1993 season, has been increasing at the rate of

$$3\left(2 + \frac{1}{2}x\right)^{-1/3} \qquad (1 \le x \le 6)$$

million viewers per year in its xth year on the air. The number of viewers of the program during its first year on the air is given by $9(5/2)^{2/3}$ million. Find how many viewers are expected in the 1998 season.

57. **Demand: Ladies' Boots**   The rate of change of the unit price p (in dollars) of Apex ladies' boots is given by

$$p'(x) = \frac{-250x}{(16 + x^2)^{3/2}},$$

where x is the quantity demanded daily in units of a hundred. Find the demand function for these boots if the quantity demanded daily is 300 pairs ($x = 3$) when the unit price is $50 per pair.

[C] 58. **Supply: Ladies' Boots**   The rate of change of the unit price p (in dollars) of Apex ladies' boots is given by

$$p'(x) = \frac{240x}{(5 - x)^2},$$

where x is the number of pairs that the supplier will make available in the market daily when the unit price is p per pair. Find the supply equation for these boots if the quantity the supplier is willing to make available is 200 pairs daily ($x = 2$) when the unit price is $50 per pair.

[C] 59. **Oil Spill**   In calm waters, the oil spilling from the ruptured hull of a grounded tanker forms an oil slick that is circular in shape. If the radius r of the circle is increasing at the rate of

$$r'(t) = \frac{30}{\sqrt{2t + 4}}$$

feet per minute t minutes after the rupture occurs, find an expression for the radius at any time t. How large is the polluted area 16 minutes after the rupture occurred? [*Hint:* $r(0) = 0$.]

C 60. **Life Expectancy of a Female**   Suppose that in a certain country the life expectancy at birth of a female is changing at the rate of

$$g'(t) = \frac{5.45218}{(1 + 1.09t)^{0.9}}$$

years per year. Here t is measured in years and $t = 0$ corresponds to the beginning of 1900. Find an expression $g(t)$ giving the life expectancy at birth (in years) of a female in that country if the life expectancy at the beginning of 1900 is 50.02 years. What is the life expectancy at birth of a female born in the year 2000 in that country?

C 61. **Average Birth Height of Boys**   Using data collected at Kaiser Hospital, pediatricians estimate that the average height of male children changes at the rate of

$$h'(t) = \frac{52.8706e^{-0.3277t}}{(1 + 2.449e^{-0.3277t})^2}$$

inches per year, where the child's height $h(t)$ is measured in inches and t, the child's age, is measured in years, with $t = 0$ corresponding to the age at birth. Find an expression $h(t)$ for the average height of a boy at age t if the height at birth of an average child is 19.4 inches. What is the height of an average eight-year-old boy?

C 62. **Learning Curves**   The average student enrolled in the 20-week Advanced Shorthand course at the American Institute of Stenography progresses according to the rule

$$N'(t) = 6e^{-0.05t} \qquad (0 \le t \le 20),$$

where $N'(t)$ measures the rate of change in the number of words per minute of dictation the student takes in shorthand after t weeks in the course. Assuming that the average student enrolled in this course begins with a dictation speed of 60 words per minute, find an expression $N(t)$ that gives the dictation speed of the student after t weeks in the course.

C 63. **Sales: Loudspeakers**   Two thousand pairs of Acrosonic model F loudspeaker systems were sold in the first year they appeared in the market. Since then, sales of these loudspeaker systems have been growing at the rate of

$$f'(t) = 2000(3 - 2e^{-t})$$

units per year, where t denotes the number of years these systems have been on the market. Determine the number of systems that were sold in the first five years after their introduction.

C 64. **Amount of Glucose in the Bloodstream**   Suppose that a patient is given a continuous intravenous infusion of glucose at a constant rate of r milligrams per minute. Then, the rate at which the amount of glucose in the bloodstream is changing at time t due to this infusion is given by

$$A'(t) = re^{-at}$$

milligrams per minute, where a is a positive constant associated with the rate at which excess glucose is eliminated from the bloodstream and is dependent on the patient's metabolism rate. Derive an expression for the amount of glucose in the bloodstream at time t. [*Hint:* $A(0) = 0$.]

C 65. **Concentration of a Drug in an Organ**   A drug is carried into an organ of volume V cm^3 by a liquid that enters the organ at the rate of a cm^3/sec and leaves it at the rate of b cm^3/sec. The concentration of the drug in the liquid entering the organ is c g/cm^3. If the concentration of the drug in the organ at time t is increasing at the rate of

$$x'(t) = \frac{1}{V}(ac - bx_0)e^{-bt/V}$$

g/cm^3 per second, and the concentration of the drug in the organ initially is x_0 g/cm^3, show that the concentration of the drug in the organ at time t is given by

$$x(t) = \frac{ac}{b} + \left(x_0 - \frac{ac}{b}\right)e^{-bt/V}.$$

SOLUTIONS TO SELF-CHECK EXERCISES 6.2

1. Let $u = 2x + 5$. Then $du = 2\,dx$, or $dx = \frac{1}{2}\,du$. Making the appropriate substitutions, we have

$$\int \sqrt{2x + 5}\,dx = \int \sqrt{u}\left(\frac{1}{2}\,du\right) = \frac{1}{2}\int u^{1/2}\,du$$

$$= \frac{1}{2}\left(\frac{2}{3}\right)u^{3/2} + C$$

$$= \frac{1}{3}(2x + 5)^{3/2} + C.$$

2. Let $u = 2x^3 + 1$, so that $du = 6x^2 \, dx$, or $x^2 \, dx = \frac{1}{6} \, du$. Making the appropriate substitutions, we have

$$\int \frac{x^2}{(2x^3 + 1)^{3/2}} \, dx = \int \frac{\left(\frac{1}{6}\right) du}{u^{3/2}} = \frac{1}{6} \int u^{-3/2} \, du$$

$$= \left(\frac{1}{6}\right)(-2)u^{-1/2} + C$$

$$= -\frac{1}{3}(2x^3 + 1)^{-1/2} + C$$

$$= -\frac{1}{3\sqrt{2x^3 + 1}} + C.$$

3. Let $u = 2x^2 - 1$, so that $du = 4x \, dx$, or $x \, dx = \frac{1}{4} \, du$. Then

$$\int xe^{2x^2-1} \, dx = \frac{1}{4} \int e^u \, du$$

$$= \frac{1}{4}e^u + C$$

$$= \frac{1}{4}e^{2x^2-1} + C.$$

4. Let $C(t)$ denote the concentration of carbon monoxide in the air due to automobile exhaust t years from now. Then

$$C'(t) = f(t) = \frac{8(0.1t + 1)}{300(0.2t^2 + 4t + 64)^{1/3}}$$

$$= \frac{8}{300}(0.1t + 1)(0.2t^2 + 4t + 64)^{-1/3}.$$

Integrating, we find

$$C(t) = \int \frac{8}{300}(0.1t + 1)(0.2t^2 + 4t + 64)^{-1/3} \, dt$$

$$= \frac{8}{300} \int (0.1t + 1)(0.2t^2 + 4t + 64)^{-1/3} \, dt.$$

Let $u = 0.2t^2 + 4t + 64$, so that $du = (0.4t + 4) \, dt = 4(0.1t + 1) \, dt$, or

$$(0.1t + 1) \, dt = \frac{1}{4} \, du.$$

Then
$$C(t) = \frac{8}{300}\left(\frac{1}{4}\right) \int u^{-1/3} \, du$$

$$= \frac{1}{150}\left(\frac{3}{2}u^{2/3}\right) + k$$

$$= 0.01(0.2t^2 + 4t + 64)^{2/3} + k,$$

where k is an arbitrary constant. To determine the value of k, we use the condition $C(0) = 0.16$, obtaining

$$C(0) = 0.16 = 0.01(64)^{2/3} + k,$$
or
$$0.16 = 0.16 + k,$$

and $k = 0$. Therefore,

$$C(t) = 0.01(0.2t^2 + 4t + 64)^{2/3}.$$

| 6.3 | **Area and the Definite Integral** |

An Intuitive Look

Suppose that a certain state's annual rate of petroleum consumption over a four-year period is constant and is given by the function

$$f(t) = 1.2 \qquad (0 \le t \le 4),$$

where t is measured in years and $f(t)$ in millions of barrels per year. Then the state's total petroleum consumption over the period of time in question is

$$(1.2)(4 - 0), \qquad \text{(Rate of consumption} \cdot \text{time elapsed)}$$

FIGURE 6.4

The total petroleum consumption is given by the area of the rectangular region.

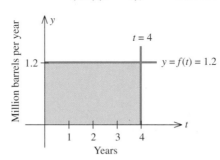

or 4.8 million barrels. If you examine the graph of f shown in Figure 6.4, you will see that this total is just the area of the rectangular region bounded above by the graph of f, below by the t-axis, and to the left and right by the vertical lines $t = 0$ (the y-axis) and $t = 4$, respectively.

Figure 6.5 shows the actual petroleum consumption of a certain New England state over a four-year period from 1990 ($t = 0$) to 1994 ($t = 4$). Observe that the rate of consumption is not constant; that is, the function f is not a constant function. What is the state's total petroleum consumption over this four-year period? It seems reasonable to conjecture that it is given by the "area" of the region bounded above by the graph of f, below by the t-axis, and to the left and right by the vertical lines $t = 0$ and $t = 4$, respectively.

This example raises two questions:

FIGURE 6.5

The daily petroleum consumption is given by the "area" of the shaded region.

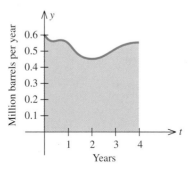

1. What is the "area" of the region shown in Figure 6.5?

2. How do we compute this area?

The Area Problem

The preceding example touches on the second fundamental problem in calculus: Calculate the area of the region bounded by the graph of a nonnegative function f, the x-axis, and the vertical lines $x = a$ and $x = b$ (Figure 6.6). This area is called the **area under the graph of f** on the interval $[a, b]$, or from a to b.

FIGURE 6.6

The area under the graph of f on $[a, b]$

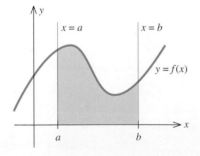

Defining Area—Two Examples

Just as we used the slopes of secant lines (quantities that we could compute) to help us define the slope of the tangent line to a point on the graph of a function, we will now adopt a parallel approach and use the areas of rectangles (quantities that we can compute) to help us define the area under the graph of a function. We begin by looking at a specific example.

EXAMPLE 1   Let $f(x) = x^2$ and consider the region R under the graph of f on the interval $[0, 1]$ (Figure 6.7a). In order to obtain an approximation of the area of R, let us construct four nonoverlapping rectangles as follows: Divide the interval $[0, 1]$ into four subintervals

$$\left[0, \frac{1}{4}\right], \left[\frac{1}{4}, \frac{1}{2}\right], \left[\frac{1}{2}, \frac{3}{4}\right], \quad \text{and} \quad \left[\frac{3}{4}, 1\right]$$

of equal length 1/4. Next, construct four rectangles with these subintervals as bases and with heights given by the values of the function at the midpoints

$$\frac{1}{8}, \frac{3}{8}, \frac{5}{8}, \quad \text{and} \quad \frac{7}{8}$$

of each subinterval. Then each of these rectangles has width 1/4 and height

$$f\left(\frac{1}{8}\right), f\left(\frac{3}{8}\right), f\left(\frac{5}{8}\right), \quad \text{and} \quad f\left(\frac{7}{8}\right),$$

respectively (Figure 6.7b).

FIGURE 6.7
The area of the region under the graph of f on [a, b] in (a) is approximated by the sum of the areas of the four rectangles in (b).

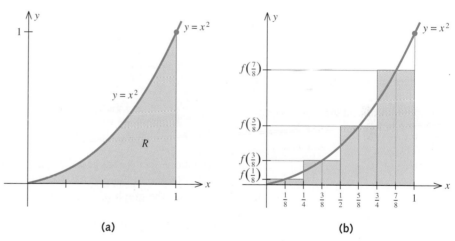

(a)   (b)

If we approximate the area A of S by the sum of the areas of the four rectangles, we obtain

$$A \approx \frac{1}{4}f\left(\frac{1}{8}\right) + \frac{1}{4}f\left(\frac{3}{8}\right) + \frac{1}{4}f\left(\frac{5}{8}\right) + \frac{1}{4}f\left(\frac{7}{8}\right)$$

$$= \frac{1}{4}\left[f\left(\frac{1}{8}\right) + f\left(\frac{3}{8}\right) + f\left(\frac{5}{8}\right) + f\left(\frac{7}{8}\right)\right]$$

$$= \frac{1}{4}\left[\left(\frac{1}{8}\right)^2 + \left(\frac{3}{8}\right)^2 + \left(\frac{5}{8}\right)^2 + \left(\frac{7}{8}\right)^2\right] \qquad \text{(Recall that } f(x) = x^2.)$$

$$= \frac{1}{4}\left(\frac{1}{64} + \frac{9}{64} + \frac{25}{64} + \frac{49}{64}\right) = \frac{21}{64},$$

or approximately 0.328125 square units.   ∎

Following the procedure of Example 1, we can obtain approximations of the area of the region R using any number n of rectangles ($n = 4$ in Example 1). Figure 6.8a shows the approximation of the area A of R using 8 rectangles ($n = 8$), and Figure 6.8b shows the approximation of the area A of R using 16 rectangles.

FIGURE 6.8

As n increases, the number of rectangles increases and the approximation improves.

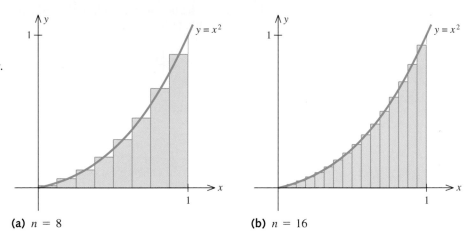

(a) $n = 8$　　　　　　　　　　　　**(b)** $n = 16$

These figures suggest that the approximations seem to get better as n increases. This is borne out by the results given in Table 6.1, which were obtained using a computer.

TABLE 6.1

Number of rectangles n	4	8	16	32	64	100	200
Approximation of A	0.328125	0.332031	0.333008	0.333252	0.333313	0.333325	0.333331

Our computations seem to suggest that the approximations approach the number $1/3$ as n gets larger and larger. This result suggests that we *define* the area of the region under the graph of $f(x) = x^2$ on the interval $[0, 1]$ to be $1/3$ square units.

In Example 1 we chose the *midpoint* of each subinterval as the point at which to evaluate $f(x)$ to obtain the height of the approximating rectangle. Let us consider another example, this time choosing the *left endpoint* of each subinterval.

EXAMPLE 2　　Let R be the region under the graph of $f(x) = 16 - x^2$ on the interval $[1, 3]$. Find an approximation of the area A of R using four subintervals of $[1, 3]$ of equal length and picking the left endpoint of each subinterval to evaluate $f(x)$ to obtain the height of the approximating rectangle.

Solution　　The graph of f is sketched in Figure 6.9a. Since the length of $[1, 3]$ is 2, we see that the length of each subinterval is $2/4$, or $1/2$. Therefore, the four subintervals are

$$\left[1, \frac{3}{2}\right], \quad \left[\frac{3}{2}, 2\right] \quad \left[2, \frac{5}{2}\right], \quad \text{and} \quad \left[\frac{5}{2}, 3\right].$$

FIGURE 6.9
*The area of R in (a) is
approximated by the sum
of the areas of the four
rectangles in (b).*

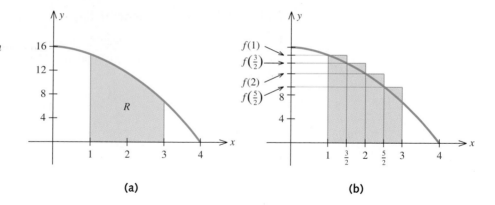

(a)   (b)

The left endpoints of these subintervals are 1, 3/2, 2, and 5/2, respectively, so the heights of the approximating rectangles are $f(1)$, $f\left(\frac{3}{2}\right)$, $f(2)$, and $f\left(\frac{5}{2}\right)$, respectively (Figure 6.9b). Therefore, the required approximation is

$$A \approx \frac{1}{2}f(1) + \frac{1}{2}f\left(\frac{3}{2}\right) + \frac{1}{2}f(2) + \frac{1}{2}f\left(\frac{5}{2}\right)$$

$$= \frac{1}{2}\left[f(1) + f\left(\frac{3}{2}\right) + f(2) + f\left(\frac{5}{2}\right)\right]$$

$$= \frac{1}{2}\left\{[16 - (1)^2] + \left[16 - \left(\frac{3}{2}\right)^2\right]\right.$$

$$\left.+ [16 - (2)^2] + \left[16 - \left(\frac{5}{2}\right)^2\right]\right\} \qquad \text{[Recall that } f(x) = 16 - x^2.\text{]}$$

$$= \frac{1}{2}\left(15 + \frac{55}{4} + 12 + \frac{39}{4}\right) = \frac{101}{4},$$

or approximately 25.25 square units.   ∎

Table 6.2 shows the approximations of the area A of the region R of Example 2 when n rectangles are used for the approximation and the heights of the approximating rectangles are found by evaluating $f(x)$ at the left endpoints.

TABLE 6.2

Number of rectangles n	4	10	100	1,000	10,000	50,000	100,000
Approximation of A	25.2500	24.1200	23.4132	23.3413	23.3341	23.3335	23.3334

Once again we see that the approximations seem to approach a unique number as n gets larger and larger—this time the number is $23\frac{1}{3}$. This result suggests that we *define* the area of the region under the graph of $f(x) = 16 - x^2$ on the interval [1, 3] to be $23\frac{1}{3}$ square units.

Defining Area—The General Case

Examples 1 and 2 point the way to defining the area A under the graph of an arbitrary but continuous and nonnegative function f on an interval $[a, b]$ (Figure 6.10a).

FIGURE 6.10

The area of the region under the graph of f on $[a, b]$ in (a) is approximated by the sum of the areas of the n rectangles shown in (b).

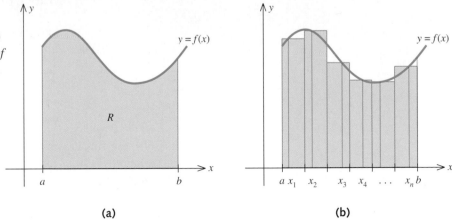

(a)   (b)

Divide the interval $[a, b]$ into n subintervals of equal length $\Delta x = (b - a)/n$. Next, pick n arbitrary points $x_1, x_2, \ldots, x_n$, called *representative points,* from the first, second, $\ldots$, and nth subintervals, respectively (Figure 6.10b). Then, approximating the area A of the region R by the n rectangles of width Δx and heights $f(x_1), f(x_2), \ldots, f(x_n)$, so that the areas of the rectangles are $f(x_1)\Delta x$, $f(x_2)\Delta x, \ldots, f(x_n)\Delta x$, we have

$$A \approx f(x_1)\Delta x + f(x_2)\Delta x + \cdots + f(x_n)\Delta x.$$

The sum on the right-hand side of this expression is called a **Riemann sum** in honor of the German mathematician Bernhard Riemann (1826–1866). Now, as the earlier examples seem to suggest, the Riemann sum will approach a unique number as n becomes arbitrarily large.[†] We define this number to be the area A of the region R.

THE AREA UNDER THE GRAPH OF A FUNCTION	Let f be a nonnegative continuous function on $[a, b]$. Then the **area** of the region under the graph of f is $$A = \lim_{n \to \infty} [f(x_1) + f(x_2) + \cdots + f(x_n)]\Delta x, \qquad (4)$$ where $x_1, x_2, \ldots, x_n$ are arbitrary points in the n subintervals of $[a, b]$ of equal width $\Delta x = (b - a)/n$.

The Definite Integral

As we have just seen, the area under the graph of a continuous *nonnegative*

[†]Even though we chose the representative points to be the midpoints of the subintervals in Example 1 and the left endpoints in Example 2, it can be shown that each of the respective sums will always approach a unique number as n approaches infinity.

function f on an interval $[a, b]$ is defined by the limit of the Riemann sum

$$\lim_{n \to \infty} [f(x_1)\Delta x + f(x_2)\Delta x + \cdots + f(x_n)\Delta x].$$

We now turn our attention to the study of limits of Riemann sums involving functions that are not necessarily nonnegative. Such limits arise in many applications of calculus.

For example, the calculation of the distance covered by a body traveling along a straight line involves evaluating a limit of this form. The computation of the total revenue realized by a company over a certain time period, the calculation of the total amount of electricity consumed in a typical home over a 24-hour period, the average concentration of a drug in a body over a certain interval of time, and the volume of a solid—all involve limits of this type.

We begin with the following definition.

DEFINITE INTEGRAL

Let f be defined on $[a, b]$. If

$$\lim_{n \to \infty} [f(x_1)\Delta x + f(x_2)\Delta x + \cdots + f(x_n)\Delta x]$$

exists for all choices of representative points $x_1, x_2, \ldots, x_n$ in the n subintervals of $[a, b]$ of equal width $\Delta x = (b - a)/n$, then this limit is called the **definite integral of f from a to b** and is denoted by $\int_a^b f(x)\, dx$. Thus,

$$\int_a^b f(x)\, dx = \lim_{n \to \infty} [f(x_1)\Delta x + f(x_2)\Delta x + \cdots + f(x_n)\Delta x]. \quad (5)$$

The number a is the **lower limit of integration** and the number b is the **upper limit of integration.**

REMARKS

1. If f is nonnegative, then the limit in (5) is the same as the limit in (4) and, therefore, the definite integral gives the area under the graph of f on $[a, b]$.
2. The limit in (5) is denoted by the integral sign $\int$ because, as we will see later, the definite integral and the antiderivative of a function f are related.
3. It is important to realize that the definite integral $\int_a^b f(x)\, dx$ is a number, whereas the indefinite integral $\int f(x)\, dx$ represents a family of functions (the antiderivatives of f).
4. If the limit in (5) exists, we say that f is **integrable** on the interval $[a, b]$.

☐

When Is a Function Integrable?

The following theorem, which we state without proof, guarantees that a continuous function is integrable.

INTEGRABILITY OF A FUNCTION

Let f be continuous on $[a, b]$. Then f is integrable on $[a, b]$; that is, the definite integral $\int_a^b f(x)\, dx$ exists.

Geometric Interpretation of the Definite Integral

If f is nonnegative and integrable on $[a, b]$, then we have the following geometric interpretation of the definite integral $\int_a^b f(x)\, dx$.

GEOMETRIC INTERPRETATION OF $\int_a^b f(x)\, dx$ FOR $f(x) \geq 0$ ON $[a, b]$	If f is nonnegative and continuous on $[a, b]$, then $$\int_a^b f(x)\, dx \qquad (6)$$ is equal to the area of the region under the graph of f on $[a, b]$ (see Figure 6.11).

FIGURE 6.11

If $f(x) \geq 0$ on $[a, b]$, then $\int_a^b f(x)\ dx$ = area under the graph of f on $[a, b]$.

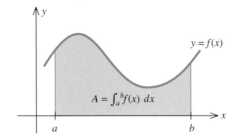

Next, let's extend our geometric interpretation of the definite integral to include the case where f assumes both positive as well as negative values on $[a, b]$. Consider a typical Riemann sum of the function f,

$$f(x_1)\Delta x + f(x_2)\Delta x + \cdots + f(x_n)\Delta x,$$

corresponding to a partition of $[a, b]$ into n subintervals of equal width $(b - a)/n$, where $x_1, x_2, \ldots, x_n$ are representative points in the subintervals. The sum consists of n terms in which a positive term corresponds to the area of a rectangle of height $f(x_k)$ [for some positive integer k] lying above the x-axis, and a negative term corresponds to the area of a rectangle of height $-f(x_k)$ lying below the x-axis. (See Figure 6.12, which depicts a situation with $n = 6$.)

FIGURE 6.12

The positive terms in the Riemann sum are associated with the areas of the rectangles that lie above the x-axis, and the negative terms are associated with the areas of those that lie below the x-axis.

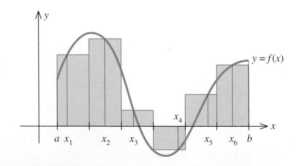

As *n* gets larger and larger, the sums of the areas of the rectangles lying above the *x*-axis seem to give a better and better approximation of the area of the region lying above the *x*-axis (see Figure 6.13). Similarly, the sums of the areas of those rectangles lying below the *x*-axis seem to give a better and better approximation of the area of the region lying below the *x*-axis.

FIGURE 6.13

As n gets larger, the approximations get better. Here n = 12, and we are approximating with twice as many rectangles as in Figure 6.12.

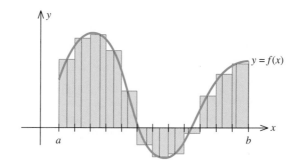

These observations suggest the following geometric interpretation of the definite integral for an arbitrary continuous function on an interval [*a*, *b*].

GEOMETRIC INTERPRETATION OF $\int_a^b f(x)\,dx$ ON [*a*, *b*]	If f is continuous on [*a*, *b*], then $$\int_a^b f(x)\,dx$$ is equal to the area of the region above [*a*, *b*] minus the area of the region below [*a*, *b*] (see Figure 6.14).

FIGURE 6.14

$\int_a^b f(x)\,dx$ = *area of* R_1 − *area of* R_2 + *area of* R_3.

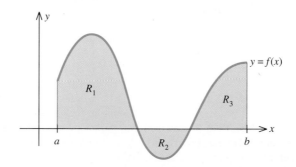

SELF-CHECK EXERCISE 6.3

Find an approximation of the area of the region *R* under the graph of $f(x) = 2x^2 + 1$ on the interval [0, 3] using four subintervals of [0, 3] of equal length and picking the midpoint of each subinterval as a representative point.

The Solution to Self-Check Exercise 6.3 can be found on page 374.

6.3 EXERCISES

In Exercises 1 and 2, find an approximation of the area of the region R under the graph of f by computing the Riemann sum of f corresponding to the partition of the interval into the subintervals shown in the accompanying figures. In each case, use the midpoints of the subintervals as the representative points.

1.

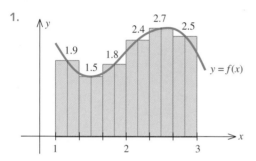

2.

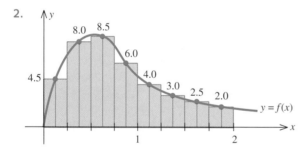

3. Let $f(x) = 3x$.
   a. Sketch the region R under the graph of f on the interval [0, 2] and find its exact area using geometry.
   b. Use a Riemann sum with four subintervals of equal length ($n = 4$) to approximate the area of R. Choose the representative points to be the left endpoints of the subintervals.
   c. Repeat (b) with eight subintervals of equal length ($n = 8$).
   d. Compare the approximations obtained in (b) and (c) with the exact area found in (a). Do the approximations improve with larger n?

4. Repeat Exercise 3 choosing the representative points to be the right endpoints of the subintervals.

5. Let $f(x) = 4 - 2x$.
   a. Sketch the region R under the graph of f on the interval [0, 2] and find its exact area using geometry.

   b. Use a Riemann sum with five subintervals of equal length ($n = 5$) to approximate the area of R. Choose the representative points to be the left endpoints of the subintervals.
   c. Repeat (b) with ten subintervals of equal length ($n = 10$).
   d. Compare the approximations obtained in (b) and (c) with the exact area found in (a). Do the approximations improve with larger n?

6. Repeat Exercise 5 choosing the representative points to be the right endpoints of the subintervals.

7. Let $f(x) = x^2$ and compute the Riemann sum of f over the interval [2, 4] using
   a. two subintervals of equal length ($n = 2$).
   b. five subintervals of equal length ($n = 5$).
   c. ten subintervals of equal length ($n = 10$).
   In each case, choose the representative points to be the midpoints of the subintervals. Can you guess at the area of the region under the graph of f on the interval [2, 4]?

8. Repeat Exercise 7 choosing the representative points to be the left endpoints of the subintervals.

9. Repeat Exercise 7 choosing the representative points to be the right endpoints of the subintervals.

10. Let $f(x) = x^3$ and compute the Riemann sum of f over the interval [0, 1] using
    a. two subintervals of equal length ($n = 2$).
    b. five subintervals of equal length ($n = 5$).
    c. ten subintervals of equal length ($n = 10$).
    In each case, choose the representative points to be the midpoints of the subintervals. Can you guess at the area of the region under the graph of f on the interval [0, 1]?

11. Repeat Exercise 10 choosing the representative points to be the left endpoints of the subintervals.

12. Repeat Exercise 10 choosing the representative points to be the right endpoints of the subintervals.

In Exercises 13–16, find an approximation of the area of the region R under the graph of the given function f on the interval [a, b]. In each case, use n subintervals and choose the representative points as indicated.

13. $f(x) = x^2 + 1$; $[0, 2]$; $n = 5$; midpoints.

14. $f(x) = 4 - x^2$; $[-1, 2]$; $n = 6$; left endpoints.

15. $f(x) = \dfrac{1}{x}$; $[1, 3]$; $n = 4$; right endpoints.

16. $f(x) = e^x$; $[0, 3]$; $n = 5$; midpoints.

17. Real Estate Figure (a) shows a vacant lot with a 100-foot frontage in a development. To estimate its area, we introduce a coordinate system so that the x-axis coincides with the edge of the straight road forming the lower boundary of the property as shown in Figure (b). Then thinking of the upper boundary of the property as the graph of a continuous function f over the interval $[0, 100]$, we see that the problem is mathematically equivalent to that of finding the area under the graph of f on $[0, 100]$. To estimate the area of the lot using a Riemann sum, we divide the interval $[0, 100]$ into five equal subintervals of length 20 feet. Then, using surveyor's equipment, we measure the distance from the midpoint of each of these subintervals to the upper boundary of the property. These measurements give the values of $f(x)$ at $x = 10, 30, 50, 70,$ and 90. What is the approximate area of the lot?

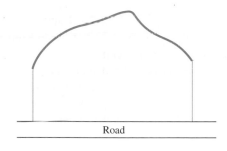

Road

(a)

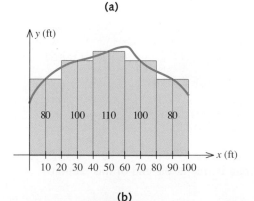

(b)

18. Real Estate Use the technique of Exercise 17 to obtain an estimate of the area of the vacant lot shown in the accompanying figure.

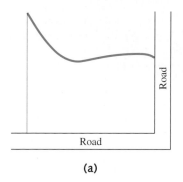

(a)

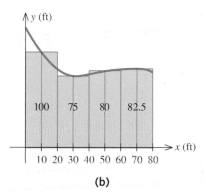

(b)

C *In Exercises 19–22, use a computer to compute the Riemann sum of f over the interval [a, b] using n subintervals and choosing the specified representative points.*

19. $f(x) = x^3 + 2x^2 + 1$; $[0, 2]$; $n = 30$; midpoints of subintervals

20. $f(x) = \dfrac{2x}{1 + x^2}$; $[1, 4]$; $n = 30$; left endpoints of subintervals

21. $f(x) = \sqrt{1 + 2x^2}$; $[0, 4]$; $n = 40$; right endpoints of subintervals

22. $f(x) = \dfrac{\sqrt{x}}{4 + x^2}$; $[0, 2]$; $n = 20$; midpoints of subintervals

SOLUTION TO
SELF-CHECK
EXERCISE 6.3

The length of each subinterval is 3/4. Therefore, the four subintervals are

$$\left[0, \frac{3}{4}\right], \quad \left[\frac{3}{4}, \frac{3}{2}\right], \quad \left[\frac{3}{2}, \frac{9}{4}\right], \quad \text{and} \quad \left[\frac{9}{4}, 3\right].$$

The representative points are $\frac{3}{8}$, $\frac{9}{8}$, $\frac{15}{8}$, and $\frac{21}{8}$, respectively. Therefore, the required approximation is

$$
\begin{aligned}
A &= \frac{3}{4}f\left(\frac{3}{8}\right) + \frac{3}{4}f\left(\frac{9}{8}\right) + \frac{3}{4}f\left(\frac{15}{8}\right) + \frac{3}{4}f\left(\frac{21}{8}\right) \\
&= \frac{3}{4}\left[f\left(\frac{3}{8}\right) + f\left(\frac{9}{8}\right) + f\left(\frac{15}{8}\right) + f\left(\frac{21}{8}\right)\right] \\
&= \frac{3}{4}\left\{\left[2\left(\frac{3}{8}\right)^2 + 1\right] + \left[2\left(\frac{9}{8}\right)^2 + 1\right] + \left[2\left(\frac{15}{8}\right)^2 + 1\right] + \left[2\left(\frac{21}{8}\right)^2 + 1\right]\right\} \\
&= \frac{3}{4}\left(\frac{41}{32} + \frac{113}{32} + \frac{257}{32} + \frac{473}{32}\right) = \frac{663}{32},
\end{aligned}
$$

or approximately 20.72 square units.

6.4   # The Fundamental Theorem of Calculus

The Fundamental Theorem of Calculus

In Section 6.3 we defined the definite integral of an arbitrary continuous function on an interval $[a, b]$ as a limit of Riemann sums. Calculating the value of a definite integral by actually taking the limit of such sums is tedious and in most cases impractical. It is important to realize that the numerical results we obtained in Examples 1 and 2 of Section 6.3 were *approximations* of the respective areas of the regions in question, even though these results enabled us to *conjecture* what the actual areas might be. Fortunately, there is a much better way of finding the exact value of a definite integral.

The following theorem shows how to evaluate the definite integral of a continuous function provided we can find an antiderivative of that function. Because of its importance in establishing the relationship between differentiation and integration, this theorem—discovered independently by Sir Isaac Newton (1642–1727) in England and Gottfried Wilhelm Leibniz (1646–1716) in Germany—is called the **Fundamental Theorem of Calculus.**

THE FUNDAMENTAL
THEOREM OF
CALCULUS

Let f be continuous on $[a, b]$. Then

$$\int_a^b f(x)\, dx = F(b) - F(a) \qquad (7)$$

where F is any antiderivative of f; that is $F'(x) = f(x)$.

We will explain why this theorem is true at the end of this section.

When applying the Fundamental Theorem of Calculus, it is convenient to use the notation

$$F(x)\Big|_a^b = F(b) - F(a).$$

For example, using this notation, (7) is written

$$\int_a^b f(x)\, dx = F(x)\Big|_a^b = F(b) - F(a).$$

EXAMPLE 1   Let R be the region under the graph of $f(x) = x$ on the interval $[1, 3]$. Use the Fundamental Theorem of Calculus to find the area A of R and verify your result by elementary means.

Solution   The region R is shown in Figure 6.15a. Since f is nonnegative on $[1, 3]$, the area of R is given by the definite integral of f from 1 to 3; that is

$$A = \int_1^3 x\, dx.$$

FIGURE 6.15

The area of R can be computed in two different ways.

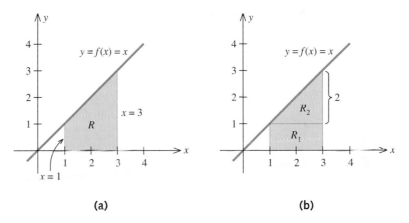

(a)   (b)

In order to evaluate the definite integral, observe that an antiderivative of $f(x) = x$ is $F(x) = \frac{1}{2}x^2 + C$, where C is an arbitrary constant. Therefore, by the Fundamental Theorem of Calculus, we have

$$A = \int_1^3 x\, dx = \frac{1}{2}x^2 + C\Big|_1^3$$

$$= \left(\frac{9}{2} + C\right) - \left(\frac{1}{2} + C\right) = 4 \text{ square units.}$$

To verify this result by elementary means, observe that the area A is the area of the rectangle R_1 (width × height) plus the area of the triangle R_2 $\left(\frac{1}{2} \text{ base} \times \text{height}\right)$ [see Figure 6.15b]; that is

$$2(1) + \frac{1}{2}(2)(2) = 2 + 2 = 4,$$

which agrees with the result obtained earlier.   ■

Observe that in evaluating the definite integral in Example 1, the constant of integration "dropped out." This is true in general, for if $F(x) + C$ denotes an antiderivative of some function f, then

$$F(x) + C \Big|_a^b = [F(b) + C] - [F(a) + C]$$
$$= F(b) + C - F(a) - C$$
$$= F(b) - F(a).$$

With this fact in mind, we may, in all future computations involving the evaluations of a definite integral, drop the constant of integration from our calculations.

Finding the Area Under a Curve

Having seen how effective the Fundamental Theorem of Calculus is in helping us find the area of simple regions, we will now use it to find the area of more complicated regions.

EXAMPLE 2   In Section 6.3, we conjectured that the area of the region R under the graph of $f(x) = x^2$ on the interval $[0, 1]$ was $1/3$ square units. Use the Fundamental Theorem of Calculus to verify this conjecture.

Solution   The region R is reproduced in Figure 6.16. Observe that f is nonnegative on $[0, 1]$, so the area of R is given by $A = \int_0^1 x^2\, dx$. Since an antiderivative of $f(x) = x^2$ is $F(x) = \frac{1}{3}x^3$, we see, using the Fundamental Theorem of Calculus, that

$$A = \int_0^1 x^2\, dx = \frac{1}{3}x^3 \Big|_0^1 = \frac{1}{3}(1) - \frac{1}{3}(0) = \frac{1}{3} \text{ square units,}$$

as we wished to show.

FIGURE 6.16
The area of R is $\int_0^1 x^2\, dx = \frac{1}{3}$.

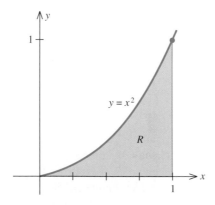

EXAMPLE 3   Find the area of the region R under the graph of $y = x^2 + 1$ from $x = -1$ to $x = 2$.

FIGURE 6.17
Area of $R = \int_{-1}^{2} (x^2 + 1)\, dx$.

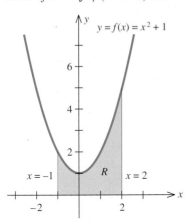

Solution The region R under consideration is shown in Figure 6.17. Using the Fundamental Theorem of Calculus, we find that the required area is

$$\int_{-1}^{2} (x^2 + 1)\, dx = \left(\frac{1}{3}x^3 + x \right)\Big|_{-1}^{2}$$

$$= \left[\frac{1}{3}(8) + 2 \right] - \left[\frac{1}{3}(-1)^3 + (-1) \right] = 6,$$

or 6 square units. ■

Evaluating Definite Integrals

In Examples 4 and 5, we make use of the rules of integration of Section 6.1 to help us evaluate the definite integrals.

EXAMPLE 4 Evaluate $\int_{1}^{3} (3x^2 + e^x)\, dx$.

Solution

$$\int_{1}^{3} (3x^2 + e^x)\, dx = x^3 + e^x \Big|_{1}^{3}$$

$$= (27 + e^3) - (1 + e) = 26 + e^3 - e. \quad ■$$

EXAMPLE 5 Evaluate $\int_{1}^{2} \left(\frac{1}{x} - \frac{1}{x^2} \right) dx$.

Solution

$$\int_{1}^{2} \left(\frac{1}{x} - \frac{1}{x^2} \right) dx = \int_{1}^{2} \left(\frac{1}{x} - x^{-2} \right) dx$$

$$= \ln |x| + \frac{1}{x}\Big|_{1}^{2}$$

$$= \left(\ln 2 + \frac{1}{2} \right) - (\ln 1 + 1)$$

$$= \ln 2 - \frac{1}{2}. \qquad \text{(Recall } \ln 1 = 0.)$$

■

Applications

EXAMPLE 6 The management of Staedtler Office Equipment has determined that the daily marginal cost function associated with producing battery-operated pencil sharpeners is given by

$$C'(x) = 0.000006x^2 - 0.006x + 4,$$

where $C'(x)$ is measured in dollars per unit and x denotes the number of units produced. Management has also determined that the daily fixed cost incurred in producing these pencil sharpeners is $100. Find Staedtler's daily total cost for producing (a) the first 500 units and (b) the 201st through 400th units.

Solution

a. Since $C'(x)$ is the marginal cost function, its antiderivative $C(x)$ is the total cost function. The daily fixed cost incurred in producing the pencil sharpeners is $C(0)$ dollars. Since the daily fixed cost is given as $100, we have $C(0) = 100$. We are required to find $C(500)$. Let us compute $C(500) - C(0)$, the net change in the total cost function $C(x)$ over the interval $[0, 500]$. Using the Fundamental Theorem of Calculus, we find

$$C(500) - C(0) = \int_0^{500} C'(x)\, dx$$

$$= \int_0^{500} (0.000006x^2 - 0.006x + 4)\, dx$$

$$= 0.000002x^3 - 0.003x^2 + 4x \Big|_0^{500}$$

$$= [0.000002(500)^3 - 0.003(500)^2 + 4(500)]$$
$$\quad - [0.000002(0)^3 - 0.003(0)^2 + 4(0)]$$

$$= 1500.$$

Therefore $C(500) = 1500 + C(0) = 1500 + 100 = 1600$, so the total cost incurred daily by Staedtler in producing 500 pencil sharpeners is $1600.

b. The daily total cost incurred by Staedtler in producing the 201st through 400th units of battery-operated pencil sharpeners is given by

$$C(400) - C(200) = \int_{200}^{400} C'(x)\, dx$$

$$= \int_{200}^{400} (0.000006x^2 - 0.006x + 4)\, dx$$

$$= 0.000002x^3 - 0.003x^2 + 4x \Big|_{200}^{400}$$

$$= 552, \text{ or } \$552. \qquad \blacksquare$$

Since $C'(x)$ is nonnegative for x in the interval $[0, \infty)$, we have the following geometrical interpretation of the two definite integrals in Example 6: $\int_0^{500} C'(x)\, dx$ is the area of the region under the graph of the function C' from $x = 0$ to $x = 500$, shown in Figure 6.18a, and $\int_{200}^{400} C'(x)\, dx$ is the area of the region shown in Figure 6.18b from $x = 200$ to $x = 400$.

FIGURE 6.18

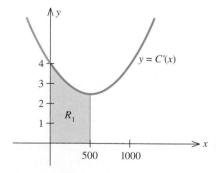

(a) Area of $R_1 = \int_0^{500} C'(x)\, dx$.

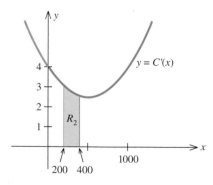

(b) Area of $R_2 = \int_{200}^{400} C'(x)\, dx$.

EXAMPLE 7 An efficiency study conducted for the Elektra Electronics Company showed that the rate at which Space Commander walkie-talkies are assembled by the average worker t hours after starting work at 8 A.M. is given by the function

$$f(t) = -3t^2 + 12t + 15 \qquad (0 \le t \le 4).$$

Determine how many walkie-talkies can be assembled by the average worker in the first hour of the morning shift. How many units can the average worker be expected to assemble in the second hour of the morning shift?

Solution Let $N(t)$ denote the number of walkie-talkies assembled by the average worker t hours after starting work in the morning shift. Then we have

$$N'(t) = f(t) = -3t^2 + 12t + 15.$$

Therefore, the number of units assembled by the average worker in the first hour of the morning shift is

$$N(1) - N(0) = \int_0^1 N'(t)\, dt = \int_0^1 (-3t^2 + 12t + 15)\, dt$$

$$= -t^3 + 6t^2 + 15t \Big|_0^1 = -1 + 6 + 15$$

$$= 20, \quad \text{or} \quad 20 \text{ units.} \qquad \blacksquare$$

EXAMPLE 8 A certain city's rate of electricity consumption is expected to grow exponentially with a growth constant of $k = 0.04$. If the present rate of consumption is 40 million kilowatt-hours (kwh) per year, what should the total production of electricity be over the next three years in order to meet the projected demand?

Solution If $R(t)$ denotes the expected rate of consumption of electricity t years from now, then

$$R(t) = 40e^{0.04t}$$

million kwh per year. Next, if $C(t)$ denotes the expected total consumption of electricity over a period of t years, then

$$C'(t) = R(t).$$

Therefore, the total consumption of electricity expected over the next three years is given by

$$\int_0^3 C'(t)\, dt = \int_0^3 40e^{0.04t}\, dt$$

$$= \frac{40}{0.04} e^{0.04t} \Big|_0^3$$

$$= 1000(e^{0.12} - 1)$$

$$= 127.5, \quad \text{or} \quad 127.5 \text{ million kwh,}$$

the amount that must be produced over the next three years in order to meet the demand. $\blacksquare$

Validity of the Fundamental Theorem of Calculus

In order to demonstrate the plausibility of the Fundamental Theorem of Calculus for the case where f is nonnegative on an interval $[a, b]$, let us define an "area function" A as follows. Let $A(t)$ denote the area of the region R under the graph of $y = f(x)$ from $x = a$ to $x = t$, where $a \leq t \leq b$ (Figure 6.19).

If h is a small positive number, then $A(t + h)$ is the area of the region under the graph of $y = f(x)$ from $x = a$ to $x = t + h$. Therefore, the difference

$$A(t + h) - A(t)$$

is the area under the graph of $y = f(x)$ from $x = t$ to $x = t + h$ (Figure 6.20).

FIGURE 6.19
$A(t) =$ the area under the graph of f from $x = a$ to $x = t$.

FIGURE 6.20
$A(t + h) - A(t) =$ the area under the graph of f from $x = t$ to $x = t + h$.

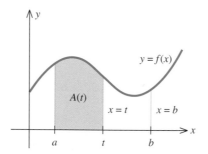

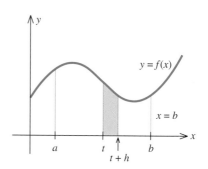

FIGURE 6.21
Area of rectangle $= h \cdot f(t)$.

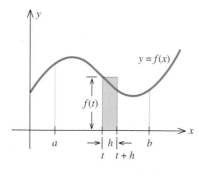

Now, the area of this last region can be approximated by the area of the rectangle of width h and height $f(t)$, that is, by the expression $h \cdot f(t)$ [Figure 6.21]. Thus,

$$A(t + h) - A(t) \approx h \cdot f(t),$$

where the approximations improve as h is taken to be smaller and smaller.

Dividing both sides of the foregoing relationship by h, we obtain

$$\frac{A(t + h) - A(t)}{h} \approx f(t).$$

Taking the limit as h approaches zero, we find, by the definition of the derivative, that the left-hand side is

$$\lim_{h \to 0} \frac{A(t + h) - A(t)}{h} = A'(t).$$

The right-hand side, which is independent of h, remains constant throughout the limiting process. Because the approximation becomes exact as h approaches zero, we find that

$$A'(t) = f(t).$$

Since the foregoing equation holds for all values of t in the interval $[a, b]$, we have shown that the *area function A* is an antiderivative of the function $f(x)$.

By Theorem 1 of Section 6.1, we conclude that $A(x)$ must have the form

$$A(x) = F(x) + C,$$

where F is any antiderivative of f and C is an arbitrary constant. To determine the value of C, observe that $A(a) = 0$. This condition implies that

$$A(a) = F(a) + C = 0,$$

FIGURE 6.22
The area of R is given by A(b).

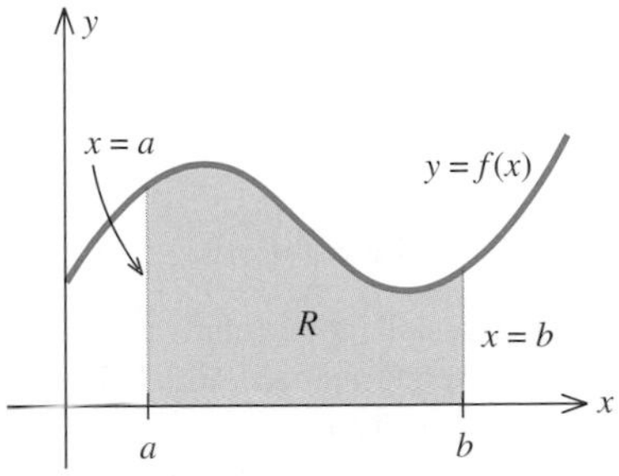

or $C = -F(a)$. Next, since the area of the region R is $A(b)$ [Figure 6.22], we see that the required area is

$$A(b) = F(b) + C$$
$$= F(b) - F(a).$$

Since the area of the region R is

$$\int_a^b f(x)\,dx,$$

we have

$$\int_a^b f(x)\,dx = F(b) - F(a),$$

as we set out to show.

**SELF-CHECK
EXERCISES 6.4**

1. Evaluate

$$\int_0^2 (x + e^x)\,dx.$$

c 2. The daily marginal profit function associated with producing and selling Texa-Pep hot sauce is

$$P'(x) = -0.000006x^2 + 6,$$

where x denotes the number of cases (each case contains 24 bottles) produced and sold daily and $P'(x)$ is measured in dollars per unit. The fixed cost is $400.
   a. What is the total profit realizable from producing and selling 1000 cases of Texa-Pep per day?
   b. What is the additional profit realizable if the production and sale of Texa-Pep is increased from 1000 to 1200 cases per day?

Solutions to Self-Check Exercises 6.4 can be found on page 383.

6.4 EXERCISES

In Exercises 1–4, find the area of the region under the graph of the given function f on the interval [a, b] using the Fundamental Theorem of Calculus. Then verify your result using geometry.

3. $f(x) = 2x$; [1, 3]   **4.** $f(x) = -\frac{1}{4}x + 1$; [1, 4]

In Exercises 5–16, find the area of the region under the graph of the given function f on the interval [a, b].

1. $f(x) = 2$; [1, 4]   **2.** $f(x) = 4$; [-1, 2]

5. $f(x) = 2x + 3$; [-1, 2]   **6.** $f(x) = 4x - 1$; [2, 4]

7. $f(x) = -x^2 + 4$; $[-1, 2]$

8. $f(x) = 4x - x^2$; $[0, 4]$

9. $f(x) = \dfrac{1}{x}$; $[1, 2]$ 10. $f(x) = \dfrac{1}{x^2}$; $[2, 4]$

11. $f(x) = \sqrt{x}$; $[1, 9]$ 12. $f(x) = x^3$; $[1, 3]$

13. $f(x) = 1 - \sqrt[3]{x}$; $[-8, -1]$

14. $f(x) = \dfrac{1}{\sqrt{x}}$; $[1, 9]$

15. $f(x) = e^x$; $[0, 2]$ 16. $f(x) = e^x - x$; $[1, 2]$

In Exercises 17–40, evaluate the given definite integral.

17. $\displaystyle\int_2^4 3\,dx$ 18. $\displaystyle\int_{-1}^2 -2\,dx$

19. $\displaystyle\int_1^3 (2x + 3)\,dx$ 20. $\displaystyle\int_{-1}^0 (4 - x)\,dx$

21. $\displaystyle\int_{-1}^3 2x^2\,dx$ 22. $\displaystyle\int_0^3 8x^3\,dx$

23. $\displaystyle\int_{-2}^2 (x^2 - 1)\,dx$ 24. $\displaystyle\int_1^4 \sqrt{u}\,du$

25. $\displaystyle\int_1^8 4x^{1/3}\,dx$ 26. $\displaystyle\int_1^4 2x^{-3/2}\,dx$

27. $\displaystyle\int_0^1 (x^3 - 2x^2 + 1)\,dx$ 28. $\displaystyle\int_1^2 (t^5 - t^3 + 1)\,dt$

29. $\displaystyle\int_2^4 \dfrac{1}{x}\,dx$ 30. $\displaystyle\int_1^3 \dfrac{2}{x}\,dx$

31. $\displaystyle\int_0^4 x(x^2 - 1)\,dx$ 32. $\displaystyle\int_0^2 (x - 4)(x - 1)\,dx$

33. $\displaystyle\int_1^3 (t^2 - t)^2\,dt$ 34. $\displaystyle\int_{-1}^1 (x^2 - 1)^2\,dx$

35. $\displaystyle\int_{-3}^{-1} \dfrac{1}{x^2}\,dx$ 36. $\displaystyle\int_1^2 \dfrac{2}{x^3}\,dx$

37. $\displaystyle\int_1^4 \left(\sqrt{x} - \dfrac{1}{\sqrt{x}}\right)dx$ 38. $\displaystyle\int_0^1 \sqrt{2x}(\sqrt{x} + \sqrt{2})\,dx$

39. $\displaystyle\int_1^4 \dfrac{3x^3 - 2x^2 + 4}{x^2}\,dx$ 40. $\displaystyle\int_1^2 \left(1 + \dfrac{1}{u} + \dfrac{1}{u^2}\right)du$

C **41. Marginal Cost**   A division of Ditton Industries manufactures a deluxe toaster oven. Management has determined that the daily marginal cost function associated with producing these toaster ovens is given by

$$C'(x) = 0.0003x^2 - 0.12x + 20,$$

where $C'(x)$ is measured in dollars per unit and x denotes the number of units produced. Management has also determined that the daily fixed cost incurred in the production is $800.

a. Find the total cost incurred by Ditton in producing the first 300 units of these toaster ovens per day.

b. What is the total cost incurred by Ditton in producing the 201st through 300th units per day?

C **42. Marginal Revenue**   The management of Ditton Industries has determined that the daily marginal revenue function associated with selling x units of their deluxe toaster ovens is given by

$$R'(x) = -0.1x + 40,$$

where $R'(x)$ is measured in dollars per unit.

a. Find the daily total revenue realized from the sale of 200 units of the toaster oven.

b. Find the additional revenue realized when the production (and sales) level is increased from 200 to 300 units.

C **43. Marginal Profit**   Refer to Exercise 41. The daily marginal profit function associated with the production and sales of the deluxe toaster ovens is known to be

$$P'(x) = -0.0003x^2 + 0.02x + 20,$$

where x denotes the number of units manufactured and sold daily and $P'(x)$ is measured in dollars per unit.

a. Find the total profit realizable from the manufacture and sale of 200 units of the toaster ovens per day. [*Hint:* $P(200) - P(0) = \int_0^{200} P'(x)\,dx$, $P(0) = -800$.]

b. What is the additional daily profit realizable if the production and sale of the toaster ovens is increased from 200 to 220 units per day?

C **44. Efficiency Studies**   Tempco Electronics, a division of Tempco Toys, Inc., manufactures an electronic football game. An efficiency study showed that the rate at which the games are assembled by the average worker t hours after starting work at 8 A.M. is

$$-\frac{3}{2}t^2 + 6t + 20 \quad (0 \le t \le 4)$$

units per hour.

a. Find the total number of games the average worker can be expected to assemble in the four-hour morning shift.

b. How many units can the average worker be expected to assemble in the first hour of the morning shift? In the second hour of the morning shift?

C 45. **Speedboat Racing** In a recent pretrial run for the world water speed record, the velocity of the *Sea Falcon II* t seconds after firing the booster rocket was given by

$$v(t) = -t^2 + 20t + 440 \quad (0 \le t \le 20)$$

feet per second. Find the distance covered by the boat over the 20-second period after the booster rocket was activated. [*Hint:* The distance is given by $\int_0^{20} v(t)\, dt$.]

C 46. **U.S. Census** According to the U.S. Census Bureau, the number of Americans aged 45 to 54 (which stood at 25 million at the beginning of 1990) will grow at the rate of

$$R(t) = 0.00933t^3 + 0.019t^2 - 0.10833t + 1.3467$$

million people per year, t years from the beginning of 1990. How many Americans aged 45 to 54 will be added to the population between 1990 and the year 2000?

C 47. **Air Purification** To test air purifiers, the engineers run a purifier in a smoke-filled 10×20-foot room. While conducting a test for a certain brand of air purifier, it was determined that the amount of smoke in the room was decreasing at the rate of

$$R(t) = 0.00032t^4 - 0.01872t^3 + 0.3948t^2$$
$$- 3.83t + 17.63 \quad (0 \le t \le 20)$$

percent of the (original) amount of the smoke per minute, t minutes after the start of the test. How much smoke was left in the room five minutes after the start of the test? Ten minutes after the start of the test?

SOLUTIONS TO SELF-CHECK EXERCISES 6.4

1.
$$\int_0^2 (x + e^x)\, dx = \frac{1}{2}x^2 + e^x \bigg|_0^2$$
$$= \left[\frac{1}{2}(2)^2 + e^2\right] - \left[\frac{1}{2}(0) + e^0\right]$$
$$= 2 + e^2 - 1$$
$$= e^2 + 1$$

2. **a.** We want $P(1000)$. But

$$P(1000) - P(0) = \int_0^{1000} P'(x)\, dx = \int_0^{1000} (-0.000006x^2 + 6)\, dx$$
$$= -0.000002x^3 + 6x \bigg|_0^{1000}$$
$$= -0.000002(1000)^3 + 6(1000)$$
$$= 4000.$$

So, $P(1000) = 4000 + P(0) = 4000 - 400$, or \$3600 per day $[P(0) = -C(0)]$.

b. The additional profit realizable is given by

$$\int_{1000}^{1200} P'(x)\, dx = -0.000002x^3 + 6x \bigg|_{1000}^{1200}$$
$$= [-0.000002(1200)^3 + 6(1200)]$$
$$- [-0.000002(1000)^3 + 6(1000)]$$
$$= 3744 - 4000$$
$$= -256;$$

that is, the company sustains a loss of \$256 per day if production is increased to 1200 cases a day.

Evaluating Definite Integrals

This section continues our discussion of the applications of the Fundamental Theorem of Calculus.

Properties of the Definite Integral

Before going on, we list the following useful properties of the definite integral, some of which parallel the rules of integration of Section 6.1.

PROPERTIES OF THE DEFINITE INTEGRAL

Let f and g be integrable functions; then

1. $\displaystyle\int_a^a f(x)\, dx = 0$

2. $\displaystyle\int_a^b f(x)\, dx = -\int_b^a f(x)\, dx$

3. $\displaystyle\int_a^b cf(x)\, dx = c\int_a^b f(x)\, dx \quad (c, \text{ a constant})$

4. $\displaystyle\int_a^b [f(x) \pm g(x)]\, dx = \int_a^b f(x)\, dx \pm \int_a^b g(x)\, dx$

5. $\displaystyle\int_a^b f(x)\, dx = \int_a^c f(x)\, dx + \int_c^b f(x)\, dx \quad (a < c < b)$

Property 5 states that if c is a number lying between a and b so that the interval $[a,b]$ is divided into the intervals $[a, c]$ and $[c, b]$, then the integral of f over the integral $[a, b]$ may be expressed as the sum of the integral of f over the interval $[a, c]$ and the integral of f over the interval $[c, b]$.

Property 5 has the following geometrical interpretation when f is nonnegative. By definition

$$\int_a^b f(x)\, dx$$

is the area of the region under the graph of $y = f(x)$ from $x = a$ to $x = b$ (Figure 6.23). Similarly, we interpret the definite integrals

$$\int_a^c f(x)\, dx \quad \text{and} \quad \int_c^b f(x)\, dx$$

as the areas of the regions under the graph of $y = f(x)$ from $x = a$ to $x = c$ and from $x = c$ to $x = b$, respectively. Since the two regions do not overlap, we see that

$$\int_a^b f(x)\, dx = \int_a^c f(x)\, dx + \int_c^b f(x)\, dx.$$

FIGURE 6.23
$\int_a^b f(x)\, dx = \int_a^c f(x)\, dx + \int_c^b f(x)\, dx$

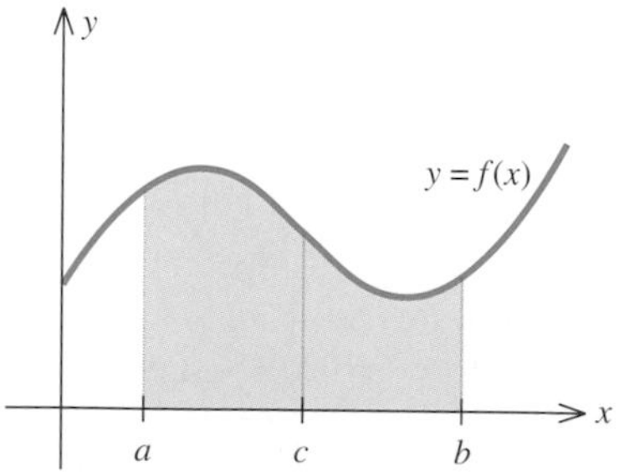

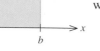

The Method of Substitution for Definite Integrals

Our first example shows two approaches generally used when evaluating a definite integral using the method of substitution.

EXAMPLE 1   Evaluate

$$\int_0^4 x\sqrt{9 + x^2} \, dx.$$

Solution

Method 1   We first find the corresponding indefinite integral

$$I = \int x\sqrt{9 + x^2} \, dx.$$

Make the substitution

$$u = 9 + x^2$$

so that $du = \dfrac{d}{dx}(9 + x^2) \, dx$

$$= 2x \, dx,$$

or $x \, dx = \dfrac{1}{2} du.$ (Dividing both sides by 2)

Then

$$I = \int \frac{1}{2} \sqrt{u} \, du = \frac{1}{2} \int u^{1/2} \, du$$

$$= \frac{1}{3} u^{3/2} + C = \frac{1}{3}(9 + x^2)^{3/2} + C. \qquad \begin{array}{l}\text{(Substituting}\\ 9 + x^2 \text{ for } u)\end{array}$$

Using this result, we now evaluate the given definite integral.

$$\int_0^4 x\sqrt{9 + x^2} \, dx = \frac{1}{3}(9 + x^2)^{3/2}\Big|_0^4$$

$$= \frac{1}{3}[(9 + 16)^{3/2} - 9^{3/2}]$$

$$= \frac{1}{3}(125 - 27) = \frac{98}{3} = 32\frac{2}{3}$$

Method 2   *Changing the Limits of Integration:* As before, we make the substitution

$$u = 9 + x^2 \qquad\qquad\qquad (8)$$

so that $du = 2x \, dx,$

or $x \, dx = \dfrac{1}{2} du.$

Next, observe that the given definite integral is evaluated *with respect to x* with the range of integration given by the interval [0, 4]. If we perform the integration *with respect to u* via the substitution (8), then we must adjust the range of integration to reflect the fact that the integration is being performed with respect to the new variable *u*. To determine the proper range of integration, note that when *x* = 0, (8) implies that

$$u = 9 + 0^2 = 9,$$

which gives the required lower limit of integration with respect to u. Similarly, when $x = 4$,

$$u = 9 + 16 = 25$$

is the required upper limit of integration with respect to u. Thus, the range of integration when the integration is performed with respect to u is given by the interval $[9, 25]$. Therefore, we have

$$\int_0^4 x\sqrt{1 + x^2}\, dx = \int_9^{25} \frac{1}{2}\sqrt{u}\, du = \frac{1}{2}\int_9^{25} u^{1/2}\, du$$

$$= \frac{1}{3}u^{3/2}\Big|_9^{25} = \frac{1}{3}(25^{3/2} - 9^{3/2})$$

$$= \frac{1}{3}(125 - 27) = \frac{98}{3} = 32\frac{2}{3},$$

which agrees with the result obtained using Method 1. ∎

EXAMPLE 2 Evaluate

$$\int_0^2 xe^{2x^2}\, dx.$$

Solution Let $u = 2x^2$ so that $du = 4x\, dx$, or $x\, dx = \frac{1}{4} du$.

When $x = 0$, $u = 0$, and when $x = 2$, $u = 8$. This gives the lower and upper limits of integration with respect to u. Making the indicated substitutions, we find

$$\int_0^2 xe^{2x^2}\, dx = \int_0^8 \frac{1}{4}e^u\, du = \frac{1}{4}e^u\Big|_0^8 = \frac{1}{4}(e^8 - 1).$$ ∎

EXAMPLE 3 Evaluate

$$\int_0^1 \frac{x^2}{x^3 + 1}\, dx.$$

Solution Let $u = x^3 + 1$ so that $du = 3x^2\, dx$, or $x^2\, dx = \frac{1}{3} du$.

When $x = 0$, $u = 1$, and when $x = 1$, $u = 2$. This gives the lower and upper limits of integration with respect to u. Making the indicated substitutions, we find

$$\int_0^1 \frac{x^2}{x^3 + 1}\, dx = \frac{1}{3}\int_1^2 \frac{du}{u} = \frac{1}{3}\ln|u|\Big|_1^2$$

$$= \frac{1}{3}(\ln 2 - \ln 1) = \frac{1}{3}\ln 2.$$ ∎

Finding the Area Under a Curve

EXAMPLE 4 Find the area of the region R under the graph of $f(x) = e^{(1/2)x}$ from $x = -1$ to $x = 1$.

Solution   The region R is shown in Figure 6.24. Its area is given by

$$A = \int_{-1}^{1} e^{(1/2)x}\, dx.$$

In order to evaluate this integral, we make the substitution

$$u = \frac{1}{2}x$$

so that $du = \dfrac{1}{2}\, dx \quad \text{or} \quad dx = 2\, du.$

When $x = -1$, $u = -\frac{1}{2}$, and when $x = 1$, $u = \frac{1}{2}$. Making the indicated substitutions, we obtain

$$A = \int_{-1}^{1} e^{(1/2)x}\, dx = 2\int_{-1/2}^{1/2} e^{u}\, du$$

$$= 2e^{u}\Big|_{-1/2}^{1/2}$$

$$= 2(e^{1/2} - e^{-1/2}),$$

or approximately 2.08 square units.

FIGURE 6.24
Area of $R = \int_{-1}^{1} e^{(1/2)x}\, dx$

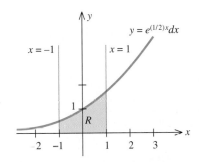

Average Value of a Function

The *average value* of a function over an interval provides us with an application of the definite integral. Recall that the average value of a set of n numbers is the number

$$\frac{y_1 + y_2 + \cdots + y_n}{n}.$$

Now, suppose that f is a continuous function defined on $[a, b]$. Let us divide the interval $[a, b]$ into n subintervals of equal length $(b - a)/n$. Choose points $x_1, x_2, \ldots, x_n$ in the first, second, $\ldots$, and nth subintervals, respectively. Then the average value of the numbers $f(x_1), f(x_2), \ldots, f(x_n)$, given by

$$\frac{f(x_1) + f(x_2) + \cdots + f(x_n)}{n},$$

is an approximation of the average of all the values of $f(x)$ on the interval

$[a, b]$. This expression can be written in the form

$$\frac{(b - a)}{(b - a)}\left[f(x_1) \cdot \frac{1}{n} + f(x_2) \cdot \frac{1}{n} + \cdots + f(x_n) \cdot \frac{1}{n}\right]$$

$$= \frac{1}{b - a}\left[f(x_1) \cdot \frac{b - a}{n} + f(x_2) \cdot \frac{b - a}{n} + \cdots + f(x_n) \cdot \frac{b - a}{n}\right]$$

$$= \frac{1}{b - a}[f(x_1)\Delta x + f(x_2)\Delta x + \cdots + f(x_n)\Delta x]. \qquad (9)$$

As n gets larger and larger, the expression (9) approximates the average value of $f(x)$ over $[a, b]$ with increasing accuracy. But the sum inside the brackets in (9) is a Riemann sum of the function f over $[a, b]$. In view of this, we have

$$\lim_{n \to \infty}\left[\frac{f(x_1) + f(x_2) + \cdots + f(x_n)}{n}\right]$$

$$= \frac{1}{b - a}\lim_{n \to \infty}[f(x_1)\Delta x + f(x_2)\Delta x + \cdots + f(x_n)\Delta x]$$

$$= \frac{1}{b - a}\int_a^b f(x)\,dx.$$

This discussion motivates the following definition.

THE AVERAGE VALUE OF A FUNCTION   Suppose that f is integrable on $[a, b]$. Then the **average value of f** over $[a, b]$ is

$$\frac{1}{b - a}\int_a^b f(x)\,dx.$$

EXAMPLE 5   Find the average value of the function $f(x) = \sqrt{x}$ over the interval $[0, 4]$.

Solution   The required average value is given by

$$\frac{1}{4 - 0}\int_0^4 \sqrt{x}\,dx = \frac{1}{4}\int_0^4 x^{1/2}\,dx$$

$$= \frac{1}{6}x^{3/2}\Big|_0^4$$

$$= \frac{4}{3}.$$

Applications

EXAMPLE 6   The interest rates charged by Madison Finance on auto loans for used cars over a certain six-month period in 1993 are approximated

by the function

$$r(t) = -\frac{1}{12}t^3 + \frac{7}{8}t^2 - 3t + 12 \quad (0 \le t \le 6),$$

where t is measured in months and $r(t)$ is the annual percentage rate. What is the average rate on auto loans extended by Madison over the six-month period?

Solution The average rate over the six-month period in question is given by

$$\frac{1}{6 - 0}\int_0^6 \left(-\frac{1}{12}t^3 + \frac{7}{8}t^2 - 3t + 12\right)dt$$

$$= \frac{1}{6}\left(-\frac{1}{48}t^4 + \frac{7}{24}t^3 - \frac{3}{2}t^2 + 12t\right)\Big|_0^6$$

$$= \frac{1}{6}\left[-\frac{1}{48}(6^4) + \frac{7}{24}(6^3) - \frac{3}{2}(6^2) + 12(6)\right]$$

$$= 9,$$

or 9 percent per year. ∎

EXAMPLE 7 The amount of a certain drug in a patient's body t days after it has been administered is

$$C(t) = 5e^{-0.2t}$$

units. Determine the average amount of the drug present in the patient's body for the first four days after the drug has been administered.

Solution The average amount of the drug present in the patient's body for the first four days after it has been administered is given by

$$\frac{1}{4 - 0}\int_0^4 5e^{-0.2t}\,dt = \frac{5}{4}\int_0^4 e^{-0.2t}\,dt$$

$$= \frac{5}{4}\left[\left(-\frac{1}{0.2}\right)e^{-0.2t}\Big|_0^4\right]$$

$$= \frac{5}{4}(-5e^{-0.8} + 5)$$

$$\approx 3.44,$$

or approximately 3.44 units. ∎

FIGURE 6.25

The average value of f over [a, b] is k.

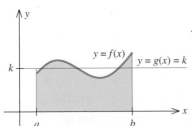

We now give a geometrical interpretation of the average value of a function f over an interval $[a, b]$. Suppose that $f(x)$ is nonnegative so that the definite integral

$$\int_a^b f(x)\,dx$$

gives the area under the graph of f from $x = a$ to $x = b$ (Figure 6.25). Observe that, in general, the "height" $f(x)$ varies from point to point. Can we replace

$f(x)$ by a constant function $g(x) = k$ (which has constant height) such that the areas under each of the two functions f and g are the same? If so, since the area under the graph of g from $x = a$ to $x = b$ is $k(b - a)$, we have

$$k(b - a) = \int_a^b f(x) \, dx$$

or
$$k = \frac{1}{b - a} \int_a^b f(x) \, dx,$$

so that k is the average value of f over $[a, b]$. Thus, the average value of a function f over an interval $[a, b]$ is the height of a rectangle with base of length $(b - a)$ that has the same area under the graph of f from $x = a$ to $x = b$.

SELF-CHECK EXERCISES 6.5

1. Evaluate
$$\int_0^2 \sqrt{2x + 5} \, dx.$$

2. Find the average value of the function $f(x) = 1 - x^2$ over the interval $[-1, 2]$.

3. The median price of a house in a southwestern state between January 1, 1989 and January 1, 1994 is approximated by the function
$$f(t) = t^3 - 7t^2 + 17t + 130 \quad (0 \le t \le 5),$$
where $f(t)$ is measured in thousands of dollars and t is expressed in years ($t = 0$ corresponds to the beginning of 1989). Determine the average median price of a house over that time interval.

Solutions to Self-Check Exercises 6.5 can be found on page 393.

6.5   EXERCISES

In Exercises 1–28, evaluate the given definite integral.

1. $\displaystyle\int_0^2 x(x^2 - 1)^3 \, dx$

2. $\displaystyle\int_0^1 x^2(2x^3 - 1)^4 \, dx$

3. $\displaystyle\int_0^1 x \sqrt{5x^2 + 4} \, dx$

4. $\displaystyle\int_1^3 x\sqrt{3x^2 - 2} \, dx$

5. $\displaystyle\int_0^2 x^2(x^3 + 1)^{3/2} \, dx$

6. $\displaystyle\int_1^5 (2x - 1)^{5/2} \, dx$

7. $\displaystyle\int_0^1 \frac{1}{\sqrt{2x + 1}} \, dx$

8. $\displaystyle\int_0^2 \frac{x}{\sqrt{x^2 + 5}} \, dx$

9. $\displaystyle\int_1^2 (2x - 1)^4 \, dx$

10. $\displaystyle\int_1^2 (2x + 4)(x^2 + 4x - 8)^3 \, dx$

11. $\displaystyle\int_{-1}^1 x^2(x^3 + 1)^4 \, dx$

12. $\displaystyle\int_{-1}^1 \left(x^3 + \tfrac{3}{4}\right)(x^4 + 3x)^{-2} \, dx$

13. $\displaystyle\int_1^5 x\sqrt{x - 1} \, dx$

14. $\displaystyle\int_1^4 x\sqrt{x + 1} \, dx$ [*Hint:* Let $u = x + 1$.]

15. $\displaystyle\int_0^2 xe^{x^2} \, dx$

16. $\displaystyle\int_0^1 e^{-x} \, dx$

17. $\int_0^1 (e^{2x} + x^2 + 1) \, dx$ **18.** $\int_0^2 (e^t - e^{-t}) \, dt$

19. $\int_{-1}^1 xe^{x^2+1} \, dx$ **20.** $\int_0^4 \frac{e^{\sqrt{x}}}{\sqrt{x}} \, dx$

21. $\int_3^6 \frac{2}{x-2} \, dx$ **22.** $\int_0^1 \frac{x}{1+2x^2} \, dx$

23. $\int_1^2 \frac{x^2 + 2x}{x^3 + 3x^2 - 1} \, dx$ **24.** $\int_0^1 \frac{e^x}{1 + e^x} \, dx$

25. $\int_1^2 \left(4e^{2u} - \frac{1}{u} \right) du$ **26.** $\int_1^2 \left(1 + \frac{1}{x} + e^x \right) dx$

27. $\int_1^2 \left(2e^{-4x} - \frac{1}{x^2} \right) dx$ **28.** $\int_1^2 \frac{\ln x}{x} \, dx$

In Exercises 29–38, find the average value of the given function f over the indicated interval [a, b].

29. $f(x) = 2x + 3$; $[0, 2]$

30. $f(x) = 8 - x$; $[1, 4]$

31. $f(x) = 2x^2 - 3$; $[1, 3]$

32. $f(x) = 4 - x^2$; $[-2, 3]$

33. $f(x) = x^2 + 2x - 3$; $[-1, 2]$

34. $f(x) = x^3$; $[-1, 1]$

35. $f(x) = \sqrt{2x + 1}$; $[0, 4]$

36. $f(x) = e^{-x}$; $[0, 4]$ **37.** $f(x) = xe^{x^2}$; $[0, 2]$

38. $f(x) = \frac{1}{x + 1}$; $[0, 2]$

C **39. World Production of Coal** A study proposed in 1980 by researchers from the major producers and consumers of the world's coal concluded that coal could and must play an important role in fueling global economic growth over the next 20 years. The world production of coal in 1980 was 3.5 billion metric tons. If output were to increase at the rate of $3.5e^{0.05t}$ billion metric tons per year in year t ($t = 0$ corresponding to 1980), determine how much coal will be produced worldwide between 1980 and the end of the century.

C **40. Newton's Law of Cooling** A bottle of white wine at room temperature (68°F) is placed in a refrigerator at 4 P.M. Its temperature after t hours is changing at the rate of

$$-18e^{-0.6t}$$

degrees Fahrenheit per hour. By how many degrees will the temperature of the wine have dropped by 7 P.M.? What will the temperature of the wine be at 7 P.M.?

C **41. Net Investment Flow** The net investment flow (rate of capital formation) of the giant conglomerate LTF Incorporated is projected to be

$$t\sqrt{\frac{1}{2}t^2 + 1}$$

million dollars per year in year t. Find the accruement on the company's capital stock in the second year. [*Hint:* The amount is given by

$$\int_1^2 t\sqrt{\frac{1}{2}t^2 + 1} \, dt.\bigg]$$

C **42. Oil Production** Based on a preliminary report by a geological survey team, it is estimated that a newly discovered oil field can be expected to produce oil at the rate of

$$R(t) = \frac{600t^2}{t^3 + 32} + 5 \quad (0 \le t \le 20)$$

thousand barrels per year, t years after production begins. Find the amount of oil that the field can be expected to yield during the first five years of production, assuming that the projection holds true.

C **43. Depreciation: Double Declining-Balance Method** Suppose that a tractor purchased at a price of $60,000 is to be depreciated by the *double declining-balance method* over a period of ten years. It can be shown that the rate at which the book value will be decreasing is given by

$$R(t) = 13388.61e^{-0.22314t} \quad (0 \le t \le 10)$$

dollars per year at year t. Find the amount by which the book value of the tractor will depreciate over the first five years of its life.

44. Velocity of a Car A car moves along a straight road in such a way that its velocity (in ft/sec) at any time t (in sec) is given by

$$v(t) = 3t\sqrt{16 - t^2} \quad (0 \le t \le 4).$$

Find the distance traveled by the car in the four seconds from $t = 0$ to $t = 4$.

C **45. Average Temperature** The temperature (in °F) in Boston over a 12-hour period on a certain December day was given by

$$T = -0.05t^3 + 0.4t^2 + 3.8t + 5.6 \quad (0 \le t \le 12),$$

where t is measured in hours with $t = 0$ corresponding to 6 A.M. Determine the average temperature on that day over the 12-hour period from 6 A.M. to 6 P.M.

C **46. Whale Population** A group of marine biologists estimate that if certain conservation measures are implemented the population of an endangered species of whale will be

$$N(t) = 3t^3 + 2t^2 - 10t + 600 \quad (0 \le t \le 10),$$

where $N(t)$ denotes the population at the end of year t. Find the average population of the whales over the next ten years.

C 47. **Cable TV Subscribers** The manager of the Tele-Star Cable Television Service estimates that the total number of subscribers to the service in a certain city t years from now will be

$$N(t) = -\frac{40{,}000}{\sqrt{1 + 0.2t}} + 50{,}000.$$

Find the average number of cable television subscribers over the next five years if this prediction holds true.

48. **Average Yearly Sales** The sales of the Universal Instruments Company in the first t years of its operation are approximated by the function

$$S(t) = t\sqrt{0.2t^2 + 4},$$

where $S(t)$ is measured in millions of dollars. What were Universal's average yearly sales over its first five years of operation?

49. **Concentration of a Drug in the Bloodstream** The concentration of a certain drug in a patient's bloodstream t hours after injection is

$$C(t) = \frac{0.2t}{t^2 + 1}$$

milligrams per cubic centimeter. Determine the average concentration of the drug in the patient's bloodstream over the first four hours after the drug is injected.

50. Refer to Exercise 44. Find the average velocity of the car over the time interval $[0, 4]$.

51. **Flow of Blood in an Artery** According to a law discovered by nineteenth-century physician Jean Louis Marie Poiseuille, the velocity of blood (in cm/sec) r centimeters from the central axis of an artery is given by

$$v(r) = k(R^2 - r^2),$$

where k is a constant and R is the radius of the artery. Find the average velocity of blood along a radius of the artery (see accompanying figure). [*Hint:* Evaluate $\frac{1}{R}\int_0^R v(r)\,dr$.]

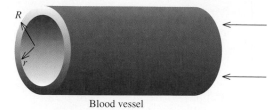

Blood vessel

Figure for Exercise 51

52. Prove Property 1 of the definite integral. [*Hint:* Let F be an antiderivative of f, and use the definition of the definite integral.]

53. Prove Property 2 of the definite integral. [*Hint:* See Exercise 52.]

54. Verify by direct computation that

$$\int_1^3 x^2\,dx = -\int_3^1 x^2\,dx.$$

55. Prove Property 3 of the definite integral. [*Hint:* See Exercise 52.]

56. Verify by direct computation that

$$\int_1^9 2\sqrt{x}\,dx = 2\int_1^9 \sqrt{x}\,dx.$$

57. Verify by direct computation that

$$\int_0^1 (1 + x - e^x)\,dx = \int_0^1 dx + \int_0^1 x\,dx - \int_0^1 e^x\,dx.$$

What properties of the definite integral are demonstrated in this exercise?

58. Verify by direct computation that

$$\int_0^3 (1 + x^3)\,dx = \int_0^1 (1 + x^3)\,dx + \int_1^3 (1 + x^3)\,dx.$$

What property of the definite integral is demonstrated here?

59. Verify by direct computation that

$$\int_0^3 (1 + x^3)\,dx = \int_0^1 (1 + x^3)\,dx$$
$$+ \int_1^2 (1 + x^3)\,dx + \int_2^3 (1 + x^3)\,dx,$$

hence showing that Property 5 may be extended.

**SOLUTIONS TO
SELF-CHECK
EXERCISES 6.5**

1. Let $u = 2x + 5$. Then $du = 2\ dx$, or $dx = \frac{1}{2}du$. Also, when $x = 0$, $u = 5$ and when $x = 2$, $u = 9$. Therefore,

$$\int_0^2 \sqrt{2x + 5}\ dx = \int_0^2 (2x + 5)^{1/2}\ dx$$

$$= \frac{1}{2}\int_5^9 u^{1/2}\ du$$

$$= \left(\frac{1}{2}\right)\left(\frac{2}{3}u^{3/2}\right)\Big|_5^9$$

$$= \frac{1}{3}[9^{3/2} - 5^{3/2}]$$

$$= \frac{1}{3}(27 - 5\sqrt{5}).$$

2. The required average value is given by

$$\frac{1}{2 - (-1)}\int_{-1}^2 (1 - x^2)\ dx = \frac{1}{3}\int_{-1}^2 (1 - x^2)\ dx$$

$$= \frac{1}{3}\left(x - \frac{1}{3}x^3\right)\Big|_{-1}^2$$

$$= \frac{1}{3}\left[\left(2 - \frac{8}{3}\right) - \left(-1 + \frac{1}{3}\right)\right]$$

$$= 0.$$

3. The average median price of a house over the stated time interval is given by

$$\frac{1}{5 - 0}\int_0^5 (t^3 - 7t^2 + 17t + 130)\ dt = \frac{1}{5}\left(\frac{1}{4}t^4 - \frac{7}{3}t^3 + \frac{17}{2}t^2 + 130t\right)\Big|_0^5$$

$$= \frac{1}{5}\left[\frac{1}{4}(5)^4 - \frac{7}{3}(5)^3 + \frac{17}{2}(5)^2 + 130(5)\right]$$

$$= 145.417,$$

or $145,417.

6.6   Area Between Two Curves

Suppose that a certain country's petroleum consumption is expected to grow at the rate of $f(t)$ million barrels per year, t years from now, for the next five years. Then the country's total petroleum consumption over the period of time in question is given by the area under the graph of f on the interval [0, 5] (Figure 6.26).

Next, suppose that because of the implementation of certain energy-conservation measures, the rate of growth of petroleum consumption is expected to be $g(t)$ million barrels per year instead. Then the country's projected total petroleum consumption over the five-year period is given by the area under the graph of g on the interval [0, 5] (Figure 6.27).

FIGURE 6.26

At a rate of consumption $f(t)$ million barrels per year, the total petroleum consumption is given by the area of the region under the graph of f.

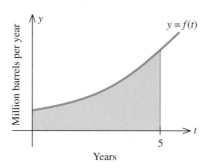

Years

FIGURE 6.27

At a rate of consumption of $g(t)$ million barrels per year, the total petroleum consumption is given by the area of the region under the graph of g.

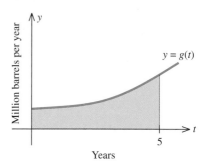

Years

Therefore, the area of the shaded region S lying between the graphs of f and g on the interval $[0, 5]$ (Figure 6.28) gives the amount of petroleum that would be saved over the five-year period because of the conservation measures.

FIGURE 6.28

The area of S gives the amount of petroleum that would be saved over the five-year period.

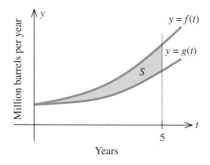

Years

But the area of S is given by

(Area under the graph of f on $[a, b]$) − (Area under the graph of g on $[a, b]$)

$$= \int_0^5 f(t)\, dt - \int_0^5 g(t)\, dt$$

$$= \int_0^5 [f(t) - g(t)]\, dt. \qquad \text{[By Property 4, Section 6.5]}$$

This example shows that some practical problems can be solved by finding the area of a region between two curves, which, in turn, can be found by evaluating an appropriate definite integral.

Finding the Area Between Two Curves

We now turn our attention to the general problem of finding the area of a plane region bounded both above and below by the graphs of functions. First, consider the situation in which the graph of one function lies above that of another.

FIGURE 6.29

Area of
$R = \int_a^b [f(x) - g(x)]\, dx.$

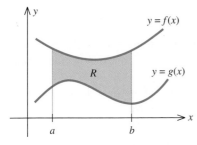

More specifically, let R be the region in the xy-plane (Figure 6.29) that is bounded above by the graph of a continuous function f, below by a continuous function g where $f(x) \geq g(x)$ on $[a, b]$, and to the left and right by the vertical lines $x = a$ and $x = b$, respectively. From the figure, we see that

$$[\text{Area of } R] = [\text{Area under } f(x)] - [\text{Area under } g(x)]$$

$$= \int_a^b f(x)\, dx - \int_a^b g(x)\, d(x)$$

$$= \int_a^b [f(x) - g(x)]\, dx,$$

upon using Property 4 of the definite integral.

THE AREA BETWEEN TWO CURVES

Let f and g be continuous functions such that $f(x) \geq g(x)$ on the interval $[a,b]$. Then the area of the region bounded above by $y = f(x)$ and below by $y = g(x)$ on $[a, b]$ is given by

$$\int_a^b [f(x) - g(x)]\, dx. \qquad (10)$$

Even though we assumed that both f and g were nonnegative in the derivation of (10), it may be shown that this equation is valid if f and g are not nonnegative (see Exercise 48). Also, observe that if $g(x)$ is 0 for all x—that is, when the lower boundary of the region R is the x-axis—Equation (10) gives the area of the region under the curve $y = f(x)$ from $x = a$ to $x = b$, as we would expect.

EXAMPLE 1   Find the area of the region bounded by the x-axis, the graph of $y = -x^2 + 4x - 8$, and the lines $x = -1$ and $x = 4$.

FIGURE 6.30

Area of $R = -\int_{-1}^4 g(x)\, dx.$

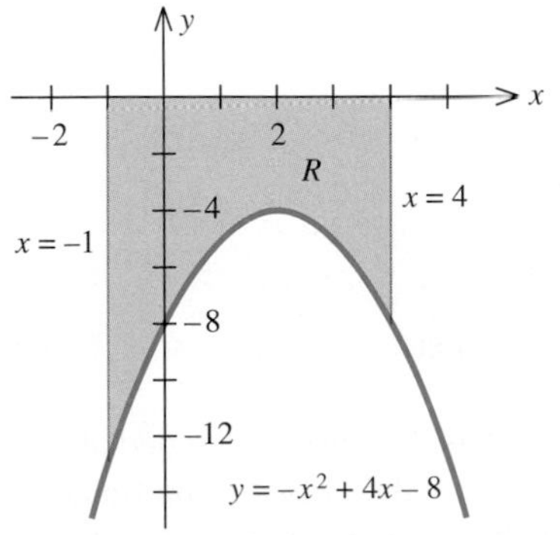

Solution   The region R under consideration is shown in Figure 6.30. We can view R as the region bounded above by the graph of $f(x) = 0$ (the x-axis) and below by the graph of $g(x) = -x^2 + 4x - 8$ on $[-1, 4]$. Therefore, the area of R is given by

$$\int_a^b [f(x) - g(x)]\, dx = \int_{-1}^4 [0 - (-x^2 + 4x - 8)]\, dx$$

$$= \int_{-1}^4 (x^2 - 4x + 8)\, dx$$

$$= \frac{1}{3}x^3 - 2x^2 + 8x \Big|_{-1}^4$$

$$= \left[\frac{1}{3}(64) - 2(16) + 8(4)\right] - \left[\frac{1}{3}(-1) - 2(1) + 8(-1)\right]$$

$$= 31\frac{2}{3}, \quad \text{or} \quad 31\frac{2}{3} \text{ square units.} \qquad \blacksquare$$

EXAMPLE 2 Find the area of the region R bounded by the graphs of

$$f(x) = 2x - 1 \quad \text{and} \quad g(x) = x^2 - 4$$

and the vertical lines $x = 1$ and $x = 2$.

FIGURE 6.31
Area of
$R = \int_1^2 [f(x) - g(x)]\, dx.$

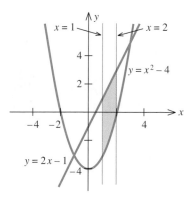

Solution We first sketch the graphs of the functions $f(x) = 2x - 1$ and $g(x) = x^2 - 4$, and the vertical lines $x = 1$ and $x = 2$, and identify the region R whose area is to be calculated (Figure 6.31).

Since the graph of f always lies above that of g for x in the interval $[1, 2]$, we see by (10) that the required area is given by

$$\int_1^2 [f(x) - g(x)]\, dx = \int_1^2 [(2x - 1) - (x^2 - 4)]\, dx$$

$$= \int_1^2 (-x^2 + 2x + 3)\, dx$$

$$= -\frac{1}{3}x^3 + x^2 + 3x \Big|_1^2$$

$$= \left(-\frac{8}{3} + 4 + 6\right) - \left(-\frac{1}{3} + 1 + 3\right) = \frac{11}{3},$$

or $3\frac{2}{3}$ square units. ■

EXAMPLE 3 Find the area of the region R that is completely enclosed by the graphs of the functions

$$f(x) = 2x - 1 \quad \text{and} \quad g(x) = x^2 - 4.$$

FIGURE 6.32
Area of
$R = \int_{-1}^3 [f(x) - g(x)]\, dx.$

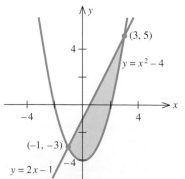

Solution The region R is shown in Figure 6.32. First we find the points of intersection of the two curves. To do this, we solve the system that comprises the two equations $y = 2x - 1$ and $y = x^2 - 4$. Equating the two values of y gives

$$x^2 - 4 = 2x - 1$$
$$x^2 - 2x - 3 = 0$$
$$(x + 1)(x - 3) = 0,$$

so $x = -1$ or $x = 3$. That is, the two curves intersect when $x = -1$ and $x = 3$.

Observe that we could also view the region R as the region bounded above by the graph of the function $f(x) = 2x - 1$, below by the graph of the function $g(x) = x^2 - 4$, and to the left and right by the vertical lines $x = -1$ and $x = 3$, respectively.

Next, since the graph of the function f always lies above that of the function g on $[-1, 3]$, we can use (10) to compute the desired area:

$$\begin{aligned}\int_a^b [f(x) - g(x)]\, dx &= \int_{-1}^{3} [(2x - 1) - (x^2 - 4)]\, dx \\ &= \int_{-1}^{3} (-x^2 + 2x + 3)\, dx \\ &= -\frac{1}{3}x^3 + x^2 + 3x\Big|_{-1}^{3} \\ &= (-9 + 9 + 9) - \left(\frac{1}{3} + 1 - 3\right) = \frac{32}{3},\end{aligned}$$

or $10\frac{2}{3}$ square units. ■

EXAMPLE 4 Find the area of the region R bounded by the graphs of the functions

$$f(x) = x^2 - 2x - 1 \quad \text{and} \quad g(x) = -e^x - 1$$

and the vertical lines $x = -1$ and $x = 1$.

FIGURE 6.33
Area of $R = \int_{-1}^{1} [f(x) - g(x)]\, dx.$

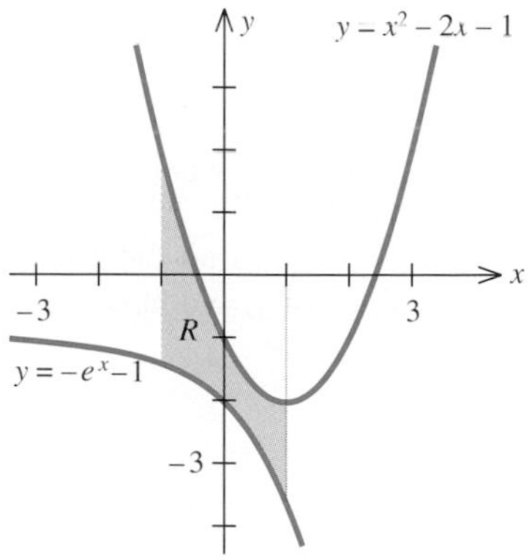

Solution The region R is shown in Figure 6.33. Since the graph of the function f always lies above that of the function g, the area of the region R is given by

$$\begin{aligned}\int_a^b [f(x) - g(x)]\, dx &= \int_{-1}^{1} [(x^2 - 2x - 1) - (-e^x - 1)]\, dx \\ &= \int_{-1}^{1} (x^2 - 2x + e^x)\, dx \\ &= \frac{1}{3}x^3 - x^2 + e^x\Big|_{-1}^{1} \\ &= \left(\frac{1}{3} - 1 + e\right) - \left(-\frac{1}{3} - 1 + e^{-1}\right) \\ &= \frac{2}{3} + e - \frac{1}{e} \text{ square units.}\end{aligned}$$

■

Equation (10), which gives the area of the region between the curves $y = f(x)$ and $y = g(x)$ for $a \le x \le b$, is valid when the graph of the function f lies above that of the function g over the interval $[a, b]$. Example 5 shows how to use (10) to find the area of a region when the latter condition does not hold.

EXAMPLE 5 Find the area of the region bounded by the graph of the function $f(x) = x^3$, the x-axis, and the lines $x = -1$ and $x = 1$.

Solution The region R under consideration can be thought of as comprising the two subregions R_1 and R_2, as shown in Figure 6.34.

FIGURE 6.34
Area of R_1 = area of R_2.

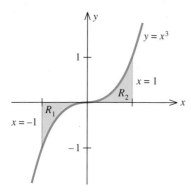

Recall that the x-axis is represented by the function $g(x) = 0$. Since $g(x) \geq f(x)$ on $[-1, 0]$, we see that the area of R_1 is given by

$$\int_a^b [g(x) - f(x)] \, dx = \int_{-1}^0 (0 - x^3) \, dx = -\int_{-1}^0 x^3 \, dx$$
$$= -\frac{1}{4}x^4 \Big|_{-1}^0 = 0 - \left(-\frac{1}{4}\right) = \frac{1}{4}.$$

To find the area of R_2, we observe that $f(x) \geq g(x)$ on $[0, 1]$, so it is given by

$$\int_a^b [f(x) - g(x)] \, dx = \int_0^1 (x^3 - 0) \, dx = \int_0^1 x^3 \, dx$$
$$= \frac{1}{4}x^4 \Big|_0^1 = \left(\frac{1}{4}\right) - 0 = \frac{1}{4}.$$

Therefore, the area of R is $\frac{1}{4} + \frac{1}{4}$, or $\frac{1}{2}$, square units.

By making use of symmetry, we could have obtained the same result by computing

$$-2 \int_{-1}^0 x^3 \, dx \quad \text{or} \quad 2 \int_0^1 x^3 \, dx,$$

as you may verify. ∎

EXAMPLE 6   Find the area of the region completely enclosed by the graphs of the functions

$$f(x) = x^3 - 3x + 3 \quad \text{and} \quad g(x) = x + 3.$$

FIGURE 6.35
Area of R_1 + area of
$R_2 = \int_{-2}^0 [f(x) - g(x)] \, dx$
$+ \int_0^2 [g(x) - f(x)] \, dx.$

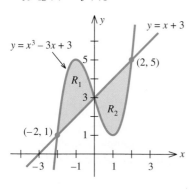

Solution   First sketch the graphs of $y = x^3 - 3x + 3$ and $y = x + 3$, and then identify the required region R. We can view the region R as being composed of the two subregions R_1 and R_2, as shown in Figure 6.35. By solving the equations $y = x + 3$ and $y = x^3 - 3x + 3$ simultaneously, we find the points of intersection of the two curves. Equating the two values of y, we have

$$x^3 - 3x + 3 = x + 3$$
$$x^3 - 4x = 0$$
or
$$x(x^2 - 4) = 0$$
$$x(x + 2)(x - 2) = 0, \quad \text{or} \quad x = 0, -2, \text{ or } 2.$$

Hence, the points of intersection of the two curves are $(-2, 1)$, $(0, 3)$, and $(2, 5)$.

For $-2 \leq x \leq 0$, we see that the graph of the function f lies above that of the function g, so the area of the region R_1 is, by virtue of (10),

$$\int_{-2}^0 [(x^3 - 3x + 3) - (x + 3)] \, dx = \int_{-2}^0 (x^3 - 4x) \, dx$$
$$= \frac{1}{4}x^4 - 2x^2 \Big|_{-2}^0$$
$$= -(4 - 8)$$
$$= 4, \quad \text{or} \quad 4 \text{ square units.}$$

For $0 \le x \le 2$, the graph of the function g lies above that of the function f, and the area of R_2 is given by

$$\int_0^2 [(x + 3) - (x^3 - 3x + 3)] \, dx = \int_0^2 (-x^3 + 4x) \, dx$$
$$= -\frac{1}{4}x^4 + 2x^2 \Big|_0^2$$
$$= -4 + 8$$
$$= 4, \quad \text{or} \quad 4 \text{ square units.}$$

Therefore, the required area is the sum of the area of the two regions $R_1 + R_2$—that is, $4 + 4$, or 8 square units.   ■

Application

EXAMPLE 7   In a 1989 study for a developing country's Economic Development Board, government economists and energy experts concluded that if the Energy Conservation Bill were implemented in 1990, the country's oil consumption for the next five years would be expected to grow in accordance with the model

$$R(t) = 20e^{0.05t},$$

where t is measured in years ($t = 0$ corresponding to the year 1990) and $R(t)$ in millions of barrels per year. Without the government-imposed conservation measures, however, the expected rate of growth of oil consumption would be given by

$$R_1(t) = 20e^{0.08t}$$

millions of barrels per year. Using these models, determine how much oil would be saved from 1990 through 1995 if the bill were implemented.

Solution   Under the Energy Conservation Bill, the total amount of oil that would have been consumed between 1990 and 1995 is given by

$$\int_0^5 R(t) \, dt = \int_0^5 20e^{0.05t} \, dt. \tag{11}$$

Without the bill, the total amount of oil that would have been consumed between 1990 and 1995 is given by

$$\int_0^5 R_1(t) \, dt = \int_0^5 20e^{0.08t} \, dt. \tag{12}$$

Equation (11) may be interpreted as the area of the region under the curve $y = R(t)$ from $t = 0$ to $t = 5$. Similarly, we interpret (12) as the area of the region under the curve $y = R_1(t)$ from $t = 0$ to $t = 5$. Furthermore, note that the graph of $y = R_1(t) = 20e^{0.08t}$ always lies on or above the graph of $y = R(t) = 20e^{0.05t}$ ($t \ge 0$). Thus, the area of the shaded region S in Figure 6.36 shows the amount of oil that would be saved from 1990 to 1995 if the Energy

FIGURE 6.36

Area of
$S = \int_0^5 [R_1(t) - R(t)] \, dt.$

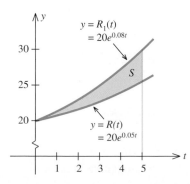

Conservation Bill were implemented. But the area of the region S is given by

$$\int_0^5 [R_1(t) - R(t)]\, dt = \int_0^5 [20e^{0.08t} - 20^{0.05t}]\, dt$$

$$= 20\int_0^5 (e^{0.08t} - e^{0.05t})\, dt$$

$$= 20\left(\frac{e^{0.08t}}{0.08} - \frac{e^{0.05t}}{0.05}\right)\Big|_0^5$$

$$= 20\left[\left(\frac{e^{0.4}}{0.08} - \frac{e^{0.25}}{0.05}\right) - \left(\frac{1}{0.08} - \frac{1}{0.05}\right)\right]$$

$$\approx 9.4,\quad \text{or}\quad \text{approximately } 9.4 \text{ square units.}$$

Thus, the amount of oil that would be saved is 9.4 million barrels. ■

SELF-CHECK EXERCISES 6.6

1. Find the area of the region bounded by the graphs of $f(x) = x^2 + 2$ and $g(x) = 1 - x$ and the vertical lines $x = 0$ and $x = 1$.

2. Find the area of the region completely enclosed by the graphs of $f(x) = -x^2 + 6x + 5$ and $g(x) = x^2 + 5$.

c 3. The management of the Kane Corporation, which operates a chain of hotels, expects its profits to grow at the rate of $1 + t^{2/3}$ million dollars per year t years from now. However, with renovations and improvements of existing hotels and proposed acquisitions of new hotels, Kane's profits are expected to grow at the rate of $t - 2\sqrt{t} + 4$ million dollars per year in the next decade. What additional profits are expected over the next ten years if the group implements the proposed plans?

Solutions to Self-Check Exercises 6.6 can be found on page 404.

6.6   EXERCISES

In Exercises 1–8, find the area of the shaded region.

1.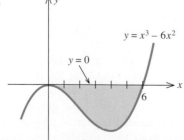

2.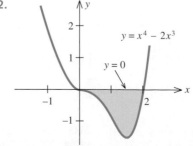

3.

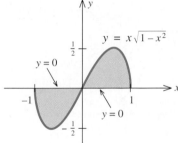

$y = x\sqrt{1-x^2}$

$y = 0$

$y = 0$

4.

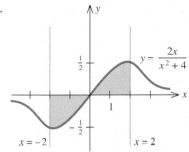

$y = \dfrac{2x}{x^2 + 4}$

$x = -2$ $x = 2$

5.

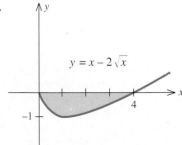

$y = x - 2\sqrt{x}$

6.

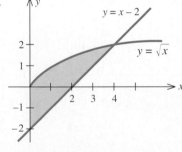

$y = x - 2$

$y = \sqrt{x}$

7.

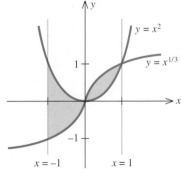

$y = x^2$

$y = x^{1/3}$

$x = -1$ $x = 1$

8.

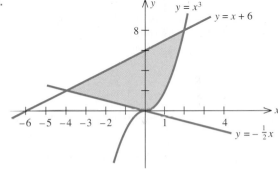

$y = x^3$

$y = x + 6$

$y = -\frac{1}{2}x$

In Exercises 9–16, sketch the graph and find the area of the region bounded below by the graph of each of the given functions and above by the x-axis from $x = a$ to $x = b$.

9. $f(x) = -x^2$; $a = -1$, $b = 2$

10. $f(x) = x^2 - 4$; $a = -2$, $b = 2$

11. $f(x) = x^2 - 5x + 4$; $a = 1$, $b = 3$

12. $f(x) = x^3$; $a = -1$, $b = 0$

13. $f(x) = -1 - \sqrt{x}$; $a = 0$, $b = 9$

14. $f(x) = \frac{1}{2}x - \sqrt{x}$; $a = 0$, $b = 4$

15. $f(x) = -e^{(1/2)x}$; $a = -2$, $b = 4$

16. $f(x) = -xe^{-x^2}$; $a = 0$, $b = 1$

In Exercises 17–26, sketch the graphs of the functions f and g and find the area of the region enclosed by these graphs and the vertical lines $x = a$ and $x = b$.

17. $f(x) = x^2 + 3$, $g(x) = 1$; $a = 1$, $b = 3$

18. $f(x) = x + 2$, $g(x) = x^2 - 4$; $a = -1$, $b = 2$

19. $f(x) = -x^2 + 2x + 3$, $g(x) = -x + 3$; $a = 0$, $b = 2$

20. $f(x) = 9 - x^2$, $g(x) = 2x + 3$; $a = -1$, $b = 1$

21. $f(x) = x^2 + 1$, $g(x) = \frac{1}{3}x^3$; $a = -1$, $b = 2$

22. $f(x) = \sqrt{x}$, $g(x) = -\frac{1}{2}x - 1$; $a = 1$, $b = 4$

23. $f(x) = \frac{1}{x}$, $g(x) = 2x - 1$; $a = 1$, $b = 4$

24. $f(x) = x^2$, $g(x) = \frac{1}{x^2}$; $a = 1$, $b = 3$

25. $f(x) = e^x$, $g(x) = \frac{1}{x}$; $a = 1$, $b = 2$

26. $f(x) = x$, $g(x) = e^{2x}$; $a = 1$, $b = 3$

In Exercises 27–34, sketch the graph and find the area of the region bounded by the graph of the function f and the lines $y = 0$, $x = a$, and $x = b$.

27. $f(x) = x$; $a = -1$, $b = 2$

28. $f(x) = x^2 - 2x$; $a = -1$, $b = 1$

29. $f(x) = -x^2 + 4x - 3$; $a = -1$, $b = 2$

30. $f(x) = x^3 - x^2$; $a = -1$, $b = 1$

31. $f(x) = x^3 - 4x^2 + 3x$; $a = 0$, $b = 2$

32. $f(x) = 4x^{1/3} + x^{4/3}$; $a = -1$, $b = 8$

33. $f(x) = e^x - 1$; $a = -1$, $b = 3$

34. $f(x) = xe^{x^2}$; $a = 0$, $b = 2$

In Exercises 35–40, sketch the graph and find the area of the region completely enclosed by the graphs of the given functions f and g.

35. $f(x) = x + 2$ and $g(x) = x^2 - 4$

36. $f(x) = -x^2 + 4x$ and $g(x) = 2x - 3$

37. $f(x) = x^2$ and $g(x) = x^3$

38. $f(x) = x^3 - 6x^2 + 9x$ and $g(x) = x^2 - 3x$

39. $f(x) = \sqrt{x}$ and $g(x) = x^2$

40. $f(x) = 2x$ and $g(x) = x\sqrt{x + 1}$

41. Effect of Advertising on Revenue   In the accompanying figure, the function f gives the rate of change of Odyssey Travel's revenue with respect to the amount x it spends on advertising with their current advertising agency. By engaging the services of a different advertising agency, it is expected that Odyssey's revenue will grow at the rate given by the function g. Give an interpretation of the area A of the region S, and find an expression for A in terms of a definite integral involving f and g.

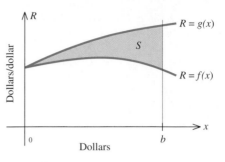

Figure for Exercise 41

42. Pulse Rate During Exercise   In the accompanying figure, the function f gives the rate of increase of an individual's pulse rate when he walked a prescribed course on a treadmill six months ago. The function g gives the rate of increase of his pulse rate when he recently walked the same prescribed course. Give an interpretation of the area A of the region S and find an expression for A in terms of a definite integral involving f and g.

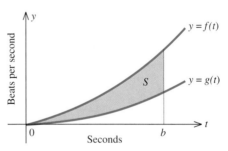

43. Air Purification   In order to study the effectiveness of air purifiers in removing smoke, engineers run each purifier in a smoke-filled 10×20-foot room. In the accompanying figure, the function f gives the rate of change of the smoke level per minute, t minutes after the start of the test, when a Brand A purifier is used. The function g gives the rate of change of the smoke level per minute when a Brand B purifier is used.
a. Give an interpretation of the area of the region S.
b. Find an expression for the area of S in terms of a definite integral involving f and g.

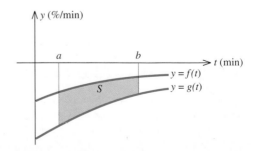

44. **Turbo-Charged Engine vs. Standard Engine** In tests conducted by Auto Test Magazine on two identical models of the Phoenix Elite—one equipped with a standard engine and the other with a turbo-charger—it was found that the acceleration of the former is given by

$$a = f(t) = 4 + 0.8t \quad (0 \leq t \leq 12)$$

feet per second per second, t seconds after starting from rest at full throttle, while the acceleration of the latter is given by

$$a = g(t) = 4 + 1.2t + 0.03t^2 \quad (0 \leq t \leq 12)$$

feet per second per second. How much faster is the turbo-charged model moving than the model with the standard engine, at the end of a ten-second test run at full throttle?

C 45. **Alternative Energy Sources** Because of the increasingly important role played by coal as a viable alternative energy source, the production of coal has been growing at the rate of

$$3.5e^{0.05t}$$

billion metric tons per year t years from 1980 (which corresponds to $t = 0$). Had it not been for the energy crisis, the rate of production of coal since 1980 might have been only

$$3.5e^{0.01t}$$

billion metric tons per year t years from 1980. Determine how much additional coal will be produced between 1980 and the end of the century as an alternate energy source.

C 46. **Effect of TV Advertising on Car Sales** Carl Williams, the new proprietor of Carl Williams Auto Sales, estimates that with extensive television advertising, car sales over the next several years could be increasing at the rate of

$$5e^{0.3t}$$

thousand cars per year t years from now, instead of at the current rate of

$$(5 + 0.5t^{3/2})$$

thousand cars per year t years from now. Find how many more cars Mr. Williams expects to sell over the next five years by implementing his advertising plans.

C 47. **Population Growth** In an endeavor to curb population growth in a Southeast Asian island state, the government has decided to launch an extensive propaganda campaign. Without curbs, the government expects the rate of population growth to have been

$$60e^{0.02t}$$

thousand people per year t years from now, over the next five years. However, successful implementation of the proposed campaign is expected to result in a population growth rate of

$$-t^2 + 60$$

thousand people per year t years from now, over the next five years. Assuming that the campaign is mounted, how many fewer people will there be in that country five years from now than there would have been if no curbs were imposed?

48. Show that the area of a region R bounded above by the graph of a function f and below by the graph of a function g from $x = a$ to $x = b$ is given by

$$\int_a^b [f(x) - g(x)] \, dx.$$

[*Hint:* The validity of the formula was verified earlier for the case when both f and g were nonnegative. Now let f and g be two functions such that $f(x) \geq g(x)$ for $a \leq x \leq b$. Then there exists some nonnegative constant c such that the curves $y = f(x) + c$ and $y = g(x) + c$ are translated in the y-direction in such a way that the region R' has the same area as the region R (see the accompanying figures). Show that the area of R' is given by

$$\int_a^b \{[f(x) + c] - [g(x) + c]\} \, dx$$
$$= \int_a^b [f(x) - g(x)] \, dx.]$$

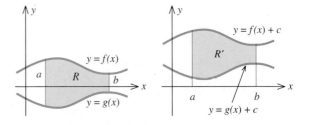

C *In Exercises 49 and 50, use a computer or graphing calculator to plot the graphs of f and g on the same set of axes. Then use these graphs to find the points of intersection of the curves. Finally, estimate the area of the region totally enclosed by the graphs of f and g.*

49. $f(x) = 2 - x^2$ and $g(x) = x^4 - 2x^3$

50. $f(x) = x^{2/3}$ and $g(x) = 3x^4 + 4x^3 + 1$

SOLUTIONS TO
SELF-CHECK
EXERCISES 6.6

1. The region in question is shown in the accompanying figure. Since the graph of the function f lies above that of the function g for $0 \le x \le 1$, we see that the

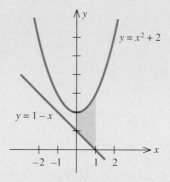

required area is given by

$$\int_0^1 [(x^2 + 2) - (1 - x)]\, dx = \int_0^1 (x^2 + x + 1)\, dx$$

$$= \left. \frac{1}{3}x^3 + \frac{1}{2}x^2 + x \right|_0^1$$

$$= \frac{1}{3} + \frac{1}{2} + 1$$

$$= \frac{11}{6},$$

or $\frac{11}{6}$ square units.

2. The region in question is shown in the accompanying figure. To find the points of intersection of the two curves, we solve the equation

$$-x^2 + 6x + 5 = x^2 + 5$$

$$2x^2 - 6x = 0$$

$$2x(x - 3) = 0,$$

giving $x = 0$ or $x = 3$. Therefore, the points of intersection are $(0, 5)$ and $(3, 14)$. Since the graph of f always lies above that of g for $0 \le x \le 3$, we see

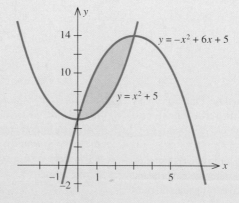

that the required area is given by

$$\int_0^3 [(-x^2 + 6x + 5) - (x^2 + 5)]\, dx = \int_0^3 (-2x^2 + 6x)\, dx$$
$$= -\frac{2}{3}x^3 + 3x^2 \Big|_0^3$$
$$= -18 + 27$$
$$= 9,$$

or 9 square units.

3. The additional profits realizable over the next ten years are given by

$$\int_0^{10} [(t - 2\sqrt{t} + 4) - (1 + t^{2/3})]\, dt$$
$$= \int_0^{10} (t - 2t^{1/2} + 3 - t^{2/3})\, dt$$
$$= \frac{1}{2}t^2 - \frac{4}{3}t^{3/2} + 3t - \frac{3}{5}t^{5/3} \Big|_0^{10}$$
$$= \frac{1}{2}(10)^2 - \frac{4}{3}(10)^{3/2} + 3(10) - \frac{3}{5}(10)^{5/3}$$
$$\approx 9.99,$$

or approximately $10 million.

6.7   Applications of the Definite Integral to Business and Economics

In this section, we consider several applications of the definite integral in the fields of business and economics.

Consumers' and Producers' Surplus

We begin by deriving a formula for computing the consumers' surplus. Suppose $p = D(x)$ is the demand function that relates the unit price p of a commodity to the quantity x demanded of it. Furthermore, suppose that a fixed unit market price has been established for the commodity and that corresponding to this unit price, the quantity demanded is $\bar{x}$ units (Figure 6.37). Then those consumers who would be willing to pay a unit price higher than $\bar{p}$ for the commodity would in effect experience a savings. This difference between what the consumers *would* be willing to pay for $\bar{x}$ units of the commodity and what they *actually* pay for them is called the **consumers' surplus.**

To derive a formula for computing the consumers' surplus, divide the interval $[0, \bar{x}]$ into n subintervals, each of length $\Delta x = \bar{x}/n$, and denote the right endpoints of these subintervals by $x_1, x_2, \ldots, x_n = \bar{x}$ (Figure 6.38).

Referring once again to Figure 6.38, we observe that there are consumers who would pay a unit price of at least $D(x_1)$ dollars for the first Δx units of the commodity instead of the market price of $\bar{p}$ dollars per unit. The savings

FIGURE 6.37

D(x) is a demand function.

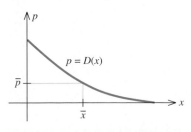

FIGURE 6.38

*Approximating consumers'
surplus by the sum of the
rectangles $r_1, r_2, \ldots, r_n$*

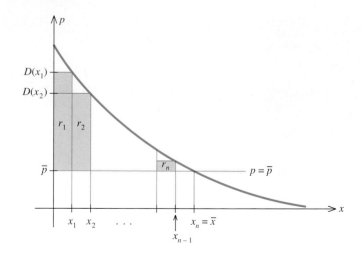

to these consumers is approximated by

$$D(x_1)\Delta x - \bar{p}\Delta x = [D(x_1) - \bar{p}\,]\Delta x,$$

which is the area of the rectangle r_1. Pursuing the same line of reasoning, we find that the savings to the consumers who would be willing to pay a unit price of at least $D(x_2)$ dollars for the next Δx units (from x_1 through x_2) of the commodity, instead of the market price of $\bar{p}$ dollars per unit, is approximated by

$$D(x_2)\Delta x - \bar{p}\Delta x = [D(x_2) - \bar{p}\,]\Delta x.$$

Continuing, we approximate the total savings to the consumers in purchasing $\bar{x}$ units of the commodity by the sum

$$[D(x_1) - \bar{p}\,]\Delta x + [D(x_2) - \bar{p}\,]\Delta x + \cdots + [D(x_n) - \bar{p}\,]\Delta x$$

$$= [D(x_1) + D(x_2) + \cdots + D(x_n)]\Delta x - \underbrace{[\bar{p}\Delta x + \bar{p}\Delta x + \cdots + \bar{p}\Delta x]}_{n \text{ terms}}$$

$$= D(x_1) + D(x_2) + \cdots + D(x_n)]\,\Delta x - n\bar{p}\Delta x$$

$$= [D(x_1) + D(x_2) + \cdots + D(x_n)]\,\Delta x - \bar{p}\bar{x}.$$

Now, the first term in the last expression is the Riemann sum of the demand function $p = D(x)$ over the interval $[0, \bar{x}]$ with representative points $x_1, x_2, \ldots, x_n$. Letting n approach infinity, we obtain the following formula for the consumers' surplus CS.

**CONSUMERS'
SURPLUS**

The **consumers' surplus** is given by

$$CS = \int_0^{\bar{x}} D(x)\,dx - \bar{p}\bar{x}, \tag{13}$$

where D is the demand function, $\bar{p}$ is the unit market price, and $\bar{x}$ is the quantity sold.

It is the area of the region bounded above by the demand curve $p = D(x)$ and below by the straight line $p = \bar{p}$ from $x = 0$ to $x = \bar{x}$ (Figure 6.39). We can also see this if we rewrite (13) in the form

$$\int_0^{\bar{x}} [D(x) - \bar{p}]\, dx$$

and interpret the result geometrically.

FIGURE 6.39
Consumers' surplus

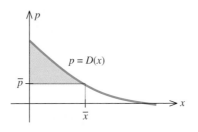

FIGURE 6.40
S(x) is a supply function.

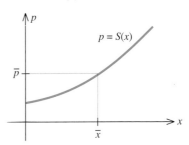

Analogously, we can derive a formula for computing the producers' surplus. Suppose that $p = S(x)$ is the supply equation that relates the unit price p of a certain commodity to the quantity x that the supplier will make available in the market at that price.

Again, suppose that a fixed market price $\bar{p}$ has been established for the commodity and that, corresponding to this unit price, a quantity of $\bar{x}$ units will be made available in the market by the supplier (Figure 6.40). Then the suppliers who would be willing to make the commodity available at a lower price stand to gain from the fact that the market price is set as such. The difference between what the suppliers actually receive and what they would be willing to receive is called the **producers' surplus.** Proceeding in a manner similar to the derivation of the equation for computing the consumers' surplus, we find that the producers' surplus PS is defined as follows:

**THE PRODUCERS'
SURPLUS**

The **producers' surplus** is given by

$$PS = \bar{p}\bar{x} - \int_0^{\bar{x}} S(x)\, dx, \qquad (14)$$

where $S(x)$ is the supply function, $\bar{p}$ is the unit market price, and $\bar{x}$ is the quantity supplied.

FIGURE 6.41
Producers' surplus

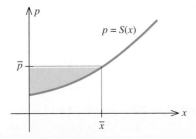

Geometrically, the producers' surplus is given by the area of the region bounded above by the straight line $p = \bar{p}$ and below by the supply curve $p = S(x)$ [Figure 6.41].

We can also show that the last statement is true by converting (14) to the form

$$\int_0^{\bar{x}} [\bar{p} - S(x)]\, dx$$

and interpreting the definite integral geometrically.

EXAMPLE 1 The demand function for a certain make of ten-speed bicycle is given by

$$p = D(x) = -0.001x^2 + 250,$$

where p is the unit price in dollars and x is the quantity demanded in units of a thousand. The supply function for these bicycles is given by

$$p = S(x) = 0.0006x^2 + 0.02x + 100,$$

where p stands for the unit price in dollars and x stands for the number of bicycles that the supplier will put on the market, in units of a thousand. Determine the consumers' surplus and the producers' surplus if the market price of a bicycle is set at the equilibrium price.

FIGURE 6.42

Consumers' surplus and producers' surplus when market price = equilibrium price

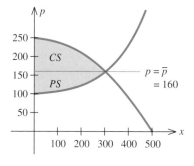

Solution Recall that the equilibrium price is the unit price of the commodity when market equilibrium occurs. We determine the equilibrium price by solving for the point of intersection of the demand curve and the supply curve (Figure 6.42). To solve the system of equations

$$p = -0.001x^2 + 250$$

and

$$p = 0.0006x^2 + 0.02x + 100,$$

we simply substitute the first equation into the second, obtaining

$$0.0006x^2 + 0.02x + 100 = -0.001x^2 + 250$$
$$0.0016x^2 + 0.02x - 150 = 0$$
$$16x^2 + 200x - 1,500,000 = 0$$
$$2x^2 + 25x - 187,500 = 0.$$

Factoring this last equation, we obtain

$$(2x + 625)(x - 300) = 0.$$

Thus, $x = -\frac{625}{2}$ or $x = 300$. The first number lies outside the interval of interest, so we are left with the solution $x = 300$, with a corresponding value of

$$p = -0.001(300)^2 + 250 = 160.$$

Thus, the equilibrium point is (300, 160); that is, the equilibrium quantity is 300,000 and the equilibrium price is \$160. Setting the market price at \$160 per unit and using (13) with $\bar{p} = 160$ and $\bar{x} = 300$, we find that the consumers' surplus is given by

$$CS = \int_0^{300} (-0.001x^2 + 250)\, dx - (160)(300)$$

$$= \left(-\frac{1}{3,000}x^3 + 250x \right) \Big|_0^{300} - 48,000$$

$$= -\frac{300^3}{3,000} + (250)(300) - 48,000$$

$$= 18,000, \quad \text{or} \quad \$18,000,000.$$

(Recall that x is measured in units of a thousand.) Next, using (14), we find that the producer's surplus is given by

$$PS = (160)(300) - \int_0^{300} (0.0006x^2 + 0.02x + 100)\, dx$$

$$= 48{,}000 - (0.0002x^3 + 0.01x^2 + 100x)\Big|_0^{300}$$

$$= 48{,}000 - [(0.0002)(300)^3 + (0.01)(300)^2 + 100(300)]$$

$$= 11{,}700, \quad \text{or} \quad \$11{,}700{,}000. \quad\blacksquare$$

The Future and Present Value of an Income Stream

To introduce the notion of the future and the present value of an income stream, suppose that a firm generates a stream of income over a period of time—the revenue generated by a large chain of retail stores over a five-year period, for example. As the income is realized, it is reinvested and earns interest at a fixed rate. The *accumulated future income stream* over the five-year period is the amount of money the firm ends up with at the end of that period.

The definite integral can be used to determine this **accumulated, or total, future income stream** over a period of time. The total future value of an income stream gives us a way to measure the value of such a stream. To find the total future value of an income stream, suppose that

$R(t) =$ rate of income generation at any time t (Dollars per year)

$r =$ interest rate compounded continuously

and $T =$ term   (In years)

Let us divide the time interval $[0, T]$ into n subintervals of equal length $\Delta t = T/n$ and denote the right endpoints of these intervals by $t_1, t_2, \ldots, t_n = T$, as shown in Figure 6.43.

If R is a continuous function on $[0, T]$, then $R(t)$ will not differ by much from $R(t_1)$ in the subinterval $[0, t_1]$ provided that the subinterval is small (which is true if n is large). Therefore, the income generated over the time interval $[0, t_1]$ is approximately

$$R(t_1)\Delta t \qquad \text{(Constant rate of income} \cdot \text{length of time)}$$

dollars. The future value of this amount, T years from now, calculated as if it were earned at time t_1, is

$$[R(t_1)\Delta t]e^{r(T-t_1)} \qquad \text{(Equation (9), Section 5.3)}$$

dollars. Similarly, the income generated over the time interval $[t_1, t_2]$ is approximately $P(t_2)\Delta t$ dollars and has a future value, T years from now, of approximately

$$[R(t_2)\Delta t]e^{r(T-t_2)}$$

dollars. Therefore, the sum of the future values of the income stream generated over the time interval $[0, T]$ is approximately

$$R(t_1)e^{r(T-t_1)}\Delta t + R(t_2)e^{r(T-t_2)}\Delta t + \cdots + R(t_n)e^{r(T-t_n)}\Delta t$$

$$= e^{rT}[R(t_1)e^{-rt_1}\Delta t + R(t_2)e^{-rt_2}\Delta t + \cdots + R(t_n)e^{-rt_n}\Delta t]$$

FIGURE 6.43

The time interval $[0, T]$ is partitioned into n subintervals.

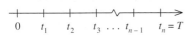

$0 \quad t_1 \quad t_2 \quad t_3 \, \cdots \, t_{n-1} \quad t_n = T$

dollars. But this sum is just the Riemann sum of the function $e^{rT}R(t)e^{-rt}$ over the interval $[0, T]$ with representative points $t_1, t_2, \ldots, t_n$. Letting n approach infinity, we obtain the following result.

ACCUMULATED OR TOTAL FUTURE VALUE OF AN INCOME STREAM

The **accumulated,** or **total, future value** after T years of an income stream of $R(t)$ dollars per year, earning interest at the rate of r per year compounded continuously, is given by

$$A = e^{rT}\int_0^T R(t)e^{-rt}\, dt. \tag{15}$$

EXAMPLE 2   Crystal Car Wash recently bought an automatic carwashing machine that is expected to generate \$40,000 in revenue per year, t years from now, for the next five years. If the income is reinvested in a business earning interest at the rate of 12 percent per year compounded continuously, find the total accumulated value of this income stream at the end of five years.

Solution   We are required to find the total future value of the given income stream after five years. Using (15) with $R(t) = 40{,}000$, $r = 0.12$, and $T = 5$, we see that the required value is given by

$$e^{0.12(5)}\int_0^5 40{,}000e^{-0.12t}\, dt$$

$$= e^{0.6}\left[-\frac{40{,}000}{0.12}e^{-0.12t}\right]\Big|_0^5 \qquad \text{(Integrate using the substitution } u = -0.12t.\text{)}$$

$$= -\frac{40{,}000e^{0.6}}{0.12}(e^{-0.6} - 1) \approx 274{,}039.60,$$

or approximately \$274,040.   ∎

Another way of measuring the value of an income stream is by considering its present value. The **present value** of an income stream of $R(t)$ dollars per year over a term of T years, earning interest at the rate of r per year compounded continuously, is the principal P that will yield the same accumulated value as the income stream itself when P is invested today for a period of T years at the same rate of interest. In other words,

$$Pe^{rT} = e^{rT}\int_0^T R(t)e^{-rt}\, dt.$$

Dividing both sides of the equation by e^{rT} gives the following result.

PRESENT VALUE OF AN INCOME STREAM

The **present value of an income stream** of $R(t)$ dollars per year, earning interest at the rate of r per year compounded continuously, is given by

$$PV = \int_0^T R(t)e^{-rt}\, dt. \tag{16}$$

EXAMPLE 3   The owner of a local cinema is considering two alternative plans for renovating and improving the theater. Plan A calls for an immediate cash outlay of $250,000, whereas plan B requires an immediate cash outlay of $180,000. It has been estimated that adopting plan A would result in a net income stream generated at the rate of

$$f(t) = 630,000$$

dollars per year, whereas adopting plan B would result in a net income stream generated at the rate of

$$g(t) = 580,000$$

dollars per year for the next three years. If the prevailing interest rate for the next five years were 10 percent per year, which plan would generate a higher net income by the end of three years?

Solution   Since the initial outlay is $250,000, we find, using (16) with $R(t) = 630,000$, $r = 0.1$, and $T = 3$, that the present value of the net income under plan A is given by

$$\int_0^3 630,000e^{-0.1t}\, dt - 250,000$$

$$= \frac{630,000}{-0.1} e^{-0.1t}\Big|_0^3 - 250,000 \qquad \text{(Integrate using the substitution } u = -0.1t.)$$

$$= -6,300,000e^{0.3} + 6,300,000 - 250,000$$

$$\approx 1,382,845,$$

or approximately $1,382,845.

To find the present value of the net income under plan B, we use (16) with $R(t) = 580,000$, $r = 0.1$, and $T = 3$, obtaining

$$\int_0^3 580,000e^{-0.1t}\, dt - 180,000$$

dollars. Proceeding as in the previous computation, we see that the required value is $1,323,254 (Exercise 8).

Comparing the present value of each plan, we conclude that plan A would generate a higher net income by the end of three years.   ■

REMARK   The function R in Example 3 is a constant function. If R is not a constant function, then we may need more sophisticated techniques of integration to evaluate the integral in (16). Exercises 7.1 and 7.2 contain problems of this type.   □

The Amount and Present Value of an Annuity

An annuity is a sequence of payments made at regular time intervals. The time period in which these payments are made is called the *term* of the annuity. Although the payments need not be equal in size, they are equal in many important applications, and we will assume that they are equal in our discussion. Examples of annuities are regular deposits to a savings account, monthly home mortgage payments, and monthly insurance payments.

The **amount of an annuity** is the sum of the payments plus the interest earned. A formula for computing the amount of an annuity A can be derived with the help of (15). Let

$$P = \text{size of each payment in the annuity}$$
$$r = \text{interest rate compounded continuously}$$
$$T = \text{term of the annuity (in years)}$$
and $$m = \text{number of payments per year.}$$

The payments into the annuity constitute a constant income stream of $R(t) = mP$ dollars per year. With this value of $R(t)$, (15) yields

$$A = e^{rT}\int_0^T R(t)e^{-rt}\,dt = e^{rT}\int_0^T mPe^{-rt}\,dt$$
$$= mPe^{rT}\left[-\frac{e^{-rt}}{r}\right]\Big|_0^T$$
$$= mPe^{rT}\left[-\frac{e^{-rT}}{r} + \frac{1}{r}\right] = \frac{mP}{r}(e^{rT} - 1).$$

This leads us to the following formula.

AMOUNT OF AN ANNUITY

The **amount of an annuity** is

$$A = \frac{mP}{r}(e^{rT} - 1), \qquad (17)$$

where P, r, T, and m are as defined earlier.

EXAMPLE 4   On January 1, 1985, Mr. Chapman deposited $2000 into an Individual Retirement Account (IRA) paying interest at the rate of 10 percent per year compounded continuously. Assuming that he deposits $2000 annually into the account, how much will he have in his IRA at the beginning of the year 2001?

Solution   We use (17), with $P = 2{,}000$, $r = 0.1$, $T = 16$, and $m = 1$, obtaining

$$A = \frac{2{,}000}{0.1}(e^{1.6} - 1)$$
$$\approx 79{,}060.65.$$

Thus, Mr. Chapman will have approximately $79,061 in his account at the beginning of the year 2001.   ∎

Using (16), we can derive the following formula for the *present value* of an annuity.

PRESENT VALUE OF AN ANNUITY	The **present value of an annuity** is given by $$PV = \frac{mP}{r}(1 - e^{-rT}),\qquad(18)$$ where P, r, T, and m are as defined earlier.

EXAMPLE 5   Mr. Carson, the proprietor of a hardware store, wants to establish a fund from which he will withdraw $1000 per month for the next ten years. If the fund earns interest at the rate of 9 percent per year compounded continuously, how much money does he need to establish the fund?

Solution   We want to find the present value of an annuity with $P = 1,000$, $r = 0.09$, $T = 10$, and $m = 12$. Using (18), we find

$$PV = \frac{12,000}{0.09}(1 - e^{-(0.09)(10)})$$

$$\approx 79,124.04.$$

Thus, Mr. Carson needs approximately $79,124 to establish the fund.   ■

Lorentz Curves and Income Distributions

One method used by economists to study the distribution of income in a society is based on the Lorentz curve, named after American statistician M. D. Lorentz. To describe the Lorentz curve, let $f(x)$ denote the proportion of the total income received by the poorest $100x$ percent of the population for $0 \leq x \leq 1$. Using this terminology, $f(0.3) = 0.1$ simply states that the lowest 30 percent of the income recipients receive 10 percent of the total income.

The function f has the following properties:

1. The domain of f is $[0, 1]$.
2. The range of f is $[0, 1]$.
3. $f(0) = 0$ and $f(1) = 1$.
4. $f(x) \leq x$ for every x in $[0, 1]$.
5. f is increasing on $[0, 1]$.

The first two properties follow from the fact that both x and $f(x)$ are fractions of a whole. Property 3 is a statement that 0 percent of the income recipients receive 0 percent of the total income and 100 percent of the income recipients receive 100 percent of the total income. Property 4 follows from the fact that the lowest $100x$ percent of the income recipients cannot receive more than $100x$ percent of the total income. A typical Lorentz curve is shown in Figure 6.44.

FIGURE 6.44

A Lorentz curve

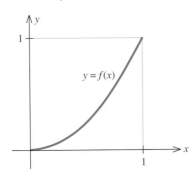

EXAMPLE 6   A developing country's income distribution is described by the function

$$f(x) = \frac{19}{20}x^2 + \frac{1}{20}x.$$

FIGURE 6.45

The Lorentz curve
$f(x) = \frac{19}{20}x^2 + \frac{1}{20}x.$

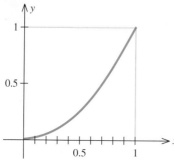

a. Sketch the Lorentz curve for the given function.

b. Compute $f(0.2)$ and $f(0.8)$ and interpret your results.

Solution

a. The Lorentz curve is shown in Figure 6.45.

b.
$$f(0.2) = \frac{19}{20}(0.2)^2 + \frac{1}{20}(0.2) = 0.048,$$

so that the lowest 20 percent of the people receive 4.8 percent of the total income.

$$f(0.8) = \frac{19}{20}(0.8)^2 + \frac{1}{20}(0.8) = 0.648,$$

so that the lowest 80 percent of the people receive 64.8 percent of the total income.   ∎

Next, let us consider the Lorentz curve described by the function $y = f(x) = x$. Since exactly $100x$ percent of the total income is received by the lowest $100x$ percent of income recipients, the line $y = x$ is called the **line of complete equality.** For example, 10 percent of the total income is received by the lowest 10 percent of income recipients, 20 percent of the total income is received by the lowest 20 percent of income recipients, and so on. Now, it is evident that the closer a Lorentz curve is to this line, the more equitable the income distribution is among the income recipients. But the proximity of a Lorentz curve to the line of complete equality is reflected by the area between the Lorentz curve and the line $y = x$ (Figure 6.46). The closer the curve is to the line, the smaller the enclosed area.

FIGURE 6.46

The closer the Lorentz curve is to the line, the more equitable the income distribution.

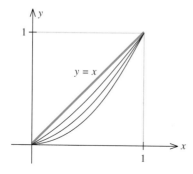

This observation suggests that we may define a number, called the **coefficient of inequality,** or **Gini Index,** of a Lorentz curve, as the ratio of the area between the line of complete equality and the Lorentz curve to the area under the line of complete equality. Since the area under the line of complete equality is $\frac{1}{2}$, we see that the coefficient of inequality is given by the following formula.

COEFFICIENT OF INEQUALITY OF A LORENTZ CURVE

The **coefficient of inequality** of a Lorentz curve is

$$L = 2\int_0^1 [x - f(x)]\, dx. \qquad (19)$$

The coefficient of inequality is a number between 0 and 1. For example, a coefficient of zero implies that the income distribution is perfectly uniform.

EXAMPLE 7   In a study conducted by a certain country's Economic Development Board with regard to the income distribution of certain segments of the country's work force, it was found that the Lorentz curves for the distribution of income of medical doctors and of movie actors and actresses are described by the functions

$$f(x) = \frac{14}{15}x^2 + \frac{1}{15}x \quad \text{and} \quad g(x) = \frac{5}{8}x^4 + \frac{3}{8}x,$$

respectively. Compute the coefficient of inequality for each Lorentz curve. Which profession has a more equitable income distribution?

Solution The required coefficients of inequality are, respectively,

$$L_1 = 2\int_0^1 \left[x - \left(\frac{14}{15}x^2 + \frac{1}{15}x \right) \right] dx = 2\int_0^1 \left(\frac{14}{15}x - \frac{14}{15}x^2 \right) dx$$

$$= \frac{28}{15}\int_0^1 (x - x^2)\, dx = \frac{28}{15}\left(\frac{1}{2}x^2 - \frac{1}{3}x^3 \right)\Big|_0^1$$

$$= \frac{14}{45} \approx 0.311$$

and

$$L_2 = 2\int_0^1 \left[x - \left(\frac{5}{8}x^4 + \frac{3}{8}x \right) \right] dx = 2\int_0^1 \left(\frac{5}{8}x - \frac{5}{8}x^4 \right) dx$$

$$= \frac{5}{4}\int_0^1 (x - x^4)\, dx = \frac{5}{4}\left(\frac{1}{2}x^2 - \frac{1}{5}x^5 \right)\Big|_0^1$$

$$= \frac{15}{40} \approx 0.375.$$

We conclude that in this country the incomes of medical doctors are more evenly distributed than the incomes of movie actors and actresses. ∎

SELF-CHECK EXERCISE 6.7

[c] The demand function for a certain make of exercise bicycle that is sold exclusively through cable television is

$$p = d(x) = \sqrt{9 - 0.02x},$$

where p is the unit price in hundreds of dollars and x is the quantity demanded per week. The corresponding supply function is given by

$$p = s(x) = \sqrt{1 + 0.02x},$$

where p has the same meaning as before and x is the number of exercise bicycles the supplier will make available at price p. Determine the consumers' surplus and the producers' surplus if the unit price is set at the equilibrium price.

The solution to Self-Check Exercise 6.7 can be found on page 417.

6.7 EXERCISES

[c] *A calculator is recommended for this exercise set.*

1. **Consumers' Surplus** The demand function for a certain make of cartridge typewriter ribbon is given by

$$p = -0.01x^2 - 0.1x + 6,$$

where p is the unit price in dollars and x is the quantity demanded each week, measured in units of a thousand. Determine the consumers' surplus if the market price is set at \$4 per cartridge.

2. **Consumers' Surplus**   The demand function for a certain brand of long-playing record is given by

$$p = -0.01x^2 - 0.2x + 8,$$

where p is the wholesale unit price in dollars and x is the quantity demanded each week, measured in units of a thousand. Determine the consumers' surplus if the wholesale market price is set at $5 per record.

3. **Consumers' Surplus**   It is known that the quantity demanded of a certain make of portable hair dryer is x hundred units per week, and the corresponding wholesale unit price is

$$p = \sqrt{225 - 5x}$$

dollars. Determine the consumers' surplus if the wholesale market price is set at $10 per unit.

4. **Producers' Surplus**   The supplier of the portable hair dryers in Exercise 3 will make x hundred units of hair dryers available in the market when the wholesale unit price is

$$p = \sqrt{36 + 1.8x}$$

dollars. Determine the producers' surplus if the wholesale market price is set at $9 per unit.

5. **Producers' Surplus**   The supply function for the long-playing records of Exercise 2 is given by

$$p = 0.01x^2 + 0.1x + 3,$$

where p is the unit wholesale price in dollars and x stands for the quantity that will be made available in the market by the supplier, measured in units of a thousand. Determine the producers' surplus if the wholesale market price is set at the equilibrium price.

6. **Consumers' and Producers' Surplus**   The management of the Titan Tire Company has determined that the quantity demanded x of their Super Titan tires per week is related to the unit price p by the relation

$$p = 144 - x^2,$$

where p is measured in dollars and x is measured in units of a thousand. Titan will make x units of the tires available in the market if the unit price is

$$p = 48 + \frac{1}{2}x^2$$

dollars. Determine the consumers' surplus and the producers' surplus when the market unit price is set at the equilibrium price.

7. **Consumers' and Producers' Surplus**   The quantity demanded x (in units of a hundred) of the Mikado miniature cameras per week is related to the unit price p (in

dollars) by

$$p = -0.2x^2 + 80,$$

and the quantity x (in units of a hundred) that the supplier is willing to make available in the market is related to the unit price p (in dollars) by

$$p = 0.1x^2 + x + 40.$$

If the market price is set at the equilibrium price, find the consumers' surplus and the producers' surplus.

8. Refer to Example 3, page 411. Verify that

$$\int_0^3 580,000e^{-0.1t}\, dt - 180,000 \approx 1,323,254.$$

9. **Present Value of an Investment**   Suppose that an investment is expected to generate income at the rate of

$$R(t) = 200,000$$

dollars per year for the next five years. Find the present value of this investment if the prevailing interest rate is 8 percent per year compounded continuously.

10. **Franchises**   Ms. Gordon purchased a 15-year franchise for a computer outlet store that is expected to generate income at the rate of

$$R(t) = 400,000$$

dollars per year. If the prevailing interest rate is 10 percent per year compounded continuously, find the present value of the franchise.

11. **The Amount of an Annuity**   Find the amount of an annuity if $250 a month is paid into it for a period of 20 years earning interest at the rate of 8 percent per year compounded continuously.

12. **The Amount of an Annuity**   Find the amount of an annuity if $400 a month is paid into it for a period of 20 years earning interest at the rate of 8 percent per year compounded continuously.

13. **The Amount of an Annuity**   Mr. Jorgenson deposits $150 per month in a savings account paying 8 percent per year compounded continuously. Estimate the amount that will be in his account after 15 years.

14. **Custodial Accounts**   The Armstrongs wish to establish a custodial account to finance their children's education. If they deposit $200 monthly for ten years in a savings account paying 9 percent per year compounded continuously, find how much their savings account will be worth at the end of this period.

15. **IRA Accounts**   Refer to Example 4, page 412. Suppose that Mr. Chapman makes his IRA payment on April 1, 1985 and annually thereafter. If interest is paid at the same initial rate, approximately how much will

Mr. Chapman have in his account at the beginning of the year 2001?

16. **Present Value of an Annuity** Estimate the present value of an annuity if payments are $800 monthly for 12 years and the account earns interest at the rate of 10 percent per year compounded continuously.

17. **Present Value of an Annuity** Estimate the present value of an annuity if payments are $1200 monthly for 15 years and the account earns interest at the rate of 10 percent per year compounded continuously.

18. **Lottery Payments** A state lottery commission pays the winner of the "Million Dollar" lottery 20 annual installments of $50,000 each. If the prevailing interest rate is 8 percent per year compounded continuously, find the present value of the winning ticket.

19. **Reverse Annuity Mortgages** Mr. Sinclair wishes to supplement his retirement income by $300 per month for the next ten years. He plans to obtain a reverse annuity mortgage (RAM) on his home to meet this need. Estimate the amount of the mortgage he will require if the prevailing interest rate is 12 percent per year compounded continuously.

20. **Reverse Annuity Mortgage** Refer to Exercise 19. Miss Santos wishes to supplement her retirement income by $400 per month for the next 15 years by obtaining a RAM. Estimate the amount of the mortgage she will require if the prevailing interest rate is 9 percent per year compounded continuously.

21. **Lorentz Curves** A certain country's income distribution is described by the function

$$f(x) = \frac{15}{16}x^2 + \frac{1}{16}x.$$

a. Sketch the Lorentz curve for this function.
b. Compute $f(0.4)$ and $f(0.9)$ and interpret your results.

22. **Lorentz Curves** In a study conducted by a certain country's Economic Development Board it was found that the Lorentz curve for the distribution of income of college teachers was described by the function

$$f(x) = \frac{13}{14}x^2 + \frac{1}{14}x$$

and that of lawyers by the function

$$g(x) = \frac{9}{11}x^4 + \frac{2}{11}x.$$

a. Compute the coefficient of inequality for each Lorentz curve.
b. Which profession has a more equitable income distribution?

23. **Lorentz Curves** A certain country's income distribution is described by the function

$$f(x) = \frac{14}{15}x^2 + \frac{1}{15}x.$$

a. Sketch the Lorentz curve for this function.
b. Compute $f(0.3)$ and $f(0.7)$.

24. **Lorentz Curves** In a study conducted by a certain country's Economic Development Board it was found that the Lorentz curve for the distribution of income of stockbrokers was described by the function

$$f(x) = \frac{11}{12}x^2 + \frac{1}{12}x$$

and that of high school teachers by the function

$$g(x) = \frac{5}{6}x^2 + \frac{1}{6}x.$$

a. Compute the coefficient of inequality for each Lorentz curve.
b. Which profession has a more equitable income distribution?

SOLUTION TO SELF-CHECK EXERCISE 6.7

We find the equilibrium price and equilibrium quantity by solving the system of equations

$$p = \sqrt{9 - 0.02x}$$

and

$$p = \sqrt{1 + 0.02x}$$

simultaneously. Substituting the first equation into the second, we have

$$\sqrt{9 - 0.02x} = \sqrt{1 + 0.02x}.$$

Squaring both sides of the equation then leads to

$$9 - 0.02x = 1 + 0.02x,$$

giving

$$x = 200.$$

Therefore,
$$p = \sqrt{9 - 0.02(200)}$$
$$= \sqrt{5} \approx 2.24,$$

and the equilibrium price is \$224 and the equilibrium quantity is 200. The consumers' surplus is given by

$$CS = \int_0^{200} \sqrt{9 - 0.02x}\, dx - (2.24)(200)$$

$$= \int_0^{200} (9 - 0.02x)^{1/2}\, dx - 448 \qquad \text{(Integrating by substitution)}$$

$$= -\frac{1}{0.02}\left(\frac{2}{3}\right)(9 - 0.02x)^{3/2}\Big|_0^{200} - 448$$

$$= -\frac{1}{0.03}(5^{3/2} - 9^{3/2}) - 448$$

$$\approx 79.32,$$

or approximately \$7,932.

Next, the producers' surplus is given by

$$PS = (2.24)(200) - \int_0^{200} \sqrt{1 + 0.02x}\, dx$$

$$= 448 - \int_0^{200} (1 + 0.02x)^{1/2}\, dx$$

$$= 448 - \frac{1}{0.02}\left(\frac{2}{3}\right)(1 + 0.02x)^{3/2}\Big|_0^{200}$$

$$= 448 - \frac{1}{0.03}(5^{3/2} - 1)$$

$$\approx 108.66,$$

or approximately \$10,866.

CHAPTER SIX SUMMARY OF PRINCIPAL FORMULAS AND TERMS

Formulas

1. Indefinite integral of a constant
$$\int k\, du = ku + C$$

2. Power Rule
$$\int u^n\, du = \frac{u^{n+1}}{n + 1} + C$$

3. Constant multiple rule
$$\int kf(u)\, du = k\int f(u)\, du$$
(k a constant)

4. Sum Rule
$$\int [f(u) \pm g(u)]\, du$$
$$= \int f(u)\, du \pm \int g(u)\, du$$

5. Indefinite integral of the exponential function
$$\int e^u\, du = e^u + C$$

6. Indefinite integral of $f(u) = \dfrac{1}{u}$
$$\int \frac{du}{u} = \ln |u| + C$$

7. Method of substitution

$$\int f'(g(x))g'(x)\, dx = \int f'(u)\, du$$

8. Fundamental Theorem of Calculus

$$\int_a^b f(x)\, dx$$
$$= F(b) - F(a),\ F'(x) = f(x)$$

9. Area between two curves

$$\int_a^b [f(x) - g(x)]\, dx,\ f(x) \geq g(x)$$

10. Definite integral as the limit of a sum

$$\int_a^b f(x)\, dx = \lim_{n \to \infty} S_n,$$
where S_n is a Riemann sum

11. Average value of f over $[a, b]$

$$\frac{1}{b-a} \int_a^b f(x)\, dx$$

12. Consumers' surplus

$$CS = \int_0^{\bar{x}} D(x)\, dx - \bar{p}\bar{x}$$

13. Producers' surplus

$$PS = \bar{p}\bar{x} - \int_0^{\bar{x}} S(x)\, dx$$

14. Accumulated (future) value of an income stream

$$A = e^{rT} \int_0^T R(t)e^{-rt}\, dt$$

15. Present value of an income stream

$$PV = \int_0^T R(t)e^{-rt}\, dt$$

16. Present value of an annuity

$$PV = \frac{mP}{r}(1 - e^{-rT})$$

17. Amount of an annuity

$$A = \frac{mP}{r}(e^{rT} - 1)$$

18. Coefficient of inequality of a Lorentz curve

$$L = 2 \int_0^1 [x - f(x)]\, dx$$

Terms

Antiderivative	Riemann sum
Integration	Definite integral
Indefinite integral	Limits of integration
Integrand	Lorentz curve
Constant of integration	

CHAPTER 6 REVIEW EXERCISES

In Exercises 1–20, evaluate each indefinite integral.

1. $\displaystyle\int (x^3 + 2x^2 - x)\, dx$

2. $\displaystyle\int \left(\tfrac{1}{3}x^3 - 2x^2 + 8\right) dx$

3. $\displaystyle\int \left(x^4 - 2x^3 + \frac{1}{x^2}\right) dx$

4. $\displaystyle\int (x^{1/3} - \sqrt{x} + 4)\, dx$

5. $\displaystyle\int x(2x^2 + x^{1/2})\, dx$

6. $\int (x^2 + 1)(\sqrt{x} - 1)\, dx$

7. $\int \left(x^2 - x + \frac{2}{x} + 5 \right) dx$ **8.** $\int \sqrt{2x + 1}\, dx$

9. $\int (3x - 1)(3x^2 - 2x + 1)^{1/3}\, dx$

10. $\int x^2(x^3 + 2)^{10}\, dx$ **11.** $\int \frac{x - 1}{x^2 - 2x + 5}\, dx$

12. $\int 2e^{-2x}\, dx$ **13.** $\int \left(x + \tfrac{1}{2} \right) e^{x^2+x+1}\, dx$

14. $\int \frac{e^{-x} - 1}{(e^{-x} + x)^2}\, dx$ **15.** $\int \frac{(\ln x)^5}{x}\, dx$

16. $\int \frac{\ln x^2}{x}\, dx$ **17.** $\int x^3(x^2 + 1)^{10}\, dx$

18. $\int x\sqrt{x + 1}\, dx$ **19.** $\int \frac{x}{\sqrt{x - 2}}\, dx$

20. $\int \frac{3x}{\sqrt{x + 1}}\, dx$

In Exercises 21–32, evaluate each definite integral.

21. $\int_0^1 (2x^3 - 3x^2 + 1)\, dx$

22. $\int_0^2 (4x^3 - 9x^2 + 2x - 1)\, dx$

23. $\int_1^4 (\sqrt{x} + x^{-3/2})\, dx$ **24.** $\int_0^1 20x(2x^2 + 1)^4\, dx$

25. $\int_{-1}^0 12(x^2 - 2x)(x^3 - 3x^2 + 1)^3\, dx$

26. $\int_4^7 x\sqrt{x - 3}\, dx$ **27.** $\int_0^2 \frac{x}{x^2 + 1}\, dx$

28. $\int_0^1 \frac{dx}{(5 - 2x)^2}$ **29.** $\int_0^2 \frac{4x}{\sqrt{1 + 2x^2}}\, dx$

30. $\int_0^2 xe^{(-1/2)x^2}\, dx$ **31.** $\int_{-1}^0 \frac{e^{-x}}{(1 + e^{-x})^2}\, dx$

32. $\int_1^e \frac{\ln x}{x}\, dx$

In Exercises 33–36, find the function f given that the slope of the tangent line to the graph at any point $(x, f(x))$ is $f'(x)$ and that the graph of f passes through the given point.

33. $f'(x) = 3x^2 - 4x + 1$; $(1, 1)$

34. $f'(x) = \frac{x}{\sqrt{x^2 + 1}}$; $(0, 1)$

35. $f'(x) = 1 - e^{-x}$; $(0, 2)$

36. $f'(x) = \frac{\ln x}{x}$; $(1, -2)$

37. Let $f(x) = -2x^2 + 1$, and compute the Riemann sum of f over the interval $[1, 2]$ by partitioning the interval into five subintervals of the same length $(n = 5)$ where the points p_i $(1 \le i \le 5)$ are taken to be the *right* endpoints of the respective subintervals.

38. The management of the National Electric Corporation has determined that the daily marginal cost function associated with producing their automatic drip coffeemakers is given by

$$C'(x) = 0.00003x^2 - 0.03x + 20,$$

where $C'(x)$ is measured in dollars per unit and x denotes the number of units produced. Management has also determined that the daily fixed cost incurred in producing these coffeemakers is $500. What is the total cost incurred by National in producing the first 400 coffeemakers per day?

39. Refer to Exercise 38. Management has also determined that the daily marginal revenue function associated with producing and selling their coffeemakers is given by

$$R'(x) = -0.03x + 60,$$

where x denotes the number of units produced and sold and $R'(x)$ is measured in dollars per unit.
a. Determine the revenue function $R(x)$ associated with producing and selling these coffeemakers.
b. What is the demand equation relating the wholesale unit price to the quantity of coffeemakers demanded?

40. The Franklin National Life Insurance Company purchased a new computer for $200,000. If the rate at which the computer's resale value changes is given by the function

$$V'(t) = 3800(t - 10),$$

where t is the length of time since the purchase date and $V'(t)$ is measured in dollars per year, find an expression $V(t)$ that gives the resale value of the computer after t years. How much would the computer cost after six years?

41. The marketing department of the Vista Vision Corporation forecasts that sales of their new line of projection television systems will grow at the rate of

$$3000 - 2000e^{-0.04t} \quad (0 \le t \le 24)$$

units per month once they are introduced into the market. Find an expression giving the total number of units of the projection television systems that Vista may expect to sell t months from the time they are put on the market. How many units of the television systems can Vista expect to sell during the first year?

42. Due to the increasing cost of fuel, the manager of the City Transit Authority estimates that the number of commuters using the city subway system will increase at the rate of

$$3000(1 + 0.4t)^{-1/2} \quad (0 \le t \le 36)$$

per month t months from now. If 100,000 commuters are currently using the system, find an expression giving the total number of commuters who will be using the subway t months from now. How many commuters will be using the subway six months from now?

43. The management of a division of Ditton Industries has determined that the daily marginal cost function associated with producing their hot-air corn poppers is given by

$$C'(x) = 0.00003x^2 - 0.03x + 10,$$

where $C'(x)$ is measured in dollars per unit and x denotes the number of units manufactured. Management has also determined that the daily fixed cost incurred in producing these corn poppers is $600. Find the total cost incurred by Ditton in producing the first 500 corn poppers.

44. In 1980 the world produced 3.5 billion metric tons of coal. If output increased at the rate of

$$3.5e^{0.04t}$$

billion metric tons per year in year t ($t = 0$ corresponds to 1980), determine how much coal was produced worldwide between 1980 and the end of 1985.

45. Find the area of the region under the curve $y = 3x^2 + 2x + 1$ from $x = -1$ to $x = 2$.

46. Find the area of the region under the curve $y = e^{2x}$ from $x = 0$ to $x = 2$.

47. Find the area of the region bounded by the graph of the function $y = 1/x^2$, the x-axis, and the lines $x = 1$ and $x = 3$.

48. Find the area of the region bounded by the curve $y = -x^2 - x + 2$ and the x-axis.

49. Find the area of the region bounded by the graphs of the functions $f(x) = e^x$ and $g(x) = x$ and the vertical lines $x = 0$ and $x = 2$.

50. Find the area of the region that is completely enclosed by the graphs of $f(x) = x^4$ and $g(x) = x$.

51. Find the area of the region between the curve $y = x(x - 1)(x - 2)$ and the x-axis.

52. Using current production techniques, the rate of oil production from a certain oil well t years from now is estimated to be

$$R_1(t) = 100e^{0.05t}$$

thousand barrels per year. Using a new production technique, however, it is estimated that the rate of oil production from that oil well t years from now will be

$$R_2(t) = 100e^{0.08t}$$

thousand barrels per year. Determine how much additional oil will be produced over the next ten years if the new technique is adopted.

53. Find the average value of the function

$$f(x) = \frac{x}{\sqrt{x^2 + 16}}$$

over the interval $[0, 3]$.

54. The demand function for a brand of blank videocassettes is given by

$$p = -0.01x^2 - 0.2x + 23,$$

where p is the wholesale unit price in dollars and x is the quantity demanded each week, measured in units of a thousand. Determine the consumers' surplus if the wholesale unit price is $8 per cassette.

55. The quantity demanded x (in units of a hundred) of the Sportsman 5×7 tents, per week, is related to the unit price p (in dollars) by the relation

$$p = -0.1x^2 - x + 40.$$

The quantity x (in units of a hundred) that the supplier is willing to make available in the market is related to the unit price by the relation

$$p = 0.1x^2 + 2x + 20.$$

If the market price is set at the equilibrium price, find the consumers' surplus and the producers' surplus.

56. Mr. Cunningham plans to deposit $4000 per year in his Keogh Retirement Account. If interest is compounded continuously at the rate of 8 percent per year, how much will he have in his retirement account after 20 years?

57. Miss Parker sold her house under an installment contract whereby the buyer gave her a down payment of $9000 and agreed to make monthly payments of $925 per month for 30 years. If the prevailing interest rate is 12 percent per year compounded continuously, find the present value of the purchase price of the house.

58. Ms. Carlson purchased a ten-year franchise for a health spa that is expected to generate income at the rate of

$$P(t) = 80,000$$

dollars per year. If the prevailing interest rate is 10 percent per year compounded continuously, find the present value of the franchise.

59. A certain country's income distribution is described by the function

$$f(x) = \frac{17}{18}x^2 + \frac{1}{18}x.$$

a. Sketch the Lorentz curve for this function.
b. Compute $f(0.3)$ and $f(0.6)$ and interpret your results.
c. Compute the coefficient of inequality for this Lorentz curve.

60. The population of a certain Sun Belt city, currently 80,000, is expected to grow exponentially in the next five years with a growth constant of 0.05. If the prediction comes true, what will the average population of the city be over the next five years?

What is the area of the oil spill caused by a grounded tanker? In Example 5, page 447, you will see how to determine the area of the oil spill.

Additional Topics in Integration

Besides the basic rules of integration developed in Chapter 6, there are more sophisticated techniques for finding the antiderivatives of functions. We begin this chapter by looking at the method of integration by parts. Another technique of integration involves using tables of integrals that have been compiled for this purpose. We will also look at numerical methods of integration that enable us to obtain approximate solutions to definite integrals, especially those whose exact value cannot be found otherwise. More specifically, we will study the Trapezoidal Rule and Simpson's Rule. Numerical integration methods are especially useful when the integrand is known only at discrete points. Finally, we learn how to evaluate integrals in which the intervals of integration are unbounded. Such integrals, called *improper integrals*, play an important role in the study of probability, the last topic of this chapter.

| 7.1 | Integration by Parts |

The Method of Integration by Parts

Integration by parts is another technique of integration that, like the method of substitution discussed in Chapter 6, is based on a corresponding rule of differentiation. In this case, the rule of differentiation is the Product Rule, which asserts that if f and g are differentiable functions, then

$$\frac{d}{dx}[f(x)g(x)] = f(x)g'(x) + g(x)f'(x). \qquad (1)$$

If we integrate both sides of (1) with respect to x, we obtain

$$\int \frac{d}{dx} f(x)g(x)\, dx = \int f(x)g'(x)\, dx + \int g(x)f'(x)\, dx$$

or

$$f(x)g(x) = \int f(x)g'(x)\, dx + \int g(x)f'(x)\, dx.$$

This last equation, which may be written in the form

$$\int f(x)g'(x)\, dx = f(x)g(x) - \int g(x)f'(x)\, dx, \qquad (2)$$

is called the formula for **integration by parts.** This formula is useful since it enables us to express one indefinite integral in terms of another that may be easier to evaluate. Formula (2) may be simplified by letting

$$u = f(x) \qquad\qquad dv = g'(x)\, dx$$
$$du = f'(x)\, dx \qquad\qquad v = g(x),$$

giving the following version of the formula for integration by parts.

| **INTEGRATION BY PARTS FORMULA** | $\displaystyle\int u\, dv = uv - \int v\, du$ | (3) |

EXAMPLE 1   Evaluate $\displaystyle\int xe^x\, dx$.

Solution   No method of integration developed thus far enables us to evaluate the given indefinite integral in its present form. Therefore, we will attempt to write it in terms of an indefinite integral that will be easier to evaluate. Let us use the integration by parts formula (3) by letting

$$u = \ x \quad \text{and} \quad dv = e^x\, dx$$

so that

$$du = dx \quad \text{and} \quad v = e^x.$$

Therefore
$$\int xe^x \, dx = \int u \, dv$$

$$= uv - \int v \, du$$

$$= xe^x - \int e^x \, dx$$

$$= xe^x - e^x + C$$
$$= (x - 1)e^x + C.$$ ∎

The success of the method of integration by parts depends on the proper choice of u and dv. For example, if we had chosen

$$u = e^x \quad \text{and} \quad dv = x \, dx$$

in the last example, then

$$du = e^x \, dx \quad \text{and} \quad v = \frac{1}{2}x^2,$$

so (3) would have yielded

$$\int xe^x \, dx = \int u \, dv$$

$$= uv - \int v \, du$$

$$= \frac{1}{2}x^2 e^x - \int \frac{1}{2}x^2 e^x \, dx.$$

Since the indefinite integral on the right-hand side of this equation is not readily evaluated (it is, in fact, more complicated than the original integral!), choosing u and dv as shown has not helped us evaluate the given indefinite integral.

In general, we can use the following guidelines.

GUIDELINES FOR CHOOSING u AND dv

Choose u and dv so that

1. du is simpler than u.
2. dv is easy to integrate.

EXAMPLE 2 Evaluate $\int x \ln x \, dx$.

Solution Letting
$$u = \ln x \quad \text{and} \quad dv = x \, dx,$$

we have
$$du = \frac{1}{x} \, dx \quad \text{and} \quad v = \frac{1}{2}x^2.$$

Therefore, $\displaystyle\int x \ln x \, dx = \int u \, dv = uv - \int v \, du$

$$= \frac{1}{2}x^2 \ln x - \int \frac{1}{2}x^2 \cdot \left(\frac{1}{x}\right) dx$$

$$= \frac{1}{2}x^2 \ln x - \frac{1}{2}\int x \, dx$$

$$= \frac{1}{2}x^2 \ln x - \frac{1}{4}x^2 + C$$

$$= \frac{1}{4}x^2(2 \ln x - 1) + C.$$ ∎

EXAMPLE 3   Evaluate

$$\int \frac{xe^x}{(x+1)^2} \, dx.$$

Solution   Let

$$u = xe^x \quad \text{and} \quad dv = \frac{1}{(x+1)^2} \, dx.$$

Then $du = (xe^x + e^x) \, dx = e^x(x+1) \, dx \quad \text{and} \quad v = -\frac{1}{x+1}.$

Therefore, $\displaystyle\int \frac{xe^x}{(x+1)^2} \, dx = \int u \, dv = uv - \int v \, du$

$$= xe^x\left(\frac{-1}{x+1}\right) - \int\left(-\frac{1}{x+1}\right)e^x(x+1) \, dx$$

$$= -\frac{xe^x}{x+1} + \int e^x \, dx$$

$$= -\frac{xe^x}{x+1} + e^x + C$$

$$= \frac{e^x}{x+1} + C.$$ ∎

The next example shows that repeated applications of the technique of integration by parts is sometimes required to evaluate an integral.

EXAMPLE 4   Evaluate

$$\int x^2 e^x \, dx.$$

Solution   Let

$$u = x^2 \qquad \text{and} \quad dv = e^x \, dx,$$
so that $\qquad\qquad du = 2x \, dx \quad \text{and} \quad v = e^x.$

Therefore, $\displaystyle\int x^2 e^x \, dx = \int u \, dv = uv - \int v \, du$

$$= x^2 e^x - \int e^x(2x) \, dx = x^2 e^x - 2\int xe^x \, dx.$$

To complete the solution of the problem, we need to evaluate the integral

$$\int xe^x \, dx.$$

But this integral may be found using integration by parts. In fact, you will recognize that this integral is precisely that of Example 1. Using the results obtained there, we now find

$$\int x^2 e^x \, dx = x^2 e^x - 2[(x-1)e^x] + C = e^x(x^2 - 2x + 2) + C.$$

∎

Application

EXAMPLE 5 The estimated rate at which oil will be produced from a certain oil well t years after production has begun is given by

$$R(t) = 100te^{-0.1t}$$

thousand barrels per year. Find an expression that describes the total production of oil at the end of year t.

Solution Let $T(t)$ denote the total production of oil from the well at the end of year t ($t \geq 0$). Then the rate of oil production will be given by $T'(t)$ thousand barrels per year. Thus,

$$T'(t) = R(t) = 100te^{-0.1t},$$

so

$$T(t) = \int 100te^{-0.1t} \, dt$$

$$= 100 \int te^{-0.1t} \, dt.$$

We use the technique of integration by parts to evaluate this integral. Let

$$u = t \quad \text{and} \quad dv = e^{-0.1t} \, dt$$

so that

$$du = dt \quad \text{and} \quad v = -\frac{1}{0.1}e^{-0.1t} = -10e^{-0.1t}.$$

Therefore,

$$T(t) = 100\left[-10te^{-0.1t} + 10 \int e^{-0.1t} \, dt \right]$$

$$= 100[-10te^{-0.1t} - 100e^{-0.1t}] + C$$

$$= -1{,}000e^{-0.1t}(t + 10) + C.$$

To determine the value of C, note that the total quantity of oil produced at the end of year 0 is nil, so $T(0) = 0$. This gives

$$T(0) = -1{,}000(10) + C = 0,$$

or

$$C = 10{,}000.$$

Thus, the required production function is given by

$$T(t) = -1{,}000e^{-0.1t}(t + 10) + 10{,}000.$$

∎

SELF-CHECK EXERCISES 7.1

1. Evaluate $\int x^2 \ln x \, dx$.

2. Since the inauguration of Ryan's Express at the beginning of 1992, the number of passengers (in millions) flying on this commuter airline has been growing at the rate of

$$R(t) = 0.1 + 0.2te^{-0.4t}$$

passengers per year ($t = 0$ corresponds to the beginning of 1992). Assuming that this trend continues through 1996, determine how many passengers will have flown on Ryan's Express by that time.

Solutions to Self-Check Exercises 7.1 can be found on page 431.

7.1   EXERCISES

In Exercises 1–26, evaluate each indefinite integral.

1. $\int xe^{2x} \, dx$

2. $\int xe^{-x} \, dx$

3. $\int xe^{x/4} \, dx$

4. $\int 6xe^{3x} \, dx$

5. $\int (e^x - x)^2 \, dx$

6. $\int (e^{-x} + x)^2 \, dx$

7. $\int (x + 1)e^x \, dx$

8. $\int (x - 3)e^{3x} \, dx$

9. $\int x(x + 1)^{-3/2} \, dx$

10. $\int x(x + 4)^{-2} \, dx$

11. $\int x\sqrt{x - 5} \, dx$

12. $\int \frac{x}{\sqrt{2x + 3}} \, dx$

13. $\int x \ln 2x \, dx$

14. $\int x^2 \ln 2x \, dx$

15. $\int x^3 \ln x \, dx$

16. $\int \sqrt{x} \ln x \, dx$

17. $\int \sqrt{x} \ln \sqrt{x} \, dx$

18. $\int \frac{\ln x}{\sqrt{x}} \, dx$

19. $\int \frac{\ln x}{x^2} \, dx$

20. $\int \frac{\ln x}{x^3} \, dx$

21. $\int \ln x \, dx$ [*Hint:* Let $u = \ln x$ and $dv = dx$.]

22. $\int \ln(x + 1) \, dx$

23. $\int x^2 e^{-x} \, dx$ [*Hint:* Integrate by parts twice.]

24. $\int e^{-\sqrt{x}} \, dx$ [*Hint:* First make the substitution $u = \sqrt{x}$; then integrate by parts.]

25. $\int x(\ln x)^2 \, dx$ [*Hint:* Integrate by parts twice.]

26. $\int x \ln(x + 1) \, dx$ [*Hint:* First make the substitution $u = x + 1$; then integrate by parts.]

In Exercises 27–32, evaluate each definite integral by using the method of integration by parts.

27. $\int_0^{\ln 2} xe^x \, dx$

28. $\int_0^2 xe^{-x} \, dx$

29. $\int_1^4 \ln x \, dx$

30. $\int_1^2 x \ln x \, dx$

31. $\int_0^2 xe^{2x} \, dx$

32. $\int_0^1 x^2 e^{-x} \, dx$

33. Find the function f given that the slope of the tangent line to the graph of f at any point $(x, f(x))$ is xe^{-2x} and that the graph passes through the point $(0, 3)$.

34. Find the function f given that the slope of the tangent line to the graph of f at any point $(x, f(x))$ is $x\sqrt{x + 1}$ and that the graph passes through the point $(3, 6)$.

35. Find the area of the region under the graph of $f(x) = \ln x$ from $x = 1$ to $x = 5$.

36. Find the area of the region under the graph of $f(x) = xe^{-x}$ from $x = 0$ to $x = 3$.

37. **Velocity of a Dragster**   The velocity of a dragster t seconds after leaving the starting line is

$$100te^{-0.2t}$$

feet per second. What is the distance covered by the dragster in the first 10 seconds of its run?

38. **Production of Steam Coal** In keeping with the projected increase in worldwide demand for steam coal, the boiler-firing fuel used for generating electricity, the management of Consolidated Mining has decided to step up its mining operations. Plans call for increasing the yearly production of steam coal by

$$2te^{-0.05t}$$

million metric tons per year for the next 20 years. The current yearly production is 20 million metric tons. Find a function that describes Consolidated's total production of steam coal at the end of t years. How much coal will Consolidated have produced over the next 20 years if this plan is carried out?

39. **Concentration of a Drug in the Bloodstream** The concentration of a certain drug in a patient's bloodstream t hours after it has been administered is given by $C(t) = 3te^{-t/3}$ mg/ml. Find the average concentration of the drug in the patient's bloodstream over the first 12 hours after administration.

C 40. **Alcohol-Related Traffic Accidents** As a result of increasingly stiff laws aimed at reducing the number of alcohol-related traffic accidents in a certain state, preliminary data indicate that the number of such accidents has been changing at the rate of

$$R(t) = -10 - te^{0.1t}$$

accidents per month t months after the laws took effect. There were 982 alcohol-related accidents for the year before the enactment of the laws. Determine how many alcohol-related accidents were expected during the first year the laws were in effect.

C 41. **Record Sales** Sales of the latest recording by Brittania, a British rock group, are currently $2te^{-0.1t}$ units per week (each unit representing 10,000 albums), where t denotes the number of weeks since the recording's release. Find an expression that gives the total number of units of record albums sold as a function of t.

42. **Rate of Return on an Investment** Suppose that an investment is expected to generate income at the rate of

$$P(t) = 30,000 + 800t$$

dollars per year for the next five years. Find the present value of this investment if the prevailing interest rate is 8 percent per year compounded continuously. [*Hint:* Use Formula (16), Section 6.7 (page 410).]

43. **Present Value of a Franchise** Ms. Hart purchased a 15-year franchise for a computer outlet store that is expected to generate income at the rate of

$$P(t) = 50,000 + 3,000t$$

dollars per year. If the prevailing interest rate is 10 percent per year compounded continuously, find the present value of the franchise. [*Hint:* Use Formula (16), Section 6.7 (page 410).]

44. **Growth of HMOs** The membership of the Cambridge Community Health Plan, Inc. (a health maintenance organization) is projected to grow at the rate of $9\sqrt{t+1}\ln\sqrt{t+1}$ thousand people per year, t years from now. If the HMO's current membership is 50,000, what will the membership be five years from now?

SOLUTIONS TO SELF-CHECK EXERCISES 7.1

1. Let $u = \ln x$ and $dv = x^2\,dx$ so that $du = \dfrac{1}{x}\,dx$ and $v = \dfrac{1}{3}x^3$.

Therefore,

$$\int x^2 \ln x\,dx = \int u\,dv = uv - \int v\,du$$

$$= \frac{1}{3}x^3 \ln x - \int \frac{1}{3}x^2\,dx$$

$$= \frac{1}{3}x^3 \ln x - \frac{1}{9}x^3 + C$$

$$= \frac{1}{9}x^3(3 \ln x - 1) + C.$$

2. Let $N(t)$ denote the total number of passengers who will have flown on Ryan's Express by the end of year t. Then $N'(t) = R(t)$, so that

$$N(t) = \int R(t)\, dt$$
$$= \int (0.1 + 0.2te^{-0.4t})\, dt$$
$$= \int 0.1\, dt + 0.2 \int te^{-0.4t}\, dt.$$

We now use the technique of integration by parts on the second integral. Letting $u = t$ and $dv = e^{-0.4t}\, dt$, we have

$$du = dt \quad \text{and} \quad v = -\frac{1}{0.4}e^{-0.4t} = -2.5e^{-0.4t}.$$

Therefore,

$$N(t) = 0.1t + 0.2\left[-2.5te^{-0.4t} + 2.5 \int e^{-0.4t}\, dt \right]$$
$$= 0.1t - 0.5te^{-0.4t} - \frac{0.5}{0.4}e^{-0.4t} + C$$
$$= 0.1t - 0.5(t + 2.5)e^{-0.4t} + C.$$

To determine the value of C, note that $N(0) = 0$, which gives

$$N(0) = -0.5(2.5) + C = 0,$$

or $$C = 1.25.$$

Therefore,

$$N(t) = 0.1t - 0.5(t + 2.5)e^{-0.4t} + 1.25.$$

The number of passengers who will have flown on Ryan's Express by the end of 1996 is given by

$$N(5) = 0.1(5) - 0.5(5 + 2.5)e^{-0.4(5)} + 1.25$$
$$= 1.242493,$$

that is, 1,242,493 passengers.

7.2   Integration Using Tables of Integrals

A Table of Integrals

We have studied several techniques for finding an antiderivative of a function. Useful as they are, these techniques are not always applicable. There are, of course, numerous other methods for finding an antiderivative of a function. These methods have been used to compile extensive lists of integration formulas.

A small sample of the integration formulas that can be found in many mathematical handbooks is given in the following table of integrals. These formulas are grouped according to the basic form of the integrand. Note that it may be necessary to modify the integrand of the integral to be evaluated in order to make use of one of these formulas.

Table of Integrals

Forms involving $a + bu$

1. $\displaystyle\int \frac{u\,du}{a + bu} = \frac{1}{b^2}[a + bu - a\ln|a + bu|] + C$

2. $\displaystyle\int \frac{u^2\,du}{a + bu} = \frac{1}{2b^3}[(a + bu)^2 - 4a(a + bu) + 2a^2\ln|a + bu|] + C$

3. $\displaystyle\int \frac{u\,du}{(a + bu)^2} = \frac{1}{b^2}\left[\frac{a}{a + bu} + \ln|a + bu|\right] + C$

4. $\displaystyle\int u\sqrt{a + bu}\,du = \frac{2}{15b^2}(3bu - 2a)(a + bu)^{3/2} + C$

5. $\displaystyle\int \frac{u\,du}{\sqrt{a + bu}} = \frac{2}{3b^2}(bu - 2a)\sqrt{a + bu} + C$

6. $\displaystyle\int \frac{du}{u\sqrt{a + bu}} = \frac{1}{\sqrt{a}}\ln\left|\frac{\sqrt{a + bu} - \sqrt{a}}{\sqrt{a + bu} + \sqrt{a}}\right| + C \qquad \text{if } a > 0$

Forms involving $\sqrt{a^2 + u^2}$

7. $\displaystyle\int \sqrt{a^2 + u^2}\,du = \frac{u}{2}\sqrt{a^2 + u^2} + \frac{a^2}{2}\ln|u + \sqrt{a^2 + u^2}| + C$

8. $\displaystyle\int u^2\sqrt{a^2 + u^2}\,du$
$= \frac{u}{8}(a^2 + 2u^2)\sqrt{a^2 + u^2} - \frac{a^4}{8}\ln|u + \sqrt{a^2 + u^2}| + C$

9. $\displaystyle\int \frac{du}{\sqrt{a^2 + u^2}} = \ln|u + \sqrt{a^2 + u^2}| + C$

10. $\displaystyle\int \frac{du}{u\sqrt{a^2 + u^2}} = -\frac{1}{a}\ln\left|\frac{\sqrt{a^2 + u^2} + a}{u}\right| + C$

11. $\displaystyle\int \frac{du}{u^2\sqrt{a^2 + u^2}} = -\frac{\sqrt{a^2 + u^2}}{a^2 u} + C$

12. $\displaystyle\int \frac{du}{(a^2 + u^2)^{3/2}} = \frac{u}{a^2\sqrt{a^2 + u^2}} + C$

Forms involving $\sqrt{u^2 - a^2}$

13. $\displaystyle\int \sqrt{u^2 - a^2}\,du = \frac{u}{2}\sqrt{u^2 - a^2} - \frac{a^2}{2}\ln|u + \sqrt{u^2 - a^2}| + C$

14. $\displaystyle\int u^2\sqrt{u^2 - a^2}\,du$
$= \frac{u}{8}(2u^2 - a^2)\sqrt{u^2 - a^2} - \frac{a^4}{8}\ln|u + \sqrt{u^2 - a^2}| + C$

15. $\displaystyle\int \frac{\sqrt{u^2 - a^2}}{u^2}\,du = -\frac{\sqrt{u^2 - a^2}}{u} + \ln|u + \sqrt{u^2 - a^2}| + C$

16. $\displaystyle\int \frac{du}{\sqrt{u^2 - a^2}} = \ln|u + \sqrt{u^2 - a^2}| + C$

17. $\displaystyle\int \frac{du}{u^2\sqrt{u^2 - a^2}} = \frac{\sqrt{u^2 - a^2}}{a^2 u} + C$

18. $\displaystyle\int \frac{du}{(u^2 - a^2)^{3/2}} = -\frac{u}{a^2\sqrt{u^2 - a^2}} + C$

Forms involving $\sqrt{a^2 - u^2}$

19. $\displaystyle\int \frac{\sqrt{a^2 - u^2}}{u}\, du = \sqrt{a^2 - u^2} - a\ln\left|\frac{a + \sqrt{a^2 - u^2}}{u}\right| + C$

20. $\displaystyle\int \frac{du}{u\sqrt{a^2 - u^2}} = -\frac{1}{a}\ln\left|\frac{a + \sqrt{a^2 - u^2}}{u}\right| + C$

21. $\displaystyle\int \frac{du}{u^2\sqrt{a^2 - u^2}} = -\frac{\sqrt{a^2 - u^2}}{a^2 u} + C$

22. $\displaystyle\int \frac{du}{(a^2 - u^2)^{3/2}} = \frac{u}{a^2\sqrt{a^2 - u^2}} + C$

Forms involving e^{au} and $\ln u$

23. $\displaystyle\int u e^{au}\, du = \frac{1}{a^2}(au - 1)e^{au} + C$

24. $\displaystyle\int u^n e^{au}\, du = \frac{1}{a}u^n e^{au} - \frac{n}{a}\int u^{n-1} e^{au}\, du$

25. $\displaystyle\int \frac{du}{1 + be^{au}} = u - \frac{1}{a}\ln(1 + be^{au}) + C$

26. $\displaystyle\int \ln u\, du = u\ln u - u + C$

27. $\displaystyle\int u^n \ln u\, du = \frac{u^{n+1}}{(n + 1)^2}[(n + 1)\ln u - 1] + C \qquad (n \neq -1)$

28. $\displaystyle\int \frac{du}{u\ln u} = \ln|\ln u| + C$

29. $\displaystyle\int (\ln u)^n\, du = u(\ln u)^n - n\int (\ln u)^{n-1}\, du$

Using a Table of Integrals

We now consider several examples that illustrate how the table of integrals can be used to evaluate an integral.

EXAMPLE 1   Use the table of integrals to evaluate $\displaystyle\int \frac{2x\, dx}{\sqrt{3 + x}}$.

Solution   We first write

$$\int \frac{2x\, dx}{\sqrt{3 + x}} = 2\int \frac{x\, dx}{\sqrt{3 + x}}.$$

Since $\sqrt{3 + x}$ is of the form $\sqrt{a + bu}$ with $a = 3$, $b = 1$, and $u = x$, we use Formula 5,

$$\int \frac{u \, du}{\sqrt{a + bu}} = \frac{2}{3b^2}(bu - 2a)\sqrt{a + bu} + C,$$

obtaining

$$2 \int \frac{x}{\sqrt{3 + x}} \, dx$$

$$= 2\left[\frac{2}{3(1)}(x - 6)\sqrt{3 + x}\right] + C$$

$$= \frac{4}{3}(x - 6)\sqrt{3 + x} + C.$$ ∎

EXAMPLE 2   Use the table of integrals to evaluate $\int x^2 \sqrt{3 + x^2} \, dx$.

Solution   Observe that if we write 3 as $(\sqrt{3})^2$, then $3 + x^2$ has the form $\sqrt{a^2 + u^2}$, with $a = \sqrt{3}$ and $u = x$. Using Formula 8,

$$\int u^2 \sqrt{a^2 + u^2} \, du = \frac{u}{8}(a^2 + 2u^2)\sqrt{a^2 + u^2} - \frac{a^4}{8} \ln|u + \sqrt{a^2 + u^2}| + C,$$

we obtain

$$\int x^2\sqrt{3 + x^2} \, dx = \frac{x}{8}(3 + 2x^2)\sqrt{3 + x^2} - \frac{9}{8} \ln|x + \sqrt{3 + x^2}| + C.$$ ∎

EXAMPLE 3   Use the table of integrals to evaluate

$$\int_3^4 \frac{dx}{x^2\sqrt{50 - 2x^2}}.$$

Solution   We first evaluate the indefinite integral

$$I = \int \frac{dx}{x^2\sqrt{50 - 2x^2}}.$$

Observe that $\sqrt{50 - 2x^2} = \sqrt{2(25 - x^2)} = \sqrt{2}\sqrt{25 - x^2}$, so we can write I as

$$I = \frac{1}{\sqrt{2}} \int \frac{dx}{x^2\sqrt{25 - x^2}} = \frac{\sqrt{2}}{2} \int \frac{dx}{x^2\sqrt{25 - x^2}} \qquad \left(\frac{1}{\sqrt{2}} = \frac{\sqrt{2}}{2}\right).$$

Next, using Formula 21,

$$\int \frac{du}{u^2\sqrt{a^2 - u^2}} = -\frac{\sqrt{a^2 - u^2}}{a^2 u} + C,$$

with $a = 5$ and $u = x$, we find

$$I = \frac{\sqrt{2}}{2}\left[-\frac{\sqrt{25 - x^2}}{25x}\right]$$

$$= -\left(\frac{\sqrt{2}}{50}\right)\frac{\sqrt{25 - x^2}}{x}.$$

Finally, using this result, we obtain

$$\int_3^4 \frac{dx}{x^2\sqrt{50 - 2x^2}} = \left. -\frac{\sqrt{2}}{50}\frac{\sqrt{25 - x^2}}{x}\right|_3^4$$

$$= -\frac{\sqrt{2}}{50}\frac{\sqrt{25 - 16}}{4} - \left(-\frac{\sqrt{2}}{50}\frac{\sqrt{25 - 9}}{3}\right)$$

$$= -\frac{3\sqrt{2}}{200} + \frac{2\sqrt{2}}{75} = \frac{7\sqrt{2}}{600}.$$ ■

As illustrated in the next example, we may need to apply a formula more than once in order to evaluate an integral.

EXAMPLE 4   Use the table of integrals to evaluate $\int x^2 e^{(-1/2)x}\, dx$.

Solution   Scanning the table of integrals for a formula involving e^{ax} in the integrand, we are led to Formula 24,

$$\int u^n e^{au}\, du = \frac{1}{a}u^n e^{au} - \frac{n}{a}\int u^{n-1} e^{au}\, du.$$

With $n = 2$, $a = -\frac{1}{2}$, and $u = x$, we have

$$\int x^2 e^{(-1/2)x}\, dx = \left(\frac{1}{-1/2}\right)x^2 e^{(-1/2)x} - \frac{2}{(-1/2)}\int x e^{(-1/2)x}\, dx$$

$$= -2x^2 e^{(-1/2)x} + 4\int x e^{(-1/2)x}\, dx.$$

If we use Formula 24 once again with $n = 1$, $a = -\frac{1}{2}$, and $u = x$ to evaluate the integral on the right, we obtain

$$\int x^2 e^{(-1/2)x}\, dx$$

$$= -2x^2 e^{(-1/2)x} + 4\left[\left(\frac{1}{-1/2}\right)x e^{(-1/2)x} - \frac{1}{(-1/2)}\int e^{(-1/2)x}\, dx\right]$$

$$= -2x^2 e^{(-1/2)x} + 4\left[-2x e^{(-1/2)x} + 2 \cdot \frac{1}{(-1/2)}e^{(-1/2)x}\right] + C$$

$$= -e^{(-1/2)x}(2x^2 + 8x + 16) + C.$$ ■

EXAMPLE 5   A study prepared for the National Association of Realtors estimated that the mortgage rate over the next t months will be

$$r(t) = \frac{8t + 100}{t + 10} \qquad (0 \le t \le 24)$$

percent per year. If the prediction holds true, what will the average mortgage rate be over the next 12 months?

Solution   The average mortgage rate over the next 12 months will be given by

$$A = \frac{1}{12 - 0} \int_0^{12} \frac{8t + 100}{t + 10} \, dt$$

$$= \frac{1}{12}\left[\int_0^{12} \frac{8t}{t + 10} \, dt + \int_0^{12} \frac{100}{t + 10} \, dt\right]$$

$$= \frac{8}{12} \int_0^{12} \frac{t}{t + 10} \, dt + \frac{100}{12} \int_0^{12} \frac{1}{t + 10} \, dt.$$

Using Formula 1,

$$\int \frac{u \, du}{a + bu} = \frac{1}{b^2}[a + bu - a \ln|a + bu|] + C \qquad (a = 10, b = 1, u = t),$$

to evaluate the first integral, we have

$$A = \left(\frac{2}{3}\right)[10 + t - 10 \ln(10 + t)]\Big|_0^{12} + \left(\frac{25}{3}\right)\ln(10 + t)\Big|_0^{12}$$

$$= \left(\frac{2}{3}\right)[(22 - 10 \ln 22) - (10 - 10 \ln 10)] + \left(\frac{25}{3}\right)[\ln 22 - \ln 10]$$

$$\approx 9.31,$$

or approximately 9.31 percent per year.   ∎

SELF-CHECK EXERCISES 7.2

1. Use the table of integrals to evaluate

$$\int_0^2 \frac{dx}{(5 - x^2)^{3/2}}.$$

c 2. During a flu epidemic, the number of children in the Easton Middle School who contracted influenza t days after the outbreak began was given by

$$N(t) = \frac{200}{1 + 9e^{-0.8t}}.$$

Determine the average number of children who contracted the flu in the first ten days of the epidemic.

Solutions to Self-Check Exercises 7.2 can be found on page 439.

7.2   EXERCISES

In Exercises 1–32, use the table of integrals in this section to evaluate each of the given integrals.

1. $\displaystyle\int \frac{2x}{2 + 3x} \, dx$

2. $\displaystyle\int \frac{x}{(1 + 2x)^2} \, dx$

3. $\displaystyle\int \frac{3x^2}{2 + 4x} \, dx$

4. $\displaystyle\int \frac{x^2}{3 + x} \, dx$

5. $\displaystyle\int x^2\sqrt{9 + 4x^2} \, dx$

6. $\displaystyle\int x^2\sqrt{4 + x^2} \, dx$

7. $\displaystyle\int \frac{dx}{x\sqrt{1 + 4x}}$

8. $\displaystyle\int_0^2 \frac{x + 1}{\sqrt{2 + 3x}}\, dx$

9. $\displaystyle\int_0^2 \frac{dx}{\sqrt{9 + 4x^2}}$

10. $\displaystyle\int \frac{dx}{x\sqrt{4 + 8x^2}}$

11. $\displaystyle\int \frac{dx}{(9 - x^2)^{3/2}}$

12. $\displaystyle\int \frac{dx}{(2 - x^2)^{3/2}}$

13. $\displaystyle\int x^2\sqrt{x^2 - 4}\, dx$

14. $\displaystyle\int_3^5 \frac{dx}{x^2\sqrt{x^2 - 9}}$

15. $\displaystyle\int \frac{\sqrt{4 - x^2}}{x}\, dx$

16. $\displaystyle\int_0^1 \frac{dx}{(4 - x^2)^{3/2}}$

17. $\displaystyle\int xe^{2x}\, dx$

18. $\displaystyle\int \frac{dx}{1 + e^{-x}}$

19. $\displaystyle\int \frac{dx}{(x + 1)\ln(1 + x)}$
[*Hint:* First use the substitution $u = x + 1$.]

20. $\displaystyle\int \frac{x}{(x^2 + 1)\ln(x^2 + 1)}\, dx$
[*Hint:* First use the substitution $u = x^2 + 1$.]

21. $\displaystyle\int \frac{e^{2x}}{(1 + 3e^x)^2}\, dx$

22. $\displaystyle\int \frac{e^{2x}}{\sqrt{1 + 3e^x}}\, dx$

23. $\displaystyle\int \frac{3e^x}{1 + e^{(1/2)x}}\, dx$

24. $\displaystyle\int \frac{dx}{1 - 2e^{-x}}$

25. $\displaystyle\int \frac{\ln x}{x(2 + 3\ln x)}\, dx$

26. $\displaystyle\int_1^e (\ln x)^2\, dx$

27. $\displaystyle\int_0^1 x^2 e^x\, dx$

28. $\displaystyle\int x^3 e^{2x}\, dx$

29. $\displaystyle\int x^2 \ln x\, dx$

30. $\displaystyle\int x^3 \ln x\, dx$

31. $\displaystyle\int (\ln x)^3\, dx$

32. $\displaystyle\int (\ln x)^4\, dx$

33. **Consumers' Surplus**   Refer to Section 6.7. The demand function for Apex ladies' boots is
$$p = \frac{250}{\sqrt{16 + x^2}},$$
where p is the wholesale unit price in dollars and x is the quantity demanded daily, in units of a hundred. Find the consumers' surplus if the wholesale price is set at $50 per pair.

34. **Producers' Surplus**   Refer to Section 6.7. The supplier of Apex ladies' boots will make x hundred pairs of the boots available in the market daily when the wholesale unit price is
$$p = \frac{30x}{5 - x}$$

dollars. Find the producers' surplus if the wholesale price is set at $50 per pair.

c 35. **Amusement Park Attendance**   The management of Astro World ("The Amusement Park of the Future") estimates that the number of visitors (in thousands) entering the amusement park t hours after opening time at 9 A.M. is given by
$$R(t) = \frac{60}{(2 + t^2)^{3/2}}$$
per hour. Determine the number of visitors admitted by noon.

36. **Voter Registration**   The number of voters in a certain district of a city is expected to grow at the rate of
$$R(t) = \frac{3000}{\sqrt{4 + t^2}}$$
people per year t years from now. If the number of voters at present is 20,000, find how many voters will be in the district five years from now.

c 37. **Growth of Fruit Flies**   Based on data collected during an experiment, a biologist found that the number of fruit flies (*Drosophila*) with a limited food supply could be approximated by the exponential model
$$N(t) = \frac{1000}{1 + 24e^{-0.02t}},$$
where t denotes the number of days since the beginning of the experiment. Find the average number of fruit flies in the colony in the first 10 days of the experiment; in the first 20 days.

c 38. **VCR Ownership**   According to estimates by Paul Kroger Associates, the percentage of households that own videocassette recorders (VCRs) is given by
$$P(t) = \frac{68}{1 + 21.67e^{-0.62t}} \qquad (0 \le t \le 12),$$
where t is measured in years, with $t = 0$ corresponding to the beginning of 1981. Find the average percentage of households owning VCRs from the beginning of 1981 to the beginning of 1993.

39. **Recycling Programs**   The commissioner of the City of Newton Department of Public Works estimates that the number of people in the city who have been recycling their magazines in year t following the introduction of the recycling program at the beginning of 1990 is
$$N(t) = \frac{100,000}{2 + 3e^{-0.2t}}.$$

Find the average number of people who will have recycled their magazines during the first five years since the program was introduced.

40. **Franchises** Ms. Lane purchased a ten-year franchise for a fast-food restaurant that is expected to generate income at the rate of $R(t) = 250,000 + 2,000t^2$ dollars per year, t years from now. If the prevailing interest rate is 10 percent per year compounded continuously, find the present value of the franchise. [*Hint:* Use Formula (16), Section 6.7.]

41. **Accumulated Value of an Income Stream** The revenue of Virtual Reality, a video-game arcade, is generated at the rate of $R(t) = 20,000t$ dollars. If the revenue is invested t years from now in a business earning interest

at the rate of 15 percent per year compounded continuously, find the accumulated value of this stream of income at the end of five years. [*Hint:* Use Formula (16), Section 6.7.]

42. **Lorentz Curves** In a study conducted by a certain country's Economic Development Board regarding the income distribution of certain segments of the country's work force, it was found that the Lorentz curve for the distribution of income of college professors is described by the function

$$g(x) = \frac{1}{3}x\sqrt{1 + 8x}.$$

Compute the coefficient of inequality of the Lorentz curve. [*Hint:* Use Formula (19), Section 6.7.]

SOLUTIONS TO SELF-CHECK EXERCISES 7.2

1. Using Formula 22, page 434, with $a^2 = 5$, we see that

$$\int_0^2 \frac{dx}{(5 - x^2)^{3/2}} = \frac{x}{5\sqrt{5 - x^2}}\bigg|_0^2$$

$$= \frac{2}{5\sqrt{5 - 4}}$$

$$= \frac{2}{5}.$$

2. The average number of children who contracted the flu in the first ten days of the epidemic is given by

$$A = \frac{1}{10}\int_0^{10} \frac{200}{1 + 9e^{-0.8t}}\,dt - 20\int_0^{10} \frac{dt}{1 + 9e^{-0.8t}}$$

$$= 20\left[t + \frac{1}{0.8}\ln(1 + 9e^{-0.8t})\right]\bigg|_0^{10} \quad \text{(Formula 25, } a = 0.8,$$
$$b = 9, u = t)$$

$$= 20\left[10 + \frac{1}{0.8}\ln(1 + 9e^{-8})\right] - 20\left(\frac{1}{0.8}\right)\ln 10$$

$$\approx 200.07536 - 57.56463$$

$$\approx 143,$$

or 143 students.

7.3 # Numerical Integration

Approximating Definite Integrals

One method of measuring cardiac output is to inject 5 to 10 mg of a dye into a vein leading to the heart. After making its way through the lungs, the dye returns to the heart and is pumped into the aorta, where its concentration is measured at equal time intervals. The graph of the function c in Figure 7.1

FIGURE 7.1

The function c gives the concentration of a dye measured at the aorta. The graph is constructed by drawing a smooth curve through a set of discrete points.

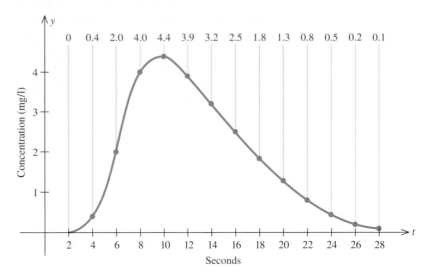

shows the concentration of dye in a person's aorta, measured at 2-second intervals after 5 mg of dye have been injected. The person's cardiac output (measured in l/min) is computed using the formula

$$R = \frac{60D}{\displaystyle\int_0^{28} c(t)\, dt}, \qquad (4)$$

where D is the quantity of dye injected (see Exercise 40).

Now, in order to use (4), we need to evaluate the definite integral

$$\int_0^{28} c(t)\, dt.$$

But we do not have the algebraic rule defining the integrand c for all values of t in [0, 28]. In fact, we are given its values only at a set of discrete points in that interval. In situations such as this, the Fundamental Theorem of Calculus proves useless because we cannot find an antiderivative of c. (We will complete the solution to this problem in Example 4.)

Other situations also arise in which an integrable function has an antiderivative that cannot be found in terms of elementary functions (functions that can be expressed as a finite combination of algebraic, exponential, logarithmic, and trigonometric functions). Examples of such functions are

$$f(x) = e^{x^2}, \qquad g(x) = x^{-1/2}e^x, \quad \text{and} \quad h(x) = \frac{1}{\ln x}.$$

Riemann sums provide us with a good approximation of a definite integral, provided the number of subintervals in the partitions is large enough. But there are better techniques and formulas, called *quadrature formulas*, that give a more efficient way of computing approximate values of definite integrals. In this section, we look at two rather simple but effective ways of approximating definite integrals.

The Trapezoidal Rule

We assume that $f(x) \geq 0$ on $[a, b]$ in order to simplify the derivation of the Trapezoidal Rule, but the result is valid without this restriction. We begin by subdividing the interval $[a, b]$ into n subintervals of equal length Δx, by means of the $(n + 1)$ points $x_0 = a, x_1, x_2, \ldots, x_n = b$, where n is a positive integer (Figure 7.2).

FIGURE 7.2

The area under the curve is equal to the sum of the n subregions $R_1, R_2, \ldots, R_n$.

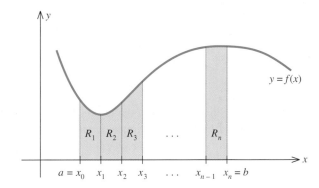

Then the length of each subinterval is given by

$$\Delta x = \frac{b - a}{n}.$$

Furthermore, as we saw earlier, we may view the definite integral

$$\int_a^b f(x) \, dx$$

as the area of the region R under the curve $y = f(x)$ between $x = a$ and $x = b$. This area is given by the sum of the areas of the n nonoverlapping subregions $R_1, R_2, \ldots, R_n$, such that R_1 represents the region under the curve $y = f(x)$ from $x = x_0$ to $x = x_1$, and so on.

The basis for the Trapezoidal Rule lies in the approximation of each of the regions $R_1, R_2, \ldots, R_n$ by a suitable trapezoid. This often leads to a much better approximation than one obtained by means of rectangles (a Riemann sum).

FIGURE 7.3

The area of R_1 is approximated by the area of the trapezoid.

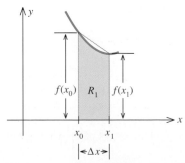

Let us consider the subregion R_1, shown magnified for the sake of clarity in Figure 7.3. Observe that the area of the region R_1 may be approximated by the trapezoid of width Δx whose parallel sides are of lengths $f(x_0)$ and $f(x_1)$. The area of the trapezoid is given by

$$\left[\frac{f(x_0) + f(x_1)}{2} \right] \Delta x \qquad \text{(Average of the lengths of the parallel sides times the width)}$$

square units. Similarly, the area of the region R_2 may be approximated by the trapezoid of width Δx and sides of lengths $f(x_1)$ and $f(x_2)$. The area of the trapezoid is given by

$$\left[\frac{f(x_1) + f(x_2)}{2} \right] \Delta x.$$

Similarly, we see that the area of the last (nth) approximating trapezoid is given by

$$\left[\frac{f(x_{n-1}) + f(x_n)}{2}\right]\Delta x.$$

Then the area of the region R is approximated by the sum of the areas of the n trapezoids, that is,

$$\left[\frac{f(x_0) + f(x_1)}{2}\right]\Delta x + \left[\frac{f(x_1) + f(x_2)}{2}\right]\Delta x + \cdots + \left[\frac{f(x_{n-1}) + f(x_n)}{2}\right]\Delta x$$

$$= \frac{\Delta x}{2}[f(x_0) + f(x_1) + f(x_1) + f(x_2) + \cdots + f(x_{n-1}) + f(x_n)]$$

$$= \frac{\Delta x}{2}[f(x_0) + 2f(x_1) + 2f(x_2) + \cdots + 2f(x_{n-1}) + f(x_n)].$$

Since the area of the region R is given by the value of the definite integral we wished to approximate, we are led to the following approximation formula, which is called the **Trapezoidal Rule.**

TRAPEZOIDAL RULE

$$\int_a^b f(x)\,dx \approx \frac{\Delta x}{2}[f(x_0) + 2f(x_1) + 2f(x_2) \tag{5}$$
$$+ \cdots + 2f(x_{n-1}) + f(x_n)]$$

where $\Delta x = \dfrac{b-a}{n}$.

The approximation generally improves with larger values of n.

EXAMPLE 1   Approximate the value of

$$\int_1^2 \frac{1}{x}\,dx$$

using the Trapezoidal Rule with $n = 10$. Compare this result with the exact value of the integral.

Solution   Here $a = 1$, $b = 2$, and $n = 10$, so

$$\Delta x = \frac{b-a}{n} = \frac{1}{10} = 0.1$$

and

$$x_0 = 1, \quad x_1 = 1.1, \quad x_2 = 1.2, \quad x_3 = 1.3, \ldots,$$
$$x_9 = 1.9, \quad \text{and} \quad x_{10} = 2.$$

The Trapezoidal Rule yields

$$\int_1^2 \frac{1}{x}\,dx \approx \frac{0.1}{2}\left[1 + 2\left(\frac{1}{1.1}\right) + 2\left(\frac{1}{1.2}\right) + 2\left(\frac{1}{1.3}\right) + \cdots + 2\left(\frac{1}{1.9}\right) + \frac{1}{2}\right]$$
$$\approx 0.693771.$$

In this case, we can easily compute the actual value of the definite integral under consideration. In fact,

$$\int_1^2 \frac{1}{x}\, dx = \ln x \Big|_1^2 = \ln 2 - \ln 1 = \ln 2$$

$$\approx 0.693147.$$

Thus, the Trapezoidal Rule with $n = 10$ yields a result with an error of 0.000624 to six decimal places. ∎

EXAMPLE 2   The demand function for a certain brand of perfume is given by

$$p = D(x) = \sqrt{10{,}000 - 0.01x^2},$$

where p is the unit price in dollars and x is the quantity demanded each week, measured in ounces. Find the consumers' surplus if the market price is set at $60 per ounce.

Solution   When $p = 60$, we have

$$\sqrt{10{,}000 - 0.01x^2} = 60$$
$$10{,}000 - 0.01x^2 = 3{,}600$$
$$x^2 = 640{,}000,$$

or $x = 800$, since x must be nonnegative. Next, using the consumers' surplus formula (page 406) with $\bar p = 60$ and $\bar x = 800$, we see that the consumers' surplus is given by

$$CS = \int_0^{800} \sqrt{10{,}000 - 0.01x^2}\, dx - (60)(800).$$

It is not easy to evaluate this definite integral by finding an antiderivative of the integrand. Instead, let us use the Trapezoidal Rule with $n = 10$.

With $a = 0$ and $b = 800$, we find that

$$\Delta x = \frac{b - a}{n} = \frac{800}{10} = 80$$

and $x_0 = 0,\qquad x_1 = 80,\qquad x_2 = 160,\qquad x_3 = 240, \ldots,$

$$x_9 = 720,\quad \text{and}\quad x_{10} = 800,$$

so

$$\int_0^{800} \sqrt{10{,}000 - 0.01x^2}\, dx$$

$$\approx \frac{80}{2}[100 + 2\sqrt{10{,}000 - (0.01)(80)^2}$$

$$+ 2\sqrt{10{,}000 - (0.01)(160)^2} + \cdots + 2\sqrt{10{,}000 - (0.01)(720)^2}$$

$$+ \sqrt{10{,}000 - (0.01)(800)^2}]$$

$$= 40[100 + 199.3590 + 197.4234 + 194.1546 + 189.4835$$

$$+ 183.3030 + 175.4537 + 165.6985$$

$$+ 153.6750 + 138.7948 + 60]$$

$$\approx 70{,}293.82.$$

Therefore, the consumers' surplus is approximately $70,294 - 48,000$, or $22,294. ∎

Simpson's Rule

Before stating Simpson's Rule, let us review the two rules we have used in approximating a definite integral. Let f be a continuous nonnegative function defined on the interval $[a, b]$. Suppose that the interval $[a, b]$ is partitioned by means of the $n + 1$ equally spaced points $x_0 = a, x_1, x_2, \ldots, x_n = b$, where n is a positive integer, so that the length of each subinterval is $\Delta x = (b - a)/n$ (Figure 7.4).

FIGURE 7.4
The area under the curve is equal to the sum of the n subregions $R_1, R_2, \ldots, R_n$.

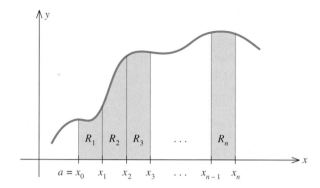

Let us concentrate on the portion of the graph of $y = f(x)$ defined on the interval $[x_0, x_2]$. In using a Riemann sum to approximate the definite integral, we are, in effect, approximating the function $f(x)$ on $[x_0, x_1]$ by the *constant* function $y = f(p_1)$ where p_1 is chosen to be a point in $[x_0, x_1]$; the function $f(x)$ on $[x_1, x_2]$ by the constant function $y = f(p_2)$ where p_2 lies in $[x_1, x_2]$; and so on. Using a Riemann sum, we see that the area of the region under the curve $y = f(x)$ between $x = a$ and $x = b$ is approximated by the area under the approximating "step" function (Figure 7.5a).

FIGURE 7.5

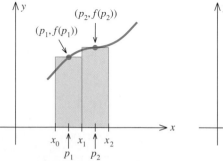

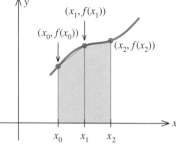

(a) The area under the curve is approximated by the area of the rectangles.

(b) The area under the curve is approximated by the area of the trapezoids.

When we use the Trapezoidal Rule, we are, in effect, approximating the function $f(x)$ on the interval $[x_0, x_1]$ by a *linear* function through the two points

$(x_0, f(x_0))$ and $(x_1, f(x_1))$; the function $f(x)$ on $[x_1, x_2]$ by a *linear* function through the two points $(x_1, f(x_1))$ and $(x_2, f(x_2))$; and so on. Thus, the Trapezoidal Rule simply approximates the actual area of the region under the curve $y = f(x)$ from $x = a$ to $x = b$ by the area under the approximating polygonal curve (Figure 7.5b).

A natural extension of the preceding idea is to approximate portions of the graph of $y = f(x)$ by means of portions of the graphs of second-degree polynomials (parts of parabolas). It can be shown that given any three noncollinear points there is a unique parabola that passes through the given points. Choose the points $(x_0, f(x_0))$, $(x_1, f(x_1))$, and $(x_2, f(x_2))$ corresponding to the first three points of the partition. Then we can approximate the function $f(x)$ on $[x_0, x_2]$ by means of a quadratic function whose graph contains these three points (Figure 7.6).

Although we will not do so here, it can be shown that the area under the parabola between $x = x_0$ and $x = x_2$ is given by

$$\frac{\Delta x}{3}[f(x_0) + 4f(x_1) + f(x_2)]$$

square units. Repeating this argument on the interval $[x_2, x_4]$, we see that the area under the curve between $x = x_2$ and $x = x_4$ is approximated by the area under the parabola between x_2 and x_4, that is, by

$$\frac{\Delta x}{3}[f(x_2) + 4f(x_3) + f(x_4)]$$

square units. Proceeding, we conclude that if n is even (Why?), then the area under the curve $y = f(x)$ from $x = a$ to $x = b$ may be approximated by the sum of the areas under the $n/2$ approximating parabolas, that is,

$$\frac{\Delta x}{3}[f(x_0) + 4f(x_1) + f(x_2)] + \frac{\Delta x}{3}[f(x_2) + 4f(x_3) + f(x_4)] + \cdots$$

$$+ \frac{\Delta x}{3}[f(x_{n-2}) + 4f(x_{n-1}) + f(x_n)]$$

$$= \frac{\Delta x}{3}[f(x_0) + 4f(x_1) + f(x_2) + f(x_2) + 4f(x_3) + f(x_4) + \cdots$$

$$+ f(x_{n-2}) + 4f(x_{n-1}) + f(x_n)]$$

$$= \frac{\Delta x}{3}[f(x_0) + 4f(x_1) + 2f(x_2) + 4f(x_3)$$

$$+ 2f(x_4) + \cdots + 4f(x_{n-1}) + f(x_n)].$$

The preceding is the derivation of the approximation formula known as **Simpson's Rule.**

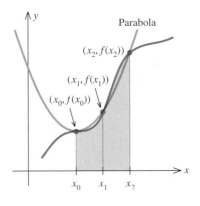

FIGURE 7.6

Simpson's Rule approximates the area under the curve by the area under the parabola.

Parabola

$(x_2, f(x_2))$

$(x_1, f(x_1))$

$(x_0, f(x_0))$

SIMPSON'S RULE

$$\int_a^b f(x)\, dx \approx \frac{\Delta x}{3}[f(x_0) + 4f(x_1) + 2f(x_2) + 4f(x_3) + 2f(x_4)$$

$$+ \cdots + 4f(x_{n-1}) + f(x_n)] \qquad (6)$$

where $\Delta x = \dfrac{b - a}{n}$ and n is even.

In using this rule, remember that n must be even.

EXAMPLE 3   Find an approximation of

$$\int_1^2 \frac{1}{x}\, dx$$

using Simpson's Rule with $n = 10$. Compare this result with that of Example 1 and also with the exact value of the integral.

Solution   We have $a = 1$, $b = 2$, $f(x) = \frac{1}{x}$, and $n = 10$, so

$$\Delta x = \frac{b - a}{n} = \frac{1}{10} = 0.1$$

and

$$x_0 = 1, \qquad x_1 = 1.1, \qquad x_2 = 1.2, \qquad x_3 = 1.3, \ldots,$$
$$x_9 = 1.9, \quad \text{and} \quad x_{10} = 2.$$

Simpson's Rule yields

$$\int_1^2 \frac{1}{x}\, dx \approx \frac{0.1}{3}[f(1) + 4f(1.1) + 2f(1.2) + \cdots + 4f(1.9) + f(2)]$$

$$= \frac{0.1}{3}\left[1 + 4\left(\frac{1}{1.1}\right) + 2\left(\frac{1}{1.2}\right) + 4\left(\frac{1}{1.3}\right) + 2\left(\frac{1}{1.4}\right) + 4\left(\frac{1}{1.5}\right)\right.$$

$$\left. + 2\left(\frac{1}{1.6}\right) + 4\left(\frac{1}{1.7}\right) + 2\left(\frac{1}{1.8}\right) + 4\left(\frac{1}{1.9}\right) + \frac{1}{2}\right]$$

$$\approx 0.693150.$$

The Trapezoidal Rule with $n = 10$ yielded an approximation of 0.693771, which is 0.000624 off of the value of $\ln 2 \approx 0.693147$ to six decimal places. Simpson's Rule yields an approximation with an error of 0.000003, a definite improvement over the Trapezoidal Rule.   ■

EXAMPLE 4   Solve the problem posed at the beginning of this section. Recall that we wished to find a person's cardiac output by using the formula

$$R = \frac{60D}{\displaystyle\int_0^{28} c(t)\, dt},$$

where D (the quantity of dye injected) is equal to 6 mg and the function c has the graph shown in Figure 7.7. Use Simpson's Rule with $n = 14$ to estimate the value of the integral.

Solution   Using Simpson's Rule with $n = 14$ and $\Delta t = 2$ so that

$$t_0 = 0, \qquad t_1 = 2, \qquad t_2 = 4, \qquad t_3 = 6, \qquad \ldots, \qquad t_{14} = 28,$$

FIGURE 7.7
The function c gives the concentration of a dye measured at the aorta. The graph is constructed by drawing a smooth curve through a set of discrete points.

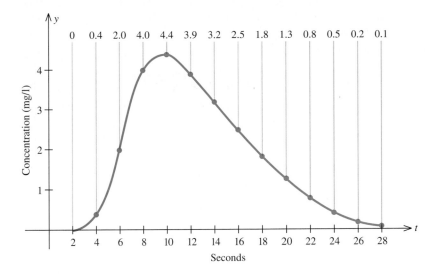

we obtain

$$\int_0^{28} c(t)\, dt \approx \frac{2}{3}[c(0) + 4c(2) + 2c(4) + 4c(6) + \cdots$$
$$+ 4c(26) + c(28)]$$
$$= \frac{2}{3}[0 + 4(0) + 2(0.4) + 4(2.0) + 2(4.0)$$
$$+ 4(4.4) + 2(3.9) + 4(3.2) + 2(2.5) + 4(1.8)$$
$$+ 2(1.3) + 4(0.8) + 2(0.5) + 4(0.2) + 0.1]$$
$$\approx 49.9.$$

Therefore, the person's cardiac output is

$$R \approx \frac{60(5)}{49.9} \approx 6.0,$$

or 6.0 l/min. ■

EXAMPLE 5 An oil spill off the coastline was caused by a ruptured tank in a grounded oil tanker. Using aerial photographs, the Coast Guard was able to obtain the dimensions of the oil spill (Figure 7.8). Using Simpson's Rule with $n = 10$, estimate the area of the oil spill.

Solution We may think of the area affected by the oil spill as the area of the plane region bounded above by the graph of the function $f(x)$ and below by the graph of the function $g(x)$ between $x = 0$ and $x = 1,000$ (Figure 7.8). Then, the required area is given by

$$A = \int_0^{1000} [f(x) - g(x)]\, dx.$$

Using Simpson's Rule with $n = 10$ and $\Delta x = 100$ so that

$$x_0 = 0, \quad x_1 = 100, \quad x_2 = 200, \quad \ldots, \quad x_{10} = 1,000,$$

FIGURE 7.8
Simpson's Rule can be used to calculate the area of the oil spill.

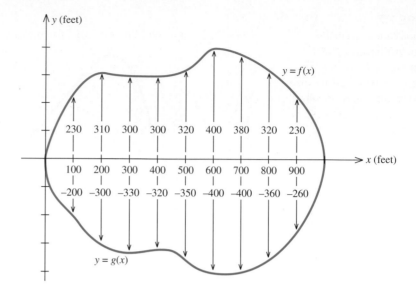

we have

$$A = \int_0^{1000} [f(x) - g(x)] \, dx$$

$$\approx \frac{\Delta x}{3} \{[f(x_0) - g(x_0)] + 4[f(x_1) - g(x_1)] + 2[f(x_2) - g(x_2)]$$

$$+ \cdots + 4[f(x_9) - g(x_9)] + [f(x_{10}) - g(x_{10})]\}$$

$$= \frac{100}{3} \{[0 - 0] + 4[230 - (-200)] + 2[310 - (-300)]$$

$$+ 4[300 - (-330)] + 2[300 - (-320)] + 4[320 - (-350)]$$

$$+ 2[400 - (-400)] + 4[380 - (-400)] + 2[320 - (-360)]$$

$$+ 4[230 - (-260)] + [0 - 0]\}$$

$$= \frac{100}{3} [0 + 4(430) + 2(610) + 4(630) + 2(620) + 4(670)$$

$$+ 2(800) + 4(780) + 2(680) + 4(490) + 0]$$

$$= \frac{100}{3} (17{,}420)$$

$$\approx 580{,}667,$$

or approximately 580,667 square feet.   ∎

Error Analysis

The following results give the bounds on the errors incurred when the Trapezoidal Rule and Simpson's Rule are used to approximate a definite integral (proof omitted).

ERRORS IN THE TRAPEZOIDAL AND SIMPSON APPROXIMATIONS

Suppose that the definite integral

$$\int_a^b f(x)\,dx$$

is approximated with n subintervals.

1. The *maximum* error incurred in using the Trapezoidal Rule is

$$\frac{M(b-a)^3}{12n^2}, \tag{7}$$

where M is a number such that $|f''(x)| \le M$ for all x in $[a, b]$.

2. The *maximum* error incurred in using Simpson's Rule is

$$\frac{M(b-a)^5}{180n^4}, \tag{8}$$

where M is a number such that $|f^{(4)}(x)| \le M$ for all x in $[a, b]$.

REMARK In many instances, the actual error is less than the upper error bounds given. ☐

EXAMPLE 6 Find bounds on the errors incurred when

$$\int_1^2 \frac{1}{x}\,dx$$

is approximated using (a) the Trapezoidal Rule and (b) Simpson's Rule with $n = 10$. Compare these with the actual errors found in Examples 1 and 3.

Solution

a. Here $a = 1$, $b = 2$, and $f(x) = 1/x$. Next, to find a value for M, we compute

$$f'(x) = -\frac{1}{x^2} \quad \text{and} \quad f''(x) = \frac{2}{x^3}.$$

Since $f''(x)$ is positive and decreasing on $(1, 2)$ [Why?], it attains its maximum value of 2 at $x = 1$, the left endpoint of the interval. Therefore, if we take $M = 2$, then $|f''(x)| \le 2$. Using (7), we see that the maximum error incurred is

$$\frac{2(2-1)^3}{12(10)^2} = \frac{2}{1200} = 0.0016667.$$

The actual error found in Example 1, 0.000624, is much less than the upper bound just found.

b. We compute

$$f'''(x) = \frac{-6}{x^4} \quad \text{and} \quad f^{(4)}(x) = \frac{24}{x^5}.$$

Since $f^{(4)}(x)$ is positive and decreasing on $(1, 2)$ [just look at $f^{(5)}$ to verify this fact], it attains its maximum at the left endpoint of $[1, 2]$. Now,

$$f^{(4)}(1) = 24,$$

and so we may take $M = 24$. Using (8), we obtain the maximum error of

$$\frac{24(2 - 1)^5}{180(10)^4} = 0.0000133.$$

The actual error is 0.000003 (see Example 3). ■

SELF-CHECK EXERCISES 7.3

c 1. Use the Trapezoidal Rule and Simpson's Rule with $n = 8$ to approximate the value of the definite integral

$$\int_0^2 \frac{1}{\sqrt{1 + x^2}}\, dx.$$

c 2. The graph in the accompanying figure shows the consumption of petroleum in the U.S. in quadrillion BTU, from 1976 to 1990. Using Simpson's Rule with $n = 14$, estimate the average consumption during the 14-year period.

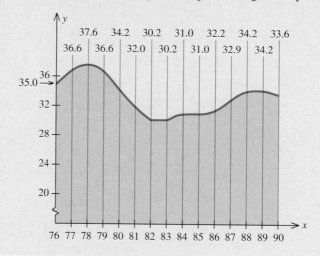

Solutions to Self-Check Exercises 7.3 can be found on page 454.

7.3   EXERCISES

c *A calculator is recommended for this exercise set. In Exercises 1–14, use the Trapezoidal Rule and Simpson's Rule to approximate the value of each definite integral. Compare your result with the exact value of the integral.*

1. $\int_0^2 x^2\, dx;\ n = 6$

2. $\int_1^3 (x^2 - 1)\, dx;\ n = 4$

3. $\int_0^1 x^3\, dx;\ n = 4$

4. $\int_1^2 x^3\, dx;\ n = 6$

5. $\int_1^2 \frac{1}{x}\, dx;\ n = 4$

6. $\int_1^2 \frac{1}{x}\, dx;\ n = 8$

7. $\int_1^2 \frac{1}{x^2}\, dx;\ n = 4$

8. $\int_0^1 \frac{1}{1 + x}\, dx;\ n = 4$

9. $\int_0^4 \sqrt{x}\, dx;\ n = 8$

10. $\int_0^2 x\sqrt{2x^2 + 1}\, dx;\ n = 6$

11. $\int_0^1 e^{-x}\, dx;\ n = 6$ 12. $\int_0^1 xe^{-x^2}\, dx;\ n = 6$

13. $\int_1^2 \ln x\, dx;\ n = 4$

14. $\int_0^1 x \ln (x^2 + 1)\, dx;\ n = 8$

Time t (hours)	0	$\frac{1}{4}$	$\frac{1}{2}$	$\frac{3}{4}$
Velocity $V(t)$ (mph)	19.5	24.3	34.2	40.5

Time t (hours)	1	$\frac{5}{4}$	$\frac{3}{2}$	$\frac{7}{4}$	2
Velocity $V(t)$ (mph)	38.4	26.2	18	16	8

In Exercises 15–22, use the Trapezoidal Rule and Simpson's Rule to approximate the value of each definite integral.

15. $\int_0^1 \sqrt{1 + x^3}\, dx;\ n = 4$

16. $\int_0^2 x\sqrt{1 + x^3}\, dx;\ n = 4$

17. $\int_0^2 \dfrac{1}{\sqrt{x^3 + 1}}\, dx;\ n = 4$

18. $\int_0^1 \sqrt{1 - x^2}\, dx;\ n = 4$

19. $\int_0^2 e^{-x^2}\, dx;\ n = 4$ 20. $\int_0^1 e^{x^2}\, dx;\ n = 6$

21. $\int_1^2 x^{-1/2}e^x\, dx;\ n = 4$ 22. $\int_2^4 \dfrac{dx}{\ln x};\ n = 6$

In Exercises 23–28, find a bound on the error in approximating the given definite integral using (a) the Trapezoidal Rule (b) Simpson's Rule with n intervals.

23. $\int_{-1}^2 x^5\, dx;\ n = 10$ 24. $\int_0^1 e^{-x}\, dx;\ n = 8$

25. $\int_1^3 \dfrac{1}{x}\, dx;\ n = 10$ 26. $\int_1^3 \dfrac{1}{x^2}\, dx;\ n = 8$

27. $\int_0^2 \dfrac{1}{\sqrt{1 + x}}\, dx;\ n = 8$ 28. $\int_1^3 \ln x\, dx;\ n = 10$

29. **Trial Run of an Attack Submarine**   In a submerged trial run of an attack submarine, a reading of the sub's velocity was made every quarter of an hour, as shown in the accompanying table. Use the Trapezoidal Rule to estimate the distance traveled by the submarine during the two-hour period.

30. **Real Estate**   Cooper Realty is considering development of a time-sharing condominium resort complex along the oceanfront property illustrated in the accompanying graph. In order to obtain an estimate of the area of this property, measurements of the distances from the edge of a straight road, which defines one boundary of the property, to the corresponding points on the shoreline are made at 100 foot intervals. Using Simpson's Rule with $n = 10$, estimate the area of the oceanfront property.

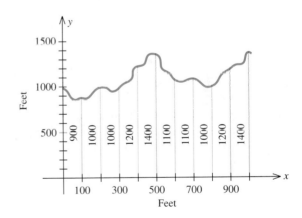

31. **Fuel Consumption of Domestic Cars**   Thanks to smaller and more fuel-efficient models, American carmakers have doubled their average fuel economy over a 13-year period from 1974 to 1987. The graph depicted in the figure on page 452 gives the average fuel consumption in miles per gallon (mpg) of domestic-built cars over the period under consideration ($t = 0$ corresponds to the beginning of 1974). Use the Trapezoidal Rule to estimate the average fuel consumption of the domestic car built during this period. [*Hint:* Approximate the integral $\frac{1}{13} \int_0^{13} f(t)\, dt$.]

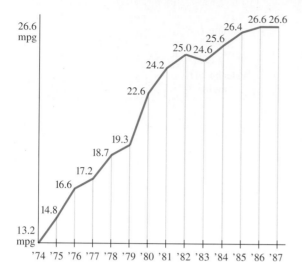

Figure for Exercise 31

32. **Average Temperature** The graph depicted in the accompanying figure shows the daily mean temperatures recorded during one September in Cameron Highlands. Using (a) the Trapezoidal Rule and (b) Simpson's Rule with $n = 10$, estimate the average temperature during that month.

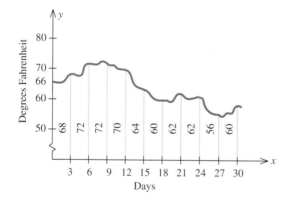

33. **Consumers' Surplus** Refer to Section 6.7. The demand equation for the Sicard wristwatch is given by

$$p = \frac{50}{0.01x^2 + 1} \qquad (0 \le x \le 20),$$

where x (measured in units of a thousand) is the quantity demanded per week and p is the unit price in dollars. Use (a) the Trapezoidal Rule and (b) Simpson's Rule (take $n = 8$) to estimate the consumers' surplus if the market price is $25 per watch.

34. **Producers' Surplus** Refer to Section 6.7. The supply function for the audio compact disc manufactured by

the Herald Record Company is given by

$$p = S(x) = \sqrt{0.01x^2 + 0.11x + 38},$$

where p is the unit wholesale price in dollars and x stands for the quantity that will be made available in the market by the supplier, measured in units of a thousand. Use (a) the Trapezoidal Rule and (b) Simpson's Rule (take $n = 8$) to estimate the producers' surplus if the wholesale price is $8 per disc.

35. **Air Pollution** The amount of nitrogen dioxide, a brown gas that impairs breathing, present in the atmosphere on a certain May day in the city of Long Beach has been approximated by

$$A(t) = \frac{136}{1 + 0.25(t - 4.5)^2} + 28 \qquad (0 \le t \le 11),$$

where $A(t)$ is measured in pollutant standard index (PSI), t is measured in hours, and $t = 0$ corresponds to 7 A.M. Use the Trapezoidal Rule with $n = 10$ to estimate the average PSI between 7 A.M. and noon. [*Hint:* $\frac{1}{5} \int_0^5 A(t) \, dt$.]

36. **Growth of Service Industries** It has been estimated that service industries, which currently make up 30 percent of the nonfarm work force in a certain country, will continue to grow at the rate of

$$R(t) = 5e^{1/(t+1)}$$

percent per decade t decades from now. Estimate the percentage of the nonfarm work force in the service industries one decade from now. [*Hint:* (a) Show that the desired answer is given by $30 + \int_0^1 5e^{1/(t+1)} \, dt$ and (b) use Simpson's Rule with $n = 10$ to approximate the definite integral.]

37. **Length of Infants at Birth** Medical records of infants delivered at the Kaiser Memorial Hospital show that the percentage of infants whose length at birth is between 19 and 21 inches is given by

$$P = 100 \int_{19}^{21} \frac{1}{2.6\sqrt{2\pi}} e^{-1/2[(x-20)/2.6]^2} \, dx.$$

Use Simpson's Rule with $n = 10$ to estimate P.

38. **Treadlives of Tires** Under normal driving conditions the percentage of Super Titan radial tires expected to have a useful treadlife of between 30,000 and 40,000 miles is given by

$$P = 100 \int_{30,000}^{40,000} \frac{1}{2,000\sqrt{2\pi}} e^{-1/2[(x-40,000)/2,000]^2} \, dx.$$

Use Simpson's Rule with $n = 10$ to estimate P.

James H. Chesebro, M.D.

Institution: Harvard Medical School, Massachusetts General Hospital

For over twenty years, James Chesebro has worked as an investigative cardiologist, diagnosing and treating patients suffering from heart and blood-vessel diseases. He specializes in researching ways to prevent blood clots from narrowing the heart's arteries. Even when patients have coronary bypass operations to correct problems, stresses Chesebro, "the piece of vein used to detour around blocked arteries is prone to plug up with clots."

Throughout his career, Chesebro has continued to investigate the body's clotting mechanisms as well as substances that may prevent, or even dissolve, clots. A study conducted in 1982 showed that bypass patients taking dipyridamole and aspirin were nearly three times less likely to develop clots in vein grafts than patients given placebos.

Recently, Chesebro has been working with a substance called hirudin derived from leech saliva. Hirudin has proved quite beneficial in blocking arterial blood-clot formations.

Determining the correct medication and dosage involves intense quantitative research. Colleagues from other branches of science such as nuclear medicine, molecular biology, and biochemistry play key roles in the research process. Chesebro and his colleagues have to understand how the heart's physiology and pharmacological variables such as distribution rates, concentration levels, elimination rates, and biological effects of particular substances dictate the choice and dosage of medication.

In considering whether a particular medication will bind effectively with specific body cells, researchers must determine bond tightness, speed, and length of time to reach optimal concentration. Researchers rely on complicated equations such as "integrating the area under a curve to find the amount of medication in the body at a given time," notes Chesebro. Without calculus, these equations can't be solved.

Whether physicians rely on medication or surgical procedures such as balloon angioplasty to open clogged passages, other factors play a major role in the outcome. Cholesterol and blood sugar levels, patient age, blood pressure, the geometry of the blockage, and the velocity and turbulence of blood flow all warrant consideration. According to Chesebro "linear modeling can be used to predict the contribution of patient variables and local blood-vessel variables in the outcome of opening blocked arteries."

The bottom line? Calculus has been a key contributor to Chesebro's success in preventing disabling arterial blood clots.

In 1992 Chesebro was appointed professor of medicine at Harvard Medical School, after having conducted cardiovascular research at the Mayo Clinic for a number of years. In addition, he was made Associate Director for Research in the Cardiac Unit at Massachusetts General Hospital, one of Harvard's teaching affiliates.

39. Measuring Cardiac Output Eight milligrams of a dye are injected into a vein leading to an individual's heart. The concentration of the dye in the aorta measured at 2-second intervals is shown in the accompanying table. Use Simpson's Rule and the formula of Example 4 to estimate the person's cardiac output.

t	0	2	4	6	8	10	12
$C(t)$	0	0	2.8	6.1	9.7	7.6	4.8

t	14	16	18	20	22	24
$C(t)$	3.7	1.9	0.8	0.3	0.1	0

40. Derive the formula

$$R = \frac{60D}{\displaystyle\int_0^T C(t)\, dt}$$

for calculating the cardiac output of a person in l/min. Here $C(t)$ is the concentration of dye in the aorta (in mg/l) at time t (in seconds) for t in $[0, T]$, and D is the amount of dye (in mg) injected into a vein leading to the heart. [*Hint:* Partition the interval $[0, T]$ into n subintervals of equal length Δt. The amount of dye that flows past the measuring point in the aorta during the time interval $[0, \Delta t]$ is approximately $C(t_1)(R\Delta t)/60$ (concentration times volume). Therefore, the total

amount of dye measured at the aorta is

$$\frac{[C(t_1)R\Delta t + C(t_2)R\Delta t + \cdots + C(t_n)R\Delta t]}{60} = D.$$

Take the limit of the Riemann sum to obtain

$$R = \frac{60D}{\displaystyle\int_0^T C(t)\, dt}.\Bigg]$$

C *In Exercises 41–44, use a computer software package with the Trapezoidal Rule to approximate the value of the given definite integral.*

41. $\displaystyle\int_0^4 \frac{1}{1 + x^2}\, dx;\ n = 20$ **42.** $\displaystyle\int_2^4 \frac{\sqrt{x}}{x^2 - 2}\, dx;\ n = 20$

43. $\displaystyle\int_0^4 \frac{e^{-x}}{\sqrt{2x + 3}}\, dx;\ n = 25$ **44.** $\displaystyle\int_1^3 \frac{\ln x}{e^x + 1}\, dx;\ n = 25$

In Exercises 45–48, use a computer software package with Simpson's Rule to approximate the value of the given definite integral.

45. $\displaystyle\int_0^4 \frac{1}{1 + x^2}\, dx;\ n = 20$ **46.** $\displaystyle\int_2^4 \frac{\sqrt{x}}{x^2 - 2}\, dx;\ n = 20$

47. $\displaystyle\int_0^4 e^{-x^2}\, dx;\ n = 30$

48. $\displaystyle\int_1^3 \frac{\ln(x + 1)}{\sqrt{x}}\, dx;\ n = 30$

SOLUTIONS TO SELF-CHECK EXERCISES 7.3

1. We have $x = 0$, $b = 2$, and $n = 8$, so

$$\Delta x = \frac{b - a}{n} = \frac{2}{8} = 0.25$$

and $x_0 = 0$, $x_1 = 0.25$, $x_2 = 0.50$, $x_3 = 0.75$, $\ldots$, $x_7 = 1.75$, and $x_8 = 2$. The Trapezoidal Rule gives

$$\int_0^2 \frac{1}{\sqrt{1 + x^2}}\, dx$$

$$\approx \frac{0.25}{2}\Bigg[1 + \frac{2}{\sqrt{1 + (0.25)^2}} + \frac{2}{\sqrt{1 + (0.5)^2}} + \cdots$$

$$+ \frac{2}{\sqrt{1 + (1.75)^2}} + \frac{1}{\sqrt{5}}\Bigg]$$

$$\approx 0.125(1 + 1.9403 + 1.7889 + 1.6000 + 1.4142 + 1.2494$$

$$+ 1.1094 + 0.9923 + 0.4472)$$

$$\approx 1.4427.$$

Using Simpson's Rule with $n = 8$ gives

$$\int_0^2 \frac{1}{\sqrt{1 + x^2}}\, dx$$

$$\approx \frac{0.25}{3}\left[1 + \frac{4}{\sqrt{1 + (0.25)^2}} + \frac{2}{\sqrt{1 + (0.5)^2}} + \frac{4}{\sqrt{1 + (0.75)^2}}\right.$$

$$\left. + \cdots + \frac{4}{\sqrt{1 + (1.75)^2}} + \frac{1}{\sqrt{5}}\right]$$

$$\approx \frac{0.25}{3}(1 + 3.8806 + 1.7889 + 3.2000 + 1.4142 + 2.4988 + 1.1094$$

$$+ 1.9846 + 0.4472)$$

$$\approx 1.4436.$$

2. The average consumption of petroleum during the 14-year period is given by

$$\frac{1}{14}\int_0^{14} f(x)\, dx,$$

where f is the function describing the given graph. Using Simpson's Rule with $a = 0$, $b = 10$, and $n = 10$ so that $\Delta x = 1$ and

$$x_0 = 0, \quad x_1 = 1, \quad x_2 = 2, \ldots, x_{10} = 10,$$

we have

$$\frac{1}{14}\int_0^{14} f(x)\, dx$$

$$\approx \left(\frac{1}{14}\right)\left(\frac{1}{3}\right)[f(x_0) + 4f(x_1) + 2f(x_2) + 4f(x_3) + \cdots + 4f(x_{13}) + f(x_{14})]$$

$$= \frac{1}{42}[35 + 4(36.6) + 2(37.6) + 4(36.6) + 2(34.2)$$

$$+ 4(32.0) + 2(30.2) + 4(30.2) + 2(31.0) + 4(31.0)$$

$$+ 2(32.2) + 4(32.9) + 2(34.2) + 4(34.2) + 33.6]$$

$$\approx 33.4,$$

or approximately 33.4 quadrillion BTU per year.

7.4   Improper Integrals

FIGURE 7.9

The area of the unbounded region R can be approximated by a definite integral.

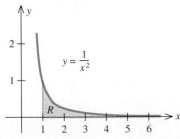

Improper Integrals

All of the definite integrals we have encountered have had finite intervals of integration. In many applications, however, we are concerned with integrals that have unbounded intervals of integration. Such integrals are called **improper integrals.**

To lead us to the definition of an improper integral of a function f over an infinite interval, consider the problem of finding the area of the region R under the curve $y = f(x) = 1/x^2$ and to the right of the vertical line $x = 1$, as shown in Figure 7.9. Because the interval over which the integration must be performed is unbounded, the method of integration presented previously cannot be applied directly in solving this problem. However, we can approximate the region R

FIGURE 7.10

Area of shaded region =
$\int_1^b \dfrac{1}{x^2} dx.$

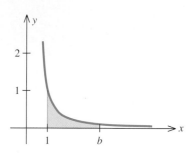

by the definite integral

$$\int_1^b \frac{1}{x^2} dx, \qquad (9)$$

which gives the area of the region under the curve $y = f(x) = 1/x^2$ from $x = 1$ to $x = b$ (Figure 7.10). You can see that the approximation of the region R by the definite integral in (9) improves as the upper limit of integration, b, becomes larger and larger. Figure 7.11 illustrates the situation for $b = 2$, 3, and 4, respectively.

This observation suggests that if we define a function $I(b)$ by

$$I(b) = \int_1^b \frac{1}{x^2} dx, \qquad (10)$$

then we can find the area of the required region R by evaluating the limit of $I(b)$ as b tends to infinity; that is, the area of R is given by

$$\lim_{b \to \infty} I(b) = \lim_{b \to \infty} \int_1^b \frac{1}{x^2} dx. \qquad (11)$$

FIGURE 7.11

As b increases, the approximation of R by the definite integral improves.

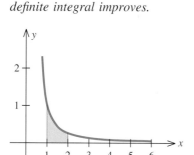

(a) Area of region under the graph of f on $[1, 2]$

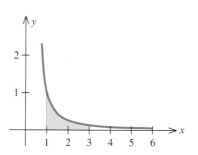

(b) Area of region under the graph of f on $[1, 3]$

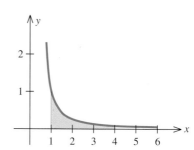

(c) Area of region under the graph of f on $[1, 4]$

EXAMPLE 1

a. Evaluate the definite integral $I(b)$ in (10).
b. Compute $I(b)$ for $b = 10, 100, 1{,}000, 10{,}000$.
c. Evaluate the limit in (11).
d. Interpret the results of (b) and (c).

Solution

a.
$$I(b) = \int_1^b \frac{1}{x^2} dx = -\frac{1}{x}\Big|_1^b = -\frac{1}{b} + 1$$

b. From the result of (a),

$$I(b) = 1 - \frac{1}{b}.$$

Therefore,

$$I(10) = 1 - \frac{1}{10} = 0.9$$

$$I(100) = 1 - \frac{1}{100} = 0.99$$

$$I(1,000) = 1 - \frac{1}{1,000} = 0.999$$

$$I(10,000) = 1 - \frac{1}{10,000} = 0.9999.$$

c. Once again, using the result of (a), we find

$$\lim_{b \to \infty} I(b) = \lim_{b \to \infty} \int_1^b \frac{1}{x^2}\, dx$$
$$= \lim_{b \to \infty} \left(1 - \frac{1}{b}\right)$$
$$= 1.$$

d. The result of (c) tells us that the area of the region R is 1 square unit. The results of the computations performed in (b) reinforce our expectation that $I(b)$ should approach 1, the area of the region R, as b approaches infinity. ∎

The preceding discussion and the results of Example 1 suggest that we define the improper integral of a continuous function f over the unbounded interval $[a, \infty)$ as follows.

IMPROPER INTEGRAL OF f OVER $[a, \infty)$

Let f be a continuous function on the unbounded interval $[a, \infty)$. Then the improper integral of f over $[a, \infty)$ is defined by

$$\int_a^\infty f(x)\, dx = \lim_{b \to \infty} \int_a^b f(x)\, dx \qquad (12)$$

if the limit exists.

If the limit exists, the improper integral is said to be **convergent**. An improper integral for which the limit in (12) fails to exist is said to be **divergent**.

EXAMPLE 2 Evaluate $\int_2^\infty \frac{1}{x}\, dx$ if it converges.

Solution
$$\int_2^\infty \frac{1}{x}\, dx = \lim_{b \to \infty} \int_2^b \frac{1}{x}\, dx$$
$$= \lim_{b \to \infty} \ln x \Big|_2^b$$
$$= \lim_{b \to \infty} (\ln b - \ln 2).$$

Since $\ln b \to \infty$ as $b \to \infty$, the limit does not exist and we conclude that the given improper integral is divergent. ∎

FIGURE 7.12

Area of $R = \displaystyle\int_0^\infty e^{-x/2}\, dx.$

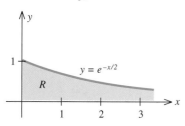

EXAMPLE 3   Find the area of the region R under the curve $y = e^{-x/2}$ for $x \geq 0$.

Solution The region R is shown in Figure 7.12. Taking $b > 0$ we compute the area of the region under the curve $y = e^{-x/2}$ from $x = 0$ to $x = b$, namely,

$$I(b) = \int_0^b e^{-x/2}\, dx = -2e^{-x/2}\Big|_0^b = -2e^{-b/2} + 2.$$

Then the area of the region R is given by

$$\lim_{b \to \infty} I(b) = \lim_{b \to \infty}(2 - 2e^{-b/2}) = 2 - 2\lim_{b \to \infty} \frac{1}{e^{b/2}}$$
$$= 2, \quad \text{or} \quad 2 \text{ square units.} \qquad ∎$$

The improper integral defined in (12) has an interval of integration that is unbounded on the right. Improper integrals with intervals of integration that are unbounded on the left also arise in practice and are defined in a similar manner.

IMPROPER INTEGRAL OF f OVER $(-\infty, b]$

Let f be a continuous function on the unbounded interval $(-\infty, b]$. Then the improper integral of f over $(-\infty, b]$ is defined by

$$\int_{-\infty}^b f(x)\, dx = \lim_{a \to -\infty} \int_a^b f(x)\, dx \qquad (13)$$

if the limit exists.

FIGURE 7.13

Area of $R = -\displaystyle\int_{-\infty}^1 -e^{2x}\, dx.$

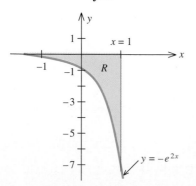

In this case, the improper integral is said to be convergent. Otherwise, the improper integral is said to be divergent.

EXAMPLE 4   Find the area of the region R bounded above by the x-axis, below by the curve $y = -e^{2x}$, and on the right by the vertical line $x = 1$.

Solution The region R is shown in Figure 7.13. Taking $a < 1$, compute the area of the region bounded above by the x-axis ($y = 0$), and below by the curve $y = -e^{2x}$ from $x = a$ to $x = 1$, namely,

$$I(a) = \int_a^1 [0 - (-e^{2x})]\, dx = \int_a^1 e^{2x}\, dx$$
$$= \frac{1}{2}e^{2x}\Big|_a^1 = \frac{1}{2}e^2 - \frac{1}{2}e^{2a}.$$

Then the area of the required region is given by

$$\lim_{a \to -\infty} I(a) = \lim_{a \to -\infty} \left(\frac{1}{2}e^2 - \frac{1}{2}e^{2a} \right)$$

$$= \frac{1}{2}e^2 - \frac{1}{2} \lim_{a \to -\infty} e^{2a}$$

$$= \frac{1}{2}e^2. \qquad\blacksquare$$

Another improper integral found in practical applications involves the integration of a function f over the unbounded interval $(-\infty, \infty)$.

IMPROPER INTEGRAL OF f OVER $(-\infty, \infty)$

Let f be a continuous function over the unbounded interval $(-\infty, \infty)$. Let c be any real number and suppose both the improper integrals

$$\int_{-\infty}^{c} f(x)\, dx \quad \text{and} \quad \int_{c}^{\infty} f(x)\, dx$$

are convergent. Then the improper integral of f over $(-\infty, \infty)$ is defined by

$$\int_{-\infty}^{\infty} f(x)\, dx = \int_{-\infty}^{c} f(x)\, dx + \int_{c}^{\infty} f(x)\, dx. \qquad (14)$$

In this case, we say that the improper integral on the left in (14) is convergent. If either one of the two improper integrals on the right in (14) is divergent, then the improper integral on the left is not defined.

REMARK   Usually we choose $c = 0$.   □

EXAMPLE 5   Evaluate the improper integral

$$\int_{-\infty}^{\infty} xe^{-x^2}\, dx$$

and give a geometrical interpretation of the results.

Solution   Take the point c in (14) to be $c = 0$. Let us first evaluate

$$\int_{-\infty}^{0} xe^{-x^2}\, dx = \lim_{a \to -\infty} \int_{a}^{0} xe^{-x^2}\, dx$$

$$= \lim_{a \to -\infty} \left. -\frac{1}{2}e^{-x^2} \right|_{a}^{0}$$

$$= \lim_{a \to -\infty} \left[-\frac{1}{2} + \frac{1}{2}e^{-a^2} \right] = -\frac{1}{2}.$$

Next, we evaluate

$$\int_0^\infty xe^{-x^2}\,dx = \lim_{b\to\infty}\int_0^b xe^{-x^2}\,dx$$

$$= \lim_{b\to\infty} -\frac{1}{2}e^{-x^2}\Big|_0^b$$

$$= \lim_{b\to\infty}\left[-\frac{1}{2}e^{-b^2} + \frac{1}{2}\right] = \frac{1}{2}.$$

Therefore,

$$\int_{-\infty}^\infty xe^{-x^2}\,dx = \int_{-\infty}^0 xe^{-x^2}\,dx + \int_0^\infty xe^{-x^2}\,dx$$

$$= -\frac{1}{2} + \frac{1}{2}$$

$$= 0.$$

FIGURE 7.14

$$\int_{-\infty}^\infty xe^{-x^2}\,dx =$$
$$\int_{-\infty}^0 xe^{-x^2}\,dx + \int_0^\infty xe^{-x^2}\,dx.$$

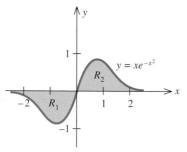

The graph of $y = xe^{-x^2}$ is sketched in Figure 7.14. A glance at the figure tells us that the improper integral

$$\int_{-\infty}^0 xe^{-x^2}\,dx$$

gives the negative of the area of the region R_1, bounded above by the x-axis, below by the curve $y = xe^{-x^2}$, and on the right by the y-axis ($x = 0$).

On the other hand, the improper integral

$$\int_0^\infty xe^{-x^2}\,dx$$

gives the area of the region R_2 under the curve $y = xe^{-x^2}$ for $x \geq 0$. Since the graph of f is symmetrical with respect to the origin, the area of R_1 is equal to the area of R_2. In other words,

$$\int_{-\infty}^0 xe^{-x^2}\,dx = -\int_0^\infty xe^{-x^2}\,dx.$$

Therefore,

$$\int_{-\infty}^\infty xe^{-x^2}\,dx = \int_{-\infty}^0 xe^{-x^2}\,dx + \int_0^\infty xe^{-x^2}\,dx$$

$$= -\int_0^\infty xe^{-x^2}\,dx + \int_0^\infty xe^{-x^2}\,dx$$

$$= 0,$$

as was shown earlier.   ■

Perpetuities

Recall from Section 6.7 that the present value of an annuity is given by

$$PV \approx mP\int_0^T e^{-rt}\,dt = \frac{mP}{r}(1 - e^{-rT}). \qquad (15)$$

Now if the payments of an annuity are allowed to continue indefinitely, we have what is called a **perpetuity.** The present value of a perpetuity may be approximated by the improper integral

$$PV \approx mP \int_0^\infty e^{-rt}\, dt,$$

obtained from (15) by allowing the term of the annuity, T, to approach infinity. Thus

$$mP \int_0^\infty e^{-rt}\, dt = \lim_{b \to \infty} mP \int_0^b e^{-rt}\, dt$$

$$= mP \lim_{b \to \infty} \int_0^b e^{-rt}\, dt$$

$$= mP \lim_{b \to \infty} \left[-\frac{1}{r} e^{-rt} \Big|_0^b \right]$$

$$= mP \lim_{b \to \infty} \left(-\frac{1}{r} e^{-rb} + \frac{1}{r} \right) = \frac{mP}{r}.$$

THE PRESENT VALUE OF A PERPETUITY

The **present value PV of a perpetuity** is given by

$$PV = \frac{mP}{r}, \qquad (16)$$

where m is the number of payments per year, P is the size of each payment, and r is the interest rate (compounded continuously).

EXAMPLE 6   The Robinson family wishes to create a scholarship fund at a college. If a scholarship in the amount of $5000 is awarded annually beginning one year from now, find the amount of the endowment they are required to make now. Assume that this fund will earn interest at a rate of 8 percent per year compounded continuously.

Solution   The amount of the endowment, A, is given by the present value of a perpetuity, with $m = 1$, $P = 5000$, and $r = 0.08$. Using (16), we find

$$A = \frac{(1)(5,000)}{0.08}$$

$$= 62,500,$$

or $62,500.

The improper integral also plays an important role in the study of probability theory, as we will see in Section 7.5.

SELF-CHECK EXERCISES 7.4

1. Evaluate $\displaystyle\int_{-\infty}^{\infty} \frac{x^3}{(1 + x^4)^{3/2}}\, dx$.

2. Suppose that an income stream is expected to continue indefinitely. Then the present value of such a stream can be calculated from the formula for the present value of an income stream by letting T approach infinity. Thus, the required present value is given by

$$PV = \int_0^{\infty} P(t)\, e^{-rt}\, dt.$$

Suppose that Marcia has an oil well in her backyard that generates a stream of income given by

$$P(t) = 20e^{-0.02t},$$

where $P(t)$ is expressed in thousands of dollars per year and t is the time in years from the present. Assuming that the prevailing interest rate in the foreseeable future is 10 percent per year compounded continuously, what is the present value of the income stream?

Solutions to Self-Check Exercises 7.4 can be found on page 463.

7.4   EXERCISES

In Exercises 1–10, find the area of the region under the given curve $y = f(x)$ over the indicated interval.

1. $f(x) = \dfrac{2}{x^2}$; $x \geq 3$

2. $f(x) = \dfrac{2}{x^3}$; $x \geq 2$

3. $f(x) = \dfrac{1}{(x - 2)^2}$; $x \geq 3$

4. $f(x) = \dfrac{2}{(x + 1)^3}$; $x \geq 0$

5. $f(x) = \dfrac{1}{x^{3/2}}$; $x \geq 1$

6. $f(x) = \dfrac{3}{x^{5/2}}$; $x \geq 4$

7. $f(x) = \dfrac{1}{(x + 1)^{5/2}}$; $x \geq 0$

8. $f(x) = \dfrac{1}{(1 - x)^{3/2}}$; $x \leq 0$

9. $f(x) = e^{2x}$; $x \leq 2$

10. $f(x) = xe^{-x^2}$; $x \geq 0$

11. Find the area of the region bounded by the x-axis and the graph of the function

$$f(x) = \frac{x}{(1 + x^2)^2}.$$

12. Find the area of the region bounded by the x-axis and the graph of the function

$$f(x) = \frac{e^x}{(1 + e^x)^2}.$$

13. Consider the improper integral

$$\int_0^{\infty} \sqrt{x}\, dx.$$

a. Evaluate $I(b) = \displaystyle\int_0^b \sqrt{x}\, dx$.

b. Show that

$$\lim_{b \to \infty} I(b) = \infty,$$

thus proving that the given improper integral is divergent.

14. Consider the improper integral

$$\int_0^{\infty} x^{-2/3}\, dx.$$

a. Evaluate $I(b) = \displaystyle\int_0^b x^{-2/3}\, dx$.

b. Show that

$$\lim_{b \to \infty} I(b) = \infty,$$

thus proving that the given improper integral is divergent.

In Exercises 15–40, evaluate each improper integral whenever it is convergent.

15. $\int_1^\infty \dfrac{3}{x^4}\,dx$

16. $\int_1^\infty \dfrac{1}{x^3}\,dx$

17. $\int_4^\infty \dfrac{2}{x^{3/2}}\,dx$

18. $\int_1^\infty \dfrac{1}{\sqrt{x}}\,dx$

19. $\int_1^\infty \dfrac{4}{x}\,dx$

20. $\int_2^\infty \dfrac{3}{x}\,dx$

21. $\int_{-\infty}^0 \dfrac{1}{(x-2)^3}\,dx$

22. $\int_2^\infty \dfrac{1}{(x+1)^2}\,dx$

23. $\int_1^\infty \dfrac{1}{(2x-1)^{3/2}}\,dx$

24. $\int_{-\infty}^0 \dfrac{1}{(4-x)^{3/2}}\,dx$

25. $\int_0^\infty e^{-x}\,dx$

26. $\int_0^\infty e^{-x/2}\,dx$

27. $\int_{-\infty}^0 e^{2x}\,dx$

28. $\int_{-\infty}^0 e^{3x}\,dx$

29. $\int_1^\infty \dfrac{e^{\sqrt{x}}}{\sqrt{x}}\,dx$

30. $\int_1^\infty \dfrac{e^{-\sqrt{x}}}{\sqrt{x}}\,dx$

31. $\int_{-\infty}^0 xe^x\,dx$

32. $\int_0^\infty xe^{-2x}\,dx$

33. $\int_{-\infty}^\infty x\,dx$

34. $\int_{-\infty}^\infty x^3\,dx$

35. $\int_{-\infty}^\infty x^2(1+x^3)^{-2}\,dx$

36. $\int_{-\infty}^\infty x(x^2+4)^{-3/2}\,dx$

37. $\int_{-\infty}^\infty xe^{1-x^2}\,dx$

38. $\int_{-\infty}^\infty \left(x-\dfrac{1}{2}\right)e^{-x^2+x-1}\,dx$

39. $\int_{-\infty}^\infty \dfrac{e^{-x}}{1+e^{-x}}\,dx$

40. $\int_{-\infty}^\infty \dfrac{xe^{-x^2}}{1+e^{-x^2}}\,dx$

C 41. **The Amount of an Endowment** A university alumni group wishes to provide an annual scholarship in the amount of $1500 beginning next year. If the scholarship fund will earn interest at the rate of 8 percent per year compounded continuously, find the amount of the endowment the alumni are required to make now.

C 42. **The Amount of an Endowment** Mr. Thompson wishes to establish a fund to provide a university medical center with an annual research grant of $50,000 beginning next year. If the fund will earn interest at the rate of 9 percent per year compounded continuously, find the amount of the endowment he is required to make now.

43. **Perpetual Net Income Streams** The present value of a perpetual stream of income that flows continually at the rate of $P(t)$ dollars per year is given by the formula

$$PV \approx \int_0^\infty P(t)e^{-rt}\,dt,$$

where r is the rate at which interest is compounded continuously. Using this formula, find the present value of a perpetual net income stream that is generated at the rate of

$$P(t) = 10,000 + 4,000t$$

dollars per year. $\left[\textit{Hint: } \lim_{b\to 0} \dfrac{b}{e^{rb}} = 0.\right]$

C 44. **Establishing a Trust Fund** Mrs. Wilkinson wants to establish a trust fund that will provide her children and heirs with a perpetual annuity in the amount of

$$P(t) = (20+t)$$

thousand dollars per year beginning next year. If the trust fund will earn interest at the rate of 10 percent per year compounded continuously, find the amount that she must place in the trust fund now. [*Hint:* Use the formula given in Exercise 43.]

C 45. **Capital Value** The capital value (present sale value) CV of property that can be rented on a perpetual basis for R dollars annually is approximated by the formula

$$CV \approx \int_0^\infty Re^{-it}\,dt,$$

where i is the prevailing continuous interest rate.
a. Show that $CV = R/i$.
b. Find the capital value of property that can be rented at $10,000 annually when the prevailing continuous interest rate is 12 percent per year.

SOLUTIONS TO SELF-CHECK EXERCISES 7.4

1. Write

$$\int_{-\infty}^\infty \dfrac{x^3}{(1+x^4)^{3/2}}\,dx = \int_{-\infty}^0 \dfrac{x^3}{(1+x^4)^{3/2}}\,dx + \int_0^\infty \dfrac{x^3}{(1+x^4)^{3/2}}\,dx.$$

Now,

$$\int_{-\infty}^{0} \frac{x^3}{(1 + x^4)^{3/2}} \, dx$$

$$= \lim_{a \to -\infty} \int_{a}^{0} x^3 (1 + x^4)^{-3/2} \, dx$$

$$= \lim_{a \to -\infty} \frac{1}{4}(-2)(1 + x^4)^{-1/2} \Big|_{a}^{0} \qquad \text{(Integrating by substitution)}$$

$$= -\frac{1}{2} \lim_{a \to -\infty} \left[1 - \frac{1}{(1 + a^4)^{1/2}} \right]$$

$$= -\frac{1}{2}.$$

Similarly, you can show that

$$\int_{0}^{\infty} \frac{x^3}{(1 + x^4)^{3/2}} \, dx = \frac{1}{2}.$$

Therefore,

$$\int_{-\infty}^{\infty} \frac{x^3}{(1 + x^4)^{3/2}} \, dx = -\frac{1}{2} + \frac{1}{2}$$

$$= 0.$$

2. The required present value is given by

$$PV = \int_{0}^{\infty} 20 e^{-0.02t} e^{-0.10t} \, dt$$

$$= 20 \int_{0}^{\infty} e^{-0.12t} \, dt$$

$$= 20 \lim_{b \to \infty} \int_{0}^{b} e^{-0.12t} \, dt$$

$$= -\frac{20}{0.12} \lim_{b \to \infty} e^{-0.12t} \Big|_{0}^{b}$$

$$= -\frac{500}{3} \lim_{b \to \infty} (e^{-0.12b} - 1)$$

$$= \frac{500}{3},$$

or approximately \$166,667.

7.5   Applications of Probability to Calculus

The systematic study of probability began in the seventeenth century, when certain aristocrats wanted to discover superior strategies to use in the gaming rooms of Europe. Some of the best mathematicians of the period were engaged in this pursuit. Since then, probability has evolved into an important branch of mathematics with widespread applications in virtually every sphere of human endeavor in which an element of uncertainty is present. In this section, we will take a brief look at the role of the integral in the study of probability theory.

Probability Density Functions

We begin by mentioning some elementary terms important in the study of probability. For our purpose, an **experiment** is an activity with observable results called **outcomes,** or **sample points.** The totality of all outcomes makes up the **sample space** of the experiment. A subset of the sample space is called an **event** of the experiment.

Now, given an event associated with an experiment, our primary objective is to determine the likelihood that this event will occur. This likelihood, or **probability of the event,** is a number between 0 and 1 and may be viewed as the proportionate number of times that the event will occur if the experiment associated with the event is repeated indefinitely under independent and similar conditions. As an example, let us consider the simple experiment of tossing an unbiased coin and observing whether it lands "heads" (H) or "tails" (T). Since the coin is *unbiased,* we see that the probability of each outcome is $\frac{1}{2}$, abbreviated

$$P(\text{H}) = \frac{1}{2} \quad \text{and} \quad P(\text{T}) = \frac{1}{2}.$$

In many situations, it is desirable to assign numerical values to the outcomes of an experiment. For example, suppose that an experiment consists of casting a die and observing the face that lands uppermost. If we let X denote the outcome of the experiment, then X assumes one of the values 1, 2, 3, 4, 5, or 6. Because the values assumed by X depend on the outcomes of a chance experiment, the outcome X is referred to as a **random variable.** In this case, the random variable X is also said to be **finite discrete** since it can assume only a finite number of integer values.

A random variable x that can assume any value in an interval is called a **continuous random variable.** Examples of continuous random variables are the life span of a light bulb, the length of a telephone call, the length of an infant at birth, the daily amount of rainfall in Boston, and the life span of a certain plant species. For the remainder of this section, we will be interested primarily in continuous, random variables.

Consider an experiment in which the associated random variable x has as its sample space the interval $[a, b]$. Then, an event of the experiment is any subset of $[a, b]$. For example, if x denotes the life span of a light bulb, then the sample space associated with the experiment is $[0, \infty)$, and the event that a light bulb selected at random has a life span between 500 and 600 hours, inclusive, is described by the interval $[500, 600]$ or, equivalently, by the inequality $500 \leq x \leq 600$. The probability that the light bulb will have a life span of between 500 and 600 hours is denoted by $P(500 \leq x \leq 600)$.

In general, we will be interested in computing $P(a \leq x \leq b)$, the probability that a random variable x assumes a value in the interval $a \leq x \leq b$. This computation is based on the notion of a probabilty density function, which we now introduce.

PROBABILITY DENSITY FUNCTION

A **probability density function** of a random variable x in an interval I, where I may be bounded or unbounded, is a nonnegative function f having the following properties:

1. The total area of the region under the graph of f is equal to 1 (see Figure 7.15a).

2. The probability that an observed value of the random variable x lies in the interval $[a, b]$ is given by

$$P(a \le x \le b) = \int_a^b f(x)\, dx \qquad \text{(Figure 7.15b)}.$$

FIGURE 7.15

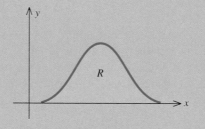

(a) Area of $R = 1$.

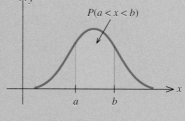

(b) $P(a < x < b)$ is the probability that an outcome of an experiment will lie between a and b.

A few comments are in order. First, a probability density function of a random variable x may be constructed using methods that range from theoretical considerations of the problem on the one extreme to an interpretation of data associated with the experiment on the other. Second, Property 1 states that the probability that a continuous random variable takes on a value lying in its range is 1, a certainty, which is expected. Third, Property 2 states that the probability that the random variable x assumes a value in an interval $a \le x \le b$ is given by the area of the region between the graph of f and the x-axis from $x = a$ to $x = b$. Because the area under one point of the graph of f is equal to zero, we see immediately that $P(a \le x \le b) = P(a < x \le b) = P(a \le x < b) = P(a < x < b)$.

EXAMPLE 1 Show that each of the following functions satisfies the nonnegativity condition and Property 1 of probability density functions.

a. $f(x) = \dfrac{2}{27}x(x - 1) \qquad (1 \le x \le 4)$

b. $f(x) = \dfrac{1}{3}e^{(-1/3)x} \qquad (0 \le x < \infty)$

Solution

a. Since the factors x and $(x - 1)$ are both nonnegative, we see that $f(x) \geq 0$ on [1, 4]. Next, we compute

$$\int_1^4 \frac{2}{27} x(x - 1)\, dx = \frac{2}{27} \int_1^4 (x^2 - x)\, dx$$

$$= \frac{2}{27} \left(\frac{1}{3} x^3 - \frac{1}{2} x^2 \right) \Big|_1^4$$

$$= \frac{2}{27} \left[\left(\frac{64}{3} - 8 \right) - \left(\frac{1}{3} - \frac{1}{2} \right) \right]$$

$$= \frac{2}{27} \left(\frac{27}{2} \right)$$

$$= 1,$$

showing that Property 1 of probability density functions holds as well.

b. First, $f(x) = \frac{1}{3} e^{(-1/3)x} \geq 0$ for all values of x in [0, ∞). Next,

$$\int_0^\infty \frac{1}{3} e^{(-1/3)x}\, dx = \lim_{b \to \infty} \int_0^b \frac{1}{3} e^{(-1/3)x}\, dx$$

$$= \lim_{b \to \infty} -e^{(-1/3)x} \Big|_0^b$$

$$= \lim_{b \to \infty} (-e^{(-1/3)b} + 1)$$

$$= 1,$$

so the area under the graph of $f(x) = \frac{1}{3} e^{(-1/3)x}$ is equal to 1, as we set out to show.   ∎

EXAMPLE 2

a. Determine the value of the constant k such that the function $f(x) = kx^2$ is a probability density function on the interval [0, 5].

b. If x is a continuous random variable with the probability density function given in (a), compute the probability that x will assume a value between $x = 1$ and $x = 2$.

Solution
a. We compute

$$\int_0^5 kx^2\, dx = k \int_0^5 x^2\, dx$$

$$= \frac{k}{3} x^3 \Big|_0^5$$

$$= \frac{125}{3} k.$$

Since this value must be equal to 1, we find that $k = \frac{3}{125}$.

FIGURE 7.16
$P(1 \le x \le 2)$ *for the probability density function*
$y = \dfrac{3}{125} x^2.$

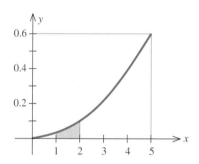

b. The required probability is given by

$$P(1 \le x \le 2) = \int_1^2 f(x)\, dx = \int_1^2 \frac{3}{125} x^2\, dx$$

$$= \frac{1}{125} x^3 \Big|_1^2 = \frac{1}{125}(8 - 1)$$

$$= \frac{7}{125}.$$

The graph of the probability density function f and the area corresponding to the probability $P(1 \le x \le 2)$ are shown in Figure 7.16.   ∎

EXAMPLE 3   The TKK Products Corporation manufactures a 200-watt electric light bulb. Laboratory tests show that the life spans of these light bulbs have a distribution described by the probability density function

$$f(x) = 0.001 e^{-0.001 x}.$$

Determine the probability that a light bulb will have a life span of

a. 500 hours or less.
b. more than 500 hours.
c. more than 1000 hours but less than 1500 hours.

Solution
Let x denote the life span of a light bulb.

a. The probability that a light bulb will have a life span of 500 hours or less is given by

$$P(0 \le x \le 500) = \int_0^{500} 0.001 e^{-0.001 x}\, dx$$

$$= -e^{-0.001 x} \Big|_0^{500} = -e^{-0.5} + 1$$

$$\approx 0.3935.$$

b. The probability that a light bulb will have a life span of more than 500 hours is given by

$$P(x > 500) = \int_{500}^{\infty} 0.001 e^{-0.001 x}\, dx$$

$$= \lim_{b \to \infty} \int_{500}^{b} 0.001 e^{-0.001 x}\, dx$$

$$= \lim_{b \to \infty} -e^{-0.001 x} \Big|_{500}^{b}$$

$$= \lim_{b \to \infty} \left(-e^{-0.001 b} + e^{-0.5} \right)$$

$$- e^{-0.5} \approx 0.6065.$$

This result may also be obtained by observing that

$$P(x > 500) = 1 - P(x \le 500)$$
$$= 1 - 0.3935 \qquad \text{[Using the result from (a)]}$$
$$\approx 0.6065.$$

c. The probability that a light bulb will have a life span of more than 1000 hours but less than 1500 hours is given by

$$
\begin{aligned}
P(1000 < x < 1500) &= \int_{1000}^{1500} 0.001 e^{-0.001x}\, dx \\
&= -e^{-0.001x}\Big|_{1000}^{1500} \\
&= -e^{-1.5} + e^{-1} \\
&\approx -0.2231 + 0.3679 \\
&= 0.1448.
\end{aligned}
$$ ∎

The probability density function of Example 3 has the form

$$f(x) = ke^{-kx},$$

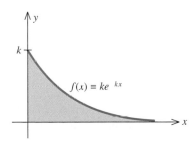

FIGURE 7.17

The area under the graph of the exponential density function is equal to 1.

where $x \ge 0$ and k is a positive constant. Its graph is shown in Figure 7.17. Such a probability function is called an **exponential density function,** and a random variable associated with such a probability density function is said to be **exponentially distributed.** Exponential random variables are used to represent the life span of electronic components, the duration of telephone calls, the waiting time in a doctor's office, and the time between successive flight arrivals and departures in an airport, to mention but a few applications.

Another probability density function, and the one most widely used, is the **normal density function,** defined by

$$f(x) = \frac{1}{\sigma\sqrt{2\pi}} e^{-(1/2)[(x-\mu)/\sigma]^2},$$

where μ and σ are constants. The graph of the normal distribution is bell-shaped (Figure 7.18). Many phenomena, such as the heights of people in a given population, the weights of newborn infants, the IQs of college students, the actual weights of 16-ounce packages of cereals, and so on, have probability distributions that are normal.

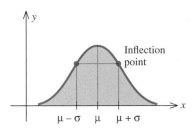

FIGURE 7.18

The area under the bell-shaped normal distribution curve

Areas under the standard normal curve (the normal curve with $\mu = 0$ and $\sigma = 1$) have been extensively computed and tabulated. Most problems involving the normal distribution can be solved with the aid of these tables.

Expected Value

The average value, or **expected value,** of a discrete variable X that takes on values $x_1, x_2, \ldots, x_n$ with associated probabilities $p_1, p_2, \ldots, p_n$ is defined by

$$E(X) = x_1 p_1 + x_2 p_2 + \cdots + x_n p_n.$$

If each of the values $x_1, x_2, \ldots, x_n$ occurs with equal frequency, then $p_1 = p_2 = \cdots = p_n = 1/n$ and

$$E(X) = x_1\left(\frac{1}{n}\right) + x_2\left(\frac{1}{n}\right) + \cdots + x_n\left(\frac{1}{n}\right)$$

$$= \frac{x_1 + x_2 + \cdots + x_n}{n},$$

giving the familiar formula for computing the average value of the n numbers $x_1, x_2, \ldots, x_n$.

Now, suppose that x is a continuous random variable and f is the probability density function associated with it. For simplicity, let us first assume that $a \le x \le b$. Divide the interval $[a, b]$ into n subintervals of equal length $\Delta x = (b - a)/n$ by means of the $(n + 1)$ points $x_0 = a, x_1, x_2, \ldots, x_n = b$ (Figure 7.19). To find an approximation of the average value, or expected value, of x on the interval $[a, b]$, let us treat x as if it were a discrete random variable that takes on the values $x_1, x_2, \ldots, x_n$ with probabilities $p_1, p_2, \ldots, p_n$. Then

$$E(x) \approx x_1 p_1 + x_2 p_2 + \cdots + x_n p_n.$$

FIGURE 7.19

Approximating the expected value of a random variable x on [a, b] by a Riemann sum

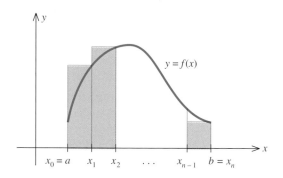

But p_1 is the probability that x is in the interval $[x_0, x_1]$, and this is just the area under the graph of f from $x = x_0$ to $x = x_1$, which may be approximated by $f(x_1)\Delta x$. The probability $p_2, \ldots, p_n$ may be approximated in a similar manner. Thus,

$$E(x) \approx x_1 f(x_1)\Delta x + x_2 f(x_2)\Delta x + \cdots + x_n f(x_n)\Delta x,$$

which is seen to be the Riemann sum of the function $g(x) = xf(x)$ over the interval $[a, b]$. Letting n approach infinity, we obtain the following formula:

EXPECTED VALUE OF A CONTINUOUS RANDOM VARIABLE

Suppose that the function f defined on the interval $[a, b]$ is the probability density function associated with a continuous random variable x. Then, the **expected value** of x is

$$E(x) = \int_a^b xf(x)\, dx. \tag{17}$$

If either $a = -\infty$ or $b = \infty$, then the integral in (17) becomes an improper integral.

The expected value of a random variable plays an important role in many practical applications. For example, if x represents the life span of a certain brand of electronic components, then the expected value of x gives the average life span of these components. If x measures the waiting time in a doctor's office, then $E(x)$ gives the average waiting time, and so on.

EXAMPLE 4   Show that if a continuous random variable x is exponentially distributed with the probability density function

$$f(x) = ke^{-kx},$$

then the expected value $E(x)$ is equal to $1/k$. Using this result, determine the average life span of a 200-watt electric light bulb manufactured by the TKK Products Corporation of Example 3.

Solution   We compute

$$E(x) = \int_0^\infty xf(x)\, dx$$

$$= \int_0^\infty kxe^{-kx}\, dx$$

$$= k \lim_{b\to\infty} \int_0^b xe^{-kx}\, dx.$$

Integrating by parts with

$$u = x \quad \text{and} \quad dv = e^{-kx}\, dx$$

so that

$$du = dx \quad \text{and} \quad v = -\frac{1}{k}e^{-kx},$$

we have

$$E(x) = k \lim_{b\to\infty} \left[-\frac{1}{k}xe^{-kx}\Big|_0^b + \frac{1}{k}\int_0^b e^{-kx}\, dx \right]$$

$$= k \lim_{b\to\infty} \left[-\left(\frac{1}{k}\right)be^{-kb} - \frac{1}{k^2}e^{-kx}\Big|_0^b \right]$$

$$= k \lim_{b\to\infty} \left[-\left(\frac{1}{k}\right)be^{-kb} - \frac{1}{k^2}e^{-kb} + \frac{1}{k^2} \right]$$

$$= -\lim_{b\to\infty}\frac{b}{e^{kb}} - \frac{1}{k}\lim_{b\to\infty}\frac{1}{e^{kb}} + \frac{1}{k}\lim_{b\to\infty}1.$$

Now, by taking a sequence of values of b that approaches infinity, for example, $b = 10, 100, 1,000, 10,000, \ldots$, we see that, for a fixed k,

$$\lim_{b\to\infty}\frac{b}{e^{kb}} = 0.$$

Therefore,

$$E(x) = \frac{1}{k},$$

as we set out to show. Next, since $k = 0.001$ in Example 3, we see that the average life span of the TKK light bulb is $1/(0.001) = 1,000$ hours.   ∎

Before considering another example, let us summarize the important result obtained in Example 4.

THE AVERAGE VALUE OF AN EXPONENTIAL DENSITY FUNCTION	If a continuous random variable x is exponentially distributed with probability density function $$f(x) = ke^{-kx},$$ then the **expected value** of x is given by $$E = \frac{1}{k}.$$

EXAMPLE 5 On a typical Monday morning, the time between successive arrivals of planes at Jackson International Airport is an exponentially distributed random variable x with expected value of 10 (minutes).

a. Find the probability density function associated with x.
b. What is the probability that between 6 and 8 minutes will elapse between successive arrivals of planes?
c. What is the probability that the time between successive arrivals of planes will be more than 15 minutes?

Solution

a. Since x is exponentially distributed, the associated probability density function has the form $f(x) = ke^{-kx}$. Next, since the expected value of x is 10, we see that

$$E(x) = \frac{1}{k}$$

or

$$k = \frac{1}{10}$$
$$= 0.1,$$

so the required probability density function is

$$f(x) = 0.1e^{-0.1x}.$$

b. The probability that between 6 and 8 minutes will elapse between successive arrivals is given by

$$P(6 \leq x \leq 8) = \int_6^8 0.1e^{-0.1x}\,dx = -e^{-0.1x}\Big|_6^8$$
$$= -e^{-0.8} + e^{-0.6}$$
$$\approx 0.100.$$

c. The probability that the time between successive arrivals will be more than 15 minutes is given by

$$P(x > 15) = \int_{15}^{\infty} 0.1e^{-0.1x}\, dx$$

$$= \lim_{b \to \infty} \int_{15}^{b} 0.1e^{-0.1x}\, dx$$

$$= \lim_{b \to \infty} \left[-e^{-0.1x} \Big|_{15}^{b} \right]$$

$$= \lim_{b \to \infty} (-e^{-0.1b} + e^{-1.5}) = e^{-1.5}$$

$$\approx 0.22.$$ ∎

**SELF-CHECK
EXERCISES 7.5**

1. Determine the value of the constant k such that the function $f(x) = k(4x - x^2)$ is a probability density function on the interval $[0, 4]$.

2. Suppose that x is a continuous random variable with the probability density function of Exercise 1. Find the probability that x will assume a value between $x = 1$ and $x = 3$.

Solutions to Self-Check Exercises 7.5 can be found on page 474.

7.5   EXERCISES

In Exercises 1–8, show that the given function is a probability density function on the specified interval.

1. $f(x) = \dfrac{2}{32}x \quad (2 \le x \le 6)$

2. $f(x) = \dfrac{2}{9}(3x - x^2) \quad (0 \le x \le 3)$

3. $f(x) = \dfrac{3}{8}x^2 \quad (0 \le x \le 2)$

4. $f(x) = \dfrac{3}{32}(x - 1)(5 - x) \quad (1 \le x \le 5)$

5. $f(x) = 20(x^3 - x^4) \quad (0 \le x \le 1)$

6. $f(x) = \dfrac{8}{7x^2} \quad (1 \le x \le 8)$

7. $f(x) = \dfrac{3}{14}\sqrt{x} \quad (1 \le x \le 4)$

8. $f(x) = \dfrac{12 - x}{72} \quad (0 \le x \le 12)$

9. a. Determine the value of the constant k such that the function $f(x) = k(4 - x)$ is a probability density function on the interval $[0, 4]$.

    b. If x is a continuous random variable with the probability density function given in (a), compute the probability that x will assume a value between $x = 1$ and $x = 3$.

10. a. Determine the value of the constant k such that the function $f(x) = k/x^2$ is a probability density function on the interval $[1, 10]$.

    b. If x is a continuous random variable with the probability density function of (a), compute the probability that x will assume a value between $x = 2$ and $x = 6$.

11. Life Span of a Plant Species The life span of a certain plant species (in days) is described by the probability density function

$$f(x) = \frac{1}{100}e^{-x/100}.$$

a. Find the probability that a plant of this species will live for 100 days or less.
b. Find the probability that a plant of this species will live more than 120 days.
c. Find the probability that a plant of this species will live more than 60 days but less than 140 days.

12. **Average Waiting Time for Patients**   The average waiting time for patients arriving at the Newtown Health Clinic between 1 P.M. and 4 P.M. on a weekday is an exponentially distributed random variable x with expected value of 15 minutes.
   a. Find the probability density function associated with x.
   b. What is the probability that a patient arriving at the clinic between 1 P.M. and 4 P.M. will have to wait between 10 and 12 minutes?
   c. What is the probability that a patient arriving at the clinic between 1 P.M. and 4 P.M. will have to wait more than 15 minutes?

13. **Expected Number of Chocolate Chips**   The number of chocolate chips in each cookie of a certain brand has a distribution described by the probability density function

$$f(x) = \frac{1}{36}(6x - x^2) \qquad (0 \leq x \leq 6).$$

Find the expected number of chips in each cookie.

14. **Reliability of Robots**   The National Welding Company uses industrial robots in some of its assembly line operations. Management has determined that, on average, a robot breaks down after 1000 hours of use and that the lengths of time between breakdowns are exponentially distributed.
   a. What is the probability that a robot selected at random will break down between 600 and 800 hours of use?

   b. What is the probability that a robot will break down after 1200 hours of use?

C 15. **Expressway Tollbooths**   Suppose that the time intervals between arrivals of successive cars at an expressway tollbooth during rush hour are exponentially distributed and that the average time interval between arrivals is 8 seconds. Find the probability that the average time interval between arrivals of successive cars is more than 8 seconds.

C 16. **Time Intervals Between Phone Calls**   A study conducted by Uni-Mart, a mail-order department store, reveals that the time intervals between incoming telephone calls on its toll-free 800 line between 10 A.M. and 2 P.M. are exponentially distributed and that the average time interval is 30 seconds. What is the probability that the time interval between successive calls is more than 2 minutes?

C 17. **Reliability of Microprocessors**   The microprocessors manufactured by the United Motor Works Company, which are used in automobiles to regulate fuel consumption, are guaranteed against defects for 20,000 miles of use. Tests conducted in the laboratory under simulated driving conditions reveal that the distances driven before the microprocessors break down are exponentially distributed and that the average distance driven before the microprocessors fail is 100,000 miles. What is the probability that a microprocessor selected at random will fail during the warranty period?

SOLUTIONS TO SELF-CHECK EXERCISES 7.5

1. We compute

$$\int_0^4 k(4x - x^2)\, dx = k\left(2x^2 - \frac{1}{3}x^3\right)\Big|_0^4$$

$$= k\left(32 - \frac{64}{3}\right)$$

$$= \frac{32}{3}k.$$

Since this value must be equal to 1, we see that $k = \frac{3}{32}$.

2. The required probability is given by

$$P(1 \leq x \leq 3) = \int_1^3 f(x)\, dx$$

$$= \int_1^3 \frac{3}{32}(4x - x^2)\, dx$$

$$= \frac{3}{32}\left(2x^2 - \frac{1}{3}x^3\right)\Big|_1^3$$

$$= \frac{3}{32}\left[(18 - 9) - \left(2 - \frac{1}{3}\right)\right] = \frac{11}{16}.$$

CHAPTER SEVEN SUMMARY OF PRINCIPAL FORMULAS AND TERMS

Formulas

1. Integration by parts

$$\int u\, dv = uv - \int v\, du$$

2. Improper integral of f over $[a, \infty)$

$$\int_a^\infty f(x)\, dx = \lim_{b \to \infty} \int_a^b f(x)\, dx$$

3. Improper integral of f over $(-\infty, b]$

$$\int_{-\infty}^b f(x)\, dx = \lim_{a \to -\infty} \int_a^b f(x)\, dx$$

4. Improper integral of f over $(-\infty, \infty)$

$$\int_{-\infty}^\infty f(x)\, dx = \int_{-\infty}^c f(x)\, dx + \int_c^\infty f(x)\, dx$$

5. Present value of a perpetuity

$$PV = \frac{mP}{r}$$

6. Trapezoidal Rule

$$\int_a^b f(x)\, dx \approx \frac{\Delta x}{2}[f(x_0) + 2f(x_1)$$
$$+ 2f(x_2) + \cdots$$
$$+ 2f(x_{n-1}) + f(x_n)],$$
where $\Delta x = \dfrac{b - a}{n}$

7. Simpson's Rule

$$\int_a^b f(x)\, dx \approx \frac{\Delta x}{3}[f(x_0) + 4f(x_1)$$
$$+ 2f(x_2) + 4f(x_3)$$
$$+ 2f(x_4) + \cdots + 4f(x_{n-1})$$
$$+ f(x_n)],$$
where $\Delta x = \dfrac{b - a}{n}$

8. Maximum error for Trapezoidal Rule

$$\frac{M(b - a)^3}{12n^2}, \text{ where } |f''(x)| \leq M,$$
$$a \leq x \leq b$$

9. Maximum error for Simpson's Rule

$$\frac{M(b - a)^5}{180n^4}, \text{ where } |f^{(4)}(x)| \leq M,$$
$$a \leq x \leq b$$

10. Probability an outcome of an experiment lies between a and b
$$P(a \le x \le b) = \int_a^b f(x)\, dx$$

11. Exponential density function
$$f(x) = ke^{-kx}$$

12. Expected value
$$E(x) = x_1 p_1 + x_2 p_2 + \cdots + x_n p_n$$

13. Expected value of a continuous random variable
$$E(x) = \int_a^b xf(x)\, dx$$

Terms

Improper integral	Probability of an event
Convergent integral	Random variable
Divergent integral	Continuous random variable
Perpetuity	Probability density function
Experiment	Normal density function
Sample space	Expected value

CHAPTER 7 REVIEW EXERCISES

In Exercises 1–6, evaluate each of the given integrals.

1. $\displaystyle\int 2xe^{-x}\, dx$

2. $\displaystyle\int xe^{4x}\, dx$

3. $\displaystyle\int \ln 5x\, dx$

4. $\displaystyle\int_1^4 \ln 2x\, dx$

5. $\displaystyle\int_0^1 xe^{-2x}\, dx$

6. $\displaystyle\int_0^2 xe^{2x}\, dx$

7. Find the function f given that the slope of the tangent line to the graph of f at any point $(x, f(x))$ is
$$f'(x) = \frac{\ln x}{\sqrt{x}}$$
and that the graph of f passes through the point $(1, -2)$.

8. Find the function f given that the slope of the tangent line to the graph of f at any point $(x, f(x))$ is
$$f'(x) = \frac{e^x}{1 + e^x}$$
and that the graph of f passes through the point $(0, 0)$.

In Exercises 9–14, use the table of integrals in Section 7.2 to evaluate each of the given integrals.

9. $\displaystyle\int \frac{x^2\, dx}{(3 + 2x)^2}$

10. $\displaystyle\int \frac{2x}{\sqrt{2x + 3}}\, dx$

11. $\displaystyle\int x^2 e^{4x}\, dx$

12. $\displaystyle\int \frac{dx}{(x^2 - 25)^{3/2}}$

13. $\displaystyle\int \frac{dx}{x^2\sqrt{x^2 - 4}}$

14. $\displaystyle\int 8x^3 \ln 2x\, dx$

In Exercises 15–20, evaluate each improper integral whenever it is convergent.

15. $\displaystyle\int_0^\infty e^{-2x}\, dx$

16. $\displaystyle\int_{-\infty}^0 e^{3x}\, dx$

17. $\displaystyle\int_3^\infty \frac{2}{x}\, dx$

18. $\displaystyle\int_2^\infty \frac{1}{(x + 2)^{3/2}}\, dx$

19. $\displaystyle\int_2^\infty \frac{dx}{(1 + 2x)^2}$

20. $\displaystyle\int_1^\infty 3e^{1-x}\, dx$

In Exercises 21–24, use the Trapezoidal Rule and Simpson's Rule to approximate the value of the definite integral.

21. $\displaystyle\int_1^3 \frac{dx}{1 + \sqrt{x}}$; $n = 4$ **22.** $\displaystyle\int_0^1 e^{x^2}\, dx$; $n = 4$

23. $\displaystyle\int_{-1}^1 \sqrt{1 + x^4}\, dx$; $n = 4$ **24.** $\displaystyle\int_1^3 \frac{e^x}{x}\, dx$; $n = 4$

25. Show that the function $f(x) = (3/128)(16 - x^2)$ is a probability density function on the interval [0, 4].

26. Show that the function $f(x) = (1/9)x\sqrt{9 - x^2}$ is a probability density function on the interval [0, 3].

27. a. Determine the value of the constant k such that the function $f(x) = kx\sqrt{4 - x^2}$ is a probability density function on the interval [0, 2].
   b. If x is a continuous random variable with the probability density function given in (a), compute the probability that x will assume a value between $x = 1$ and $x = 2$.

28. a. Determine the value of the constant k such that the function $f(x) = k/\sqrt{x}$ is a probability density function on the interval [1, 4].
   b. If x is a continuous random variable with the probability density function given in (a), compute the probability that x will assume a value between $x = 2$ and $x = 3$.

29. a. Determine the value of the constant k such that the function $f(x) = kx^2(3 - x)$ is a probability density function on the interval [0, 3].
   b. If x is a continuous random variable with the probability density function given in (a), compute the probability that x will assume a value between $x = 1$ and $x = 2$.

30. Records at the Centerville Hospital indicate that the length of time in days that a maternity patient stays in the hospital has a probability density function given by

$$P(t) = \frac{1}{4}e^{(-1/4)t}.$$

   a. What is the probability that a woman entering the maternity wing will be there more than six days?
   b. What is the probability that a woman entering the maternity wing will be there less than two days?
   c. What is the average length of time that a woman entering the maternity wing stays in the hospital?

31. The supply equation for the GTC Slim-Phone is given by

$$p = 2\sqrt{25 + x^2},$$

where p is the unit price in dollars and x is the quantity demanded per month in units of ten thousand. Find the producers' surplus if the market price is $26. Use the table of integrals in Section 7.2 to evaluate the definite integral.

32. The sales of the Starr Communication Company's newest video cartridge game, Laser Beams, are currently

$$te^{-0.05t}$$

units per month (each unit representing 1000 cartridges), where t denotes the number of months since the release of the cartridge. Find an expression that gives the total number of units of video cartridge games sold as a function of t. How many video games will be sold by the end of the first year?

33. The demand equation for a computer software program is given by

$$p = 2\sqrt{325 - x^2},$$

where p is the unit price in dollars and x is the quantity demanded per month in units of a thousand. Find the consumers' surplus if the market price is $30. Evaluate the definite integral using Simpson's Rule with $n = 10$.

34. Using aerial photographs, the Coast Guard was able to determine the dimensions of an oil spill along an embankment on a coastline, as shown in the accompanying figure. Using (a) the Trapezoidal Rule and (b) Simpson's Rule with $n = 10$, estimate the area of the oil spill.

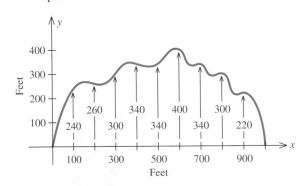

35. Mr. Lindsey wishes to establish a memorial fund at the Newtown Hospital in the amount of $10,000 per year beginning next year. If the fund earns interest at a rate of 9 percent per year compounded continuously, find the amount of endowment that he is required to make now.

What should the dimensions of the swimming pool be? The operators of the Viking Princess, *a luxury cruise ship, are thinking about adding another swimming pool to the* Princess. *The chief engineer has suggested that an area in the form of an ellipse, located in the rear of the promenade deck, would be suitable for this purpose. Subject to this constraint, what are the dimensions of the largest pool that can be built? See Example 5, page 530, to see how to solve this problem.*

Calculus of Several Variables

Up to now we have dealt with functions involving one variable. In many real-life situations, however, we encounter quantities that depend on two or more quantities. For example, the Consumer Price Index (CPI) compiled every month by the Bureau of Labor Statistics depends on the price of more than 95,000 consumer items from gas to groceries. In order to study such relationships, we need the notion of a function of several variables. Next, generalizing the concept of the derivative of a function of one variable, we are led to the idea of the *partial derivatives* of a function of two or more variables. Partial derivatives enable us to study the rate of change of a function with respect to one variable while holding all the other variables constant. We will also learn how to find the maximum and/or minimum values of a function of several variables. For example, we will learn how a manufacturer can maximize her profits by producing the right amount of each of her products. As another example of the application of optimization theory, we will learn how to find an equation of the straight line that "best" fits a set of data points scattered about a straight line.

—————————————

8.1   Functions of Several Variables

Up to now, our study of calculus has been restricted to functions of one variable. In many practical situations, however, the formulation of a problem results in a mathematical model that involves a function of two or more variables. For example, suppose that the Ace Novelty Company determines that the profits are $6, $5, and $4 for three types of souvenirs it produces. Let x, y, and z denote the number of type A, type B, and type C souvenirs to be made; then the company's profit is given by

$$P = 6x + 5y + 4z,$$

and P is a function of the three variables x, y, and z.

Functions of Two Variables

Although this chapter deals with real-valued functions of several variables, most of our definitions and results are stated in terms of a function of two variables. One reason for adopting this approach, as you will soon see, is that there is a geometrical interpretation for this special case, which serves as an important visual aid. We can then draw upon the experience gained from studying the two-variable case to help us understand the concepts and results connected with the more general case, which, by and large, is just a simple extension of the lower-dimensional case.

A FUNCTION OF TWO VARIABLES

A **real-valued function of two variables,** f, consists of

1. a set A of ordered pairs of real numbers (x, y) called the **domain** of the function and

2. a rule that associates with each ordered pair in the domain of f one and only one real number, denoted by $z = f(x, y)$.

The variables x and y are called **independent variables,** and the variable z, which is dependent on the values of x and y, is referred to as a **dependent variable.**

As in the case of a real-valued function of one real variable, the number $z = f(x, y)$ is called the **value of** f at the point (x, y). And, unless specified, the domain of the function f will be taken to be the largest possible set for which the rule defining f is meaningful.

EXAMPLE 1   Let f be the function defined by

$$f(x, y) = x + xy + y^2 + 2.$$

Compute $f(0, 0)$, $f(1, 2)$, and $f(2, 1)$.

Solution   We have

$$f(0, 0) = 0 + (0)(0) + 0^2 + 2 = 2$$
$$f(1, 2) = 1 + (1)(2) + 2^2 + 2 = 9$$

and

$$f(2, 1) = 2 + (2)(1) + 1^2 + 2 = 7.$$ ∎

The domain of a function of two variables $f(x, y)$ is a set of ordered pairs of real numbers and may therefore be viewed as a subset of the xy-plane.

EXAMPLE 2   Find the domain of each of the following functions.

a. $f(x, y) = x^2 + y^2$ **b.** $g(x, y) = \dfrac{2}{x - y}$

c. $h(x, y) = \sqrt{1 - x^2 - y^2}$

Solution

a. $f(x, y)$ is defined for all real values of x and y, so the domain of the function f is the set of all points (x, y) in the xy-plane.

b. $g(x, y)$ is defined for all $x \neq y$, so the domain of the function g is the set of all points in the xy-plane except those lying on the line $y = x$ (Figure 8.1a).

c. We require that $1 - x^2 - y^2 \geq 0$ or $x^2 + y^2 \leq 1$, which is just the set of all points (x, y) lying on and inside the circle of radius 1 with center at the origin (Figure 8.1b).

FIGURE 8.1

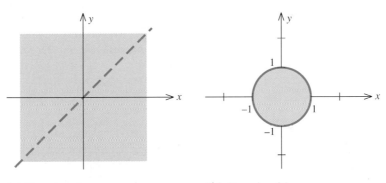

(a) Domain of g **(b)** Domain of h ■

Applications

EXAMPLE 3   The Acrosonic Company manufactures a bookshelf loudspeaker system that may be bought fully assembled or in a kit. The demand equations that relate the unit prices, p and q, to the quantities demanded weekly, x and y, of the assembled and kit versions of the loudspeaker systems are given by

$$p = 300 - \frac{1}{4}x - \frac{1}{8}y \quad \text{and} \quad q = 240 - \frac{1}{8}x - \frac{3}{8}y.$$

a. What is the weekly total revenue function $R(x, y)$?

b. What is the domain of the function R?

Solution

a. The weekly revenue realizable from the sale of x units of the assembled speaker systems at p dollars per unit is given by xp dollars. Similarly, the weekly revenue realizable from the sale of y units of the kits at q dollars per unit is given by yq dollars. Therefore, the weekly total revenue function

R is given by

$$R(x, y) = xp + yq$$

$$= x\left(300 - \frac{1}{4}x - \frac{1}{8}y\right) + y\left(240 - \frac{1}{8}x - \frac{3}{8}y\right)$$

$$= -\frac{1}{4}x^2 - \frac{3}{8}y^2 - \frac{1}{4}xy + 300x + 240y.$$

FIGURE 8.2

The domain of $R(x, y)$

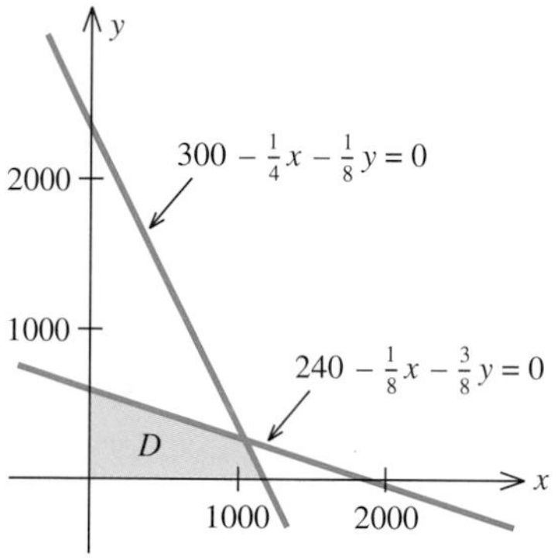

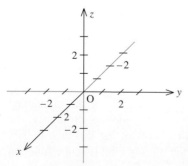

b. To find the domain of the function R, let us observe that the quantities x, y, p, and q must be nonnegative. This observation leads to the following system of linear inequalities:

$$300 - \frac{1}{4}x - \frac{1}{8}y \geq 0$$

$$240 - \frac{1}{8}x - \frac{3}{8}y \geq 0$$

$$x \geq 0$$

$$y \geq 0.$$

The domain of the function R is sketched in Figure 8.2.   ∎

EXAMPLE 4   The monthly payment that amortizes a loan of A dollars in t years when the interest rate is r per year is given by

$$P = f(A, r, t) = \frac{Ar}{12\left[1 - \left(1 + \dfrac{r}{12}\right)^{-12t}\right]}.$$

Find the monthly payment for a home mortgage of $90,000 to be amortized over 30 years when the interest rate is 10 percent per year.

Solution   Letting $A = 90,000$, $r = 0.1$, and $t = 30$, we find the required monthly payment to be

$$P = f(90,000, 0.1, 30) = \frac{90,000(0.1)}{12\left[1 - \left(1 + \dfrac{0.1}{12}\right)^{-360}\right]}$$

$$\approx 789.81,$$

or approximately $789.81.   ∎

FIGURE 8.3

The three-dimensional Cartesian coordinate system

Graphs of Functions of Two Variables

In order to graph a function of two variables, we need a three-dimensional coordinate system. This is readily constructed by adding a third axis to the plane Cartesian coordinate system in such a way that the three resulting axes are mutually perpendicular and intersect at O. Observe that, by construction, the zeros of the three number scales coincide at the origin of the **three-dimensional Cartesian coordinate system** (Figure 8.3).

A point in three-dimensional space can now be represented uniquely in this coordinate system by an **ordered triple** of numbers (x, y, z), and, conversely, every ordered triple of real numbers (x, y, z) represents a point in three-dimensional space (Figure 8.4a). For example, the points $A(2, 3, 4)$, $B(1, -2, -2)$, $C(2, 4, 0)$, and $D(0, 0, 4)$ are shown in Figure 8.4b.

FIGURE 8.4

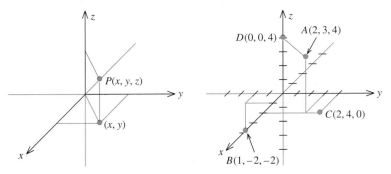

(a) A point in three-dimensional space

(b) Some sample points in three-dimensional space

FIGURE 8.5

The graph of a function in three-dimensional space

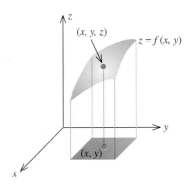

Now, if $f(x, y)$ is a function of two variables x and y, the domain of f is a subset of the xy-plane. Let $z = f(x, y)$ so that there is one and only one point $(x, y, z) \equiv (x, y, f(x, y))$ associated with each point (x, y) in the domain of f. The totality of all such points makes up the **graph** of the function f and is, except for certain degenerate cases, a surface in three-dimensional space (Figure 8.5).

In interpreting the graph of a function $f(x, y)$ one often thinks of the value $z = f(x, y)$ of the function at the point (x, y) as the "height" of the point (x, y, z) on the graph of f. If $f(x, y) > 0$, then the point (x, y, z) is $f(x, y)$ units above the xy-plane; if $f(x, y) < 0$, then the point (x, y, z) is $|f(x, y)|$ units below the xy-plane.

In general, it is quite difficult to draw the graph of a function of two variables. But techniques have been developed that enable us to generate such graphs with minimum effort using a computer. Figure 8.6 shows the computer-generated graphs of two functions.

FIGURE 8.6

Two computer-generated graphs of functions of two variables

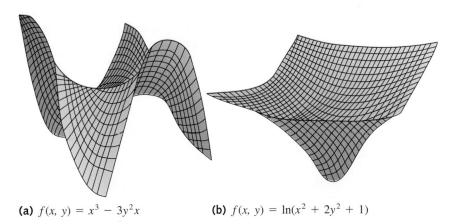

(a) $f(x, y) = x^3 - 3y^2 x$

(b) $f(x, y) = \ln(x^2 + 2y^2 + 1)$

Level Curves

As mentioned earlier, the graph of a function of two variables is often difficult to sketch, and we will not develop a systematic procedure for sketching it. Instead, we will describe a method that is used in constructing topographical maps. This method is relatively easy to apply and conveys sufficient information to enable one to obtain a feel for the graph of the function.

Suppose that $f(x, y)$ is a function of two variables x and y with a graph as shown in Figure 8.7. If c is some value of the function f, then the equation $f(x, y) = c$ describes a curve lying on the plane $z = c$ called the **trace** of the graph of f in the plane $z = c$. If this trace is projected onto the xy-plane, the resulting curve in the xy-plane is called a **level curve.** By drawing the level curves corresponding to several admissible values of c, we obtain a **contour map.** Observe that, by construction, every point on a particular level

FIGURE 8.7

The graph of the function $z = f(x, y)$ and its intersection with the plane $z = c$

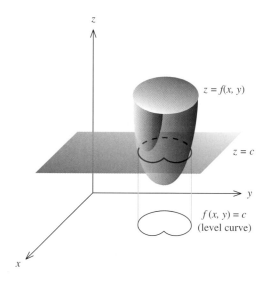

curve corresponds to a point in the surface $z = f(x, y)$ that is a certain fixed distance from the xy-plane. Thus, by elevating or depressing the level curves that make up the contour map in one's mind, it is possible to obtain a feel for the general shape of the surface represented by the function f. Figure 8.8a shows a part of a mountain range with one peak, and Figure 8.8b is the associated contour map.

FIGURE 8.8

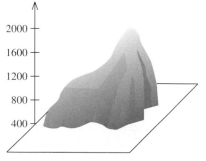

(a) A peak on a mountain range

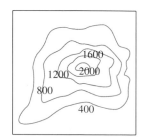

(b) A contour map for the mountain peak

EXAMPLE 5   Sketch a contour map for the function $f(x, y) = x^2 + y^2$.

Solution   The level curves are the graphs of the equation $x^2 + y^2 = c$ for nonnegative numbers c. Taking $c = 0, 1, 4, 9$, and 16, for example, we obtain

$$c = 0 : x^2 + y^2 = 0$$
$$c = 1 : x^2 + y^2 = 1$$
$$c = 4 : x^2 + y^2 = 4 = 2^2$$
$$c = 9 : x^2 + y^2 = 9 = 3^2$$
$$c = 16: x^2 + y^2 = 16 = 4^2.$$

FIGURE 8.9

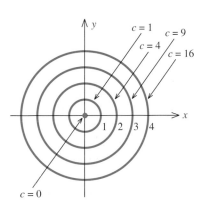
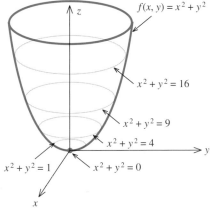

(a) Level curves of $f(x, y) = x^2 + y^2$

(b) The graph of $f(x, y) = x^2 + y^2$

The five level curves are concentric circles with center at the origin and radius given by $r = 0, 1, 2, 3$, and 4, respectively (Figure 8.9a). A sketch of the graph of $f(x, y) = x^2 + y^2$ is included for your reference in Figure 8.9b.

FIGURE 8.10

Level curves for $f(x, y) = 2x^2 - y$

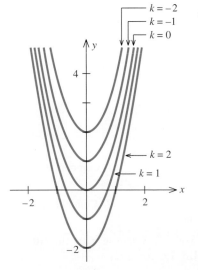

EXAMPLE 6   Sketch the level curves for the function $f(x, y) = 2x^2 - y$ corresponding to $z = -2, -1, 0, 1$, and 2.

Solution   The level curves are the graphs of the equation $2x^2 - y = k$ or $y = 2x^2 - k$ for $k = -2, -1, 0, 1$, and 2. The required level curves are shown in Figure 8.10.

Level curves of functions of two variables are found in many practical applications. For example, if $f(x, y)$ denotes the temperature at a location within the continental United States with longitude x and latitude y at a certain time of day, then the temperature at the point (x, y) is given by the "height" of the surface, represented by $z = f(x, y)$. In this situation, the level curve $f(x, y) = k$ is a curve superimposed on a map of the United States connecting points having the same temperature at a given time (Figure 8.11). These level curves are called **isotherms.**

Similarly, if $f(x, y)$ gives the barometric pressure at the location (x, y), then the level curves of the function f are called **isobars,** lines connecting points having the same barometric pressure at a given time.

As a final example, suppose that $P(x, y, z)$ is a function of three variables x, y, and z giving the profit realized when x, y, and z units of three products

FIGURE 8.11
Isotherms: curves connecting points that have the same temperature

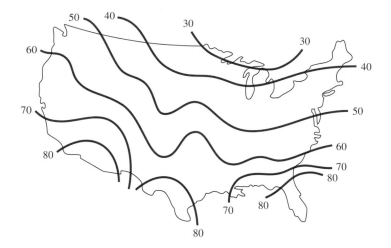

A, B, and C, respectively, are produced and sold. Then the equation $P(x, y, z) = k$, where k is a constant, represents a surface in three-dimensional space called a **level surface** of P. In this situation, the level surface represented by $P(x, y, z) = k$ represents the product mix that results in a profit of exactly k dollars. Such a level surface is called an **isoprofit surface.**

SELF-CHECK EXERCISES 8.1

1. Let $f(x, y) = x^2 - 3xy + \sqrt{x + y}$. Compute $f(1, 3)$ and $f(-1, 1)$. Is the point $(-1, 0)$ in the domain of f?

2. Find the domain of $f(x, y) = \dfrac{1}{x} + \dfrac{1}{x - y} - e^{x+y}$.

c 3. The Odyssey Travel Agency has a monthly advertising budget of $20,000. Odyssey's management estimates that if they spend x dollars on newspaper advertising and y dollars on television advertising, then the monthly revenue will be

$$f(x, y) = 30x^{1/4}y^{3/4}$$

dollars. What will the monthly revenue be if Odyssey spends $5,000 per month on newspaper ads and $15,000 per month on television ads? If Odyssey spends $4,000 per month on newspaper ads and $16,000 per month on television ads?

Solutions to Self-Check Exercises 8.1 can be found on page 489.

8.1 EXERCISES

1. Let $f(x, y) = 2x + 3y - 4$. Compute $f(0, 0), f(1, 0)$, $f(0, 1), f(1, 2)$, and $f(2, -1)$.

2. Let $g(x, y) = 2x^2 - y^2$. Compute $g(1, 2), g(2, 1), g(1, 1)$, $g(-1, 1)$, and $g(2, -1)$.

3. Let $f(x, y) = x^2 + 2xy - x + 3$. Compute $f(1, 2)$, $f(2, 1), f(-1, 2)$, and $f(2, -1)$.

4. Let $h(x, y) = (x + y)/(x - y)$. Compute $h(0, 1)$, $h(-1, 1), h(2, 1)$, and $h(\pi, -\pi)$.

5. Let $g(s, t) = 3s\sqrt{t} + t\sqrt{s} + 2$. Compute $g(1, 2)$, $g(2, 1), g(0, 4)$, and $g(4, 9)$.

6. Let $f(x, y) = xye^{x^2+y^2}$. Compute $f(0, 0), f(0, 1)$, $f(1, 1)$, and $f(-1, -1)$.

7. Let $h(s, t) = s \ln t - t \ln s$. Compute $h(1, e)$, $h(e, 1)$, and $h(e, e)$.

8. Let $f(u, v) = (u^2 + v^2)e^{uv^2}$. Compute $f(0, 1), f(-1, -1)$, $f(a, b)$, and $f(b, a)$.

9. Let $g(r, s, t) = re^{s/t}$. Compute $g(1, 1, 1)$, $g(1, 0, 1)$, and $g(-1, -1, -1)$.

10. Let $g(u, v, w) = (ue^{vw} + ve^{uw} + we^{uv})/(u^2 + v^2 + w^2)$. Compute $g(1, 2, 3)$ and $g(3, 2, 1)$.

In Exercises 11–18, find the domain of the given function.

11. $f(x, y) = 2x + 3y$

12. $g(x, y, z) = x^2 + y^2 + z^2$

13. $h(u, v) = \dfrac{uv}{u - v}$ 14. $f(s, t) = \sqrt{s^2 + t^2}$

15. $g(r, s) = \sqrt{rs}$ 16. $f(x, y) = e^{-xy}$

17. $h(x, y) = \ln(x + y - 5)$

18. $h(u, v) = \sqrt{4 - u^2 - v^2}$

In Exercises 19–24, sketch the level curves of the given function corresponding to the given values of z.

19. $f(x, y) = 2x + 3y; z = -2, -1, 0, 1, 2$

20. $f(x, y) = -x^2 + y; z = -2, -1, 0, 1, 2$

21. $f(x, y) = 2x^2 + y; z = -2, -1, 0, 1, 2$

22. $f(x, y) = xy; z = -4, -2, 2, 4$

23. $f(x, y) = \sqrt{16 - x^2 - y^2}; z = 0, 1, 2, 3, 4$

24. $f(x, y) = e^x - y; z = -2, -1, 0, 1, 2$

25. The volume of a cylindrical tank of radius r and height h is given by

$$V = f(r, h) = \pi r^2 h.$$

Find the volume of a cylindrical tank of radius 1.5 ft and height 4 ft.

26. **IQs**   The IQ (intelligence quotient) of a person whose mental age is m years and whose chronological age is c years is defined as

$$f(m, c) = \frac{100m}{c}.$$

What is the IQ of a nine-year-old child who has a mental age of 13.5 years?

C 27. **Poiseuille's Law**   Poiseuille's law states that the resistance R, measured in dynes, of blood flowing in a blood vessel of length l and radius r (both in centimeters) is

given by

$$R = f(l, r) = \frac{kl}{r^4},$$

where k is the viscosity of blood (in dyne-sec/cm^2). What is the resistance, in terms of k, of blood flowing through an arteriole 4 cm long and of radius 0.1 cm?

28. **Revenue Functions**   The Country Workshop manufactures both finished and unfinished furniture for the home. The estimated quantities demanded each week of its roll-top desks in the finished and unfinished versions are x and y units when the corresponding unit prices are

$$p = 200 - \frac{1}{5}x - \frac{1}{10}y$$

$$q = 160 - \frac{1}{10}x - \frac{1}{4}y$$

dollars, respectively.
a. What is the weekly total revenue function $R(x, y)$?
b. Find the domain of the function R.

29. For the total revenue function $R(x, y)$ of Exercise 28, compute $R(100, 60)$ and $R(60, 100)$. Interpret your results.

30. **Revenue Functions**   The Weston Publishing Company publishes a deluxe edition and a standard edition of its English language dictionary. Weston's management estimates that the number of deluxe editions demanded is x copies per day, and the number of standard editions demanded is y copies per day when the unit prices are

$$p = 20 - 0.005x - 0.001y$$
$$q = 15 - 0.001x - 0.003y$$

dollars, respectively.
a. Find the daily total revenue function $R(x, y)$.
b. Find the domain of the function R.

31. For the total revenue function $R(x, y)$ of Exercise 30, compute $R(300, 200)$ and $R(200, 300)$. Interpret your results.

C 32. **Volume of a Gas**   The volume of a certain mass of gas is related to its pressure and temperature by the formula

$$V = \frac{30.9T}{P},$$

where the volume V is measured in liters, the temperature T is measured in degrees Kelvin (obtained by adding 273° to the Celsius temperature), and the pressure P is measured in millimeters of mercury pressure.
a. Find the domain of the function V.

b. Calculate the volume of the gas at standard temperature and pressure, that is, when $T = 273°K$ and $P = 760$ mm of mercury.

c **33. Surface Area of a Human Body** An empirical formula by E. F. Dubois relates the surface area S of a human body (in square meters) to its weight W in kilograms and its height H in centimeters. The formula, given by

$$S = 0.007184W^{0.425}H^{0.725},$$

is used by physiologists in metabolism studies.
a. Find the domain of the function S.
b. What is the surface area of a human body that weighs 70 kilograms and has a height of 178 centimeters?

c **34. Arson for Profit** A study of arson for profit was conducted by a team of paid civilian experts and police detectives appointed by the mayor of a large city. It was found that the number of suspicious fires in that city in 1992 was very closely related to the concentration of tenants in the city's public housing and to the level of reinvestment in the area in conventional mortgages by the ten largest banks. In fact, the number of fires was closely approximated by the formula

$$N(x, y) = \frac{100(1000 + 0.03x^2 y)^{1/2}}{(5 + 0.2y)^2} \quad \begin{array}{l} (0 \le x \le 150; \\ 5 \le y \le 35), \end{array}$$

where x denotes the number of persons per census tract and y denotes the level of reinvestment in the area in cents per dollar deposited. Using this formula, estimate the total number of suspicious fires in the districts of the city where the concentration of public housing tenants was 100 per census tract and the level of reinvestment was 20 cents per dollar deposited.

c **35. Continuously Compounded Interest** If a principal of P dollars is deposited in an account earning interest at the rate of r per year compounded continuously, then the accumulated amount at the end of t years is given by

$$A = f(P, r, t) = Pe^{rt}$$

dollars. Find the accumulated amount at the end of three years if a sum of $10,000 is deposited in an account earning interest at the rate of 10 percent per year.

c **36. Home Mortgages** The monthly payment that amortizes a loan of A dollars in t years when the interest rate is r per year is given by

$$P = f(A, r, t) = \frac{Ar}{12\left[1 - \left(1 + \dfrac{r}{12}\right)^{-12t}\right]}.$$

a. Find the monthly payment for a home mortgage of $100,000 that will be amortized over 30 years with an interest rate of 8 percent per year. With an interest rate of 10 percent per year.
b. Find the monthly payment for a home mortgage of $100,000 that will be amortized over 20 years with an interest rate of 8 percent per year.

c **37. Home Mortgages** Suppose that a home buyer secures a bank loan of A dollars to purchase a house. If the interest rate charged is r per year and the loan is to be amortized in t years, then the principal repayment at the end of i months is given by

$$B = f(A, r, t, i)$$
$$= A\left[\frac{\left(1 + \dfrac{r}{12}\right)^{i} - 1}{\left(1 + \dfrac{r}{12}\right)^{12t} - 1}\right] \quad (0 \le i \le 12t).$$

Suppose that the Blakelys borrow a sum of $80,000 from a bank to help finance the purchase of a house and the bank charges interest at a rate of 9 percent per year. If the Blakelys agree to repay the loan in equal installments over 30 years, how much will they owe the bank after the 60th payment (5 years)? After the 240th payment (20 years)?

c **38. Force Generated by a Centrifuge** A centrifuge is a machine designed for the specific purpose of subjecting materials to a sustained centrifugal force. The actual amount of centrifugal force, F, expressed in dynes (1 gram of force = 980 dynes) is given by

$$F = f(M, S, R) = \frac{\pi^2 S^2 MR}{900},$$

where S is in revolutions per minute (rpm), M is in grams, and R is in centimeters. Show that an object revolving at the rate of 600 rpm in a circle with radius of 10 cm generates a centrifugal force that is approximately 40 times gravity.

39. Wilson Lot Size Formula The Wilson lot size formula in economics states that the optimal quantity Q of goods for a store to order is given by

$$Q = f(C, N, h) = \sqrt{\frac{2CN}{h}},$$

where C is the cost of placing an order, N is the number of items the store sells per week, and h is the weekly holding cost for each item. Find the most economical quantity of ten-speed bicycles to order if it costs the store $20 to place an order, $5 to hold a bicycle for a week, and the store expects to sell 40 bicycles a week.

SOLUTIONS TO
SELF-CHECK
EXERCISES 8.1

1. $f(1, 3) = 1^2 - 3(1)(3) + \sqrt{1 + 3} = -6$

$f(-1, 1) = (-1)^2 - 3(-1)(1) + \sqrt{-1 + 1} = 4$

The point $(-1, 0)$ is not in the domain of f because the term $\sqrt{x + y}$ is not defined when $x = -1$ and $y = 0$. In fact, the domain of f consists of all real values of x and y that satisfy the inequality $x + y \geq 0$, the shaded half-plane shown in the accompanying figure.

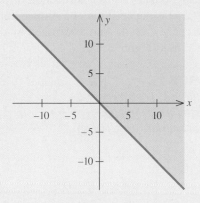

2. Since division by zero is not permitted, we see that $x \neq 0$ and $x - y \neq 0$. Therefore, the domain of f is the set of all points in the xy-plane not containing the y-axis ($x = 0$) and the straight line $x = y$.

3. If Odyssey spends \$5,000 per month on newspaper ads ($x = 5,000$) and \$15,000 per month on television ads ($y = 15,000$), then its monthly revenue will be given by

$$f(5,000, 15,000) = 30(5,000)^{1/4}(15,000)^{3/4}$$
$$\approx 341,925.13,$$

or approximately \$341,925. If the agency spends \$4,000 per month on newspaper ads and \$16,000 per month on television ads, then its monthly revenue will be given by

$$f(4,000, 16,000) = 30(4,000)^{1/4}(16,000)^{3/4}$$
$$\approx 339,410.53,$$

or approximately \$339,411.

8.2   Partial Derivatives

Partial Derivatives

For a function $f(x)$ of one variable x, there is no ambiguity when we speak about the rate of change of $f(x)$ with respect to x, since x must be constrained to move along the x-axis. The situation becomes more complicated, however, when we study the rate of change of a function of two or more variables. For example, the domain D of a function of two variables $f(x, y)$ is a subset of the

FIGURE 8.12
We can approach a point in the plane from infinitely many directions.

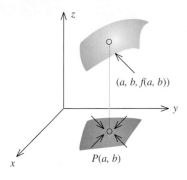

plane (Figure 8.12), so if $P(a, b)$ is any point in the domain of f, there are infinitely many directions from which one can approach the point P. We may therefore ask for the rate of change of f at P along any of these directions.

However, we will not deal with this general problem. Instead, we will restrict ourselves to studying the rate of change of the function $f(x, y)$ at a point $P(a, b)$ in each of two *preferred directions,* namely, the direction parallel to the x-axis and the direction parallel to the y-axis. Let $y = b$, where b is a constant, so that $f(x, b)$ is a function of the one variable x. Since the equation $z = f(x, y)$ is the equation of a surface, the equation $z = f(x, b)$ is the equation of the curve C on the surface formed by the intersection of the surface and the plane $y = b$ (Figure 8.13).

FIGURE 8.13
The curve C is formed by the intersection of the plane $y = b$ with the surface $z = f(x, y)$.

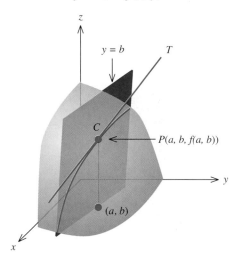

Because $f(x, b)$ is a function of one variable x, we may compute the derivative of f with respect to x at $x = a$. This derivative, obtained by keeping the variable y fixed and differentiating the resulting function $f(x, y)$ with respect to x, is called the **first partial derivative of f with respect to x at (a, b),** written

$$\frac{\partial z}{\partial x}(a, b) \quad \text{or} \quad \frac{\partial f}{\partial x}(a, b) \quad \text{or} \quad f_x(a, b).$$

Thus, $\dfrac{\partial z}{\partial x}(a, b) = \dfrac{\partial f}{\partial x}(a, b) = f_x(a, b) = \lim\limits_{h \to 0} \dfrac{f(a + h, b) - f(a, b)}{h},$

provided that the limit exists. The first partial derivative of f with respect to x at (a, b) measures both the slope of the tangent line T to the curve C and the rate of change of the function f in the x-direction when $x = a$ and $y = b$. We also write

$$\left.\frac{\partial f}{\partial x}\right|_{(a,b)} \equiv f_x(a, b).$$

Similarly, we define the **first partial derivative of f with respect to y** at (a, b), written

$$\frac{\partial z}{\partial y}(a, b) \quad \text{or} \quad \frac{\partial f}{\partial y}(a, b) \quad \text{or} \quad f_y(a, b),$$

as the derivative obtained by keeping the variable x fixed and differentiating the resulting function $f(x, y)$ with respect to y. That is,

$$\frac{\partial z}{\partial y}(a, b) = \frac{\partial f}{\partial y}(a, b) = f_y(a, b)$$

$$= \lim_{k \to 0} \frac{f(a, b + k) - f(a, b)}{k},$$

if the limit exists. The first partial derivative of f with respect to y at (a, b) measures both the slope of the tangent line T to the curve C, obtained by holding x constant (Figure 8.14), and the rate of change of the function f in the y-direction when $x = a$ and $y = b$. We write

$$\left.\frac{\partial f}{\partial y}\right|_{(a,b)} \equiv f_y(a, b).$$

FIGURE 8.14

The first partial derivative of f with respect to y at (a, b) measures the slope of the tangent line T to the curve C with x held constant.

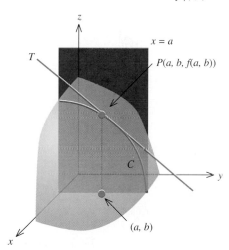

Before looking at some examples, let us summarize these definitions.

FIRST PARTIAL DERIVATIVES OF $f(x, y)$

Suppose that $f(x, y)$ is a function of the two variables x and y. Then, the first partial derivative of f with respect to x at the point (x, y) is

$$\frac{\partial f}{\partial x} = \lim_{h \to 0} \frac{f(x + h, y) - f(x, y)}{h},$$

provided the limit exists. The first partial derivative of f with respect to y at the point (x, y) is

$$\frac{\partial f}{\partial y} = \lim_{k \to 0} \frac{f(x, y + k) - f(x, y)}{k},$$

provided the limit exists.

EXAMPLE 1   Find the partial derivatives $\partial f/\partial x$ and $\partial f/\partial y$ of the function

$$f(x, y) = x^2 - xy^2 + y^3.$$

What is the rate of change of the function f in the x-direction at the point $(1, 2)$? What is the rate of change of the function f in the y-direction at the point $(1, 2)$?

Solution   To compute $\partial f/\partial x$, think of the variable y as a constant and differentiate the resulting function of x with respect to x. Let us write

$$f(x, y) = x^2 - xy^2 + y^3,$$

where the variable y to be treated as a constant is shown in color. Then,

$$\frac{\partial f}{\partial x} = 2x - y^2.$$

To compute $\partial f/\partial y$, think of the variable x as being fixed, that is, as a constant, and differentiate the resulting function of y with respect to y. In this case,

$$f(x, y) = x^2 - xy^2 + y^3$$

so that

$$\frac{\partial f}{\partial y} = -2xy + 3y^2.$$

The rate of change of the function f in the x-direction at the point $(1, 2)$ is given by

$$f_x(1, 2) = \left.\frac{\partial f}{\partial x}\right|_{(1,2)} = 2(1) - 2^2 = -2;$$

that is, f decreases 2 units for each unit increase in the x-direction, y being kept constant ($y = 2$). The rate of change of the function f in the y-direction at the point $(1, 2)$ is given by

$$f_y(1, 2) = \left.\frac{\partial f}{\partial y}\right|_{(1,2)} = -2(1)(2) + 3(2)^2 = 8;$$

that is, f increases 8 units for each unit increase in the y-direction, x being kept constant ($x = 1$).   ∎

EXAMPLE 2   Compute the first partial derivatives of each of the following functions.

a. $f(x, y) = \dfrac{xy}{x^2 + y^2}$ **b.** $g(s, t) = (s^2 - st + t^2)^5$

c. $h(u, v) = e^{u^2 - v^2}$ **d.** $f(x, y) = \ln(x^2 + 2y^2)$

Solution
a. To compute $\partial f/\partial x$, think of the variable y as a constant. Thus,

$$f(x, y) = \frac{xy}{x^2 + y^2},$$

so that, upon using the Quotient Rule, we have

$$\frac{\partial f}{\partial x} = \frac{(x^2 + y^2)y - xy(2x)}{(x^2 + y^2)^2}$$

$$= \frac{y(y^2 - x^2)}{(x^2 + y^2)^2}$$

upon simplification and factorization. To compute $\partial f/\partial y$, think of the variable x as a constant. Thus,

$$f(x,\ y) = \frac{xy}{x^2 + y^2},$$

so that, upon using the Quotient Rule once again, we obtain

$$\frac{\partial f}{\partial y} = \frac{(x^2 + y^2)x - xy(2y)}{(x^2 + y^2)^2}$$

$$= \frac{x(x^2 - y^2)}{(x^2 + y^2)^2}.$$

b. To compute $\partial g/\partial s$, we treat the variable t as if it were a constant. Thus,

$$g(s,\ t) = (s^2 - st + t^2)^5.$$

Using the General Power Rule, we find

$$\frac{\partial g}{\partial s} = 5(s^2 - st + t^2)^4 \cdot (2s - t)$$

$$= 5(2s - t)(s^2 - st + t^2)^4.$$

To compute $\partial g/\partial t$, we treat the variable s as if it were a constant. Thus,

$$g(s,\ t) = (s^2 - st + t^2)^5$$

$$\frac{\partial g}{\partial t} = 5(s^2 - st + t^2)^4(-s + 2t)$$

$$= 5(2t - s)(s^2 - st + t^2)^4.$$

c. To compute $\partial h/\partial u$, think of the variable v as a constant. Thus,

$$h(u,\ v) = e^{u^2 - v^2}.$$

Using the Chain Rule for exponential functions, we have

$$\frac{\partial h}{\partial u} = e^{u^2 - v^2} \cdot 2u$$

$$= 2u e^{u^2 - v^2}.$$

Next, we treat the variable u as if it were a constant,

$$h(u,\ v) = e^{u^2 - v^2},$$

and we obtain

$$\frac{\partial h}{\partial v} = e^{u^2 - v^2} \cdot (-2v)$$

$$= -2v e^{u^2 - v^2}.$$

d. To compute $\partial f/\partial x$, think of the variable y as a constant. Thus,

$$f(x,\ y) = \ln(x^2 + 2y^2),$$

so that the Chain Rule for logarithmic functions gives

$$\frac{\partial f}{\partial x} = \frac{2x}{x^2 + 2y^2}.$$

Next, treating the variable x as if it were a constant, we find

$$f(x,\ y) = \ln(x^2 + 2y^2)$$
$$\frac{\partial f}{\partial y} = \frac{4y}{x^2 + 2y^2}.$$ ∎

To compute the partial derivative of a function of several variables with respect to one variable, say x, we think of the other variables as if they were constants and differentiate the resulting function with respect to x.

EXAMPLE 3   Compute the first partial derivatives of the function

$$w = f(x,\ y,\ z) = xyz - xe^{yz} + x \ln y.$$

Solution   Here we have a function of three variables x, y, and z, and we are required to compute

$$\frac{\partial f}{\partial x},\quad \frac{\partial f}{\partial y},\quad \text{and}\quad \frac{\partial f}{\partial z}.$$

To compute f_x, we think of the other two variables, y and z, as fixed, and we differentiate the resulting function of x with respect to x, thereby obtaining

$$f_x = yz - e^{yz} + \ln y.$$

To compute f_y, we think of the other two variables, x and z, as constants, and we differentiate the resulting function of y with respect to y. We then obtain

$$f_y = xz - xze^{yz} + \frac{x}{y}.$$

Finally, to compute f_z, we treat the variables x and y as constants and differentiate the function f with respect to z, obtaining

$$f_z = xy - xye^{yz}.$$ ∎

The Cobb-Douglas Production Function

For an economic interpretation of the first partial derivatives of a function of two variables, let us turn our attention to the function

$$f(x,\ y) = ax^b y^{1-b}, \tag{1}$$

where a and b are positive constants with $0 < b < 1$. This function is called the **Cobb-Douglas production function.** Here x stands for the amount of money expended for labor, y stands for the cost of capital equipment (buildings, machinery, and other tools of production), and the function f measures the output of the finished product (in suitable units) and is called, accordingly, the **production function.**

The partial derivative f_x is called the **marginal productivity of labor.** It measures the rate of change of production with respect to the amount of money expended for labor, with the level of capital expenditure held constant. Similarly, the partial derivative f_y, called the **marginal productivity of capital,** measures the rate of change of production with respect to the amount expended on capital, with the level of labor expenditure held fixed.

EXAMPLE 4   A certain country's production in the early years following World War II is described by the function

$$f(x,\ y) = 30x^{2/3}y^{1/3}$$

units, when x units of labor and y units of capital were used.

a. Compute f_x and f_y.
b. What is the marginal productivity of labor and the marginal productivity of capital when the amounts expended on labor and capital are 125 units and 27 units, respectively?
c. Should the government have encouraged capital investment rather than increasing expenditure on labor to increase the country's productivity?

Solution

a.
$$f_x = 30 \cdot \frac{2}{3}x^{-1/3}y^{1/3} = 20\left(\frac{y}{x}\right)^{1/3}$$

$$f_y = 30x^{2/3} \cdot \frac{1}{3}y^{-2/3} = 10\left(\frac{x}{y}\right)^{2/3}$$

b. The required marginal productivity of labor is given by

$$f_x(125,\ 27) = 20\left(\frac{27}{125}\right)^{1/3} = 20\left(\frac{3}{5}\right),$$

or 12 units per unit increase in labor expenditure (capital expenditure is held constant at 27 units). The required marginal productivity of capital is given by

$$f_y(125,\ 27) = 10\left(\frac{125}{27}\right)^{2/3} = 10\left(\frac{25}{9}\right),$$

or $27\frac{7}{9}$ units per unit increase in capital expenditure (labor outlay is held constant at 125 units).

c. From the results of (b), we see that a unit increase in capital expenditure resulted in a much faster increase in productivity than a unit increase in labor expenditure would have. Therefore, the government should have encouraged increased spending on capital rather than on labor during the early years of reconstruction.   ∎

Substitute and Complementary Commodities

For another application of the first partial derivatives of a function of two variables in the field of economics, let us consider the relative demands of two commodities. We say that the two commodities are **substitute** (competitive) **commodities** if a decrease in the demand for one results in an increase in the demand for the other. Examples of competitive commodities are coffee and tea. On the other hand, two commodities are referred to as **complementary commodities** if a decrease in the demand for one results in a decrease in the demand for the other as well. Examples of complementary commodities are automobiles and tires.

We will now derive a criterion for determining whether two commodities A and B are substitute or complementary. Suppose that the demand equations that relate the quantities demanded, x and y, to the unit prices, p and q, of the two commodities are given by

$$x = f(p, q) \quad \text{and} \quad y = g(p, q).$$

Let us consider the partial derivative $\partial f/\partial p$. Since f is the demand function for commodity A, we see that, for fixed q, f is typically a decreasing function of p; that is, $\partial f/\partial p < 0$. Now, if the two commodities were substitute commodities, then the quantity demanded of commodity B would increase with respect to p, that is $\partial g/\partial p > 0$. A similar argument with p fixed shows that if A and B are substitute commodities, then $\partial f/\partial q > 0$. Thus, the two commodities A and B are substitute commodities if

$$\frac{\partial f}{\partial q} > 0 \quad \text{and} \quad \frac{\partial g}{\partial p} > 0.$$

Similarly A and B are complementary commodities if

$$\frac{\partial f}{\partial q} < 0 \quad \text{and} \quad \frac{\partial g}{\partial p} < 0.$$

SUBSTITUTE AND COMPLEMENTARY COMMODITIES

Two commodities A and B are **substitute commodities** if

$$\frac{\partial f}{\partial q} > 0 \quad \text{and} \quad \frac{\partial g}{\partial p} > 0. \tag{2}$$

Two commodities A and B are **complementary commodities** if

$$\frac{\partial f}{\partial q} < 0 \quad \text{and} \quad \frac{\partial g}{\partial p} < 0. \tag{3}$$

EXAMPLE 5   Suppose that the daily demand for butter is given by

$$x = f(p, q) = \frac{3q}{1 + p^2}$$

and the daily demand for margarine is given by

$$y = g(p, q) = \frac{2p}{1 + \sqrt{q}} \quad (p > 0, q > 0),$$

where p and q denote the prices per pound (in dollars) of butter and margarine, respectively, and x and y are measured in millions of pounds. Determine whether these two commodities are substitute, complementary, or neither.

where p and q denote the unit prices, respectively, and x and y denote the number of VCRs and the number of blank VCR tapes demanded per week. Determine whether these two products are substitute, complementary, or neither.

46. **Complementary and Substitute Commodities**   Refer to Problem 28, Exercises 8.1. Show that the finished and unfinished home furniture manufactured by the Country Workshop are substitute commodities. [*Hint:* Solve the system of equations for x and y in terms of p and q.]

C 47. **Revenue Functions**   The total weekly revenue (in dollars) of the Country Workshop associated with manufacturing and selling their roll-top desks is given by the function

$$R(x, y) = -0.2x^2 - 0.25y^2 - 0.2xy + 200x + 160y,$$

where x denotes the number of finished units and y denotes the number of unfinished units manufactured and sold per week. Compute $\partial R/\partial x$ and $\partial R/\partial y$ when $x = 300$ and $y = 250$. Interpret your results.

C 48. **Profit Functions**   The monthly profit (in dollars) of the Bond and Barker Department Store depends on the level of inventory x (in thousands of dollars) and the

floor space y (in thousands of square feet) available for display of the merchandise, as given by the equation

$$P(x, y) = -0.02x^2 - 15y^2 + xy$$
$$+ 39x + 25y - 20,000.$$

Compute $\partial P/\partial x$ and $\partial P/\partial y$ when $x = 4000$ and $y = 150$. Interpret your results. Repeat with $x = 5000$ and $y = 150$.

C 49. **Volume of a Gas**   The volume V (in liters) of a certain mass of gas is related to its pressure P (in millimeters of mercury) and its temperature T (in degrees Kelvin) by the law

$$V = \frac{30.9T}{P}.$$

Compute $\partial V/\partial T$ and $\partial V/\partial P$ when $T = 300$ and $P = 800$. Interpret your results.

50. **Surface Area of a Human Body**   The formula

$$S = 0.007184W^{0.425}H^{0.725}$$

gives the surface area S of a human body (in square meters) in terms of its weight W in kilograms and its height H in centimeters. Compute $\partial S/\partial W$ and $\partial S/\partial H$ when $W = 70$ kg and $H = 180$ cm. Interpret your results.

SOLUTIONS TO SELF-CHECK EXERCISES 8.2

1.
$$f_x = \frac{\partial f}{\partial x} = 3x^2 - 2y^2;$$

$$f_y = \frac{\partial f}{\partial y} = -2x(2y) + 2y$$
$$= 2y(1 - 2x)$$

2.
$$f_x = \ln y + ye^x - 2x; f_y = \frac{x}{y} + e^x$$

In particular,
$$f_x(0, 1) = \ln 1 + 1e^0 - 2(0) = 1$$

$$f_y(0, 1) = \frac{0}{1} + e^0 = 1.$$

The results tell us that at the point $(0, 1)$, $f(x, y)$ increases 1 unit for each unit increase in the x-direction, y being kept constant; $f(x, y)$ also increases 1 unit for each unit increase in the y-direction, x being kept constant.

3. From the results of Exercise 1,

$$f_x = 3x^2 - 2y^2.$$

Therefore,

$$f_{xx} = \frac{\partial}{\partial x}(3x^2 - 2y^2) = 6x$$

and
$$f_{xy} = \frac{\partial}{\partial y}(3x^2 - 2y^2) = -4y.$$

Also, from the results of Exercise 1,

$$f_y = 2y(1 - 2x),$$

and so

$$f_{yx} = \frac{\partial}{\partial x}[2y(1 - 2x)] = -4y$$

and
$$f_{yy} = \frac{\partial}{\partial y}[2y(1 - 2x)] = 2(1 - 2x).$$

4. a. The marginal productivity of labor when the amount expended on labor and capital are x and y units, respectively, is given by

$$f_x(x, y) = 60\left(\frac{1}{3}x^{-2/3}\right)y^{2/3} = 20\left(\frac{y}{x}\right)^{2/3}.$$

In particular, the required marginal productivity of labor is given by

$$f_x(125, 8) = 20\left(\frac{8}{125}\right)^{2/3} = 20\left(\frac{4}{25}\right),$$

or 3.2 units per unit increase in labor expenditure, capital expenditure being held constant at 8 units. Next, we compute

$$f_y(x, y) = 60x^{1/3}\left(\frac{2}{3}y^{-1/3}\right) = 40\left(\frac{x}{y}\right)^{1/3}$$

and deduce that the required marginal productivity of capital is given by

$$f_y(125, 8) = 40\left(\frac{125}{8}\right)^{1/3} = 40\left(\frac{5}{2}\right),$$

or 100 units per unit increase in capital expenditure, labor expenditure being held constant at 125 units.

b. The results of (a) tell us that the government should encourage increased spending on capital rather than on labor.

8.3   Maxima and Minima of Functions of Several Variables

Maxima and Minima

In Chapter 4 we saw how solving a practical problem formulated in terms of a function f of one variable often centers on determining an extreme value of f with respect to that variable. For example, we can solve the problem of finding a firm's production level x that will yield a maximum profit by finding the absolute maximum value of the profit function $P(x)$ with respect to x. On the other hand, to find a manufacturer's production level x that will result in a minimal cost of operation, we find the absolute minimum value of the cost function $C(x)$ with respect to x.

The notion of an extreme value of a function plays an equally important role in the case of a function of several variables. As in the case of a function of one variable, it is important to distinguish between the concept of a relative maximum (relative minimum) of f and that of an absolute maximum (absolute minimum) of f. More specifically, a function $f(x, y)$ of two variables has a **relative maximum** at a point (a, b) in the domain of f if $f(x, y) \leq f(a, b)$ for all points (x, y) that are sufficiently close to the point (a, b). Similarly, f has a **relative minimum** at (a, b) if $f(x, y) \geq f(a, b)$ for all points (x, y) that are sufficiently close to the point (a, b). Geometrically, this means that no point on the graph of f corresponding to points (x, y) lying sufficiently close to the point (a, b) can be above (or below) the point $(a, b, f(a, b))$. Figure 8.15 depicts the graph of a function f of two variables that has a relative maximum at the point (a, b) and a relative minimum at the point (c, d).

FIGURE 8.15

f has a relative maximum at (a, b) and a relative minimum at (c, d).

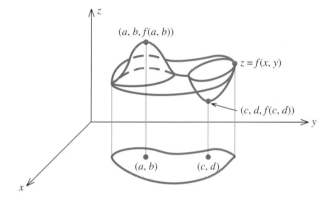

On the other hand, a function $f(x, y)$ of two variables has an **absolute maximum** at a point (a, b) if

$$f(x, y) \leq f(a, b)$$

for *all* points (x, y) in the domain of f. Similarly, a function $f(x, y)$ has an **absolute minimum** at a point (a, b) if

$$f(x, y) \geq f(a, b)$$

for *all* points (x, y) in the domain of f. Geometrically, the absolute maximum and the absolute minimum of a function f are, respectively, the highest and the lowest points on the graph of f.

Just as in the case of a function of one variable, a relative extremum (relative maximum or relative minimum) may or may not be an absolute extremum. However, in order to simplify matters, we will assume that whenever an absolute extremum exists, it will occur at a point where f has a relative extremum.

Just as the first and second derivatives play an important role in determining the relative extrema of a function of one variable, the first and second partial derivatives are powerful tools for locating and classifying the relative extrema of functions of several variables.

Suppose now that a differentiable function $f(x, y)$ of two variables has a relative maximum (relative minimum) at a point (a, b) in the domain of f. From Figure 8.16 it is clear that at the point (a, b) the slope of the "tangent lines"

FIGURE 8.16

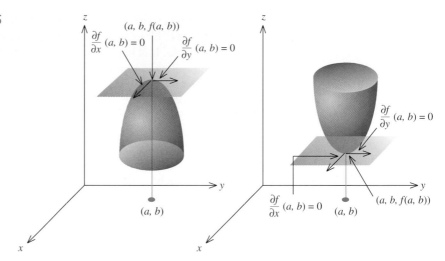

(a) f has a relative maximum at (a, b) **(b)** f has a relative minimum at (a, b)

to the surface in any direction must be zero. In particular, this implies that both

$$\frac{\partial f}{\partial x}(a,\ b) \quad \text{and} \quad \frac{\partial f}{\partial y}(a,\ b)$$

must be zero.

The point $(a,\ b)$ is called a **critical point** of the function f. In case we are tempted to conclude that a critical point of a function f must automatically be a relative extremum of f, let us consider the graph of the function f in Figure 8.17. Here we have both

$$\frac{\partial f}{\partial x}(a,\ b) = 0 \quad \text{and} \quad \frac{\partial f}{\partial y}(a,\ b) = 0.$$

FIGURE 8.17
The point $(a,\ b,\ f(a,\ b))$ is called a saddle point.

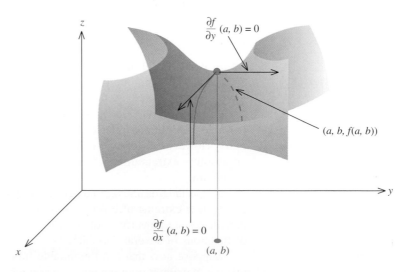

The point $(a,\ b,\ f(a,\ b))$ is neither a relative maximum nor a relative minimum of the function f, since there are points nearby that are higher and others that are lower than it. Such a point is called a **saddle point.** Thus, as in the case of a function of one variable, we may conclude that a critical point of a function of two (or more) variables is only a candidate for a relative extremum of f.

To determine the nature of a critical point of a function $f(x, y)$ of two variables, we make use of the second partial derivatives of f. The resulting test, which aids us in classifying these points, is called the **Second Derivative Test** and is incorporated in the following procedure for finding and classifying the relative extrema of f.

DETERMINING
RELATIVE EXTREMA

1. Find the critical points of $f(x, y)$ by solving the system of simultaneous equations

$$f_x = 0$$
$$f_y = 0.$$

2. The Second Derivative Test: Let

$$D(x, y) = f_{xx}f_{yy} - f_{xy}^2.$$

Then

   a. $D(a, b) > 0$ and $f_{xx}(a, b) < 0$ implies that $f(x, y)$ has a **relative maximum** at the point (a, b).
   b. $D(a, b) > 0$ and $f_{xx}(a, b) > 0$ implies that $f(x, y)$ has a **relative minimum** at the point (a, b).
   c. $D(a, b) < 0$ implies that $f(x, y)$ has neither a relative maximum nor a relative minimum at the point (a, b).
   d. $D(a, b) = 0$ implies that the test is inconclusive, so some other technique must be used to solve the problem.

EXAMPLE 1   Find the relative extrema of the function

$$f(x, y) = x^2 + y^2.$$

Solution   We have

$$f_x = 2x$$

and

$$f_y = 2y.$$

To find the critical point(s) of f, we set $f_x = 0$ and $f_y = 0$ and solve the resulting system of simultaneous equations

$$2x = 0$$
$$2y = 0,$$

obtaining $x = 0$, $y = 0$, or $(0, 0)$, as the sole critical point of f. Next, we apply the Second Derivative Test to determine the nature of the critical point $(0, 0)$. We compute

$$f_{xx} = 2, \qquad f_{xy} = 0, \qquad f_{yy} = 2$$

and

$$D(x, y) = f_{xx}f_{yy} - f_{xy}^2 = (2)(2) - 0 = 4.$$

FIGURE 8.18
The graph of $f(x, y) = x^2 + y^2$

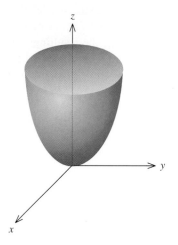

In particular, $D(0, 0) = 4$. Since $D(0, 0) > 0$ and $f_{xx}(0, 0) = 2 > 0$, we conclude that $f(x, y)$ has a relative minimum at the point $(0, 0)$. The relative minimum value, 0, also happens to be the absolute minimum of f. The graph of the function f, shown in Figure 8.18, confirms these results. ■

EXAMPLE 2   Find the relative extrema of the function

$$f(x, y) = 3x^2 - 4xy + 4y^2 - 4x + 8y + 4.$$

Solution   We have

$$f_x = 6x - 4y - 4$$

and

$$f_y = -4x + 8y + 8.$$

To find the critical points of f, we set $f_x = 0$ and $f_y = 0$ and solve the resulting system of simultaneous equations

$$6x - 4y = 4$$
$$-4x + 8y = -8.$$

Multiplying the first equation by 2 and the second equation by 3, we obtain the equivalent system

$$12x - 8y = 8$$
$$-12x + 24y = -24.$$

Adding the two equations gives $16y = -16$, or $y = -1$. We substitute this value for y into either equation in the system to get $x = 0$. Thus, the only critical point of f is the point $(0, -1)$. Next, we apply the Second Derivative Test to determine whether the point $(0, -1)$ gives rise to a relative extremum of f. We compute

$$f_{xx} = 6, \qquad f_{xy} = -4, \qquad f_{yy} = 8$$

and

$$D(x, y) = f_{xx}f_{yy} - f_{xy}^2 = (6)(8) - (-4)^2 = 32.$$

Since $D(0, -1) = 32 > 0$ and $f_{xx}(0, -1) = 6 > 0$, we conclude that $f(x, y)$ has a relative minimum at the point $(0, -1)$. The value of $f(x, y)$ at the point $(0, -1)$ is given by

$$f(0, -1) = 3(0)^2 - 4(0)(-1) + 4(-1)^2 - 4(0) + 8(-1) + 4 = 0.$$
■

EXAMPLE 3   Find the relative extrema of the function

$$f(x, y) = 4y^3 + x^2 - 12y^2 - 36y + 2.$$

Solution   To find the critical points of f, we set $f_x = 0$ and $f_y = 0$ simultaneously, obtaining

$$f_x = 2x = 0$$

$$f_y = 12y^2 - 24y - 36 = 0.$$

The first equation implies that $x = 0$. The second equation implies that

$$y^2 - 2y - 3 = 0$$
$$(y + 1)(y - 3) = 0,$$

that is, $y = -1$ or 3. Therefore, there are two critical points of the function f, namely, $(0, -1)$ and $(0, 3)$.

Next, we apply the Second Derivative Test to determine the nature of each of the two critical points. We compute

$$f_{xx} = 2, \qquad f_{xy} = 0, \qquad f_{yy} = 24y - 24 = 24(y - 1).$$

Therefore,

$$D(x, y) = f_{xx}f_{yy} - f_{xy}^2 = 48(y - 1).$$

For the point $(0, -1)$,

$$D(0, -1) = 48(-1 - 1) = -96 < 0.$$

Since $D(0, -1) < 0$, we conclude that the point $(0, -1)$ gives a saddle point of f. For the point $(0, 3)$,

$$D(0, 3) = 48(3 - 1) = 96 > 0.$$

Since $D(0, 3) > 0$ and $f_{xx}(0, 3) > 0$, we conclude that the function f has a relative minimum at the point $(0, 3)$. Furthermore, since

$$f(0, 3) = 4(3)^3 + (0)^2 - 12(3)^2 - 36(3) + 2$$
$$= -106,$$

we see that the relative minimum value of f is -106.   ■

Applications

As in the case of a practical optimization problem involving a function of one variable, the solution to an optimization problem involving a function of several variables calls for finding the *absolute* extremum of the function. Determining the absolute extremum of a function of several variables is more difficult than merely finding the relative extrema of the function. However, in many situations, the absolute extremum of a function actually coincides with the largest relative extremum of the function that occurs in the interior of its domain. We will assume that the problems considered here belong to this category. Furthermore, the existence of the absolute extremum (solution) of a practical problem is often deduced from the geometric or physical nature of the problem.

EXAMPLE 4   The total weekly revenue (in dollars) that the Acrosonic Company realizes in producing and selling its bookshelf loudspeaker systems is given by

$$R(x, y) = -\frac{1}{4}x^2 - \frac{3}{8}y^2 - \frac{1}{4}xy + 300x + 240y,$$

where x denotes the number of fully assembled units and y denotes the number of kits produced and sold per week. The total weekly cost attributable to the production of these loudspeakers is

$$C(x, y) = 180x + 140y + 5000$$

dollars, where x and y have the same meaning as before. Determine how many assembled units and how many kits Acrosonic should produce per week to maximize its profit.

Solution The contribution to Acrosonic's weekly profit stemming from the production and sale of the bookshelf loudspeaker systems is given by

$$P(x, y) = R(x, y) - C(x, y)$$

$$= \left(-\frac{1}{4}x^2 - \frac{3}{8}y^2 - \frac{1}{4}xy + 300x + 240y\right) - (180x + 140y + 5000)$$

$$= -\frac{1}{4}x^2 - \frac{3}{8}y^2 - \frac{1}{4}xy + 120x + 100y - 5000.$$

To find the relative maximum of the profit function $P(x, y)$, we first locate the critical point(s) of P. Setting $P_x(x, y)$ and $P_y(x, y)$ equal to zero, we obtain

$$P_x = -\frac{1}{2}x - \frac{1}{4}y + 120 = 0$$

$$P_y = -\frac{3}{4}y - \frac{1}{4}x + 100 = 0.$$

Solving the first of these equations for y yields

$$y = -2x + 480,$$

which, upon substitution into the second equation, yields

$$-\frac{3}{4}(-2x + 480) - \frac{1}{4}x + 100 = 0$$

$$6x - 1440 - x + 400 = 0$$

or $$x = 208.$$

We substitute this value of x into the equation $y = -2x + 480$ to get

$$y = 64.$$

Therefore, the function P has the sole critical point $(208, 64)$. To show that the point $(208, 64)$ is a solution to our problem, we use the Second Derivative Test. We compute

$$P_{xx} = -\frac{1}{2}, \qquad P_{xy} = -\frac{1}{4}, \quad \text{and} \quad P_{yy} = -\frac{3}{4}.$$

So

$$D(x, y) = \left(-\frac{1}{2}\right)\left(-\frac{3}{4}\right) - \left(-\frac{1}{4}\right)^2 = \frac{3}{8} - \frac{1}{16} = \frac{5}{16}.$$

In particular, $D(208, 64) = \frac{5}{16} > 0$.

Since $D(208, 64) > 0$ and $P_{xx}(208, 64) < 0$, the point $(208, 64)$ yields a relative maximum of P. This relative maximum is also the absolute maximum of P. We conclude that Acrosonic can maximize its weekly profit by manufacturing 208 assembled units and 64 kits of their bookshelf loudspeaker systems. The maximum weekly profit realizable from the production and sale of these loudspeaker systems is given by

$$P(208, 64) = -\frac{1}{4}(208)^2 - \frac{3}{8}(64)^2 - \frac{1}{4}(208)(64)$$
$$+ 120(208) + 100(64) - 5{,}000$$
$$= 10{,}680,$$

or $10,680. ∎

FIGURE 8.19

Locating a site for a television relay station

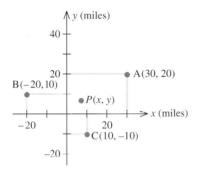

EXAMPLE 5   A television relay station will serve towns A, B, and C, whose relative locations are shown in Figure 8.19. Determine a site for the location of the station if the sum of the squares of the distances from each town to the site is minimized.

Solution   Suppose the required site is located at the point $P(x, y)$. With the aid of the distance formula, we find that the square of the distance from town A to the site is

$$(x - 30)^2 + (y - 20)^2.$$

The respective distances from towns B and C to the site are found in a similar manner, so that the sum of the squares of the distances from each town to the site is given by

$$f(x, y) = (x - 30)^2 + (y - 20)^2 + (x + 20)^2$$
$$+ (y - 10)^2 + (x - 10)^2 + (y + 10)^2.$$

To find the relative minimum of $f(x, y)$, we first find the critical point(s) of f. Using the Chain Rule to find $f_x(x, y)$ and $f_y(x, y)$ and setting each equal to zero, we obtain

$$f_x = 2(x - 30) + 2(x + 20) + 2(x - 10) = 6x - 40 = 0$$

and $$f_y = 2(y - 20) + 2(y - 10) + 2(y + 10) = 6y - 40 = 0,$$

from which we deduce that $\left(\frac{20}{3}, \frac{20}{3}\right)$ is the sole critical point of f. Since

$$f_{xx} = 6, \qquad f_{xy} = 0, \quad \text{and} \quad f_{yy} = 6,$$

we have

$$D(x, y) = f_{xx} f_{yy} - f_{xy}^2 = (6)(6) - 0 = 36.$$

Since $D\left(\frac{20}{3}, \frac{20}{3}\right) > 0$ and $f_{xx}\left(\frac{20}{3}, \frac{20}{3}\right) > 0$, we conclude that the point $\left(\frac{20}{3}, \frac{20}{3}\right)$ yields a relative minimum of f. Thus, the required site has coordinates $x = \frac{20}{3}$ and $y = \frac{20}{3}$. ∎

SELF-CHECK EXERCISES 8.3

1. Let $f(x, y) = 2x^2 + 3y^2 - 4xy + 4x - 2y + 3$.

    a. Find the critical point of f.
    b. Use the Second Derivative Test to classify the nature of the critical point.
    c. Find the relative extremum of f if it exists.

2. The Robertson Controls Company manufactures two basic models of setback thermostats: a standard mechanical thermostat and a deluxe electronic thermostat. Robertson's monthly revenue (in hundreds of dollars) is

$$R(x, y) = -\frac{1}{8}x^2 - \frac{1}{2}y^2 - \frac{1}{4}xy + 20x + 60y,$$

where x (in units of a hundred) denotes the number of mechanical thermostats manufactured and y (in units of a hundred) denotes the number of electronic thermostats manufactured per month. The total monthly cost incurred in producing these thermostats is

$$C(x, y) = 7x + 20y + 280$$

hundred dollars. Find how many thermostats of each model Robertson should manufacture per month in order to maximize its profits. What is the maximum profit?

Solutions to Self-Check Exercises 8.3 can be found on page 512.

8.3   EXERCISES

In Exercises 1–20, find the critical point(s) of the given functions. Then use the Second Derivative Test to classify the nature of each of these point(s), if possible. Finally, determine the relative extrema of each function.

1. $f(x, y) = 1 - 2x^2 - 3y^2$

2. $f(x, y) = x^2 - xy + y^2 + 1$

3. $f(x, y) = x^2 - y^2 - 2x + 4y + 1$

4. $f(x, y) = 2x^2 + y^2 - 4x + 6y + 3$

5. $f(x, y) = x^2 + 2xy + 2y^2 - 4x + 8y - 1$

6. $f(x, y) = x^2 - 4xy + 2y^2 + 4x + 8y - 1$

7. $f(x, y) = 2x^3 + y^2 - 9x^2 - 4y + 12x - 2$

8. $f(x, y) = 2x^3 + y^2 - 6x^2 - 4y + 12x - 2$

9. $f(x, y) = x^3 + y^2 - 2xy + 7x - 8y + 4$

10. $f(x, y) = 2y^3 - 3y^2 - 12y + 2x^2 - 6x + 2$

11. $f(x, y) = x^3 - 3xy + y^3 - 2$

12. $f(x, y) = x^3 - 2xy + y^2 + 5$

13. $f(x, y) = xy + \dfrac{4}{x} + \dfrac{2}{y}$

14. $f(x, y) = \dfrac{x}{y^2} + xy$ 15. $f(x, y) = x^2 - e^{y^2}$

16. $f(x, y) = e^{x^2 - y^2}$ 17. $f(x, y) = e^{x^2 + y^2}$

18. $f(x, y) = e^{xy}$

19. $f(x, y) = \ln(1 + x^2 + y^2)$

20. $f(x, y) = xy + \ln x + 2y^2$

21. **Maximizing Profit**   The total weekly revenue (in dollars) of the Country Workshop realized in manufacturing and selling its roll-top desks is given by

$$R(x, y) = -0.2x^2 - 0.25y^2 - 0.2xy + 200x + 160y,$$

where x denotes the number of finished units and y denotes the number of unfinished units manufactured and sold per week. The total weekly cost attributable to the manufacture of these desks is given by

$$C(x, y) = 100x + 70y + 4000$$

dollars. Determine how many finished units and how many unfinished units the company should manufacture per week in order to maximize its profit. What is the maximum profit realizable?

22. **Maximizing Profit**   The total daily revenue (in dollars) that the Weston Publishing Company realizes in publishing and selling its English language dictionaries is given by

$$R(x, y) = -0.005x^2 - 0.003y^2 - 0.002xy + 20x + 15y,$$

where x denotes the number of deluxe copies and y denotes the number of standard copies published and sold daily. The total daily cost of publishing these dictionaries is given by

$$C(x, y) = 6x + 3y + 200$$

dollars. Determine how many deluxe copies and how many standard copies Weston should publish per day to maximize its profits. What is the maximum profit realizable?

23. **Maximum Price**   The rectangular region R shown in the accompanying figure represents the financial district of a city. The price of land within the district is approximated by the function

$$p(x, y) = 200 - 10\left(x - \frac{1}{2}\right)^2 - 15(y - 1)^2,$$

where $p(x, y)$ is the price of land at the point (x, y) in dollars per square foot and x and y are measured in miles. At what point within the financial district is the price of land highest?

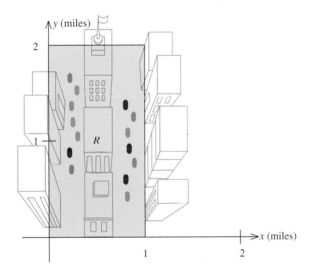

24. **Maximizing Profit**   C & G Imports, Inc., imports two brands of white wine, one from Germany and the other from Italy. The German wine costs $4 a bottle, and the Italian wine can be obtained for $3 a bottle. It has been estimated that if the German wine retails at p dollars per bottle and the Italian wine is sold for q dollars per bottle, then

$$2000 - 150p + 100q$$

bottles of the German wine and

$$1000 + 80p - 120q$$

bottles of the Italian wine will be sold per week. Determine the unit price for each brand that will allow C & G to realize the largest possible weekly profit.

25. **Determining the Optimal Site**   An auxiliary electric power station will serve three communities, A, B, and

C, whose relative locations are shown in the accompanying figure. Determine where the power station should be located if the sum of the squares of the distances from each community to the site is minimized.

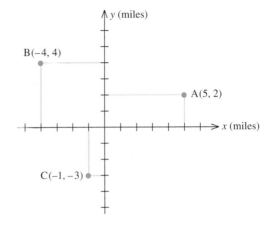

26. **Packaging**   An open rectangular box having a volume of 108 cubic inches is to be constructed from a tin sheet. Find the dimensions of such a box if the amount of material used in its construction is to be minimal. [*Hint:* Let the dimensions of the box be x'' by y'' by z''. Then

$$xyz = 108$$

and the amount of material used is given by

$$S = xy + 2yz + 2xz.$$

Show that

$$S = f(x, y) = xy + \frac{216}{x} + \frac{216}{y}.$$

Minimize $f(x, y)$.]

27. **Packaging**   Postal regulations specify that the combined length and girth of a parcel sent by parcel post may not exceed 108 inches. Find the dimensions of the rectangular package that would have the greatest possible volume under these regulations. [*Hint:* Let the dimensions of the box be x'' by y'' by z'' (see the figure on page 512). Then

$$2x + 2z + y = 108$$

and the volume

$$V = xyz.$$

Show that

$$V = f(x, z) = 108xz - 2x^2z - 2xz^2.$$

Maximize $f(x, z)$.]

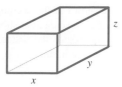

Figure for Exercise 27

28. **Minimizing Heating and Cooling Costs**   A building in the shape of a rectangular box is to have a volume of 12,000 cubic feet (see figure). It is estimated that the annual heating and cooling costs will be $2 per square foot for the top, $4 per square foot for the front and back, and $3 per square foot for the sides. Find the dimensions of the building that will result in a minimal annual heating and cooling cost. What is the minimal annual heating and cooling cost?

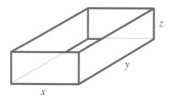

29. **Packaging**   An open box having a volume of 48 cubic inches is to be constructed. If the box is to include a partition that is parallel to a side of the box, as shown in the figure, and the amount of material used is to be minimal, what should the dimensions of the box be?

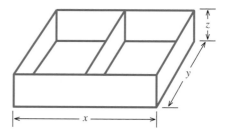

SOLUTIONS TO SELF-CHECK EXERCISES 8.3

1. a. To find the critical point(s) of f, we solve the system of equations

$$f_x = 4x - 4y + 4 = 0$$
$$f_y = -4x + 6y - 2 = 0,$$

obtaining $x = -2$ and $y = -1$. Thus, the only critical point of f is the point $(-2, -1)$.

b. We have $f_{xx} = 4$, $f_{xy} = -4$, and $f_{yy} = 6$, so

$$D(x, y) = f_{xx} f_{yy} - f_{xy}^2$$
$$= (4)(6) - (-4)^2 = 8.$$

Since $D(-2, -1) > 0$ and $f_{xx}(-2, -1) > 0$, we conclude that f has a relative minimum at the point $(-2, -1)$.

c. The relative minimum value of $f(x, y)$ at the point $(-2, -1)$ is

$$f(-2, -1) = 2(-2)^2 + 3(-1)^2 - 4(-2)(-1) + 4(-2) - 2(-1) + 3$$
$$= 0.$$

2. Robertson's monthly profit is

$$P(x, y) = R(x, y) - C(x, y)$$
$$= \left(-\frac{1}{8}x^2 - \frac{1}{2}y^2 - \frac{1}{4}xy + 20x + 60y \right) - (7x + 20y + 280)$$
$$= -\frac{1}{8}x^2 - \frac{1}{2}y^2 - \frac{1}{4}xy + 13x + 40y - 280.$$

The critical point of P is found by solving the system

$$P_x = -\frac{1}{4}x - \frac{1}{4}y + 13 = 0$$

$$P_y = -\frac{1}{4}x - y + 40 = 0,$$

giving $x = 16$ and $y = 36$. Thus $(16, 36)$ is the critical point of P. Next,

$$P_{xx} = -\frac{1}{4}, \qquad P_{xy} = -\frac{1}{4}, \quad \text{and} \quad P_{yy} = -1$$

and

$$D(x, y) = f_{xx}f_{yy} - f_{xy}^2$$

$$= \left(-\frac{1}{4}\right)(-1) - \left(-\frac{1}{4}\right)^2 = \frac{3}{16}.$$

Since $D(16, 36) > 0$ and $P_{xx}(16, 36) < 0$, the point $(16, 36)$ yields a relative maximum of P. We conclude that the monthly profit is maximized by manufacturing 1600 mechanical and 3600 electronic setback thermostats per month. The maximum monthly profit realizable is

$$P(16, 36) = -\frac{1}{8}(16)^2 - \frac{1}{2}(36)^2 - \frac{1}{4}(16)(36) + 13(16) + 40(36) - 280$$

$$= 544,$$

or $54,400.

8.4   The Method of Least Squares

The Method of Least Squares

In Section 1.4, Example 10, we saw how a linear equation can be used to approximate the sales trend for a local sporting goods store. As we saw there, one use of a **trend line** is to predict a store's future sales. Recall that we obtained the line by requiring that it pass through two data points, the rationale being that such a line seems to *fit* the data reasonably well.

In this section we describe a general method, known as the **method of least squares**, for determining a straight line that, in some sense, *best* fits a set of data points when the points are scattered about a straight line. In order to illustrate the principle behind the method of least squares, suppose, for simplicity, that we are given five data points,

$$P_1(x_1, y_1), \quad P_2(x_2, y_2), \quad P_3(x_3, y_3), \quad P_4(x_4, y_4), \quad P_5(x_5, y_5),$$

that describe the relationship between the two variables x and y. By plotting these data points we obtain a graph called a **scatter diagram** (Figure 8.20).

If we try to fit a straight line to these data points, the line will miss the first, second, third, fourth, and fifth data points by the amounts d_1, d_2, d_3, d_4, and d_5, respectively (Figure 8.21).

FIGURE 8.20
A scatter diagram

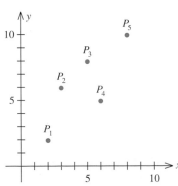

FIGURE 8.21

The approximating line misses each point by the amounts d_1, d_2, ..., d_5.

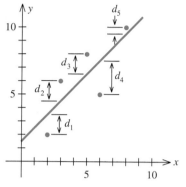

The **principle of least squares** states that the straight line L that fits the data points best is the one chosen by requiring that the sum of the squares of d_1, d_2, ..., d_5, that is,

$$d_1^2 + d_2^2 + d_3^2 + d_4^2 + d_5^2,$$

be made as small as possible. If we think of the amount d_1 as the error made when the value y_1 is approximated by the corresponding value of y lying on the straight line L, and d_2 as the error made when the value y_2 is approximated by the corresponding value of y, and so on, then it can be seen that the least-squares criterion calls for minimizing the sum of the squares of the errors. The line L obtained in this manner is called the **least-squares line**, or **regression line**.

To find a method for computing the regression line L, suppose L has representation $y = f(x) = mx + b$, where m and b are to be determined. Observe that

$$
\begin{aligned}
&d_1^2 + d_2^2 + d_3^2 + d_4^2 + d_5^2 \\
&= [f(x_1) - y_1]^2 + [f(x_2) - y_2]^2 + [f(x_3) - y_3]^2 \\
&\quad + [f(x_4) - y_4]^2 + [f(x_5) - y_5]^2 \\
&= (mx_1 + b - y_1)^2 + (mx_2 + b - y_2)^2 + (mx_3 + b - y_3)^2 \\
&\quad + (mx_4 + b - y_4)^2 + (mx_5 + b - y_5)^2
\end{aligned}
$$

and may be viewed as a function of the two variables m and b. Thus, the least-squares criterion is equivalent to minimizing the function

$$
\begin{aligned}
f(m, b) &= (mx_1 + b - y_1)^2 + (mx_2 + b - y_2)^2 + (mx_3 + b - y_3)^2 \\
&\quad + (mx_4 + b - y_4)^2 + (mx_5 + b - y_5)^2
\end{aligned}
$$

with respect to m and b. Using the Chain Rule, we compute

$$
\begin{aligned}
\frac{\partial f}{\partial m} &= 2(mx_1 + b - y_1)x_1 + 2(mx_2 + b - y_2)x_2 + 2(mx_3 + b - y_3)x_3 \\
&\quad + 2(mx_4 + b - y_4)x_4 + 2(mx_5 + b - y_5)x_5 \\
&= 2[mx_1^2 + bx_1 - x_1y_1 + mx_2^2 + bx_2 - x_2y_2 + mx_3^2 + bx_3 - x_3y_3 \\
&\quad + mx_4^2 + bx_4 - x_4y_4 + mx_5^2 + bx_5 - x_5y_5] \\
&= 2[(x_1^2 + x_2^2 + x_3^2 + x_4^2 + x_5^2)m + (x_1 + x_2 + x_3 + x_4 + x_5)b \\
&\quad - (x_1y_1 + x_2y_2 + x_3y_3 + x_4y_4 + x_5y_5)]
\end{aligned}
$$

and

$$
\begin{aligned}
\frac{\partial f}{\partial b} &= 2(mx_1 + b - y_1) + 2(mx_2 + b - y_2) + 2(mx_3 + b - y_3) \\
&\quad + 2(mx_4 + b - y_4) + 2(mx_5 + b - y_5) \\
&= 2[(x_1 + x_2 + x_3 + x_4 + x_5)m + 5b - (y_1 + y_2 + y_3 + y_4 + y_5)].
\end{aligned}
$$

Setting

$$\frac{\partial f}{\partial m} = 0 \quad \text{and} \quad \frac{\partial f}{\partial b} = 0$$

gives

$$(x_1^2 + x_2^2 + x_3^2 + x_4^2 + x_5^2)m + (x_1 + x_2 + x_3 + x_4 + x_5)b$$
$$= x_1 y_1 + x_2 y_2 + x_3 y_3 + x_4 y_4 + x_5 y_5$$

and

$$(x_1 + x_2 + x_3 + x_4 + x_5)m + 5b = y_1 + y_2 + y_3 + y_4 + y_5.$$

Solving these two simultaneous equations for m and b then leads to an equation $y = mx + b$ of a straight line.

Before looking at an example, we state a more general result whose derivation is identical to the special case involving the five data points just discussed.

THE METHOD OF LEAST SQUARES

Suppose that we are given n data points

$$P_1(x_1, y_1), \quad P_2(x_2, y_2), \quad P_3(x_3, y_3), \quad \ldots, \quad P_n(x_n, y_n).$$

Then, the least-squares (regression) line for the data is given by the linear equation

$$y = f(x) = mx + b,$$

where the constants m and b satisfy the equations

$$(x_1^2 + x_2^2 + x_3^2 + \cdots + x_n^2)m + (x_1 + x_2 + x_3 + \cdots + x_n)b$$
$$= x_1 y_1 + x_2 y_2 + x_3 y_3 + \cdots + x_n y_n \tag{4}$$

and

$$(x_1 + x_2 + x_3 + \cdots + x_n)m + nb$$
$$= y_1 + y_2 + y_3 + \cdots + y_n \tag{5}$$

simultaneously. Equations (4) and (5) are called **normal equations.**

EXAMPLE 1   Find an equation of the least-squares line for the data

$$P_1(1, 1), \qquad P_2(2, 3), \qquad P_3(3, 4), \qquad P_4(4, 3), \quad \text{and} \quad P_5(5, 6).$$

Solution   Here we have $n = 5$ and

$$x_1 = 1, \qquad x_2 = 2, \qquad x_3 = 3, \qquad x_4 = 4, \qquad x_5 = 5$$
$$y_1 = 1, \qquad y_2 = 3, \qquad y_3 = 4, \qquad y_4 = 3, \qquad y_5 = 6,$$

so (4) becomes

$$(1 + 4 + 9 + 16 + 25)m + (1 + 2 + 3 + 4 + 5)b$$
$$= 1 + 6 + 12 + 12 + 30$$

or $$55m + 15b = 61, \tag{6}$$

and (5) becomes

$$(1 + 2 + 3 + 4 + 5)m + 5b = 1 + 3 + 4 + 3 + 6$$

or $$15m + 5b = 17. \tag{7}$$

Solving (7) for b gives

$$b = -3m + \frac{17}{5}, \tag{8}$$

which, upon substitution into (6), gives

$$15\left(-3m + \frac{17}{5}\right) + 55m = 61$$
$$-45m + 51 + 55m = 61$$
$$10m = 10$$

or $$m = 1.$$

Substituting this value of m into (8) gives

$$b = -3 + \frac{17}{5}$$

or $$b = \frac{2}{5} = 0.4.$$

Therefore, the required least-squares line is

$$y = x + 0.4.$$

The scatter diagram and the regression line are shown in Figure 8.22.

◼

FIGURE 8.22

The scatter diagram and the least-squares line $y = x + 0.4$

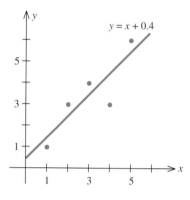

Applications

EXAMPLE 2   The proprietor of the Leisure Travel Service compiled the following data relating the firm's annual profit to its annual advertising expenditure (both measured in thousands of dollars).

Annual advertising expenditure (x)	12	14	17	21	26	30
Annual profit (y)	60	70	90	100	100	120

a. Determine an equation of the least-squares line for these data.
b. Draw a scatter diagram and the least-squares line for these data.
c. Use the result obtained in (a) to predict Leisure Travel's annual profit if the annual advertising budget is $20,000.

Solution

a. The calculations required for obtaining the normal equations may be summarized as follows:

	x	y	x^2	xy
	12	60	144	720
	14	70	196	980
	17	90	289	1,530
	21	100	441	2,100
	26	100	676	2,600
	30	120	900	3,600
Sum	120	540	2,646	11,530

The normal equations are

$$6b + 120m = 540 \qquad\qquad (9)$$
$$120b + 2{,}646m = 11{,}530. \qquad\qquad (10)$$

Solving (9) for b gives

$$b = -20m + 90, \qquad\qquad (11)$$

which, upon substitution into (10), gives

$$120(-20m + 90) + 2{,}646m = 11{,}530$$
$$-2{,}400m + 10{,}800 + 2{,}646m = 11{,}530$$
$$246m = 730$$
or
$$m = 2.97.$$

Substituting this value of m into (11) gives

$$b = -20(2.97) + 90$$
or
$$b = 30.6.$$

Therefore, the required least-squares line is given by

$$y = f(x) = 2.97x + 30.6.$$

b. The scatter diagram and the least-squares line are shown in Figure 8.23.

c. Leisure Travel's predicted annual profit corresponding to an annual budget of \$20,000 is given by

$$f(20) = 2.97(20) + 30.6$$
$$= 90,$$

or \$90,000.

FIGURE 8.23

The scatter diagram and the least-squares line $y = 2.97x + 30.6$

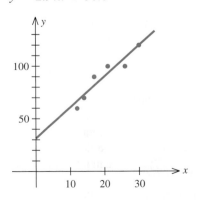

EXAMPLE 3   A market research study conducted for the Century Communications Company provided the following data based on the projected monthly sales x (in thousands) of Century's videocassette version of a box-office hit adventure movie with a proposed wholesale unit price of p dollars.

p	38	36	34.5	30	28.5
x	2.2	5.4	7.0	11.5	14.6

a. Find the demand equation if the demand curve is the least-squares line for these data.

b. Suppose the total monthly cost function associated with producing and distributing the videocassette movies is given by

$$C(x) = 4x + 25,$$

where x denotes the number of units (in thousands) produced and sold and $C(x)$ is in thousands of dollars. Determine the unit wholesale price that will maximize Century's monthly profit.

Solution

a. The calculations required for obtaining the normal equations may be summarized as follows:

	x	p	x^2	xp
	2.2	38	4.84	83.6
	5.4	36	29.16	194.4
	7.0	34.5	49	241.5
	11.5	30	132.25	345
	14.6	28.5	213.16	416.1
Sum	40.7	167	428.41	1280.6

The normal equations are

$$5b + 40.7m = 167$$
$$40.7b + 428.41m = 1280.6.$$

Solving this system of linear equations simultaneously, we find that

$$m = -0.81 \quad \text{and} \quad b = 39.99.$$

Therefore, the required least-squares is given by

$$p = f(x) = -0.81x + 39.99,$$

which is the required demand equation, provided $0 \le x \le 49.37$.

b. The total revenue function in this case is given by

$$R(x) = xp = -0.81x^2 + 39.99x,$$

and since the total cost function is

$$C(x) = 4x + 25,$$

we see that the profit function is

$$P(x) = -0.81x^2 + 39.99x - (4x + 25)$$
$$= -0.81x^2 + 35.99x - 25.$$

To find the absolute maximum of $P(x)$ over the closed interval [0, 49.37], we compute

$$P'(x) = -1.62x + 35.99.$$

Since $P'(x) = 0$, we find $x = 22.22$ as the only critical point of P. Finally, from the table

x	0	22.22	49.37
$P(x)$	-25	374.78	-222.47

we see that the optimal wholesale price is $22.22 dollars per videocassette.

■

SELF-CHECK EXERCISES 8.4

1. Find an equation of the least-squares line for the data

$$P_1(0, 3), \quad P_2(2, 6.5), \quad P_3(4, 10),$$
$$P_4(6, 16), \quad \text{and} \quad P_5(7, 16.5).$$

2. The following data obtained from the U.S. Department of Commerce give the percentage of people over the age of 65 who have high school diplomas.

Year (x)	0	6	11	16	22	26
Percent with diploma (y)	19	25	30	35	44	48

Here $x = 0$ corresponds to the beginning of the year 1959.

a. Find an equation of the least-squares line for the given data.
b. Assuming that this trend continues, what percentage of people over 65 years of age will have high school diplomas at the beginning of the year 1999 ($x = 40$)?

Solutions to Self-Check Exercises 8.4 can be found on page 523.

8.4   EXERCISES

C *A calculator is recommended for this exercise set. In Exercises 1–6, (a) find the equation of the least-squares line for the given data and (b) draw a scatter diagram for the given data and graph the least-squares line.*

1.

x	1	2	3	4
y	4	6	8	11

2.

x	1	3	5	7	9
y	9	8	6	3	2

3.

x	1	2	3	4	4	6
y	4.5	5	3	2	3.5	1

4.

x	1	1	2	3	4	4	5
y	2	3	3	3.5	3.5	4	5

5. $P_1(1, 3)$, $P_2(2, 5)$, $P_3(3, 5)$, $P_4(4, 7)$, $P_5(5, 8)$

6. $P_1(1, 8)$, $P_2(2, 6)$, $P_3(5, 6)$, $P_4(7, 4)$, $P_5(10, 1)$

7. **College Admissions**   The following data were compiled by the admissions office at Faber College during the past five years. The data relate the number of college brochures and follow-up letters (x) sent to a preselected list of high-school juniors who had taken the PSAT and the number of completed applications (y) received from these students (both measured in units of 1000).

x	4	4.5	5	5.5	6
y	0.5	0.6	0.8	0.9	1.2

a. Determine the equation of the least-squares line for these data.
b. Draw a scatter diagram and the least-squares line for these data.

c. Use the result obtained in (a) to predict the number of completed applications that might be expected if 6400 brochures and follow-up letters are sent out during the next year.

8. **Net Sales**   The management of Kaldor, Inc., a manufacturer of electric motors, submitted the following data in the annual report to its stockholders. The table shows the net sales (in millions of dollars) during the five years that have elapsed since the new management team took over. (The first year the firm operated under the new management corresponds to the time period $x = 1$, and the four subsequent years correspond to $x = 2, 3, 4, 5$.)

Year (x)	1	2	3	4	5
Net sales (y)	426	437	460	473	477

a. Determine the equation of the least-squares line for these data.
b. Draw a scatter diagram and the least-squares line for these data.
c. Use the result obtained in (a) to predict the net sales for the upcoming year.

9. **SAT Verbal Scores**   The following data were compiled by the superintendent of schools in a large metropolitan area. The table shows the average SAT verbal scores of high-school seniors during the five years since the district implemented the "back-to-basics" program.

Year (x)	1	2	3	4	5
Average score (y)	436	438	428	430	426

a. Determine the equation of the least-squares line for these data.
b. Draw a scatter diagram and the least-squares line for these data.
c. Use the result obtained in (a) to predict the average SAT verbal score of high-school seniors two years from now ($x = 7$).

10. **Auto Operating Costs**   The following figures were compiled by Clarke, Kingsley, and Company, a consulting firm that specializes in auto operating costs, relat-

ing the annual mileage (in thousands of miles) that an average new compact car is driven to the cost per mile (in cents) of operating the car.

Annual mileage (x)	5	10	15	20	25	30
Cost per mile (y)	50.3	34.8	30.1	27.4	25.6	23.5

   a. Determine an equation of the least-squares line for these data.
   b. Draw a scatter diagram and the least-squares line for these data.
   c. Use the result obtained in (a) to estimate the cost per mile of operating a new company car if it is driven 8000 miles during the first year of ownership.

11. **Size of Average Farm**   The size of the average farm in the United States has been growing steadily over the years. The following data, obtained from the U.S. Department of Agriculture, give the size of the average farm y (in acres) from 1940 through 1991. (Here $x = 0$ corresponds to the beginning of the year 1940.)

Year (x)	0	10	20	30	40	51
Size of farm (y)	168	213	297	374	427	467

   a. Find the equation of the least-squares line for these data.
   b. Use the result of (a) to estimate the size of the average farm in the year 2000.

12. **Welfare Costs**   According to the Massachusetts Department of Welfare, the spending (in billions of dollars) by Medicaid, the national health-care plan for the poor, over the five-year period from 1988 to 1992 is summarized in the following table. (Here $x = 0$ represents the beginning of the year 1988.)

Year (x)	0	1	2	3	4
Expenditure (y)	1.550	1.662	1.786	1.888	2.009

   a. Find an equation of the least-squares line for these data.

   b. Use the result of (a) to estimate Medicaid spending for the year 1996, assuming the trend continues.

13. **Mass Transit Subsidies**   The following table gives the projected state subsidies (in millions of dollars) to the Massachusetts Bay Transit Authority (MBTA) over a five-year period.

Year (x)	1	2	3	4	5
Subsidy (y)	20	24	26	28	32

   a. Find an equation of the least-squares line for these data.
   b. Use the result of (a) to estimate the state subsidy to the MBTA for the eighth year ($x = 8$).

14. **Social Security Wage Base**   The Social Security (FICA) wage base (in thousands of dollars) from 1987 to 1992 is given in the following table.

Year	1987	1988	1989
Wage base (y)	43.8	45.0	48.0

Year	1990	1991	1992
Wage base (y)	51.3	53.4	55.5

   a. Find an equation of the least-squares line for these data. (Let $x = 1$ represent the year 1987.)
   b. Use your result of (a) to estimate the FICA wage base in the year 1996.

15. **Production of All-Aluminum Cans**   Steel has been playing a decreasing role in the manufacture of beverage cans in the United States. According to the Can Manufacture Institute, the use of bimetallic cans has been dwindling while the use of all-aluminum cans has been growing steadily. The accompanying table gives the production of all-aluminum cans over the period from 1975 through 1989.

Year	1975	1977	1979	1981
No. of cans (in billions)	16.7	26	33.3	48.3

Year	1983	1985	1987	1989
No. of cans (in billions)	57	65.8	74.2	83.3

a. Find an equation of the least-squares line for these data. (Let $x = 1$ represent 1975.)
b. Use the result of (a) to estimate the number of cans produced in 1993, assuming the trend continues.

16. **Health-Care Spending** The following data, compiled by the Organization for Economic Cooperation and Development (OECD) in 1990, gives the per capita Gross Domestic Product (GDP) and the corresponding per capita spending on health care for selected countries.

Country	Turkey	Spain	Netherlands
GDP (per capita in thousands of dollars)	4.25	10	14
Health-care spending (per capita in dollars)	178	667	1194

Country	Sweden	Switzerland	Canada
GDP (per capita in thousands of dollars)	15.5	17.8	19.5
Health-care spending (per capita in dollars)	1500	1388	1640

a. Letting x denote a country's GDP (in thousands of dollars per capita) and y denote the per capita health-care spending (in dollars), find an equation of the least-squares line for these data giving the typical relationship between GDP and health-care spending for the selected countries.

b. The per capita GDP of the United States is $20,000. If the health-care spending of the United States were in line with that of these sample OECD countries, what would it be? [*Note:* The actual per capita health-care spending of the United States in 1990 was $2,444.]

C *Use a computer or programmable calculator to solve Exercises 17 and 18.*

17. According to data from the Council on Environmental Quality, the amount of waste (in millions of tons per year) generated in the United States from 1960 to 1990 was:

Year	1960	1965	1970	1975
Amount (y)	81	100	120	124

Year	1980	1985	1990
Amount (y)	140	152	164

a. Find an equation of the least-squares line for these data. (Let x be in units of 5 and let $x = 1$ represent 1960.)
b. Use the result of (a) to estimate the amount of waste generated in the year 2000, assuming the trend continues.

18. According to data from the Association of Realtors, the median price (in thousands of dollars) of existing homes in a certain metropolitan area from 1982 to 1992 was:

Year	1982	1983	1984	1985	1986	1987
Price (y)	66.4	69.8	72.8	76.0	79.6	83.1

Year	1988	1989	1990	1991	1992
Price (y)	86.3	89.5	92.3	96.0	99.5

a. Find an equation of the least-squares line for these data. (Let $x = 1$ represent 1982.)
b. Use the result of (a) to estimate the median price of a house in the year 1997, assuming the trend continues.

SOLUTIONS TO
SELF-CHECK
EXERCISES 8.4

1. We first construct the table:

	x	y	x^2	xy
	0	3	0	0
	2	6.5	4	13
	4	10	16	40
	6	16	36	96
	7	16.5	49	115.5
Sum	19	52	105	264.5

The normal equations are

$$5b + 19m = 52$$
$$19b + 105m = 264.5.$$

Solving the first equation for b gives

$$b = -3.8m + 10.4,$$

which, upon substitution into the second equation, gives

$$19(-3.8m + 10.4) + 105m = 264.5$$
$$-72.2m + 197.6 + 105m = 264.5$$
$$32.8m = 66.9$$

or $\qquad\qquad\qquad\qquad\qquad m = 2.04.$

Substituting this value of m into the expression for b found earlier gives

$$b = -3.8(2.04) + 10.4 = 2.65.$$

Therefore, the required least-squares line has equation given by

$$y = 2.04x + 2.65.$$

2. a. The calculations required for obtaining the normal equations may be summarized as follows:

	x	y	x^2	xy
	0	19	0	0
	6	25	36	150
	11	30	121	330
	16	35	256	560
	22	44	484	968
	26	48	676	1248
Sum	81	201	1573	3256

The normal equations are

$$6b + 81m = 201$$
$$81b + 1573m = 3256.$$

Solving this system of linear equations simultaneously, we find

$$m = 1.13 \quad \text{and} \quad b = 18.23.$$

Therefore, the required least-squares line has equation given by

$$y = f(x) = 1.13x + 18.23.$$

b. The percentage of people over the age of 65 who will have high school diplomas at the beginning of the year 1999 is given by

$$f(40) = 1.13(40) + 18.23$$
$$= 63.43,$$

or approximately 63.4 percent.

8.5   Constrained Maxima and Minima and the Method of Lagrange Multipliers

Constrained Relative Extrema

In Section 8.3 we studied the problem of determining the relative extremum of a function $f(x, y)$ without placing any restrictions on the independent variables x and y except, of course, that the point (x, y) lie in the domain of f. Such a relative extremum of a function f is referred to as an **unconstrained relative extremum of f.** However, in many practical optimization problems, we must maximize or minimize a function in which the independent variables are subjected to certain further constraints.

In this section we discuss a powerful method for determining the relative extrema of a function $f(x, y)$ whose independent variables x and y are required to satisfy one or more constraints of the form $g(x, y) = 0$. Such a relative extremum of a function f is called a **constrained relative extremum of f.** We can see the difference between an unconstrained extremum of a function $f(x, y)$ of two variables and a constrained extremum of f, where the independent variables x and y are subjected to a constraint of the form $g(x, y) = 0$, by considering the geometry of the two cases. Figure 8.24a depicts the graph of

FIGURE 8.24

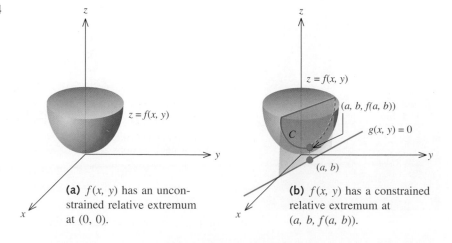

(a) $f(x, y)$ has an unconstrained relative extremum at $(0, 0)$.

(b) $f(x, y)$ has a constrained relative extremum at $(a, b, f(a, b))$.

a function $f(x, y)$ that has an unconstrained relative minimum at the point $(0, 0)$. However, when the independent variables x and y are subjected to an equality constraint of the form $g(x, y) = 0$, the points (x, y, z) that satisfy both $z = f(x, y)$ and the constraint equation $g(x, y) = 0$ lie on a curve C. Therefore, the constrained relative minimum of f must also lie on C (Figure 8.24b).

Our first example involves an equality constraint $g(x, y) = 0$ in which we solve for the variable y explicitly in terms of x. In this case, we may apply the technique used in Chapter 4 to find the relative extrema of a function of one variable.

EXAMPLE 1   Find the relative minimum of the function

$$f(x, y) = 2x^2 + y^2$$

subject to the constraint $g(x, y) = x + y - 1 = 0$.

Solution   Solving the constraint equation for y explicitly in terms of x, we obtain $y = -x + 1$. Substituting this value of y into the function $f(x, y) = 2x^2 + y^2$ results in a function of x,

$$h(x) = 2x^2 + (-x + 1)^2 = 3x^2 - 2x + 1.$$

The function h describes the curve C lying on the graph of f on which the constrained relative minimum of f occurs. To find this point, use the technique developed in Chapter 4 to determine the relative extrema of a function of one variable:

$$h'(x) = 6x - 2 = 2(3x - 1).$$

Setting $h'(x) = 0$ gives $x = \frac{1}{3}$ as the sole critical point of the function h. Next, we find

$$h''(x) = 6$$

and, in particular, $$h''\left(\frac{1}{3}\right) = 6 > 0.$$

Therefore, by the Second Derivative Test, the point $x = \frac{1}{3}$ gives rise to a relative minimum of h. Substitute this value of x into the constraint equation $x + y - 1 = 0$ to get $y = \frac{2}{3}$. Thus, the point $\left(\frac{1}{3}, \frac{2}{3}\right)$ gives rise to the required constrained relative minimum of f. Since

$$f\left(\frac{1}{3}, \frac{2}{3}\right) = 2\left(\frac{1}{3}\right)^2 + \left(\frac{2}{3}\right)^2 = \frac{2}{3},$$

the required constrained relative minimum value of f is $\frac{2}{3}$ at the point $\left(\frac{1}{3}, \frac{2}{3}\right)$.

It may be shown that $\frac{2}{3}$ is, in fact, a constrained absolute minimum value of f (Figure 8.25).

FIGURE 8.25
*f has a constrained absolute
minimum of $\frac{2}{3}$ at $\left(\frac{1}{3}, \frac{2}{3}\right)$.*

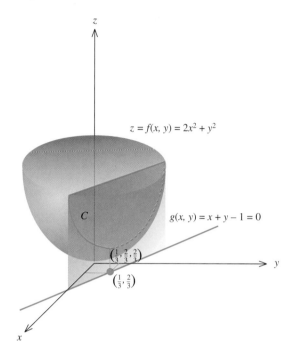

$$z = f(x, y) = 2x^2 + y^2$$

$$g(x, y) = x + y - 1 = 0$$

$$\left(\frac{1}{3}, \frac{2}{3}, \frac{2}{3}\right)$$

$$\left(\frac{1}{3}, \frac{2}{3}\right)$$

The Method of Lagrange Multipliers

The major drawback of the technique used in Example 1 is that it relies on our ability to solve the constraint equation $g(x, y) = 0$ for y explicitly in terms of x. This is not always an easy task. Moreover, even when we can solve the constraint equation $g(x, y) = 0$ for y explicitly in terms of x, the resulting function of one variable that is to be optimized may turn out to be unnecessarily complicated. Fortunately, an easier method exists. This method, called the **Method of Lagrange Multipliers** (Joseph Lagrange, 1736–1813), is as follows:

THE METHOD OF LAGRANGE MULTIPLIERS

To find the relative extremum of the function $f(x, y)$ subject to the constraint $g(x, y) = 0$ (assuming that these extreme values exist),

1. form an auxiliary function

$$F(x, y, \lambda) = f(x, y) + \lambda g(x, y),$$

called the Lagrangian function (the variable λ is called the Lagrange multiplier).

2. solve the system that comprises the equations

$$F_x = 0, \qquad F_y = 0, \quad \text{and} \quad F_\lambda = 0,$$

for all values of x, y, and λ.

3. evaluate f at each of the points (x, y) found in Step 2. The largest (smallest) of these values is the maximum (minimum) value of f.

Let us re-solve Example 1 using the Method of Lagrange Multipliers.

EXAMPLE 2   Using the Method of Lagrange Multipliers, find the relative minimum of the function

$$f(x, y) = 2x^2 + y^2$$

subject to the constraint $x + y = 1$.

Solution Write the constraint equation $x + y = 1$ in the form $g(x, y) = x + y - 1 = 0$. Then form the Lagrangian function

$$F(x, y, \lambda) = f(x, y) + \lambda g(x, y)$$
$$= 2x^2 + y^2 + \lambda(x + y - 1).$$

To find the critical point(s) of the function F, solve the system that comprises the equations

$$F_x = 4x + \lambda = 0$$
$$F_y = 2y + \lambda = 0$$
$$F_\lambda = x + y - 1 = 0.$$

Solving the first and second equations in this system for x and y in terms of λ, we obtain

$$x = -\frac{1}{4}\lambda, \qquad y = -\frac{1}{2}\lambda,$$

which, upon substitution into the third equation, yields

$$-\frac{1}{4}\lambda - \frac{1}{2}\lambda - 1 = 0 \quad \text{or} \quad \lambda = -\frac{4}{3}.$$

Therefore, $x = \frac{1}{3}$ and $y = \frac{2}{3}$ and $\left(\frac{1}{3}, \frac{2}{3}\right)$ affords a constrained minimum of the function f, in agreement with the result obtained earlier. ∎

The Method of Lagrange Multipliers may be used to solve a problem involving a function of three or more variables, as illustrated in the next example.

EXAMPLE 3   Use the Method of Lagrange Multipliers to find the minimum of the function

$$f(x, y, z) = 2xy + 6yz + 8xz$$

subject to the constraint

$$xyz = 12,000.$$

(*Note:* The existence of the minimum is suggested by the geometry of the problem.)

Solution Write the constraint equation $xyz = 12,000$ in the form $g(x, y, z) = xyz - 12,000$. Then the Lagrangian function is

$$F(x, y, z, \lambda) = f(x, y, z) + \lambda g(x, y, z)$$
$$= 2xy + 6yz + 8xz + \lambda(xyz - 12,000).$$

To find the critical point(s) of the function F, we solve the system that comprises the equations

$$F_x = 2y + 8z + \lambda yz = 0$$

$$F_y = 2x + 6z + \lambda xz = 0$$

$$F_z = 6y + 8x + \lambda xy = 0$$

$$F_\lambda = xyz - 12,000 = 0.$$

Solving the first three equations of the system for λ in terms of x, y, and z, we have

$$\lambda = -\frac{2y + 8z}{yz}$$

$$\lambda = -\frac{2x + 6z}{xz}$$

$$\lambda = -\frac{6y + 8x}{xy}.$$

Equating the first two expressions for λ leads to

$$\frac{2y + 8z}{yz} = \frac{2x + 6z}{xz}$$

or

$$2xy + 8xz = 2xy + 6yz$$

and

$$x = \frac{3}{4}y.$$

Next, equating the second and third expressions for λ in the same system yields

$$\frac{2x + 6z}{xz} = \frac{6y + 8x}{xy}$$

$$2xy + 6yz = 6yz + 8xz$$

and

$$z = \frac{1}{4}y.$$

Finally, substituting these values of x and z into the equation $xyz - 12,000 = 0$, the fourth equation of the first system of equations, we have

$$\left(\frac{3}{4}y\right)(y)\left(\frac{1}{4}y\right) - 12,000 = 0$$

$$y^3 = \frac{(12,000)(4)(4)}{3} = 64,000,$$

or $y = 40$. The corresponding values of x and z are given by $x = \frac{3}{4}(40) = 30$ and $z = \frac{1}{4}(40) = 10$. Therefore, we see that the point $(30, 40, 10)$ gives the constrained minimum of f. The minimum value is

$$f(30, 40, 10) = 2(30)(40) + 6(40)(10) + 8(30)(10) = 7200. \quad \blacksquare$$

Applications

EXAMPLE 4   Refer to Example 3, Section 8.1. The total weekly profit (in dollars) that the Acrosonic Company realized in producing and selling its bookshelf loudspeaker systems is given by the profit function

$$P(x, y) = -\frac{1}{4}x^2 - \frac{3}{8}y^2 - \frac{1}{4}xy + 120x + 100y - 5000,$$

where x denotes the number of fully assembled units and y denotes the number of kits produced and sold per week. Acrosonic's management decides that production of these loudspeaker systems should be restricted to a total of exactly 230 units per week. Under this condition, how many fully assembled units and how many kits should be produced per week to maximize Acrosonic's weekly profit?

Solution   The problem is equivalent to the problem of maximizing the function

$$P(x, y) = -\frac{1}{4}x^2 - \frac{3}{8}y^2 - \frac{1}{4}xy + 120x + 100y - 5000$$

subject to the constraint

$$g(x, y) = x + y - 230 = 0.$$

The Lagrangian function is

$$\begin{aligned} F(x, y, \lambda) &= P(x, y) + \lambda g(x, y) \\ &= -\frac{1}{4}x^2 - \frac{3}{8}y^2 - \frac{1}{4}xy + 120x + 100y \\ &\quad -5000 + \lambda(x + y - 230). \end{aligned}$$

To find the critical point(s) of F, solve the following system of equations:

$$F_x = -\frac{1}{2}x - \frac{1}{4}y + 120 + \lambda = 0$$

$$F_y = -\frac{3}{4}y - \frac{1}{4}x + 100 + \lambda = 0$$

$$F_\lambda = x + y - 230 = 0.$$

Solving the first equation of this system for λ, we obtain

$$\lambda = \frac{1}{2}x + \frac{1}{4}y - 120,$$

which, upon substitution into the second equation, yields

$$-\frac{3}{4}y - \frac{1}{4}x + 100 + \frac{1}{2}x + \frac{1}{4}y - 120 = 0$$

or $$-\frac{1}{2}y + \frac{1}{4}x - 20 = 0.$$

Solving the last equation for y gives

$$y = \frac{1}{2}x - 40.$$

When we substitute this value of y into the third equation of the system, we have

$$x + \frac{1}{2}x - 40 - 230 = 0$$

or $x = 180$.

The corresponding value of y is $\left(\frac{1}{2}\right)(180) - 40$, or 50. Thus, the required constrained relative maximum of P occurs at the point (180, 50). Again, we can show that the point (180, 50) in fact yields a constrained absolute maximum for P. Thus, Acrosonic's profit is maximized by producing 180 assembled and 50 kit versions of their bookshelf loudspeaker systems. The maximum weekly profit realizable is given by

$$P(180, 50) = -\frac{1}{4}(180)^2 - \frac{3}{8}(50)^2 - \frac{1}{4}(180)(50)$$
$$+ 120(180) + 100(50) - 5{,}000$$
$$= 10{,}312.5,$$

or \$10,312.50.   ∎

FIGURE 8.26

A rectangular-shaped pool will be built in the elliptical-shaped poolside area.

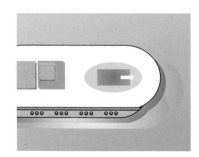

EXAMPLE 5   The operators of the *Viking Princess*, a luxury cruise liner, are contemplating the addition of another swimming pool to the ship. The chief engineer has suggested that an area in the form of an ellipse located in the rear of the promenade deck would be suitable for this purpose. This location would provide a poolside area with sufficient space for passenger movement and placement of deck chairs (Figure 8.26). It has been determined that the shape of the ellipse may be described by the equation $x^2 + 4y^2 = 3600$, where x and y are measured in feet. *Viking's* operators would like to know the dimensions of the rectangular pool with the largest possible area that would meet these requirements.

Solution   In order to solve this problem, we need to find the rectangle inscribed in the ellipse with equation $x^2 + 4y^2 = 3600$ and having the largest area. Letting the sides of the rectangle be $2x$ and $2y$ feet, we see that the area of the rectangle is $A = 4xy$ (Figure 8.27). Furthermore, the point (x, y) must be constrained to move along the ellipse so that it satisfies the equation $x^2 + 4y^2 = 3600$. Thus, the problem is equivalent to the problem of maximizing the function

$$f(x, y) = 4xy$$

subject to the constraint $g(x, y) = x^2 + 4y^2 - 3600 = 0$. The Lagrangian function is

$$F(x, y, \lambda) = f(x, y) + \lambda g(x, y)$$
$$= 4xy + \lambda(x^2 + 4y^2 - 3600).$$

To find the critical point(s) of F, we solve the following system of equations:

$$F_x = 4y + 2\lambda x = 0$$
$$F_y = 4x + 8\lambda y = 0$$
$$F_\lambda = x^2 + 4y^2 - 3600 = 0.$$

Solving the first equation of this system for λ, we obtain

$$\lambda = -\frac{2y}{x},$$

FIGURE 8.27

We want to find the largest rectangle that can be inscribed in the ellipse described by $x^2 + 4y^2 = 3600$.

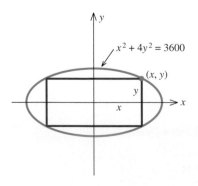

which, upon substitution into the second equation, yields

$$4x + 8\left(-\frac{2y}{x}\right)(y) = 0 \quad \text{or} \quad x^2 - 4y^2 = 0,$$

that is, $x = \pm 2y$. Substituting these values of x into the third equation of the system, we have

$$4y^2 + 4y^2 - 3600 = 0,$$

or, upon solving $y = \pm\sqrt{450} = \pm 15\sqrt{2}$. The corresponding values of x are $\pm 30\sqrt{2}$. Because both x and y must be nonnegative, we have $x = 30\sqrt{2}$ and $y = 15\sqrt{2}$. Thus, the dimensions of the pool with maximum area are $30\sqrt{2}$ ft by $60\sqrt{2}$ ft, or approximately 42 ft × 85 ft. ■

EXAMPLE 6 Suppose that x units of labor and y units of capital are required to produce

$$f(x, y) = 100x^{3/4}y^{1/4}$$

units of a certain product (recall that this is a Cobb-Douglas production function). If each unit of labor costs \$200 and each unit of capital costs \$300, and a total of \$60,000 is available for production, determine how many units of labor and how many units of capital should be used in order to maximize production.

Solution The total cost of x units of labor at \$200 per unit and y units of capital at \$300 per unit is equal to $200x + 300y$ dollars. But \$60,000 is budgeted for production, so $200x + 300y = 60{,}000$, which we rewrite as

$$g(x, y) = 200x + 300y - 60{,}000 = 0.$$

To maximize $f(x, y) = 100x^{3/4}y^{1/4}$ subject to the constraint $g(x, y) = 0$, we form the Lagrangian function

$$F(x, y, \lambda) = f(x, y) + \lambda g(x, y)$$
$$= 100x^{3/4}y^{1/4} + \lambda(200x + 300y - 60{,}000).$$

To find the critical point(s) of F, we solve the system that comprises the equations

$$F_x = 75x^{-1/4}y^{1/4} + 200\lambda = 0$$
$$F_y = 25x^{3/4}y^{-3/4} + 300\lambda = 0$$
$$F_\lambda = 200x + 300y - 60{,}000 = 0.$$

Solving the first equation for λ, we have

$$\lambda = -\frac{75x^{-1/4}y^{1/4}}{200} = -\frac{3}{8}\left(\frac{y}{x}\right)^{1/4},$$

which, when substituted into the second equation, yields

$$25\left(\frac{x}{y}\right)^{3/4} + 300\left(-\frac{3}{8}\right)\left(\frac{y}{x}\right)^{1/4} = 0.$$

Multiplying the last equation by $(x/y)^{1/4}$ then gives

$$25\left(\frac{x}{y}\right) - \frac{900}{8} = 0$$

or

$$x = \left(\frac{900}{8}\right)\left(\frac{1}{25}\right)y = \frac{9}{2}y.$$

Substituting this value of x into the third equation of the first system of equations, we have

$$200\left(\frac{9}{2}y\right) + 300y - 60,000 = 0,$$

from which we deduce that $y = 50$. Hence, $x = 225$. Thus, maximum production is achieved when 225 units of labor and 50 units of capital are used. ∎

When used in the context of Example 6, the negative of the Lagrange multiplier λ is called the **marginal productivity of money.** That is, if one additional dollar is available for production, then approximately $-\lambda$ units of a product can be produced. Here

$$\lambda = -\frac{3}{8}\left(\frac{y}{x}\right)^{1/4} = -\frac{3}{8}\left(\frac{50}{225}\right)^{1/4} \approx -0.258,$$

so, in this case, the marginal productivity of money is 0.258. For example, if $65,000 is available for production instead of the originally budgeted figure of $60,000, then the maximum production may be boosted from the original

$$f(225, 50) = 100(225)^{3/4}(50)^{1/4},$$

or 15,448 units, to

$$15,448 + 5,000(0.258),$$

or 16,738 units.

SELF-CHECK
EXERCISES 8.5

1. Use the Method of Lagrange Multipliers to find the relative maximum of the function

$$f(x, y) = -2x^2 - y^2$$

subject to the constraint $3x + 4y = 12$.

2. The total monthly profit of the Robertson Controls Company in manufacturing and selling x hundred of its standard mechanical setback thermostats and y hundred of its deluxe electronic setback thermostats per month is given by the total profit function

$$P(x, y) = -\frac{1}{8}x^2 - \frac{1}{2}y^2 - \frac{1}{4}xy + 13x + 40y - 280,$$

where P is in hundreds of dollars. If the production of setback thermostats is to be restricted to a total of exactly 4000 per month, how many of each model should Robertson manufacture in order to maximize its monthly profits? What is the maximum monthly profit?

Solutions to Self-Check Exercises 8.5 can be found on page 534.

8.5 EXERCISES

In Exercises 1–16, use the Method of Lagrange Multipliers to optimize the given function subject to the given constraint.

1. Minimize the function $f(x, y) = x^2 + 3y^2$ subject to the constraint $x + y - 1 = 0$.

2. Minimize the function $f(x, y) = x^2 + y^2 - xy$ subject to the constraint $x + 2y - 14 = 0$.

3. Maximize the function $f(x, y) = 2x + 3y - x^2 - y^2$ subject to the constraint $x + 2y = 9$.

4. Maximize the function $f(x, y) = 16 - x^2 - y^2$ subject to the constraint $x + y - 6 = 0$.

5. Minimize the function $f(x, y) = x^2 + 4y^2$ subject to the constraint $xy = 1$.

6. Minimize the function $f(x, y) = xy$ subject to the constraint $x^2 + 4y^2 = 4$.

7. Maximize the function $f(x, y) = x + 5y - 2xy - x^2 - 2y^2$ subject to the constraint $2x + y = 4$.

8. Maximize the function $f(x, y) = xy$ subject to the constraint $2x + 3y - 6 = 0$.

9. Maximize the function $f(x, y) = xy^2$ subject to the constraint $9x^2 + y^2 = 9$.

10. Minimize the function $f(x, y) = \sqrt{y^2 - x^2}$ subject to the constraint $x + 2y - 5 = 0$.

11. Find the maximum and minimum values of the function $f(x, y) = xy$ subject to the constraint $x^2 + y^2 = 16$.

12. Find the maximum and minimum values of the function $f(x, y) = e^{xy}$ subject to the constraint $x^2 + y^2 = 8$.

13. Find the maximum and minimum values of the function $f(x, y) = xy^2$ subject to the constraint $x^2 + y^2 = 1$.

14. Maximize the function $f(x, y, z) = xyz$ subject to the constraint $2x + 2y + z = 84$.

15. Minimize the function $f(x, y, z) = x^2 + y^2 + z^2$ subject to the constraint $3x + 2y + z = 6$.

16. Find the maximum and minimum values of the function $f(x, y, z) = x + 2y - 3z$ subject to the constraint $z = 4x^2 + y^2$.

17. **Maximizing Profit** The total weekly profit (in dollars) realized by the Country Workshop in manufacturing and selling its roll-top desks is given by the profit function

$$P(x, y) = -0.2x^2 - 0.25y^2 - 0.2xy \\ + 100x + 90y - 4000,$$

where x stands for the number of finished units and y denotes the number of unfinished units manufactured and sold per week. The company's management decides to restrict the manufacture of these desks to a total of exactly 200 units per week. How many finished and how many unfinished units should be manufactured per week to maximize the company's weekly profit?

18. **Maximizing Profit** The total daily profit (in dollars) realized by the Weston Publishing Company in publishing and selling its dictionaries is given by the profit function

$$P(x, y) = -0.005x^2 - 0.003y^2 - 0.002xy \\ + 14x + 12y - 200,$$

where x stands for the number of deluxe editions and y denotes the number of standard editions sold daily. Weston's management decides that publication of these dictionaries should be restricted to a total of exactly 400 copies per day. How many deluxe copies and how many standard copies should be published per day to maximize Weston's daily profit?

19. **Minimizing Construction Costs** The management of UNICO Department Store decides to enclose an 800-ft^2 area outside their building to display potted plants. The enclosed area will be a rectangle, one side of which is provided by the external walls of the store. Two sides of the enclosure will be made of pine board, and the fourth side will be made of galvanized steel fencing material. If the pine board fencing costs $6 per running foot and the steel fencing costs $3 per running foot, determine the dimensions of the enclosure that will cost the least to erect.

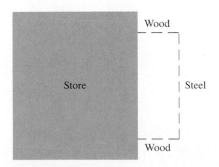

20. **Minimizing Container Costs** The Betty Moore Company requires that its corned beef hash containers have a capacity of 64 cubic inches, be right circular cylinders, and be made of a tin alloy. Find the radius and height of the least expensive container that can be made if the

metal for the side and bottom costs 4 cents per square inch and the metal for the pull-off lid costs 2 cents per square inch. [*Hint:* Let the radius and height of the container be r and h inches, respectively. Then, the volume of the container is $\pi r^2 h = 64$, and the cost is given by

$$C(r, h) = 8\pi rh + 6\pi r^2.]$$

21. **Minimizing Construction Costs** An open rectangular box is to be constructed from material that costs \$3 per square foot for the bottom and \$1 per square foot for its sides. Find the dimensions of the box of greatest volume that can be constructed for \$36.

22. **Minimizing Construction Costs** A closed rectangular box having a volume of 4 cubic feet is to be constructed. If the material for the sides costs \$1 per square foot and the material for the top and bottom costs \$1.50 per square foot, find the dimensions of the box that can be constructed with minimum cost.

23. **Maximizing Sales** The Ross-Simons Company has a monthly advertising budget of \$60,000. Their marketing department estimates that if they spend x dollars on newspaper advertising and y dollars on television advertising, then the monthly sales will be given by

$$z = f(x, y) = 90x^{1/4}y^{3/4}$$

dollars. Determine how much money Ross-Simons should spend on newspaper ads and on television ads per month to maximize its monthly sales.

24. **Maximizing Production** John Mills, the proprietor of the Mills Engine Company, a manufacturer of model airplane engines, finds that it takes x units of labor and y units of capital to produce

$$f(x, y) = 100x^{3/4}y^{1/4}$$

units of the product. If a unit of labor costs \$100 and a unit of capital costs \$200, and \$200,000 is budgeted for production, determine how many units should be expended on labor and how many units should be expended on capital in order to maximize production.

25. Use the Method of Lagrange Multipliers to solve Problem 28, Exercises 8.3.

SOLUTIONS TO SELF-CHECK EXERCISES 8.5

1. Write the constraint equation in the form $g(x, y) = 3x + 4y - 12 = 0$. Then the Lagrangian function is

$$F(x, y, \lambda) = -2x^2 - y^2 + \lambda(3x + 4y - 12).$$

To find the critical point(s) of F, we solve the system

$$F_x = -4x + 3\lambda = 0$$
$$F_y = -2y + 4\lambda = 0$$
$$F_\lambda = 3x + 4y - 12 = 0.$$

Solving the first two equations for x and y in terms of λ, we find $x = \frac{3}{4}\lambda$ and $y = 2\lambda$. Substituting these values of x and y into the third equation of the system yields

$$3\left(\frac{3}{4}\lambda\right) + 4(2\lambda) - 12 = 0,$$

or $\lambda = \frac{48}{41}$. Therefore, $x = \frac{3}{4}\left(\frac{48}{41}\right) = \frac{36}{41}$ and $y = 2\left(\frac{48}{41}\right) = \frac{96}{41}$, and we see that the point $\left(\frac{36}{41}, \frac{96}{41}\right)$ gives the constrained maximum of f. The maximum value is

$$f\left(\frac{36}{41}, \frac{96}{41}\right) = -2\left(\frac{36}{41}\right)^2 - \left(\frac{96}{41}\right)^2$$

$$= -\frac{11{,}808}{1{,}681} = -\frac{288}{41}.$$

2. We want to maximize

$$P(x, y) = -\frac{1}{8}x^2 - \frac{1}{2}y^2 - \frac{1}{4}xy + 13x + 40y - 280$$

subject to the constraint

$$g(x, y) = x + y - 40 = 0.$$

The Lagrangian function is

$$F(x, y, \lambda) = P(x, y) + \lambda g(x, y)$$
$$= -\frac{1}{8}x^2 - \frac{1}{2}y^2 - \frac{1}{4}xy + 13x$$
$$+ 40y - 280 + \lambda(x + y - 40).$$

To find the critical points of F, solve the following system of equations:

$$F_x = -\frac{1}{4}x - \frac{1}{4}y + 13 + \lambda = 0$$

$$F_y = -\frac{1}{4}x - y + 40 + \lambda = 0$$

$$F_\lambda = x + y - 40 = 0.$$

Subtracting the first equation from the second gives

$$-\frac{3}{4}y + 27 = 0, \quad \text{or} \quad y = 36.$$

Substituting this value of y into the third equation yields $x = 4$. Therefore, in order to maximize its monthly profits, Robertson should manufacture 400 standard and 3600 deluxe thermostats. The maximum monthly profit is given by

$$P(4, 36) = -\frac{1}{8}(4)^2 - \frac{1}{2}(36)^2 - \frac{1}{4}(4)(36)$$
$$+ 13(4) + 40(36) - 280$$
$$= 526,$$

or $52,600.

CHAPTER EIGHT SUMMARY OF PRINCIPAL TERMS

Terms

Three-dimensional Cartesian
  coordinate system
Function of two variables:
  Domain
  First partial derivative
  Second-order partial derivative
  Relative maximum
  Relative minimum
  Absolute maximum
  Absolute minimum
  Critical point
  Saddle point
  Second Derivative Test

Level curve
Constrained relative extremum
Cobb-Douglas production function
Marginal productivity of labor
Marginal productivity of capital
Substitute commodities
Complementary commodities
Method of least squares
Least-squares line
Regression line
Normal equations
Method of Lagrange Multipliers

CHAPTER 8 REVIEW EXERCISES

1. Let $f(x, y) = xy/(x^2 + y^2)$. Compute $f(0, 1)$, $f(1, 0)$, and $f(1, 1)$. Does $f(0, 0)$ exist?

2. Let $f(x, y) = \dfrac{xe^y}{1 + \ln xy}$. Compute $f(1, 1)$, $f(1, 2)$, and $f(2, 1)$. Does $f(1, 0)$ exist?

3. Let $h(x, y, z) = xye^z + (x/y)$. Compute $h(1, 1, 0)$, $h(-1, 1, 1)$, and $h(1, -1, 1)$.

4. Find the domain of the function $f(u, v) = \dfrac{\sqrt{u}}{u - v}$.

5. Find the domain of the function $f(x, y) = \dfrac{x - y}{x + y}$.

6. Find the domain of the function $f(x, y) = x\sqrt{y} + y\sqrt{1 - x}$.

7. Find the domain of the function $f(x, y, z) = \dfrac{xy\sqrt{z}}{(1 - x)(1 - y)(1 - z)}$.

In Exercises 8–11, sketch the level curves of the given function corresponding to the given values of z.

8. $z = f(x, y) = 2x + 3y$; $z = -2, -1, 0, 1, 2$

9. $z = f(x, y) = y - x^2$; $z = -2, -1, 0, 1, 2$

10. $z = f(x, y) = \sqrt{x^2 + y^2}$; $z = 0, 1, 2, 3, 4$

11. $z = f(x, y) = e^{xy}$; $z = 1, 2, 3$

In Exercises 12–21, compute the first partial derivatives of the given function.

12. $f(x, y) = x^2y^3 + 3xy^2 + \dfrac{x}{y}$

13. $f(x, y) = x\sqrt{y} + y\sqrt{x}$

14. $f(u, v) = \sqrt{uv^2 - 2u}$

15. $f(x, y) = \dfrac{x - y}{y + 2x}$ 16. $g(x, y) = \dfrac{xy}{x^2 + y^2}$

17. $h(x, y) = (2xy + 3y^2)^5$

18. $f(x, y) = (xe^y + 1)^{1/2}$

19. $f(x, y) = (x^2 + y^2)e^{x^2+y^2}$

20. $f(x, y) = \ln(1 + 2x^2 + 4y^4)$

21. $f(x, y) = \ln\left(1 + \dfrac{x^2}{y^2}\right)$

In Exercises 22–27, compute the second-order partial derivatives of the given function.

22. $f(x, y) = x^3 - 2x^2y + y^2 + x - 2y$

23. $f(x, y) = x^4 + 2x^2y^2 - y^4$

24. $f(x, y) = (2x^2 + 3y^2)^3$ 25. $g(x, y) = \dfrac{x}{x + y^2}$

26. $g(x, y) = e^{x^2+y^2}$ 27. $h(s, t) = \ln\left(\dfrac{s}{t}\right)$

28. Let $f(x, y, z) = x^3y^2z + xy^2z + 3xy - 4z$. Compute $f_x(1, 1, 0)$, $f_y(1, 1, 0)$, and $f_z(1, 1, 0)$ and interpret your results.

In Exercises 29–34, find the critical point(s) of the given functions. Then use the Second Derivative Test to classify the nature of each of these points, if possible. Finally, determine the relative extrema of each function.

29. $f(x, y) = 2x^2 + y^2 - 8x - 6y + 4$

30. $f(x, y) = x^2 + 3xy + y^2 - 10x - 20y + 12$

31. $f(x, y) = x^3 - 3xy + y^2$

32. $f(x, y) = x^3 + y^2 - 4xy + 17x - 10y + 8$

33. $f(x, y) = e^{2x^2+y^2}$

34. $f(x, y) = \ln(x^2 + y^2 - 2x - 2y + 4)$

In Exercises 35–38, use the Method of Lagrange Multipliers to optimize the given functions subject to the given constraints.

35. Maximize the function $f(x, y) = -3x^2 - y^2 + 2xy$ subject to the constraint $2x + y = 4$.

36. Minimize the function $f(x, y) = 2x^2 + 3y^2 - 6xy + 4x - 9y + 10$ subject to the constraint $x + y = 1$.

37. Find the maximum and minimum values of the function $f(x, y) = 2x - 3y + 1$ subject to the constraint $2x^2 + 3y^2 - 125 = 0$.

38. Find the maximum and minimum values of the function $f(x, y) = e^{x-y}$ subject to the constraint $x^2 + y^2 = 1$.

39. A division of Ditton Industries makes a 16-speed and a 10-speed electric blender. The company's management estimates that x units of the 16-speed model and y units of the 10-speed model are demanded daily when the

unit prices are

$$p = 80 - 0.02x - 0.1y$$

and $\qquad q = 60 - 0.1x - 0.05y$

dollars, respectively.

a. Find the daily total revenue function $R(x, y)$.
b. Find the domain of the function R.
c. Compute $R(100, 300)$ and interpret your result.

40. In a survey conducted by *Home Entertainment* magazine it was determined that the demand equation for compact disc (CD) players is given by

$$x = f(p, q) = 900 - 9p - e^{0.4q},$$

whereas the demand equation for compact audio discs is given by

$$y = g(p, q) = 20,000 - 3,000q - 4p,$$

where p and q denote the unit prices (in dollars) for the CD players and audio discs, respectively, and x and y denote the number of CD players and audio discs demanded per week. Determine whether these two products are substitute, complementary, or neither.

41. The Odyssey Travel Agency's monthly revenue depends on the amount of money x (in thousands of dollars) spent on advertising per month and the number of agents y in its employ in accordance with the rule

$$R(x, y) = -x^2 - 0.5y^2 + xy + 8x + 3y + 20.$$

Determine the amount of money the agency should spend per month and the number of agents it should employ in order to maximize its monthly revenue.

42. The following data were compiled by the Bureau of Television Advertising in a large metropolitan area, giving the average daily TV viewing time per household in that area over the years 1986 to 1994.

Year	1986	1988
Daily viewing time y	6 hr 9 min	6 hr 30 min

Year	1990	1992	1994
Daily viewing time y	6 hr 36 min	7 hr	7 hr 16 min

a. Find the least-squares line for these data. (Let $x = 1$ represent the year 1986.)
b. Estimate the average daily TV viewing time per household in the year 1996.

43. The owner of the Rancho Grande wants to enclose a rectangular piece of grazing land along the straight portion of a river and then subdivide it using a fence running parallel to the sides. No fencing is required along the river. If the material for the sides costs $3 per running yard and the material for the divider costs $2 per running yard, what will the dimensions of a 303,750-yard pasture be if the cost of fencing material is kept to a minimum?

44. The production of Q units of a commodity is related to the amount of labor x and the amount of capital y (in suitable units) expended by the equation

$$Q = f(x, y) = x^{3/4}y^{1/4}.$$

If an expenditure of 100 units is available for production, how should it be apportioned between labor and capital so that Q is maximized? [*Hint:* Use the Method of Lagrange Multipliers to maximize the function Q subject to the constraint $x + y = 100$.]

TABLES

TABLE 1
Exponential Functions

x	e^x	e^{-x}	x	e^x	e^{-x}
0.00	1.0000	1.0000	1.5	4.4817	0.2231
0.01	1.0101	0.9901	1.6	4.9530	0.2019
0.02	1.0202	0.9802	1.7	5.4739	0.1827
0.03	1.0305	0.9705	1.8	6.0496	0.1653
0.04	1.0408	0.9608	1.9	6.6859	0.1496
0.05	1.0513	0.9512	2.0	7.3891	0.1353
0.06	1.0618	0.9418	2.1	8.1662	0.1225
0.07	1.0725	0.9324	2.2	9.0250	0.1108
0.08	1.0833	0.9331	2.3	9.9742	0.1003
0.09	1.0942	0.9139	2.4	11.023	0.0907
0.10	1.1052	0.9048	2.5	12.182	0.0821
0.11	1.1163	0.8958	2.6	13.464	0.0743
0.12	1.1275	0.8869	2.7	14.880	0.0672
0.13	1.1388	0.8781	2.8	16.445	0.0608
0.14	1.1503	0.8694	2.9	18.174	0.0550
0.15	1.1618	0.8607	3.0	20.086	0.0498
0.16	1.1735	0.8521	3.1	22.198	0.0450
0.17	1.1853	0.8437	3.2	24.533	0.0408
0.18	1.1972	0.8353	3.3	27.113	0.0369
0.19	1.2092	0.8270	3.4	29.964	0.0334
0.20	1.2214	0.8187	3.5	33.115	0.0302
0.21	1.2337	0.8106	3.6	36.598	0.0273
0.22	1.2161	0.8025	3.7	40.447	0.0247
0.23	1.2586	0.7945	3.8	44.701	0.0224
0.24	1.2712	0.7866	3.9	49.402	0.0202
0.25	1.2840	0.7788	4.0	54.598	0.0183
0.30	1.3499	0.7408	4.1	60.340	0.0166
0.35	1.4191	0.7047	4.2	66.686	0.0150
0.40	1.4918	0.6703	4.3	73.700	0.0136
0.45	1.5683	0.6376	4.4	81.451	0.0123
0.50	1.6487	0.6065	4.5	90.017	0.0111
0.55	1.7333	0.5769	4.6	99.484	0.0101
0.60	1.8221	0.5488	4.7	109.95	0.0091
0.65	1.9155	0.5220	4.8	121.51	0.0082
0.70	2.0138	0.4966	4.9	134.29	0.0074
0.75	2.1170	0.4724	5.0	148.41	0.0067
0.80	2.2255	0.4493	5.5	244.69	0.0041
0.85	2.3396	0.4274	6.0	403.43	0.0025
0.90	2.4596	0.4066	6.5	665.14	0.0015
0.95	2.5857	0.3867	7.0	1096.6	0.0009
1.0	2.7183	0.3679	7.5	1808.0	0.0006
1.1	3.0042	0.3329	8.0	2981.0	0.0003
1.2	3.3201	0.3012	8.5	4914.8	0.0002
1.3	3.6693	0.2725	9.0	8103.1	0.0001
1.4	4.0552	0.2466	10.0	22026	0.00005

From MODERN COLLEGE ALGEBRA AND TRIGONOMETRY, Fourth Edition, by Edwin F. Beckenbach, Irving Drooyan, and Michael D. Grady. © 1981 by Wadsworth, Inc. Reprinted by permission of Wadsworth Publishing Company, Belmont, California 94002.

TABLE 2
Common Logarithms

x	0	1	2	3	4	5	6	7	8	9
1.0	.0000	.0043	.0086	.0128	.0170	.0212	.0253	.0294	.0334	.0374
1.1	.0414	.0453	.0492	.0531	.0569	.0607	.0645	.0682	.0719	.0755
1.2	.0792	.0828	.0864	.0899	.0934	.0969	.1004	.1038	.1072	.1106
1.3	.1139	.1173	.1206	.1239	.1271	.1303	.1335	.1367	.1399	.1430
1.4	.1461	.1492	.1523	.1553	.1584	.1614	.1644	.1673	.1703	.1732
1.5	.1761	.1790	.1818	.1847	.1875	.1903	.1931	.1959	.1987	.2014
1.6	.2041	.2068	.2095	.2122	.2148	.2175	.2201	.2227	.2253	.2279
1.7	.2304	.2330	.2355	.2380	.2405	.2430	.2455	.2480	.2504	.2529
1.8	.2553	.2577	.2601	.2625	.2648	.2672	.2695	.2718	.2742	.2765
1.9	.2788	.2810	.2833	.2856	.2878	.2900	.2923	.2945	.2967	.2989
2.0	.3010	.3032	.3054	.3075	.3096	.3118	.3139	.3160	.3181	.3201
2.1	.3222	.3243	.3263	.3284	.3304	.3324	.3345	.3365	.3385	.3404
2.2	.3424	.3444	.3464	.3483	.3502	.3522	.3541	.3560	.3579	.3598
2.3	.3617	.3636	.3655	.3674	.3692	.3711	.3729	.3747	.3766	.3784
2.4	.3802	.3820	.3838	.3856	.3874	.3892	.3909	.3927	.3945	.3962
2.5	.3979	.3997	.4014	.4031	.4048	.4065	.4082	.4099	.4116	.4133
2.6	.4150	.4166	.4183	.4200	.4216	.4232	.4249	.4265	.4281	.4298
2.7	.4314	.4330	.4346	.4362	.4378	.4393	.4409	.4425	.4440	.4456
2.8	.4472	.4487	.4502	.4518	.4533	.4548	.4564	.4579	.4594	.4609
2.9	.4624	.4639	.4654	.4669	.4683	.4698	.4713	.4728	.4742	.4757
3.0	.4771	.4786	.4800	.4814	.4829	.4843	.4857	.4871	.4886	.4900
3.1	.4914	.4928	.4942	.4955	.4969	.4983	.4997	.5011	.5024	.5038
3.2	.5051	.5065	.5079	.5092	.5105	.5119	.5132	.5145	.5159	.5172
3.3	.5185	.5198	.5211	.5224	.5237	.5250	.5263	.5276	.5289	.5302
3.4	.5315	.5328	.5340	.5353	.5366	.5378	.5391	.5403	.5416	.5428
3.5	.5441	.5453	.5465	.5478	.5490	.5502	.5514	.5527	.5539	.5551
3.6	.5563	.5575	.5587	.5599	.5611	.5623	.5635	.5647	.5658	.5670
3.7	.5682	.5694	.5705	.5717	.5729	.5740	.5752	.5763	.5775	.5786
3.8	.5798	.5809	.5821	.5832	.5843	.5855	.5866	.5877	.5888	.5899
3.9	.5911	.5922	.5933	.5944	.5955	.5966	.5977	.5988	.5999	.6010
4.0	.6021	.6031	.6042	.6053	.6064	.6075	.6085	.6096	.6107	.6117
4.1	.6128	.6138	.6149	.6160	.6170	.6180	.6191	.6201	.6212	.6222
4.2	.6232	.6243	.6253	.6263	.6274	.6284	.6294	.6304	.6314	.6325
4.3	.6335	.6345	.6355	.6365	.6375	.6385	.6395	.6405	.6415	.6425
4.4	.6435	.6444	.6454	.6464	.6474	.6484	.6493	.6503	.6513	.6522
4.5	.6532	.6542	.6551	.6561	.6571	.6580	.6590	.6599	.6609	.6618
4.6	.6628	.6637	.6646	.6656	.6665	.6675	.6684	.6693	.6702	.6712
4.7	.6721	.6730	.6739	.6749	.6758	.6767	.6776	.6785	.6794	.6803
4.8	.6812	.6821	.6830	.6839	.6848	.6857	.6866	.6875	.6884	.6893
4.9	.6902	.6911	.6920	.6928	.6937	.6946	.6955	.6964	.6972	.6981
5.0	.6990	.6998	.7007	.7016	.7024	.7033	.7042	.7050	.7059	.7067
5.1	.7076	.7084	.7093	.7101	.7110	.7118	.7126	.7135	.7143	.7152
5.2	.7160	.7168	.7177	.7185	.7193	.7202	.7210	.7218	.7226	.7235
5.3	.7243	.7251	.7259	.7267	.7275	.7284	.7292	.7300	.7308	.7316
5.4	.7324	.7332	.7340	.7348	.7356	.7364	.7372	.7380	.7388	.7396
x	0	1	2	3	4	5	6	7	8	9

(continued)

From MODERN COLLEGE ALGEBRA AND TRIGONOMETRY, Fourth Edition, by Edwin F. Beckenbach, Irving Drooyan, and Michael D. Grady. © 1981 by Wadsworth, Inc. Reprinted by permission of Wadsworth Publishing Company, Belmont, California 94002.

TABLE 2
Common Logarithms (continued)

x	0	1	2	3	4	5	6	7	8	9
5.5	.7404	.7412	.7419	.7427	.7435	.7443	.7451	.7459	.7466	.7474
5.6	.7482	.7490	.7497	.7505	.7513	.7520	.7528	.7536	.7543	.7551
5.7	.7559	.7566	.7574	.7582	.7589	.7597	.7604	.7612	.7619	.7627
5.8	.7634	.7642	.7649	.7657	.7664	.7672	.7679	.7686	.7694	.7701
5.9	.7709	.7716	.7723	.7731	.7738	.7745	.7752	.7760	.7767	.7774
6.0	.7782	.7789	.7796	.7803	.7810	.7818	.7825	.7832	.7839	.7846
6.1	.7853	.7860	.7868	.7875	.7882	.7889	.7896	.7903	.7910	.7917
6.2	.7924	.7931	.7938	.7945	.7952	.7959	.7966	.7973	.7980	.7987
6.3	.7993	.8000	.8007	.8014	.8021	.8028	.8035	.8041	.8048	.8055
6.4	.8062	.8069	.8075	.8082	.8089	.8096	.8102	.8109	.8116	.8122
6.5	.8129	.8136	.8142	.8149	.8156	.8162	.8169	.8176	.8182	.8189
6.6	.8195	.8202	.8209	.8215	.8222	.8228	.8235	.8241	.8248	.8254
6.7	.8261	.8267	.8274	.8280	.8287	.8293	.8299	.8306	.8312	.8319
6.8	.8325	.8331	.8338	.8344	.8351	.8357	.8363	.8370	.8376	.8382
6.9	.8388	.8395	.8401	.8407	.8414	.8420	.8426	.8432	.8439	.8445
7.0	.8451	.8457	.8463	.8470	.8476	.8482	.8488	.8494	.8500	.8506
7.1	.8513	.8519	.8525	.8531	.8537	.8543	.8549	.8555	.8561	.8567
7.2	.8573	.8579	.8585	.8591	.8597	.8603	.8609	.8615	.8621	.8627
7.3	.8633	.8639	.8645	.8651	.8657	.8663	.8669	.8675	.8681	.8686
7.4	.8692	.8698	.8704	.8710	.8716	.8722	.8727	.8733	.8739	.8745
7.5	.8751	.8756	.8762	.8768	.8774	.8779	.8785	.8791	.8797	.8802
7.6	.8808	.8814	.8820	.8825	.8831	.8837	.8842	.8848	.8854	.8859
7.7	.8865	.8871	.8876	.8882	.8887	.8893	.8899	.8904	.8910	.8915
7.8	.8921	.8927	.8932	.8938	.8943	.8949	.8954	.8960	.8965	.8971
7.9	.8976	.8982	.8987	.8993	.8998	.9004	.9009	.9015	.9020	.9025
8.0	.9031	.9036	.9042	.9047	.9053	.9058	.9063	.9069	.9074	.9079
8.1	.9085	.9090	.9096	.9101	.9106	.9112	.9117	.9122	.9128	.9133
8.2	.9138	.9143	.9149	.9154	.9159	.9165	.9170	.9175	.9180	.9186
8.3	.9191	.9196	.9201	.9206	.9212	.9217	.9222	.9227	.9232	.9238
8.4	.9243	.9248	.9253	.9258	.9263	.9269	.9274	.9279	.9284	.9289
8.5	.9294	.9299	.9304	.9309	.9315	.9320	.9325	.9330	.9335	.9340
8.6	.9345	.9350	.9355	.9360	.9365	.9370	.9375	.9380	.9385	.9390
8.7	.9395	.9400	.9405	.9410	.9415	.9420	.9425	.9430	.9435	.9440
8.8	.9445	.9450	.9455	.9460	.9465	.9469	.9474	.9479	.9484	.9489
8.9	.9494	.9499	.9504	.9509	.9513	.9518	.9523	.9528	.9533	.9538
9.0	.9542	.9547	.9552	.9557	.9562	.9566	.9571	.9576	.9581	.9586
9.1	.9590	.9595	.9600	.9605	.9609	.9614	.9619	.9624	.9628	.9633
9.2	.9638	.9643	.9647	.9652	.9657	.9661	.9666	.9671	.9675	.9680
9.3	.9685	.9689	.9694	.9699	.9703	.9708	.9713	.9717	.9722	.9727
9.4	.9731	.9736	.9741	.9745	.9750	.9754	.9759	.9763	.9768	.9773
9.5	.9777	.9782	.9786	.9791	.9795	.9800	.9805	.9809	.9814	.9818
9.6	.9823	.9827	.9832	.9836	.9841	.9845	.9850	.9854	.9859	.9863
9.7	.9868	.9872	.9877	.9881	.9886	.9890	.9894	.9899	.9903	.9908
9.8	.9912	.9917	.9921	.9926	.9930	.9934	.9939	.9943	.9948	.9952
9.9	.9956	.9961	.9965	.9969	.9974	.9978	.9983	.9987	.9991	.9996
x	0	1	2	3	4	5	6	7	8	9

TABLE 3
Natural Logarithms

n	$\log_e n$	n	$\log_e n$	n	$\log_e n$
0.1	7.6974 *	4.5	1.5041	9.0	2.1972
0.2	8.3906	4.6	1.5261	9.1	2.2083
0.3	8.7960	4.7	1.5476	9.2	2.2192
0.4	9.0837	4.8	1.5686	9.3	2.2300
		4.9	1.5892	9.4	2.2407
0.5	9.3069	5.0	1.6094	9.5	2.2513
0.6	9.4892	5.1	1.6292	9.6	2.2618
0.7	9.6433	5.2	1.6487	9.7	2.2721
0.8	9.7769	5.3	1.6677	9.8	2.2824
0.9	9.8946	5.4	1.6864	9.9	2.2925
1.0	0.0000	5.5	1.7047	10	2.3026
1.1	0.0953	5.6	1.7228	11	2.3979
1.2	0.1823	5.7	1.7405	12	2.4849
1.3	0.2624	5.8	1.7579	13	2.5649
1.4	0.3365	5.9	1.7750	14	2.6391
1.5	0.4055	6.0	1.7918	15	2.7081
1.6	0.4700	6.1	1.8083	16	2.7726
1.7	0.5306	6.2	1.8245	17	2.8332
1.8	0.5878	6.3	1.8405	18	2.8904
1.9	0.6419	6.4	1.8563	19	2.9444
2.0	0.6931	6.5	1.8718	20	2.9957
2.1	0.7419	6.6	1.8871	25	3.2189
2.2	0.7885	6.7	1.9021	30	3.4012
2.3	0.8329	6.8	1.9169	35	3.5553
2.4	0.8755	6.9	1.9315	40	3.6889
2.5	0.9163	7.0	1.9459	45	3.8067
2.6	0.9555	7.1	1.9601	50	3.9120
2.7	0.9933	7.2	1.9741	55	4.0073
2.8	1.0296	7.3	1.9879	60	4.0943
2.9	1.0647	7.4	2.0015	65	4.1744
3.0	1.0986	7.5	2.0149	70	4.2485
3.1	1.1314	7.6	2.0281	75	4.3175
3.2	1.1632	7.7	2.0412	80	4.3820
3.3	1.1939	7.8	2.0541	85	4.4427
3.4	1.2238	7.9	2.0669	90	4.4998
3.5	1.2528	8.0	2.0794	100	4.6052
3.6	1.2809	8.1	2.0919	110	4.7005
3.7	1.3083	8.2	2.1041	120	4.7875
3.8	1.3350	8.3	2.1163	130	4.8676
3.9	1.3610	8.4	2.1282	140	4.9416
4.0	1.3863	8.5	2.1401	150	5.0106
4.1	1.4110	8.6	2.1518	160	5.0752
4.2	1.4351	8.7	2.1633	170	5.1358
4.3	1.4586	8.8	2.1748	180	5.1930
4.4	1.4816	8.9	2.1861	190	5.2470

*Subtract 10 for $n < 1$. Thus $\log_e 0.1 = 7.6974 - 10 = -2.3026$.

From MODERN COLLEGE ALGEBRA AND TRIGONOMETRY, Fourth Edition, by Edwin F. Beckenbach, Irving Drooyan, and Michael D. Grady. © 1981 by Wadsworth, Inc. Reprinted by permission of Wadsworth Publishing Company, Belmont, California 94002.

ANSWERS TO ODD-NUMBERED EXERCISES

CHAPTER 1

Exercises 1.1, page 11

1. False **3.** False

5. number line with marks at 0, 3, 6

7. number line with marks at -1, 0, 4

9. number line with mark at 0

11. $(-\infty, 2)$ **13.** $(-\infty, -5]$
15. $(-4, 6)$ **17.** $(-\infty, -3) \cup (3, \infty)$
19. $(-2, 3)$ **21.** 4 **23.** 2
25. $5\sqrt{3}$ **27.** $\pi + 1$ **29.** 2
31. False **33.** False **35.** True
37. True **39.** True **41.** False
43. 9 **45.** 1 **47.** 4
49. 7 **51.** $\frac{1}{5}$ **53.** 2
55. 2 **57.** 1 **59.** True
61. False **63.** False **65.** False

67. False **69.** $\dfrac{1}{(xy)^2}$ **71.** $\dfrac{1}{x^{5/6}}$

73. $\dfrac{1}{(s+t)^3}$ **75.** $x^{13/3}$ **77.** $\dfrac{1}{x^3}$

79. x **81.** $\dfrac{9}{x^2 y^4}$ **83.** $\dfrac{y^8}{x^{10}}$

85. $2x^{11/6}$ **87.** $-2xy^2$ **89.** $2x^{4/3}y^{1/2}$
91. 2.828 **93.** 5.196 **95.** 31.62

97. 316.2 **99.** $\dfrac{3\sqrt{x}}{2x}$ **101.** $\dfrac{2\sqrt{3y}}{3}$

103. $\dfrac{\sqrt[3]{x^2}}{x}$ **105.** $\dfrac{2x}{3\sqrt{x}}$ **107.** $\dfrac{2y}{\sqrt{2xy}}$

109. $\dfrac{xz}{y\sqrt[3]{xz^2}}$ **111.** $[362, 488.7]$

113. \$12,300 **115.** \$18,666.67
117. Between 1000 and 4000 units

Exercises 1.2, page 23

1. $9x^2 + 3x + 1$ **3.** $4y^2 + y + 8$

5. $-x - 1$ **7.** $\frac{2}{3} + e - e^{-1}$
9. $6\sqrt{2} + 8 + \frac{1}{2}\sqrt{x} - \frac{11}{4}\sqrt{y}$
11. $x^2 + 6x - 16$ **13.** $a^2 + 10a + 25$
15. $x^2 + 4xy + 4y^2$ **17.** $4x^2 - y^2$
19. $-2x$ **21.** $2t(2\sqrt{t} + 1)$
23. $2x^3(2x^2 - 6x - 3)$ **25.** $7a^2(a^2 + 7ab - 6b^2)$
27. $e^{-x}(1 - x)$ **29.** $\frac{1}{2}x^{-5/2}(4 - 3x)$
31. $(2a + b)(3c - 2d)$ **33.** $(2a + b)(2a - b)$
35. $-2(3x + 5)(2x - 1)$ **37.** $3(x - 4)(x + 2)$
39. $2(3x - 5)(2x + 3)$ **41.** $(3x - 4y)(3x + 4y)$
43. $(x^2 + 5)(x^4 - 5x^2 + 25)$
45. $x^3 - xy^2$
47. $4(x - 1)(3x - 1)(2x + 2)^3$
49. $4(x - 1)(3x - 1)(2x + 2)^3$
51. $2x(x^2 + 2)^2(5x^4 + 20x^2 + 17)$
53. -4 and 3 **55.** -1 and $\frac{1}{2}$
57. 2 and 2 **59.** -2 and $\frac{3}{4}$
61. $\frac{1}{2} + \frac{1}{4}\sqrt{10}$ and $\frac{1}{2} - \frac{1}{4}\sqrt{10}$
63. $-1 + \frac{1}{2}\sqrt{10}$ and $-1 - \frac{1}{2}\sqrt{10}$

65. $\dfrac{x - 1}{x - 2}$ **67.** $\dfrac{3(2t + 1)}{2t - 1}$

69. $-\dfrac{7}{(4x - 1)^2}$ **71.** -8

73. $\dfrac{3x - 1}{2}$ **75.** $\dfrac{t + 20}{3t + 2}$

77. $-\dfrac{x(2x - 13)}{(2x - 1)(2x + 5)}$ **79.** $-\dfrac{x + 27}{(x - 3)^2(x + 3)}$

81. $\dfrac{x + 1}{x - 1}$ **83.** $\dfrac{4x^2 + 7}{\sqrt{2x^2 + 7}}$

85. $\dfrac{x - 1}{x^2\sqrt{x + 1}}$ **87.** $\dfrac{x - 1}{(2x + 1)^{3/2}}$

89. $\dfrac{\sqrt{3} + 1}{2}$ **91.** $\dfrac{\sqrt{x} + \sqrt{y}}{x - y}$

93. $\dfrac{(\sqrt{a} + \sqrt{b})^2}{a - b}$ **95.** $\dfrac{x}{3\sqrt{x}}$

97. $-\dfrac{2}{3(1 + \sqrt{3})}$

99. $-\dfrac{x + 1}{\sqrt{x + 2}\,(1 - \sqrt{x + 2})}$

Exercises 1.3, page 28

1. (3, 3); Quadrant I   **3.** (2, −2); Quadrant IV
5. (−4, −6); Quadrant III   **7.** *A*
9. *E*, *F*, and *G*   **11.** *F*
13.–19. See accompanying figure.

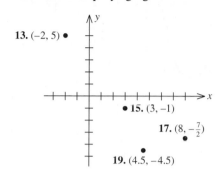

13. (−2, 5) •

• **15.** (3, −1)

17. $(8, -\frac{7}{2})$

19. (4.5, −4.5)

21. 5   **23.** $\sqrt{61}$
25. (−8, −6) and (8, −6)   **29.** 3400 miles
31. Route 1   **33.** Model C

Exercises 1.4, page 40

1. $\frac{1}{2}$ **3.** Not defined
5. 5   **7.** 5/6
9. $\dfrac{d - b}{c - a}\,(a \neq c)$ **11. a.** 4   **b.** −8
13. Parallel   **15.** Perpendicular
17. −5   **19.** $y = -3$
21. $y = 2x - 10$ **23.** $y = 2$
25. $y = 3x - 2$ **27.** $y = x + 1$
29. $y = 3x + 4$ **31.** $y = 5$
33. $y = \frac{1}{2}x;\ m = \frac{1}{2};\ b = 0$
35. $y = \frac{2}{3}x - 3;\ m = \frac{2}{3};\ b = -3$
37. $y = -\frac{1}{2}x + \frac{7}{2};\ m = -\frac{1}{2};\ b = \frac{7}{2}$
39. $y = \frac{1}{2}x + 3$ **41.** $y = -6$
43. $y = b$ **45.** $y = \frac{2}{3}x - \frac{2}{3}$
47. $k = 8$
49.

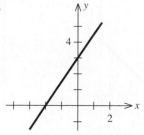

51.

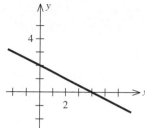

53.

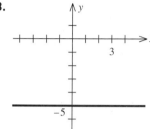

57. $y = -2x - 4$ **59.** $y = \frac{1}{8}x - \frac{1}{2}$
61. Yes
63. a. $y = 0.55x$ **b.** 2000
65. a. and **b.**

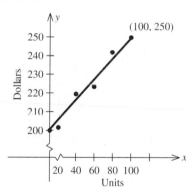

c. $y = \frac{1}{2}x + 200$ **d.** \$227
67. a. and **b.**

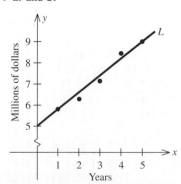

c. $y = 0.8x + 5$ **d.** \$12.2 million

Chapter 1 Review Exercises, page 45

1. $[-2, \infty)$ **3.** $(-\infty, -4) \cup (5, \infty)$

5. 4   **7.** $\pi - 6$ **9.** $\frac{27}{8}$

11. $\frac{1}{144}$ **13.** $\frac{1}{4}$ **15.** $4(x^2 + y)^2$

17. $\dfrac{2x}{3z}$ **19.** $6xy^7$

21. $2vw(v^2 + w^2 + u^2)$ **23.** $6t(t + 1)(2t - 3)$

25. -2 and $\frac{1}{3}$ **27.** $\dfrac{\sqrt{2}}{2}$ and $-\dfrac{\sqrt{2}}{2}$

29. $-2 + \dfrac{\sqrt{2}}{2}$ and $-2 - \dfrac{\sqrt{2}}{2}$

31. $\dfrac{15x^2 + 24x + 2}{4(x + 2)(3x^2 + 2)}$ **33.** $\dfrac{2(x + 2)}{\sqrt{x + 1}}$

35. $\dfrac{x - \sqrt{x}}{2x}$ **37.** 2

39. $y = 4$ **41.** $y = -\frac{4}{5}x + \frac{12}{5}$

43. $y = \frac{3}{4}x + \frac{11}{2}$ **45.** $y = -\frac{3}{5}x + \frac{12}{5}$

47.
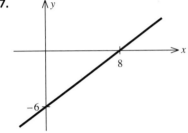

49. $100

c.
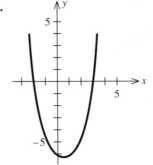

33.
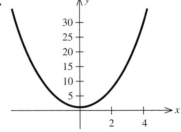

35.
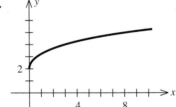

37.
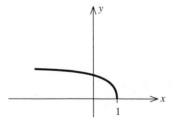

39.
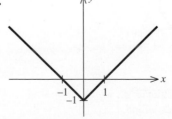

CHAPTER 2

Exercises 2.1, page 57

1. $21, -9, 5a + 6, -5a + 6, 5a + 21$

3. $-3, 6, 3a^2 - 6a - 3, 3a^2 + 6a - 3, 3x^2 - 6$

5. $\dfrac{8}{15}, 0, \dfrac{2a}{a^2 - 1}, \dfrac{2(2 + a)}{a^2 + 4a + 3}, \dfrac{2(t + 1)}{t(t + 2)}$

7. $8, \dfrac{2a^2}{\sqrt{a - 1}}, \dfrac{2(x + 1)^2}{\sqrt{x}}, \dfrac{2(x - 1)^2}{\sqrt{x - 2}}$

9. $5, 1, 1$ **11.** $\frac{5}{2}, 3, 3, 9$

13. Yes   **15.** Yes

17. $(-\infty, \infty)$ **19.** $(-\infty, 0) \cup (0, \infty)$

21. $(-\infty, \infty)$ **23.** $(-\infty, 5]$

25. $(-\infty, -1) \cup (-1, 1) \cup (1, \infty)$

27. $[-3, \infty)$ **29.** $(-\infty, -2) \cup (-2, 1]$

31. a. $(-\infty, \infty)$

    b. $6, 0, -4, -6, -\frac{25}{4}, -6, -4, 0$

41.

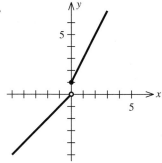

43.

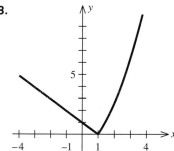

45. Yes   **47.** No   **49.** Yes
51. Yes   **53.** 10π in.
55. $6 billion; $43.5 billion; $81 billion
57. $S(r) = 4\pi r^2$
59. a. $I(x) = 1.053x$ **b.** $652.86

61. a. $V = -12,000n + 120,000$
b.

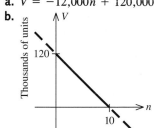

c. $48,000
d. $12,000

63. The domain of P is $(0, \infty)$.

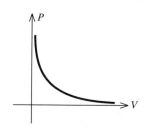

65. 990; 2240
67. 37.5 words per minute; 42 words per minute
69. $0.1 billion; $4.5 billion

Exercises 2.2, page 67

1. $f(x) + g(x) = x^3 + x^2 + 3$
3. $f(x)g(x) = x^5 - 2x^3 + 5x^2 - 10$
5. $\dfrac{f(x)}{g(x)} = \dfrac{x^3 + 5}{x^2 - 2}$
7. $\dfrac{f(x)g(x)}{h(x)} = \dfrac{x^5 - 2x^3 + 5x^2 - 10}{2x + 4}$
9. $f(x) + g(x) = x - 1 + \sqrt{x + 1}$
11. $f(x)g(x) = (x - 1)\sqrt{x + 1}$
13. $\dfrac{g(x)}{h(x)} = \dfrac{\sqrt{x + 1}}{2x^3 - 1}$
15. $\dfrac{f(x)g(x)}{h(x)} = \dfrac{(x - 1)\sqrt{x + 1}}{2x^3 - 1}$
17. $\dfrac{f(x) - h(x)}{g(x)} = \dfrac{x - 2x^3}{\sqrt{x + 1}}$
19. $f(x) + g(x) = x^2 + \sqrt{x} + 3$;
$f(x) - g(x) = x^2 - \sqrt{x} + 7$;
$f(x)g(x) = (x^2 + 5)(\sqrt{x} - 2)$;
$\dfrac{f(x)}{g(x)} = \dfrac{x^2 + 5}{\sqrt{x} - 2}$
21. $f(x) + g(x) = \dfrac{(x - 1)\sqrt{x + 3} + 1}{x - 1}$;
$f(x) - g(x) = \dfrac{(x - 1)\sqrt{x + 3} - 1}{x - 1}$;
$f(x)g(x) = \dfrac{\sqrt{x + 3}}{x - 1}$;
$\dfrac{f(x)}{g(x)} = (x - 1)\sqrt{x + 3}$
23. $f(x) + g(x) = \dfrac{2(x^2 - 2)}{(x - 1)(x - 2)}$;
$f(x) - g(x) = \dfrac{-2x}{(x - 1)(x - 2)}$;
$f(x)g(x) = \dfrac{(x + 1)(x + 2)}{(x - 1)(x - 2)}$;
$\dfrac{f(x)}{g(x)} = \dfrac{(x + 1)(x - 2)}{(x - 1)(x + 2)}$
25. $f(g(x)) = x^4 + x^2 + 1$;
$g(f(x)) = (x^2 + x + 1)^2$
27. $f(g(x)) = \sqrt{x^2 - 1} + 1$; $g(f(x)) = x + 2\sqrt{x}$
29. $f(g(x)) = \dfrac{x}{x^2 + 1}$; $g(f(x)) = \dfrac{x^2 + 1}{x}$
31. 49   **33.** $\dfrac{\sqrt{5}}{5}$
35. $f(x) = 2x^3 + x^2 + 1$ and $g(x) = x^5$
37. $f(x) = x^2 - 1$ and $g(x) = \sqrt{x}$
39. $f(x) = x^2 - 1$ and $g(x) = \dfrac{1}{x}$
41. $f(x) = 3x^2 + 2$ and $g(x) = \dfrac{1}{x^{3/2}}$
43. $3h$ **45.** $-h(2a + h)$ **47.** $2a + h$
49. $C(x) = 0.6x + 12,100$
51. a. $P(x) = -0.000003x^3 - 0.07x^2 + 300x - 100,000$
b. $182,375

53. a. $N(t) = \dfrac{7}{1 + 0.02\left[\dfrac{10t + 150}{t + 10}\right]^2}$

b. 1.27 million units; 1.74 million units; 1.85 million units

55. $N(x(t)) = 9.94\left[\dfrac{(t + 10)^2}{(t + 10)^2 + 2(t + 15)^2}\right]$

2.24 million jobs; 2.48 million jobs

Exercises 2.3, page 76

1. Polynomial function, degree 6

3. Polynomial function, degree 6

5. Some other function

7. $28,800   **9.** 104 mg   **11.** $400,000

13. a. $R(x) = \dfrac{100x}{40 + x}$ **b.** 60%

15. $72,000

17. $\dfrac{110}{\frac{1}{2}t + 1} - 26\left(\frac{1}{4}t^2 - 1\right)^2 - 52$; $32, $6.71, $3; the gap was closing

19. a.

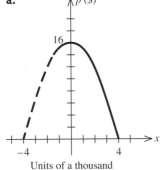

Units of a thousand

b. 3000 units

21. a.

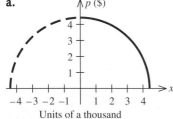

Units of a thousand

b. 3000

23. $p = \sqrt{-x^2 + 100}$; 6614 units

25. a.

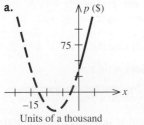

Units of a thousand

b. $76

27. a.

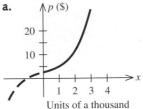

Units of a thousand

b. $15

29. $p = \frac{1}{10}\sqrt{x} + 10$; $30

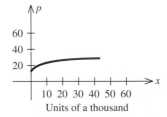

Units of a thousand

31. Equilibrium quantity is 5000; equilibrium price is $65

33. 2000 units; $52   **35.** 8000 units; $80

Exercises 2.4, page 92

1. $\lim\limits_{x \to -2} f(x) = 3$ **3.** $\lim\limits_{x \to 3} f(x) = 3$

5. $\lim\limits_{x \to -2} f(x) = 3$

7. The limit does not exist.

9.

x	1.9	1.99	1.999
$f(x)$	4.61	4.9601	4.9960

x	2.001	2.01	2.1
$f(x)$	5.004	5.0401	5.41

$\lim\limits_{x \to 2} (x^2 + 1) = 5$

11.

x	−0.1	−0.01	−0.001
$f(x)$	−1	−1	−1

x	0.001	0.01	0.1
$f(x)$	1	1	1

The limit does not exist.

13.

x	0.9	0.99	0.999
f(x)	100	10,000	1,000,000

x	1.001	1.01	1.1
f(x)	1,000,000	10,000	100

The limit does not exist.

15.

x	0.9	0.99	0.999	1.001	1.01	1.1
f(x)	2.9	2.99	2.999	3.001	3.01	3.1

$$\lim_{x \to 1} \frac{x^2 + x - 2}{x - 1} = 3$$

17.

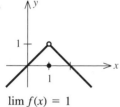

$$\lim_{x \to 0} f(x) = -1$$

19.

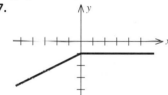

$$\lim_{x \to 1} f(x) = 1$$

21.

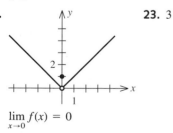

23. 3

$$\lim_{x \to 0} f(x) = 0$$

25. 3   **27.** −1   **29.** 2
31. −4   **33.** $\frac{5}{4}$ **35.** 2
37. $\sqrt{171} = 3\sqrt{19}$
39. $\frac{3}{2}$ **41.** −1   **43.** −6
45. 2   **47.** $\frac{1}{6}$ **49.** 2
51. −1   **53.** −10
55. The limit does not exist.

57. $\frac{5}{3}$ **59.** $\frac{1}{2}$ **61.** $\frac{1}{3}$
63. $\lim_{x \to \infty} f(x) = \infty$; $\lim_{x \to -\infty} f(x) = \infty$
65. 0; 0
67. $\lim_{x \to \infty} f(x) = -\infty$; $\lim_{x \to -\infty} f(x) = -\infty$

69.

x	1	10	100	1000
f(x)	0.5	0.009901	0.0001	0.000001

x	−1	−10	−100	−1000
f(x)	0.5	0.009901	0.0001	0.000001

$\lim_{x \to \infty} f(x) = 0$ and $\lim_{x \to -\infty} f(x) = 0$

71.

x	1	5	10	100
f(x)	12	360	2910	2.99×10^6

x	1000	−1	−5
f(x)	2.999×10^9	6	−390

x	−10	−100	−1000
f(x)	−3090	-3.01×10^6	-3.0×10^9

$\lim_{x \to \infty} f(x) = \infty$ and $\lim_{x \to -\infty} f(x) = -\infty$

73. 3   **75.** 3
77. $\lim_{x \to \infty} f(x) = -\infty$ **79.** 0
81. **a.** $0.5 million; $0.75 million; $1,166,667; $2 million;
     $4.5 million; $9.5 million
   b. The limit does not exist; as the percentage of pol-
     lutant to be removed approaches 100, the cost
     becomes astronomical.
83. $\lim_{t \to \infty} C(t) = 0$; in the long run, the concentration of drug
   in the bloodstream decreases to zero.
85. **a.** 5000   **b.** 25,000
87. $\lim_{x \to \infty} V(x) = a$; as the amount of substrate increases indefinitely, the initial speed approaches the constant a moles
   per liter per second.

89. $\lim\limits_{x \to -1} f(x) = -\frac{1}{2}$

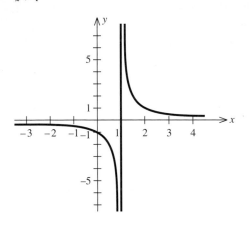

Exercises 2.5, page 103

1. 3; 2; the limit does not exist.
3. The limit does not exist; 2; the limit does not exist.
5. 0; 2; the limit does not exist.
7. -2; 2; the limit does not exist.
9. 6   **11.** $-\frac{1}{4}$
13. The limit does not exist.
15. -1 **17.** 0   **19.** -4
21. The limit does not exist.
23. 4   **25.** 0
27. 0; 0   **29.** 2; 3
31. $x = 0$; conditions 2 and 3
33. Continuous everywhere
35. $x = 0$; condition 3   **37.** $x = 0$; condition 3
39. $(-\infty, \infty)$ **41.** $(-\infty, \infty)$
43. $\left(-\infty, \frac{1}{2}\right) \cup \left(\frac{1}{2}, \infty\right)$
45. $(-\infty, -2) \cup (-2, 1) \cup (1, \infty)$
47. $(-\infty, \infty)$ **49.** $(-\infty, \infty)$ **51.** $(-\infty, \infty)$
53. $(-\infty, \infty)$ **55.** -1 and 1   **57.** 1 and 2
59. f is discontinuous at $x = 1, 2, \ldots, 11$
61. Michael makes progress toward solving the problem until $x = x_1$. Between $x = x_1$ and $x = x_2$, he makes no further progress. But at $x = x_2$ he suddenly achieves a breakthrough, and at $x = x_3$ he proceeds to complete the problem.
63. Conditions 2 and 3 are not satisfied at each of these points.

65.

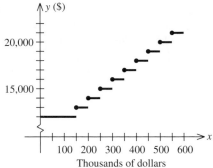

f is discontinuous at $x = 150{,}000, 200{,}000, 250{,}000, \ldots$

67.

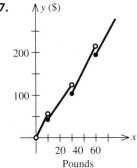

C is discontinuous at $x = 0, 10, 30, 60$
69. a. f is a polynomial of degree 2
     b. $f(1) = 3$ and $f(3) = -1$
71. a. f is a polynomial of degree 3
     b. $f(0) = 14$ and $f(1) = -23$
73. c. $\pm \dfrac{\sqrt{2}}{2}$

77.

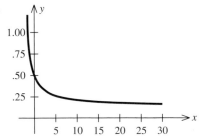

$\lim\limits_{x \to 0^+} f(x) = \frac{1}{2}$; $\lim\limits_{x \to 0^-} f(x) = \frac{1}{2}$; $\lim\limits_{x \to 0} f(x) = \frac{1}{2}$

Exercises 2.6, page 120

1. 1.5; 0.5833; 1.25   **3.** 3.075; -21.15
5. 0
7. 2   **9.** $6x$
11. $-2x + 3$ **13.** 2; $y = 2x + 7$
15. 6; $y = 6x - 3$ **17.** $\frac{1}{9}$; $y = \frac{1}{9}x - \frac{2}{3}$

19. a. $4x$ **b.** $y = 4x - 1$
c.

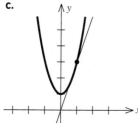

21. a. $2x - 2$ **b.** $(1, 0)$
c.

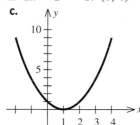

d. 0
23. a. $6; 5.5; 5.1$ **b.** 5
c. The computations in (a) show that as h approaches zero, the average velocity approaches the instantaneous velocity.
25. a. 130 ft/sec; 128.2 ft/sec; 128.02 ft/sec
b. 128 ft/sec
c. The computations in (a) show that as the time intervals over which the average velocity are computed become smaller and smaller, the average velocity approaches the instantaneous velocity of the car at $t = 20$.
27. a. $C'(x) = -20x + 300$
b. $100/surfboard   **c.** $200
29. a. $-$2.01/1000$ tents; $-$2/1000$ tents
b. $-$2/1000$ tents   **31.** 62 bacteria/min
33. Average rate of change of the prime rate over the time interval $[a, a + h]$; instantaneous rate of change of the prime rate at time a
35. Average rate of change of the total cost of production when the level is between a and $a + h$ units; instantaneous rate of change of the total cost of production when the level of production is a units
37. a. Yes   **b.** Yes   **c.** Yes
39. a. No   **b.** No   **c.** No
41. a. No   **b.** No   **c.** No
43.

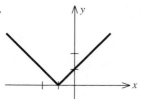

45.

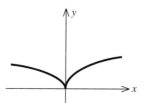

Yes; no; the graph of f has a kink at $x = 0$
49. $5.06060, 5.06006, 5.060006, 5.0600006, 5.06000006$; 5.06

Chapter 2 Review Exercises, page 125

1. a. $(-\infty, 9]$
b. $(-\infty, -1) \cup \left(-1, \frac{3}{2}\right) \cup \left(\frac{3}{2}, \infty\right)$

3. a.   **b.** No   **c.** Yes

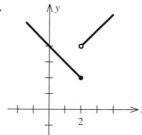

5. a. $\dfrac{2x + 3}{x}$ **b.** $\dfrac{1}{x(2x + 3)}$
c. $\dfrac{1}{2x + 3}$ **d.** $\dfrac{2}{x} + 3$
7. 2 **9.** 0
11. The limit does not exist.
13. $\frac{9}{2}$ **15.** $\frac{1}{2}$ **17.** 1
19. The limit does not exist.
21.

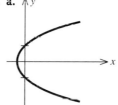

$4; 2;$ the limit does not exist.
23. $x = -\frac{1}{2}, 1$ **25.** $x = 0$
27. 3 **29.** $\frac{3}{2}; y = \frac{3}{2}x + 5$
31. a. Yes   **b.** No
33. a. $S(t) = t + 2.4$ **b.** 5.4 million
35. $\left(6, \frac{21}{2}\right)$ **37.** $6000; $22   **39.** $45,000

41.

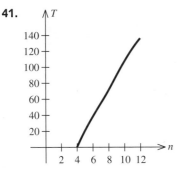

As the length of the list increases, the time taken to learn the list increases by a very large amount.

43. $C(x) = \begin{cases} 5 & \text{if} & 1 \le x \le 100 \\ 9 & \text{if} & 100 < x \le 200 \\ 12.50 & \text{if} & 200 < x \le 300 \\ 15.00 & \text{if} & 300 < x \le 400 \\ 7 + 0.02x & \text{if} & x > 400 \end{cases}$

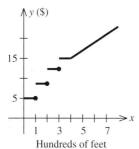

Hundreds of feet

The function is discontinuous at $x = 100$, 200, and 300.

CHAPTER 3

Exercises 3.1, page 136

1. 0   **3.** $5x^4$ **5.** $2.1x^{1.1}$

7. $6x$ **9.** $2\pi r$ **11.** $\dfrac{3}{x^{2/3}}$

13. $\dfrac{3}{2\sqrt{x}}$ **15.** $-84x^{-13}$ **17.** $10x - 3$

19. $-3x^2 + 4x$ **21.** $0.06x - 0.4$

23. $2x - 4 - \dfrac{3}{x^2}$ **25.** $16x^3 - 7.5x^{3/2}$

27. $-\dfrac{3}{x^2} - \dfrac{8}{x^3}$ **29.** $-\dfrac{16}{t^5} + \dfrac{9}{t^4} - \dfrac{2}{t^2}$

31. $2 - \dfrac{5}{2\sqrt{x}}$ **33.** $-\dfrac{4}{x^3} + \dfrac{1}{x^{4/3}}$

35. a. 20   **b.** -4 **c.** 20
37. $m = 5; y = 5x - 4$ **39.** $m = -2; y = -2x + 2$

41. a. $(0, 0)$ **b.**

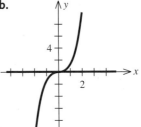

43. a $(-2, -7), (2, 9)$
   b. $y = 12x + 17$ and $y = 12x - 15$
   c.

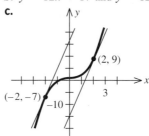

45. a. $(0, 0); \left(1, -\frac{13}{12}\right)$
   b. $(0, 0); \left(2, -\frac{8}{3}\right); \left(-1, -\frac{5}{12}\right)$
   c. $(0, 0); \left(4, \frac{80}{3}\right); \left(-3, \frac{81}{4}\right)$
47. a. 30 cm/sec
   b. -200 cm/sec; the velocity is decreasing at the rate of 200 cm/sec.
49. a. 15 pts/yr; 12.6 pts/yr; 0 pts/yr   **b.** 10 pts/yr
51. 155/month; 200/month
53. 32 turtles/yr; 428 turtles/hr; 3260 turtles
55. a. $120 - 30t$ **b.** 120 ft/sec   **c.** 240 ft
57. a. $\dfrac{1}{20\sqrt{x}}$ **b.** 0.025 cents/radio
59. a. $0.08508t^2 - 0.10334t + 9.60881$
   b. \$27.20 billion/yr; \$110.22 billion/yr
   c. \$270.12 billion/yr; \$1530.85 billion/yr

61. $x \approx 1.5; y \approx 21$

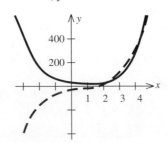

63. $x \approx 4$; $y \approx 1/4$

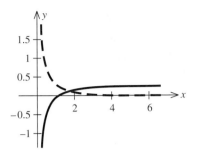

55. $(1.7, -10.4)$; $(-1.7, 10.4)$

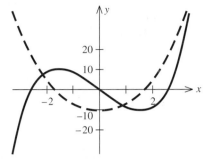

57. $\left(-1, -\frac{3}{2}\right)$; $\left(1, \frac{3}{2}\right)$

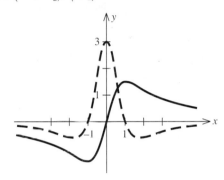

Exercises 3.2, page 146

1. $6x^2 + 2$ **3.** $4t - 1$

5. $9x^2 + 2x - 6$ **7.** $4x^3 + 3x^2 - 1$

9. $5w^4 - 4w^3 + 9w^2 - 6w + 2$

11. $\dfrac{25x^2 - 10x^{3/2} + 1}{x^{1/2}}$ **13.** $\dfrac{3x^4 - 10x^3 + 4}{x^2}$

15. $\dfrac{-1}{(x - 2)^2}$ **17.** $\dfrac{3}{(2x + 1)^2}$

19. $-\dfrac{2x}{(x^2 + 1)^2}$ **21.** $\dfrac{s^2 + 2s + 4}{(s + 1)^2}$

23. $\dfrac{1 - 3x^2}{2\sqrt{x}(x^2 + 1)^2}$ **25.** $\dfrac{x^2 - 2x - 2}{(x^2 + x + 1)^2}$

27. $\dfrac{2x^3 - 5x^2 - 4x - 3}{(x - 2)^2}$

29. $\dfrac{-2(x^5 + 12x^4 - 8x^3 + 16x + 64)}{(x^4 - 16)^2}$

31. $2(3x^2 - x + 3)$; 10

33. $\dfrac{-3x^4 + 2x^2 - 1}{(x^4 - 2x^2 - 1)^2}$; $-\dfrac{1}{2}$

35. 60; $y = 60x - 102$ **37.** $-\frac{1}{2}$; $y = -\frac{1}{2}x + \frac{3}{2}$

39. $y = 7x - 5$ **41.** $\left(\frac{1}{3}, \frac{50}{27}\right)$; $(1, 2)$

43. $\left(\frac{4}{3}, -\frac{770}{27}\right)$; $(2, -30)$

45. a. $\dfrac{0.2(1 - t^2)}{(t^2 + 1)^2}$

  b. 0.096 percent/hr; 0 percent/hr; -0.024 percent/hr

47. $\dfrac{6000}{(t + 12)^2}$; 18.5 mg/yr; 12.4 mg/yr

49. a. $-\dfrac{x}{(0.01x^2 + 1)^2}$ **b.** -3.2; -2.5; -1.4

51. $\$38.4$ million/yr; $\$17.04$ million/yr; $\$5.71$ million/yr

Exercises 3.3, page 155

1. $8(2x - 1)^3$ **3.** $10x(x^2 + 2)^4$

5. $6x^2(1 - x)(2 - x)^2$ **7.** $\dfrac{-4}{(2x + 1)^3}$

9. $3x\sqrt{x^2 - 4}$ **11.** $\dfrac{3}{2\sqrt{3x - 2}}$

13. $\dfrac{-2x}{3(1 - x^2)^{2/3}}$ **15.** $-\dfrac{6}{(2x + 3)^4}$

17. $\dfrac{-1}{(2t - 3)^{3/2}}$ **19.** $-\dfrac{3(16x^3 + 1)}{2(4x^4 + x)^{5/2}}$

21. $\dfrac{-4(3x + 1)}{(3x^2 + 2x + 1)^3}$ **23.** $6x(2x^2 - x + 1)$

25. $3(t^{-1} - t^{-2})^2(-t^{-2} + 2t^{-3})$

27. $\dfrac{1}{2\sqrt{x - 1}} + \dfrac{1}{2\sqrt{x + 1}}$

29. $-12x(4x - 1)(3 - 4x)^3$

31. $6(x - 1)(2x - 1)(2x + 1)^3$

33. $-\dfrac{15(x + 3)^2}{(x - 2)^4}$ **35.** $\dfrac{3\sqrt{t}}{2(2t + 1)^{5/2}}$

37. $\dfrac{-1}{2\sqrt{u + 1}\,(3u + 2)^{3/2}}$ **39.** $-\dfrac{2x(3x^2 + 1)}{(x^2 - 1)^5}$

41. $-\dfrac{2x(3x^2 + 1)^2(3x^2 + 13)}{(x^2 - 1)^5}$

43. $-\dfrac{3x^2 + 2x + 1}{\sqrt{2x + 1}\,(x^2 - 1)^2}$ **45.** $-\dfrac{t^2 + 2t - 1}{2\sqrt{t + 1}\,(t^2 + 1)^{3/2}}$

47. $\frac{4}{3}u^{1/3}$; $6x$; $8x(3x^2 - 1)^{1/3}$

49. $-\dfrac{2(6x^2 - 1)}{3(2x^3 - x + 1)^{5/3}}$

51. $\dfrac{(3x^2 - 1)(x^3 - x - 1)}{2(x^3 - x)^{3/2}}$

53. $y = -33x + 57$ **55.** $y = \frac{43}{5}x - \frac{54}{5}$

57. 0.333 million/wk; 0.305 million/wk; 16 million; 22.7 million

59. a. $0.027(0.2t^2 + 4t + 64)^{-1/3}(0.1t + 1)$
    b. 0.0091 parts per million

61. a. $0.21t^2(t - 3)(t - 7)^3$
    b. 90.72; 0; −90.72; at 8 A.M. the level of nitrogen dioxide is increasing, at 10 A.M. the level stops increasing, and at 11 A.M. the level is decreasing.

63. $\dfrac{3450t}{(t + 25)^2\sqrt{\frac{1}{2}t^2 + 2t + 25}}$; 2.9 beats/min; 0.7 beats/min; 0.2 beats/min; 179 beats/min

65. 160π ft²/sec

67. a. $\dfrac{dr}{dt} = \dfrac{10}{27}t^2 - \dfrac{20}{3}t + \dfrac{200}{9}$
    b. $\dfrac{dR}{dr} = -\dfrac{9}{5000}r^2 + \dfrac{9}{25}r$
    c. \$336,000/month; −\$80,000/month

69. 18.7 units/month

73. (0.7, 0.5); (−0.7, 0.5)

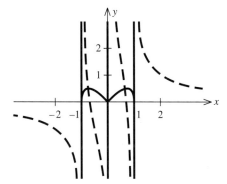

75. (−0.7, −1.4)

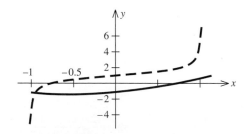

Exercises 3.4, page 169

1. a. \$1.80/record; \$1.60/record
    b. \$1.80/record; \$1.60/record

3. a. $100 + \dfrac{200,000}{x}$ **b.** $-\dfrac{200,000}{x^2}$
    c. $\lim\limits_{x \to \infty}\left(100 + \dfrac{200,000}{x}\right) = 100$

5. $\dfrac{2000}{x} + 2 - 0.0001x$; $-\dfrac{2000}{x^2} - 0.0001$

7. a. $8000 - 200x$
    b. 200; 0; −200; revenue seems to be maximized when the fare is approximately \$40/passenger

9. a. \$750
    b. \$760; the marginal profit approximates the actual profit

11. a. $600x - 0.05x^2$; $-0.000002x^3 - 0.02x^2 + 200x - 80,000$
    b. $0.000006x^2 - 0.06x + 400$; $600 - 0.1x$; $-0.000006x^2 - 0.04x + 200$
    c. 304; 400; 96
    d.

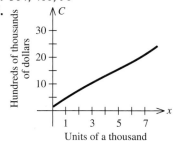

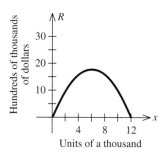

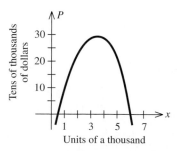

13. $0.000002x^2 - 0.03x + 400 + \dfrac{80,000}{x}$
    a. $0.000004x - 0.03 - \dfrac{80,000}{x^2}$
    b. −0.0132; 0.0092; the marginal average cost is negative (average cost is decreasing) when 5000 units

are produced and positive (average cost is increasing)
when 10,000 units are produced

c.

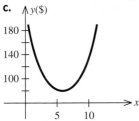

y($)

180

140

100

5   10

Units of a thousand

15. a. $\dfrac{50x}{0.01x^2 + 1}$ **b.** $\dfrac{50 - 0.5x^2}{(0.01x^2 + 1)^2}$

  c. \$44,380; when the level of production is 2000 units,
  the revenue increases at the rate of \$44,380 per additional 1000 units produced.

17. \$1.21 billion per billion dollars

19. \$0.288 billion per billion dollars

21. Inelastic   **23.** Unitary   **25.** Unitary

27. a. 0.39; 1; 1.5   **b.** Inelastic; unitary; elastic

29. a. Inelastic; elastic   **b.** $p = 53\frac{1}{3}$ **c.** Increase

  d. Increase

31. $E = \dfrac{2p^2}{9 - p^2}$; inelastic in $[0, \sqrt{3})$, unitary when $p = \sqrt{3}$, and elastic in $(\sqrt{3}, 3]$

33. (10, 250)

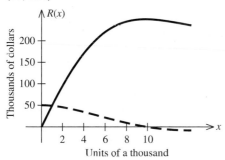

R(x)

Thousands of dollars

200

150

100

50

2   4   6   8   10

Units of a thousand

The revenue seems to be maximized when 10,000
watches are sold.

Exercises 3.5, page 177

1. $8x - 2$; 8   **3.** $6x^2 - 6x$; $6(2x - 1)$

5. $4t^3 - 6t^2 + 12t - 3$; $12(t^2 - t + 1)$

7. $10x(x^2 + 2)^4$; $10(x^2 + 2)^3(9x^2 + 2)$

9. $6t(2t^2 - 1)(6t^2 - 1)$; $6(60t^4 - 24t^2 + 1)$

11. $14x(2x^2 + 2)^{5/2}$; $28(2x^2 + 2)^{3/2}(6x + 1)$

13. $(x^2 + 1)(5x^2 + 1)$; $4x(5x^2 + 3)$

15. $\dfrac{1}{(2x + 1)^2}$; $-\dfrac{4}{(2x + 1)^3}$

17. $\dfrac{2}{(s + 1)^2}$; $-\dfrac{4}{(s + 1)^3}$

19. $-\dfrac{3}{2(4 - 3u)^{1/2}}$; $-\dfrac{9}{4(4 - 3u)^{3/2}}$

21. $72x - 24$ **23.** $-\dfrac{6}{x^4}$

25. $\frac{81}{8}(3s - 2)^{-5/2}$ **27.** $192(2x - 3)$

29. 128 ft/sec; 32 ft/sec²

31. a. and b.

t	0	1	2	3
$N'(t)$	0	2.7	4.8	6.3
$N''(t)$				

t	4	5	6	7
$N'(t)$	7.2	7.5	7.2	6.3
$N''(t)$	0.6	0	-0.6	-1.2

33. a. $\frac{1}{4}t(t^2 - 12t + 32)$

  b. 0 ft/sec; 0 ft/sec; 0 ft/sec

  c. $\frac{3}{4}t^2 - 6t + 8$

  d. 8 ft/sec²; −4 ft/sec²; 8 ft/sec²

  e. 0 ft; 16 ft; 0 ft

35. −3.09; 0.35; ten minutes after the test began, the smoke
  was being removed at the rate of 3 percent/min and the
  rate of purification was also increasing at the rate of
  0.35/sec².

37. $f(x) = x^{n+1/2}$

Exercises 3.6, page 187

1. a. $-\frac{1}{2}$ **b.** $-\frac{1}{2}$ **3. a.** $-\dfrac{1}{x^2}$ **b.** $-\dfrac{y}{x}$

5. a. $2x - 1 + \dfrac{4}{x^2}$ **b.** $3x - 2 - \dfrac{y}{x}$

7. a. $\dfrac{1 - x^2}{(1 + x^2)^2}$ **b.** $-2y^2 + \dfrac{y}{x}$

9. $-\dfrac{x}{y}$ **11.** $\dfrac{x}{2y}$ **13.** $1 - \dfrac{y}{x}$

15. $-\dfrac{y}{x}$ **17.** $-\dfrac{\sqrt{y}}{\sqrt{x}}$

19. $2\sqrt{x} + y - 1$ **21.** $-\dfrac{y^3}{x^3}$

23. $\dfrac{2\sqrt{xy} - y}{x - 2\sqrt{xy}}$ **25.** $\dfrac{6x - 3y - 1}{3x + 1}$

27. $\dfrac{2(2x - y^{3/2})}{3x\sqrt{y} - 4y}$ **29.** $-\dfrac{2x^2 + 2xy + y^2}{x^2 + 2xy + 2y^2}$

31. $y = 2$ **33.** $y = -\frac{3}{2}x + \frac{5}{2}$

35. $\dfrac{2y}{x^2}$ **37.** $\dfrac{2y(y - x)}{(2y - x)^3}$

39. Dropping at the rate of 111 tires/wk

41. Dropping at the rate of 44 ten-packs/week
43. Dropping at the rate of 3.7 cents/carton/week
45. Inelastic   **47.** 160π ft^2/sec
49. 17 ft/sec   **51.** 7.69 ft/sec
53. Sliding down at the rate of 1.5 ft/sec

Exercises 3.7, page 195

1. $4x\,dx$ **3.** $(3x^2 - 1)\,dx$ **5.** $\dfrac{dx}{2\sqrt{x+1}}$

7. $\dfrac{6x+1}{2\sqrt{x}}\,dx$ **9.** $\dfrac{x^2-2}{x^2}\,dx$

11. $\dfrac{-x^2+2x+1}{(x^2+1)^2}\,dx$ **13.** $\dfrac{6x-1}{2\sqrt{3x^2-x}}\,dx$

15. a. $2x\,dx$ **b.** 0.04 **c.** 0.0404

17. a. $-\dfrac{dx}{x^2}$ **b.** -0.05 **c.** -0.05263

19. 3.167   **21.** 7.0358   **23.** 1.983
25. 0.298   **27.** ± 8.64 cm^3
29. ± 0.076 cm^3 **31.** 274 seconds
33. 111,595 fewer housing starts
35. $-\$1.33$ **37.** $\pm\$64,800$/yr
39. 11 fewer serious crimes

Chapter 3 Review Exercises, page 199

1. $15x^4 - 8x^3 + 6x - 2$ **3.** $\dfrac{6}{x^4} - \dfrac{3}{x^2}$

5. $-\dfrac{1}{t^{3/2}} - \dfrac{6}{t^{5/2}}$ **7.** $1 - \dfrac{2}{t^2} - \dfrac{6}{t^3}$

9. $2x + \dfrac{3}{x^{5/2}}$ **11.** $\dfrac{2t}{(2t^2+1)^2}$

13. $\dfrac{1}{\sqrt{x}(\sqrt{x}+1)^2}$ **15.** $\dfrac{2x(x^4-2x^2-1)}{(x^2-1)^2}$

17. $72x^2(3x^3-2)^7$ **19.** $\dfrac{2t}{\sqrt{2t^2+1}}$

21. $\dfrac{-4(3t-1)}{(3t^2-2t+5)^3}$ **23.** $\dfrac{2(x^2+1)(x^2-1)}{x^3}$

25. $4t^2(5t+3)(t^2+t)^3$, or $4t^5(5t+3)(t+1)^3$

27. $\dfrac{1}{2\sqrt{x}}(x^2-1)^2(13x^2-1)$

29. $-\dfrac{12x+25}{2\sqrt{3x+2}\,(4x-3)^2}$

31. $2(12x^2-9x+2)$ **33.** $\dfrac{2t(t^2-12)}{(t^2+4)^3}$

35. $\dfrac{2}{(2x^2+1)^{3/2}}$ **37.** $\dfrac{2x}{y}$

39. $-\dfrac{2x}{y^2-1}$ **41.** $\dfrac{x-2y}{2x+y}$

43. a. $(2, -25)$ and $(-1, 14)$
    b. $y = -4x - 17$; $y = -4x + 10$

45. $y = -\dfrac{\sqrt{3}}{3}x + \dfrac{4}{3}\sqrt{3}$

47. $-\dfrac{48}{(2x-1)^4}$; $\left(-\infty; \tfrac{1}{2}\right) \cup \left(\tfrac{1}{2}, \infty\right)$

49. a. \$2.20; \$2.20   **b.** $\dfrac{2500}{x} + 2.2$; $-\dfrac{2500}{x^2}$

    c. $\displaystyle\lim_{x\to\infty}\left(\dfrac{2500}{x} + 2.2\right) = 2.2$

51. a. $2000x - 0.04x^2$; $-0.000002x^3 - 0.02x^2 + 1000x -$
     $120,000$; $0.000002x^2 - 0.02x + 1000 + \dfrac{120,000}{x}$

    b. $0.000006x^2 - 0.04x + 1000$; $2000 - 0.08x$;
     $-0.000006x^2 - 0.04x + 1000$; $0.000004x -$
     $0.02 - \dfrac{120,000}{x^2}$

    c. 934; 1760; 826
    d. -0.0048; 0.010125; at a production level of 5000, the average cost is decreasing by 0.48 cents/unit. At a production level of 8000, the average cost is increasing by 1.0125 cents/unit.

CHAPTER 4

Exercises 4.1, page 210

1. Decreasing on $(-\infty, 0)$ and increasing on $(0, \infty)$
3. Increasing on $(-\infty, -1) \cup (1, \infty)$ and decreasing on $(-1, 1)$
5. Decreasing on $(-\infty, 0) \cup (2, \infty)$ and increasing on $(0, 2)$
7. Decreasing on $(-\infty, -1) \cup (1, \infty)$ and increasing on $(-1, 1)$
9. Increasing on $(-\infty, \infty)$
11. Decreasing on $\left(-\infty, \tfrac{3}{2}\right)$ and increasing on $\left(\tfrac{3}{2}, \infty\right)$
13. Decreasing on $\left(-\infty, -\dfrac{\sqrt{3}}{3}\right) \cup \left(\dfrac{\sqrt{3}}{3}, \infty\right)$ and increasing on $\left(-\dfrac{\sqrt{3}}{3}, \dfrac{\sqrt{3}}{3}\right)$
15. Increasing on $(-\infty, -2) \cup (0, \infty)$ and decreasing on $(-2, 0)$
17. Increasing on $(-\infty, 0) \cup (2, \infty)$ and decreasing on $(0, 2)$
19. Increasing on $(-\infty, 3) \cup (3, \infty)$
21. Decreasing on $(-\infty, 0) \cup (0, 3)$ and increasing on $(3, \infty)$
23. Decreasing on $(-\infty, 2) \cup (2, \infty)$
25. Decreasing on $(-\infty, 1) \cup (1, \infty)$
27. Increasing on $(2, \infty)$; decreasing on $(-\infty, 0) \cup (0, 2)$
29. Increasing on $(-\infty, 0) \cup (0, \infty)$
31. Increasing on $(-1, \infty)$
33. Increasing on $(-4, 0)$; decreasing on $(0, 4)$
35. Increasing on $(-\infty, 0) \cup (0, \infty)$

37. Increasing on $(-\infty, 1)$; decreasing on $(1, \infty)$

39.

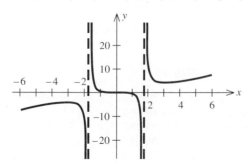

Rising in the time interval $(0, 2)$; falling in the time interval $(2, 5)$; when $t = 5$ sec

41. Rising on $(0, 33)$; descending on $(33, T)$

43. Decreasing on $(0, 1)$; increasing on $(1, 4)$; the average speed of a vehicle on Route 134 was decreasing from 6 A.M. to 7 A.M. and increasing from 7 A.M. to 10 A.M.

45. Increasing from 7 A.M. to 10 A.M.; decreasing from 10 A.M. to 2 P.M.

47. $N'(t) = \dfrac{180}{(t + 6)^2}$ is always positive.

49. Increasing on $(0, 4.5)$; decreasing on $(4.5, 11)$; the air pollution was increasing from 7 A.M. to 11:30 A.M. and decreasing from 11:30 A.M. to 6 P.M.

51. Decreasing on $(0, 2)$; increasing on $(2, \infty)$; after organic waste is dumped into the pond, the pond's oxygen content decreases for 2 days and increases thereafter.

53. For $a > 0$: f decreasing on $\left(-\infty, -\dfrac{b}{2a}\right)$ and increasing on $\left(-\dfrac{b}{2a}, \infty\right)$

55. Increasing on $(-\infty, -3) \cup (3, \infty)$; decreasing on $(-3, -\sqrt{3}) \cup (-\sqrt{3}, 0) \cup (0, \sqrt{3}) \cup (\sqrt{3}, 3)$

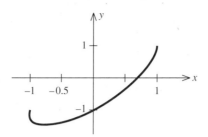

57. Decreasing on $\left(-1, -\dfrac{\sqrt{2}}{2}\right)$; increasing on $\left(-\dfrac{\sqrt{2}}{2}, 1\right)$

Exercises 4.2, page 219

1. Relative maximum: $f(0) = 1$; relative minimum: $f(-1) = 0$ and $f(1) = 0$

3. Relative maximum: $f(0) = 0$; relative minimum: $f(4) = -32$

5. Relative minimum: $f(-1) = 0$

7. Relative maximum: $f(-3) = -\frac{9}{2}$; relative minimum: $f(3) = \frac{9}{2}$

9. Relative minimum: $f(2) = -4$

11. Relative minimum: $f(2) = 2$

13. Relative minimum: $f(0) = 2$

15. Relative maximum: $g(0) = 4$; relative minimum: $g(2) = 0$

17. Relative minimum: $F(3) = -5$; relative maximum: $F(-1) = \frac{17}{3}$

19. Relative maximum: $f\left(-\frac{1}{3}\right) = \frac{119}{54}$; relative minimum: $f\left(\frac{1}{2}\right) = \frac{13}{8}$

21. Relative minimum: $g(3) = -19$

23. Relative minimum: $f\left(\frac{1}{2}\right) = \frac{63}{16}$

25. None

27. Relative maximum: $f(-3) = -4$; relative minimum: $f(3) = 8$

29. Relative maximum: $f(1) = \frac{1}{2}$; relative minimum: $f(-1) = -\frac{1}{2}$

31. Relative maximum: $f(0) = 0$

33. Relative minimum: $f(1) = 0$

37. Relative minimum: $f(0.4) = 2.6$

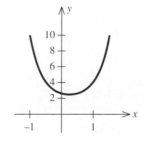

39. Relative maximum: $f(1) = 1$; relative minimum: $f(-0.7) = -1.4$

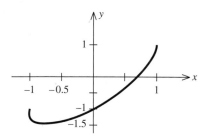

Exercises 4.3, page 230

1. Concave downward on $(-\infty, 0)$ and concave upward on $(0, \infty)$; inflection point: $(0, 0)$

3. Concave downward on $(-\infty, 0) \cup (0, \infty)$

5. Concave upward on $(-\infty, 0) \cup (1, \infty)$ and concave downward on $(0, 1)$; inflection points: $(0, 0)$ and $(1, -1)$

7. Concave downward on $(-\infty, -2) \cup (-2, 2) \cup (2, \infty)$

9. a　　　　**11.** b

19. Concave upward on $(-\infty, \infty)$

21. Concave downward on $(-\infty, 0)$; concave upward on $(0, \infty)$

23. Concave downward on $(-\infty, 2)$; concave upward on $(2, \infty)$

25. Concave upward on $(-\infty, 0) \cup (3, \infty)$; concave downward on $(0, 3)$

27. Concave downward on $(-\infty, 0) \cup (0, \infty)$

29. Concave downward on $(-\infty, 4)$

31. Concave downward on $(-\infty, 2)$; concave upward on $(2, \infty)$

33. Concave upward on $\left(-\infty, -\frac{\sqrt{6}}{3}\right) \cup \left(\frac{\sqrt{6}}{3}, \infty\right)$; concave downward on $\left(-\frac{\sqrt{6}}{3}, \frac{\sqrt{6}}{3}\right)$

35. Concave downward on $(-\infty, 1)$; concave upward on $(1, \infty)$

37. Concave upward on $(-\infty, 0) \cup (0, \infty)$

39. Concave upward on $(-\infty, -1) \cup (1, \infty)$; concave downward on $(-1, 1)$

41. Concave upward on $(-\infty, 2)$; concave downward on $(2, \infty)$

43. $(0, -2)$　　　　**45.** $(1, -15)$

47. $(0, 1)$ and $\left(\frac{2}{3}, \frac{11}{27}\right)$　　　　**49.** $(0, 0)$

51. $(1, 2)$

53. $\left(-\frac{\sqrt{3}}{3}, \frac{3}{2}\right)$ and $\left(\frac{\sqrt{3}}{3}, \frac{3}{2}\right)$

55. Relative maximum: $f(1) = 5$

57. None

59. Relative maximum: $f(-1) = -\frac{22}{3}$; relative minimum: $f(5) = -\frac{130}{3}$

61. Relative maximum: $f(-3) = -6$; relative minimum: $f(3) = 6$

63. None

65. Relative minimum: $f(-2) = 12$

67. Relative maximum: $g(1) = \frac{1}{2}$; relative minimum: $g(-1) = -\frac{1}{2}$

69. Relative maximum: $f(0) = 0$; relative minimum: $f\left(\frac{4}{3}\right) = \frac{256}{27}$

71. Relative minimum: $g(4) = -\frac{1}{108}$

73. **a.** The number of drug-related crimes is projected to be increasing over the next 12 months.
　b. $N_1''(t) < 0$; $N_2''(t) > 0$
　c. If the budget is cut, the rate of growth of the number of drug-related crimes will increase; without the budget cuts the rate will decrease.

75. $(100, 4600)$; the sales increase rapidly until \$100,000 is spent on advertising; after that any additional expenditure results in increased sales but at a slower rate of increase.

77. 10 A.M.

79. $(18, 20{,}310)$; 1452 ft/sec

85. Concave downward on $(-\infty, 0) \cup (1.6, 4.4)$; concave upward on $(0, 1.6) \cup (4.4, \infty)$; inflection points: $(0, 0)$, $(1.6, 3483)$, $(4.4, 3893)$

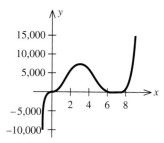

87. Concave upward on $(0, 3)$; concave downward on $(3, \infty)$; inflection point: $(3, 2.3)$

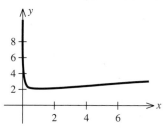

Exercises 4.4, page 242

1. y-axis　　　**3.** y-axis　　　**5.** Origin
7. y-axis　　　**9.** y-axis　　　**11.** y-axis
13. Neither　　**15.** y-axis　　**17.** y-axis
19. Neither　　**21.** y-axis

23. $y = 0$ is a horizontal asymptote
25. $y = 0$ is a horizontal asymptote and $x = 0$ is a vertical asymptote
27. $y = 0$ is a horizontal asymptote; $x = -1$ and $x = 1$ are vertical asymptotes
29. $y = 3$ is a horizontal asymptote and $x = 0$ is a vertical asymptote
31. $y = 0$ is a horizontal asymptote and $x = 0$ is a vertical asymptote
33. $y = 0$ is a horizontal asymptote and $x = 0$ is a vertical asymptote
35. $y = 1$ is a horizontal asymptote and $x = -1$ is a vertical asymptote
37. None
39. Horizontal asymptote: $y = 1$; vertical asymptotes: $t = -3$ and $t = 3$
41. Horizontal asymptote: $y = 0$; vertical asymptotes: $x = -2$ and $x = 3$
43. Horizontal asymptote: $y = 2$; vertical asymptote: $t = 2$
45. Horizontal asymptote: $y = 1$; vertical asymptotes: $x = -2$ and $x = 2$
47. Vertical asymptote: $x = -1$
49. a. $x = 100$;   **b.** No
51. a. $y = 0$
   b. As time passes, the concentration of the drug decreases and approaches zero.

Exercises 4.5, page 251

1.

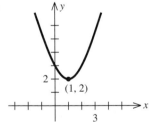

3.

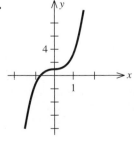

5.

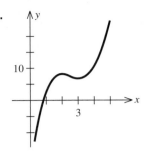

7.

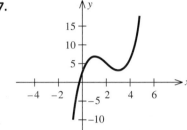

9.

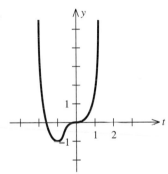

11.

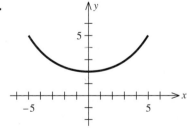

13.

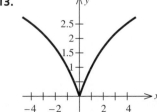

15.

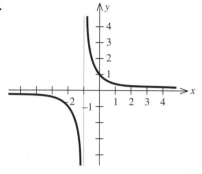

17.

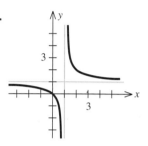

19.

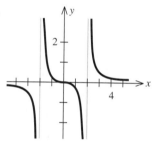

21.

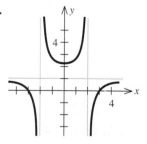

23.

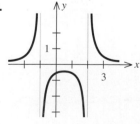

25.

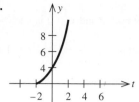

27.

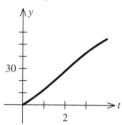

29.

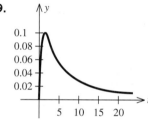

31.

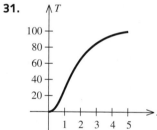

33.

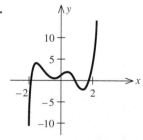

35.

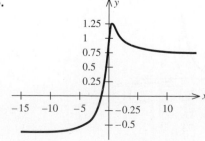

Exercises 4.6, page 261

1. None
3. Absolute minimum value: 0
5. Absolute maximum value: 3; absolute minimum value: -2
7. Absolute maximum value: 3; absolute minimum value: $-\frac{27}{16}$
9. Absolute minimum value: $-\frac{41}{8}$
11. No absolute extremum
13. Absolute maximum value: 1
15. Absolute maximum value: 5; absolute minimum value: -4
17. Absolute maximum value: 10; absolute minimum value: 1
19. Absolute maximum value: 19; absolute minimum value: -1
21. Absolute maximum value: 16; absolute minimum value: -1
23. Absolute maximum value: 2; absolute minimum value: -1
25. Absolute maximum value: $\frac{37}{3}$; absolute minimum value: 5
27. Absolute maximum value ≈ 1.04; absolute minimum value: -1.5
29. No absolute extremum
31. Absolute maximum value: 3; absolute minimum value: 0
33. Absolute maximum value: 0; absolute minimum value: -3
35. Absolute maximum value: $\dfrac{\sqrt{2}}{4} \approx 0.35$; absolute minimum value: $-\frac{1}{3}$
37. Absolute maximum value: $\dfrac{\sqrt{2}}{2}$; absolute minimum value: $-\dfrac{\sqrt{2}}{2}$
39. 144 ft   **41.** 3000   **43.** $3600
45. 6000   **47.** 110 units
49. a. $0.000002x^2 + 5 + \dfrac{400}{x}$ **b.** 464 cases
    c. 464 cases   **d.** Same
51. 10,000   **53.** 11:30 A.M.

63. c.

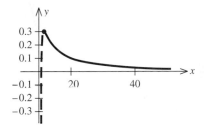

65. Absolute minimum value: 0; absolute maximum value: $\frac{26}{9}$

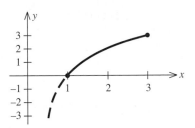

67. Absolute maximum value $= 0.3$

Exercises 4.7, page 273

1. 750 yd $\times$ 1500 yd; 1,125,000 sq yd
3. $10\sqrt{2}$ ft $\times$ $40\sqrt{2}$ ft
5. $\frac{16}{3}$ inches $\times$ $\frac{16}{3}$ inches $\times$ $\frac{4}{3}$ inches
7. 5.04 inches $\times$ 5.04 inches $\times$ 5.04 inches
9. 18 inches $\times$ 18 inches $\times$ 36 inches; 11,664 cu in.
11. $r = \frac{36}{\pi}$ inches; $l = 36$ inches; $\frac{46,656}{\pi}$ cu in.
13. $\frac{2}{3}\sqrt[3]{9}$ ft $\times$ $\sqrt[3]{9}$ ft $\times$ $\frac{2}{5}\sqrt[3]{9}$
15. 250; $62,500; $250
17. $w \approx 13.86$ inches; $h \approx 19.60$ inches
19. $x = 750\sqrt{2}$ ft   **21.** 56.57 mph
23. 45; 44,721

Chapter 4 Review Exercises, page 279

1. a. f is increasing on $(-\infty, 1) \cup (1, \infty)$
   b. No relative extrema
   c. Concave down on $(-\infty, 1)$; concave up on $(1, \infty)$
   d. $\left(1, -\frac{17}{3}\right)$
3. a. f is increasing on $(-1, 0) \cup (1, \infty)$; decreasing on $(-\infty, -1) \cup (0, 1)$
   b. Relative maximum: 0; relative minimum: -1
   c. Concave up on $\left(-\infty, -\dfrac{\sqrt{3}}{3}\right) \cup \left(\dfrac{\sqrt{3}}{3}, \infty\right)$; concave down on $\left(-\dfrac{\sqrt{3}}{3}, \dfrac{\sqrt{3}}{3}\right)$
   d. $\left(-\dfrac{\sqrt{3}}{3}, -\dfrac{5}{9}\right); \left(\dfrac{\sqrt{3}}{3}, -\dfrac{5}{9}\right)$
5. a. f is increasing on $(-\infty, 0) \cup (2, \infty)$; decreasing on $(0, 1) \cup (1, 2)$

b. Relative maximum: 0; relative minimum: 4
c. Concave up on $(1, \infty)$; concave down on $(-\infty, 1)$
d. None
7. a. f is decreasing on $(-\infty, 1) \cup (1, \infty)$
   b. No relative extrema
   c. Concave down on $(-\infty, 1)$; concave up on $(1, \infty)$
   d. $(1, 0)$
9. a. f is increasing on $(-\infty, -1) \cup (-1, \infty)$
   b. No relative extrema
   c. Concave down on $(-1, \infty)$; concave up on $(-\infty, -1)$
   d. No inflection point

11.
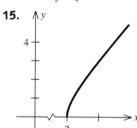

13.
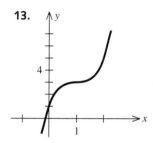

15.
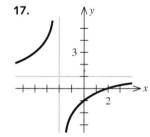

17.
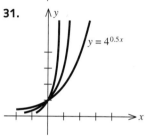

19. Vertical asymptote: $x = -\frac{3}{2}$; horizontal asymptote: $y = 0$
21. Vertical asymptotes: $x = -2$, $x = 4$; horizontal asymptote: $y = 0$
23. Absolute minimum value: $-\frac{25}{8}$
25. Absolute maximum value: 5; absolute minimum value: 0
27. Absolute maximum value: -16; absolute minimum value: -32
29. Absolute maximum value: $\frac{8}{3}$; absolute minimum value: 0
31. Absolute maximum value: $\frac{1}{2}$; absolute minimum value: $-\frac{1}{2}$
33. $4000
35. 168
37. 10 A.M.
39. 20,000 cases

CHAPTER 5

Exercises 5.1, page 288

1. a. 16   **b.** 27   **3. a.** 3   **b.** $\sqrt{5}$
5. a. -3 **b.** 8   **7. a.** 25   **b.** $4^{1.8}$

9. a. $4x^3$ **b.** $5xy^2\sqrt{x}$
11. a. $\dfrac{2}{a^2}$ **b.** $\frac{1}{3}b^2$
13. a. $8x^9y^6$ **b.** $16x^4y^4z^6$
15. a. $\dfrac{64x^6}{y^4}$ **b.** $(x - y)(x + y)$
17. 2   **19.** 3   **21.** 3
23. $\frac{5}{4}$ **25.** 1 or 2

27.
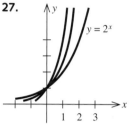

29.
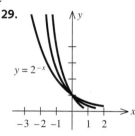

31.
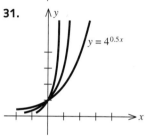

33.
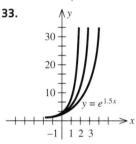

35.
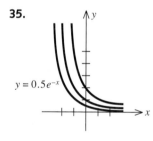

37.
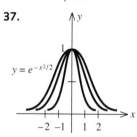

Exercises 5.2, page 295

1. $\log_2 64 = 6$ **3.** $\log_3 \frac{1}{9} = -2$
5. $\log_{1/3} \frac{1}{3} = 1$ **7.** $\log_{32} 8 = \frac{3}{5}$
9. $\log_{10} 0.001 = -3$ **11.** 1.0792
13. 1.2042   **15.** 1.6813
17. $\log x + 4 \log (x + 1)$
19. $\frac{1}{2} \log (x + 1) - \log (x^2 + 1)$
21. $\ln x - x^2$
23. $-\frac{3}{2} \ln x - \frac{1}{2} \ln (1 + x^2)$
25. $x \ln x$

27.

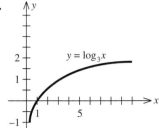

$y = \log_3 x$

29.

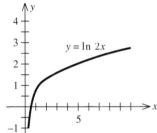

$y = \ln 2x$

31.

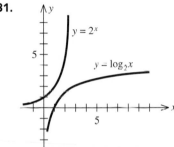

$y = 2^x$

$y = \log_2 x$

33. 5.1986   **35.** −0.0912   **37.** −8.0472

39. −4.9041   **41.** $-2 \ln \left(\dfrac{A}{B}\right)$ **43.** 105.7 mm

45. a. $10^3 I_0$ **b.** 100,000 times greater
c. 10,000,000 times greater
47. $34\frac{1}{2}$ hrs earlier, at 1:30 P.M.

Exercises 5.3, page 306

1. $4,974.47   **3.** $223,403.11
5. a. 10.25 percent per year   **b.** 9.31 percent per year
7. a. $29,227.61   **b.** $29,137.83
9. $6,885.64   **11.** $112,926.52
13. $3.795 million   **15.** $23,329.49
17. 9.58 years   **19.** 12.75 percent
21. $40,000
23. a. $33,885.14   **b.** $33,565.38
25. a. $16,262.79   **b.** $12,047.77   **c.** $6611.96
29. Bank B   **31.** 9.531 percent

Exercises 5.4, page 315

1. $3e^{3x}$ **3.** $-e^{-t}$ **5.** $e^x + 1$

7. $x^2 e^x (x + 3)$ **9.** $\dfrac{2e^x(x - 1)}{x^2}$ **11.** $3(e^x - e^{-x})$

13. $-\dfrac{1}{e^w}$ **15.** $6e^{3x-1}$ **17.** $-2xe^{-x^2}$

19. $\dfrac{3e^{-1/x}}{x^2}$ **21.** $25e^x(e^x + 1)^{24}$

23. $\dfrac{e^{\sqrt{x}}}{2\sqrt{x}}$ **25.** $e^{3x+2}(3x - 2)$

27. $\dfrac{2e^x}{(e^x + 1)^2}$ **29.** $2(8e^{-4x} + 9e^{3x})$

31. $6e^{3x}(3x + 2)$ **33.** $y = 2x - 2$
35. f is increasing on $(-\infty, 0)$ and decreasing on $(0, \infty)$
37. Concave downward on $(-\infty, 0)$; concave upward on $(0, \infty)$
39. $(1, e^{-2})$
41. Abs. max. value: 1; abs. min. value: e^{-1}
43. Abs. min. value: −1; abs. max. value: $2e^{-3/2}$
45.   **47.**

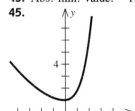
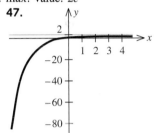

49. −$6,065/day; −$3,679/day; −$2,231/day;
−$1,353/day
51. b. 4505 per year; 273 cases per year
53. a. −1.68 cents/case   **b.** $40.36 per case
55. Tenth year
57. 1.8; −0.1084; −0.2286; −0.1279; initially, the number
of barrels of oil needed to fuel productivity ($1000 worth
of economic output) was increasing, but after one decade
the number of barrels required was decreasing.
59.

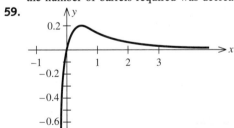

61. a. 69.63 percent   **b.** 5.09 percent/yr

Exercises 5.5, page 323

1. $\dfrac{5}{x}$ **3.** $\dfrac{1}{x + 1}$

5. $\dfrac{8}{x}$ **7.** $\dfrac{1}{2x}$

9. $\dfrac{2}{x}$ **11.** $\dfrac{2(4x - 3)}{4x^2 - 6x + 3}$

13. $\dfrac{1}{x(x + 1)}$ **15.** $x(1 + 2 \ln x)$

17. $\dfrac{2(1 - \ln x)}{x^2}$

19. $\dfrac{3}{u - 2}$

21. $\dfrac{1}{2x\sqrt{\ln x}}$

23. $\dfrac{3(\ln x)^2}{x}$

25. $\dfrac{3x^2}{x^3 + 1}$

27. $\dfrac{(x \ln x + 1)e^x}{x}$

29. $\dfrac{e^{2t}[2(t + 1)\ln (t + 1) + 1]}{t + 1}$

31. $\dfrac{1 - \ln x}{x^2}$

33. $-\dfrac{1}{x^2}$

35. $\dfrac{2(2 - x^2)}{(x^2 + 2)^2}$

37. $(x + 1)(5x + 7)(x + 2)^2$

39. $(x - 1)(x + 1)^2(x + 3)^3(9x^2 + 14x - 7)$

41. $\dfrac{(2x^2 - 1)^4(38x^2 + 40x + 1)}{2(x + 1)^{3/2}}$

43. $3^x \ln 3$

45. $(x^2 + 1)^{x-1}[2x^2 + (x^2 + 1)\ln (x^2 + 1)]$

47. $y = x - 1$

49. f is decreasing on $(-\infty, 0)$ and increasing on $(0, \infty)$

51. Concave up: $(-\infty, -1) \cup (1, \infty)$; concave down: $(-1, 0) \cup (0, 1)$

53. $(-1, \ln 2)$ and $(1, \ln 2)$

55. Abs. min. value: 1; abs. max. value: $3 - \ln 3$

57.

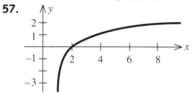

61.

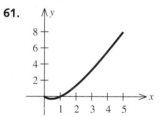

11. 13,412 years ago

13. a. 60 words per minute   **b.** 107 words per minute
    c. 136 words per minute

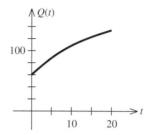

15. a. 573 computers; 1177 computers; 1548 computers; 1925 computers
    b. 2000 computers
    c. 46 computers/month

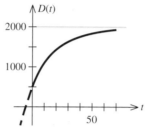

17. a. 11   **b.** 937   **c.** 1000

19. 3 percent; 65 percent

21. a. $\dfrac{r}{k}$

  b.

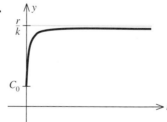

Exercises 5.6, page 332

1. a. 0.05   **b.** 400
  c.

t	0	10	20	100	1000
Q	400	660	1087	59,365	2.07×10^{24}

3. $Q(t) = 100e^{0.035t}$; 266 minutes; $Q(t) = 1000e^{0.035t}$

5. a. 55.5 years   **b.** 14.25 billion

7. 8.7 lb/sq. in.

9. $Q(t) = 100e^{-0.049t}$; 70.6 grams

Chapter 5 Review Exercises, page 336

1. a.–b.   **3.** $\log_{16} 0.125 = -\frac{3}{4}$

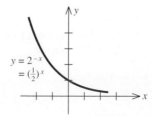

5. 2   **7.** $x + 2y - z$

9.

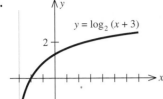

$y = \log_2 (x + 3)$

11. $(2x + 1)e^{2x}$

13. $\dfrac{(1 - 4t)}{2\sqrt{t}e^{2t}}$

15. $\dfrac{2(e^{2x} + 2)}{(1 + e^{-2x})^2}$

17. $(1 - 2x^2)e^{-x^2}$

19. $(x + 1)^2 e^x$

21. $\dfrac{2xe^{x^2}}{e^{x^2} + 1}$

23. $\dfrac{x + 1 - x \ln x}{x(x + 1)^2}$

25. $\dfrac{4e^{4x}}{e^{4x} + 3}$

27. $\dfrac{1 + e^x(1 - x \ln x)}{x(1 + e^x)^2}$

29. $-\dfrac{9}{(3x + 1)^2}$

31. 0

33. $6x(x^2 + 2)^2(3x^3 + 2x + 1)$

35. $y = -(2x - 3)e^{-2}$

37.

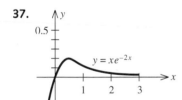

$y = xe^{-2x}$

39. Abs. max. value: $\dfrac{1}{e}$

41. 12 percent per year

43. a. $Q(t) = 2000e^{0.01831t}$ **b.** 161,992

45.

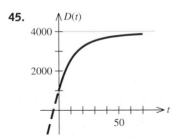

$D(t)$

a. 1175; 2540; 3289   **b.** 4000

CHAPTER 6

Exercises 6.1, page 350

5. b. $y = 2x + C$

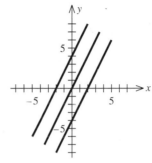

7. b. $y = \frac{1}{3}x^3 + C$

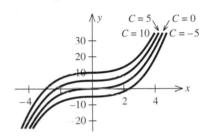

9. $6x + C$

11. $\frac{1}{4}x^4 + C$

13. $-\dfrac{1}{3x^3} + C$

15. $\frac{3}{5}x^{5/3} + C$

17. $-\dfrac{4}{x^{1/4}} + C$

19. $-\dfrac{2}{x} + C$

21. $\frac{2}{3}\pi t^{3/2} + C$

23. $3x - x^2 + C$

25. $\frac{1}{3}x^3 + \frac{1}{2}x^2 - \dfrac{1}{2x^2} + C$

27. $4e^x + C$

29. $x + \frac{1}{2}x^2 + e^x + C$

31. $x^4 + \dfrac{2}{x} - x + C$

33. $\frac{2}{7}x^{7/2} + \frac{4}{5}x^{5/2} - \frac{1}{2}x^2 + C$

35. $\frac{2}{3}x^{3/2} + 6\sqrt{x} + C$

37. $\frac{1}{9}u^3 + \frac{1}{3}u^2 - \frac{1}{3}u + C$

39. $\frac{2}{3}t^3 - \frac{3}{2}t^2 - 2t + C$

41. $\frac{1}{3}x^3 - 2x - \dfrac{1}{x} + C$

43. $\frac{1}{3}s^3 + s^2 + s + C$

45. $e^t + \dfrac{t^{e+1}}{e + 1} + C$

47. $\frac{1}{2}x^2 + x - \ln |x| - \dfrac{1}{x} + C$

49. $\ln |x| + \dfrac{4}{\sqrt{x}} - \dfrac{1}{x} + C$

51. $x^3 + 2x^2 - x - 5$

53. $x + \ln |x|$

55. $\sqrt{x}$

57. $e^x + \frac{1}{2}x^2 + 2$

59. $f(t) = \frac{4}{3}t^{3/2}$

61. $R(x) = -0.0045x^2 + 12x$; $p = -0.0045x + 12$
63. $0.001x^2 + 100x + 4000$
65. a. $-16t + 400$ **b.** 5 sec   **c.** -160 ft/sec
67. 3.375 parts per million   **69.** $1.0974t^3 - 0.0915t^4$
71. $-t^3 + 96t^2 + 120t$; 63,000 ft

Exercises 6.2, page 360

1. $\frac{1}{5}(4x + 3)^5 + C$ **3.** $\frac{1}{3}(x^3 - 2x)^3 + C$

5. $-\dfrac{1}{2(2x^2 + 3)^2} + C$ **7.** $\frac{2}{3}(t^3 + 2)^{3/2} + C$

9. $\frac{1}{20}(x^2 - 1)^{10} + C$ **11.** $-\frac{1}{5}\ln|1 - x^5| + C$

13. $\ln(x - 2)^2 + C$

15. $\frac{1}{2}\ln(0.3x^2 - 0.4x + 2) + C$

17. $\frac{1}{6}\ln|3x^2 - 1| + C$ **19.** $-\frac{1}{2}e^{-2x} + C$

21. $-e^{2-x} + C$ **23.** $-\frac{1}{2}e^{-x^2} + C$

25. $e^x + e^{-x} + C$ **27.** $\ln(1 + e^x) + C$

29. $2e^{\sqrt{x}} + C$ **31.** $-\dfrac{1}{6(e^{3x} + x^3)^2} + C$

33. $\frac{1}{8}(e^{2x} + 1)^4 + C$ **35.** $\frac{1}{2}(\ln 5x)^2 + C$

37. $\ln|\ln x| + C$ **39.** $\frac{2}{3}(\ln x)^{3/2} + C$

41. $\frac{1}{2}e^{x^2} - \frac{1}{2}\ln(x^2 + 2) + C$

43. $\frac{2}{3}(\sqrt{x} - 1)^3 + 3(\sqrt{x} - 1)^2 + 8(\sqrt{x} - 1)$
     $+ 4\ln|\sqrt{x} - 1| + C$

45. $\dfrac{(6x + 1)(x - 1)^6}{42} + C$

47. $5 + 4\sqrt{x} - x - 4\ln(1 + \sqrt{x}) + C$

49. $-\frac{1}{252}(1 - v)^7(28v^2 + 7v + 1) + C$

51. $\frac{1}{2}[(2x - 1)^5 + 5]$ **53.** $e^{-x^2+1} - 1$

55. $21,000 - \dfrac{20,000}{\sqrt{1 + 0.2t}}$; 6858

57. $p(x) = \dfrac{250}{\sqrt{16 + x^2}}$

59. $30(\sqrt{2t + 4} - 2)$; $14,400\pi$ sq ft

61. $\dfrac{71.86887}{1 + 2.449e^{-0.3227t}} - 1.43760$; 59.6 inches

63. $6000t + 4000e^{-t} - 5472$; 24,555 pairs

Exercises 6.3, page 372

1. 4.27 sq units
3. a. 6 sq units   **b.** 4.5 sq units
   c. 5.25 sq units   **d.** Yes
5. a. 4 sq units   **b.** 4.8 sq units
   c. 4.4 sq units   **d.** Yes
7. a. 18.5 sq units   **b.** 18.64 sq units
   c. 18.66 sq units   **d.** ≈ 18.7 sq units
9. a. 25 sq units   **b.** 21.12 sq units
   c. 19.88 sq units   **d.** ≈ 19.9 sq units
11. a. 0.0625 sq units   **b.** 0.16 sq units
   c. 0.2025 sq units   **d.** ≈ 0.2 sq units

13. 4.64 sq units   **15.** 0.95 sq units
17. 9400 sq ft   **19.** 11.32963 sq units
21. 12.58797 sq units

Exercises 6.4, page 381

1. 6 sq units   **3.** 8 sq units   **5.** 12 sq units
7. 9 sq units   **9.** $\ln 2$ sq units   **11.** $17\frac{1}{3}$ sq units
13. $18\frac{1}{4}$ sq units   **15.** $(e^2 - 1)$ sq units
17. 6   **19.** 14   **21.** $18\frac{2}{3}$
23. $\frac{4}{3}$ **25.** 45   **27.** $\frac{7}{12}$
29. $\ln 2$ **31.** 56   **33.** $\frac{981}{10}$
35. $\frac{2}{3}$ **37.** $2\frac{2}{3}$ **39.** $19\frac{1}{2}$

41. a. \$4100   **b.** \$900
43. a. \$2800   **b.** \$219.20
45. $10,133\frac{1}{3}$ ft
47. a. 46 percent of the smoke
   b. 24 percent of the smoke

Exercises 6.5, page 390

1. 10   **3.** $\frac{19}{15}$ **5.** $32\frac{4}{15}$
7. $\sqrt{3} - 1$ **9.** $24\frac{1}{5}$ **11.** $\frac{32}{15}$
13. $18\frac{2}{15}$ **15.** $\frac{1}{2}(e^4 - 1)$ **17.** $\frac{1}{2}e^2 + \frac{5}{6}$
19. 0   **21.** $2\ln 4$
23. $\frac{1}{3}(\ln 19 - \ln 3)$ **25.** $2e^4 - 2e^2 - \ln 2$
27. $\frac{1}{2}(e^{-4} - e^{-8} - 1)$ **29.** 5
31. $\frac{17}{3}$ **33.** -1
35. $\frac{13}{6}$ **37.** $\frac{1}{4}(e^4 - 1)$
39. 120.3 billion metric tons
41. $\approx$ \$2.24 million   **43.** \$40,339.50
45. 26°F   **47.** 16,863
49. 0.071 mg/cu cm/hr   **51.** $\frac{2k}{3}R^2$ cm/sec

Exercises 6.6, page 400

1. 108 sq units   **3.** $\frac{2}{3}$ sq units
5. $2\frac{2}{3}$ sq units   **7.** $1\frac{1}{2}$ sq units
9. 3   **11.** $3\frac{1}{3}$ **13.** 27
15. $2(e^2 - e^{-1})$ **17.** $12\frac{2}{3}$ **19.** $3\frac{1}{3}$
21. $4\frac{3}{4}$ **23.** $12 - \ln 4$
25. $e^2 - e - \ln 2$ **27.** $2\frac{1}{2}$
29. $7\frac{1}{3}$ **31.** $\frac{3}{2}$
33. $e^3 - 4 + \dfrac{1}{e}$ **35.** $20\frac{5}{6}$
37. $\frac{1}{12}$ **39.** $\frac{1}{3}$

41. S is the additional revenue that Odyssey Travel could realize by switching to the new agency;

$$S = \int_a^b [g(x) - f(x)]\, dx$$

43. S is the additional amount of smoke that brand B will remove over brand A in the time interval $[a, b]$;

$$S = \int_a^b [f(t) - g(t)]\, dt$$

45. 42.79 million metric tons
47. 57,179 people
49. ≈ 4.3

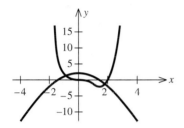

Exercises 6.7, page 415

1. $\$11,667$ **3.** $\$6,667$ **5.** $\$11,667$
7. C.S. $\$13,333$; P.S. $\$11,667$
9. $\$824,200$ **11.** $\$148,239$ **13.** $\$52,203$
15. $\$76,615$ **17.** $\$111,869$ **19.** $\$20,964$
21. a.

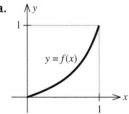

    b. 0.175; 0.816
23. a.

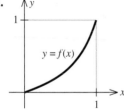

    b. 0.104; 0.504

Chapter 6 Review Exercises, page 419

1. $\frac{1}{4}x^4 + \frac{2}{3}x^3 - \frac{1}{2}x^2 + C$ **3.** $\frac{1}{5}x^5 - \frac{1}{2}x^4 - \frac{1}{x} + C$
5. $\frac{1}{2}x^4 + \frac{2}{5}x^{5/2} + C$
7. $\frac{1}{3}x^3 - \frac{1}{2}x^2 + 2\ln|x| + 5x + C$
9. $\frac{3}{8}(3x^2 - 2x + 1)^{4/3} + C$
11. $\frac{1}{2}\ln(x^2 - 2x + 5) + C$ **13.** $\frac{1}{2}e^{x^2 + x + 1} + C$

15. $\frac{1}{6}(\ln x)^6 + C$
17. $\dfrac{(11x^2 - 1)(x^2 + 1)^{11}}{264} + C$
19. $\frac{2}{3}(x + 4)\sqrt{x - 2} + C$ **21.** $\frac{1}{2}$
23. $5\frac{2}{3}$ **25.** -80 **27.** $\frac{1}{2}\ln 5$
29. 4 **31.** $\dfrac{e - 1}{2(1 + e)}$
33. $f(x) = x^3 - 2x^2 + x + 1$
35. $f(x) = x + e^{-x} + 1$ **37.** -4.28
39. $-0.015x^2 + 60x;\ p = -0.015x + 60$
41. $3000t - 50,000(1 - e^{-0.04t})$; 16,939
43. $\$3100$ **45.** 15 sq units
47. $\frac{2}{3}$ sq units   **49.** $e^2 - 3$ sq units
51. $\frac{1}{2}$ sq units   **53.** $\frac{1}{3}$ sq units
55. $\$2,083$; $\$3,333$ **57.** $\$98,973$
59. a.

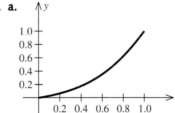

    b. 0.1017; 0.3733   **c.** 0.315

CHAPTER 7

Exercises 7.1, page 430

1. $\frac{1}{4}e^{2x}(2x - 1) + C$ **3.** $4(x - 4)e^{x/4} + C$
5. $\frac{1}{2}e^{2x} - 2(x - 1)e^x + \frac{1}{3}x^3 + C$
7. $xe^x + C$ **9.** $\dfrac{2(x + 2)}{\sqrt{x + 1}} + C$
11. $\frac{2}{3}x(x - 5)^{3/2} - \frac{4}{15}(x - 5)^{5/2} + C$
13. $\dfrac{x^2}{4}(2\ln 2x - 1) + C$ **15.** $\dfrac{x^4}{16}(4\ln x - 1) + C$
17. $\frac{2}{9}x^{3/2}(3\ln\sqrt{x} - 1) + C$
19. $-\dfrac{1}{x}(\ln x + 1) + C$ **21.** $x(\ln x - 1) + C$
23. $-(x^2 + 2x + 2)e^{-x} + C$
25. $\frac{1}{4}x^2[2(\ln x)^2 - 2\ln x + 1] + C$
27. $2\ln 2 - 1$ **29.** $4\ln 4 - 3$
31. $\frac{1}{4}(3e^4 + 1)$
33. $-\frac{1}{2}xe^{-2x} - \frac{1}{4}e^{-2x} + \frac{13}{4}$
35. $5\ln 5 - 4$ **37.** 1485 ft   **39.** 2.04 mg/ml
41. $-20e^{-0.1t}(t + 10) + 200$
43. $\$521,087$

Exercises 7.2, page 437

1. $\frac{2}{9}[2 + 3x - 2\ln|2 + 3x|] + C$

3. $\frac{3}{32}[(1 + 2x)^2 - 4(1 + 2x) + 2\ln|1 + 2x|] + C$

5. $2\left[\frac{x}{8}\left(\frac{9}{4} + 2x^2\right)\sqrt{\frac{9}{4} + x^2} - \frac{81}{128}\ln\left(x + \sqrt{\frac{9}{4} + x^2}\right)\right] + C$

7. $\ln\left(\frac{\sqrt{1 + 4x} - 1}{\sqrt{1 + 4x} + 1}\right) + C$

9. $\frac{1}{2}\ln 3$

11. $\frac{x}{9\sqrt{9 - x^2}} + C$

13. $\frac{x}{8}(2x^2 - 4)\sqrt{x^2 - 4} - 2\ln|x + \sqrt{x^2 - 4}| + C$

15. $\sqrt{4 - x^2} - 2\ln\left|\frac{2 + \sqrt{4 - x^2}}{x}\right| + C$

17. $\frac{1}{4}(2x - 1)e^{2x} + C$ **19.** $\ln|\ln(1 + x)| + C$

21. $\frac{1}{9}\left[\frac{1}{1 + 3e^x} + \ln(1 + 3e^x)\right] + C$

23. $6[e^{(1/2)x} - \ln(1 + e^{(1/2)x}] + C$

25. $\frac{1}{9}(2 + 3\ln x - 2\ln|2 + 3\ln x|) + C$

27. $e - 2$ **29.** $\frac{x^3}{9}(3\ln x - 1) + C$

31. $x[(\ln x)^3 - 3(\ln x)^2 + 6\ln x - 6] + C$

33. $\approx \$2329$ **35.** $27{,}136$ **37.** $44; 49$

39. $26{,}157$ **41.** $\$418{,}444$

Exercises 7.3, page 450

1. $2.7037; 2.6667; 2\frac{2}{3}$ **3.** $0.2656; 0.2500; \frac{1}{4}$

5. $0.6970; 0.6933; \approx 0.6931$

7. $0.5090; 0.5004; \frac{1}{2}$ **9.** $5.2650; 5.3046; \frac{16}{3}$

11. $0.6336; 0.6321; \approx 0.6321$

13. $0.3837; 0.3863; \approx 0.3863$

15. $1.1170; 1.1114$ **17.** $1.3973; 1.4052$

19. $0.8806; 0.8818$ **21.** $3.7757; 3.7625$

23. a. 3.6 **b.** 0.0324

25. a. 0.013 **b.** 0.00043

27. a. 0.0078125 **b.** 0.0002848

29. 52.84 miles   **31.** 21.65 mpg

33. a. $\$142{,}374$ **b.** $\$142{,}698$

35. 103.9 PSI   **37.** ≈ 30 percent

39. ≈ 6.42 l/min   **41.** ≈ 1.32573

43. ≈ 0.46420 **45.** ≈ 1.32582

47. ≈ 0.88623

Exercises 7.4, page 462

1. $\frac{2}{3}$ sq unit   **3.** 1 sq unit

5. 2 sq units   **7.** $\frac{2}{3}$ sq unit

9. $\frac{1}{2}e^4$ sq units   **11.** 1 sq unit

13. a. $\frac{2}{3}b^{3/2}$ **15.** 1 **17.** 2

19. Divergent   **21.** $-\frac{1}{8}$ **23.** 1

25. 1 **27.** $\frac{1}{2}$ **29.** Divergent

31. -1 **33.** Divergent   **35.** 0

37. 0 **39.** Divergent   **41.** $\$18{,}750$

43. $\frac{\$10{,}000r + 4000}{r^2}$ dollars

45. b. $\$83{,}333$

Exercises 7.5, page 473

9. a. $k = \frac{1}{8}$ **b.** $\frac{1}{2}$

11. a. 0.63 **b.** 0.30 **c.** 0.30

13. 3 **15.** 0.37 **17.** 0.18

Chapter 7 Review Exercises, page 476

1. $-2(1 + x)e^{-x} + C$ **3.** $x(\ln 5x - 1) + C$

5. $\frac{1}{4}(1 - 3e^{-2})$ **7.** $2\sqrt{x}(\ln x - 2) + 2$

9. $\frac{1}{8}\left[3 + 2x - \frac{9}{3 + 2x} - 6\ln|3 + 2x|\right] + C$

11. $\frac{1}{32}e^{4x}(8x^2 - 4x + 1) + C$

13. $\frac{1}{4}\frac{\sqrt{x^2 - 4}}{x} + C$ **15.** $\frac{1}{2}$

17. Divergent   **19.** $\frac{1}{10}$

21. $0.8421; 0.8404$ **23.** $2.2379; 2.1791$

27. a. $k = \frac{3}{8}$ **b.** 0.6495

29. a. $k = \frac{4}{27}$ **b.** 0.4815

31. $\$1{,}157{,}641$ **33.** $\$41{,}100$

35. $\$111{,}111$

CHAPTER 8

Exercises 8.1, page 486

1. $f(0, 0) = -4; f(1, 0) = -2; f(0, 1) = -1; f(1, 2) = 4; f(2, -1) = -3$

3. $f(1, 2) = 7; f(2, 1) = 9; f(-1, 2) = 1; f(2, -1) = 1$

5. $g(1, 2) = 4 + 3\sqrt{2}; g(2, 1) = 8 + \sqrt{2}; g(0, 4) = 2; g(4, 9) = 56$

7. $h(1, e) = 1; h(e, 1) = -1; h(e, e) = 0$

9. $g(1, 1, 1) = e; g(1, 0, 1) = 1; g(-1, -1, -1) = -e$

11. All real values of x and y

13. All real values of u and v except those satisfying the equation $u = v$

15. All real values of r and s satisfying $rs \geq 0$

17. All real values of x and y satisfying $x + y > 5$

19. **21.**

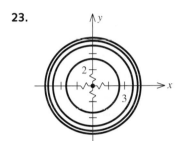

23.

25. 9π cu ft

27. 40,000k dynes

29. \$25,500; \$23,580

31. \$8310; \$7910

33. a. The set of all nonnegative values of W and H

  b. 1.871 sq meters

35. \$13,498.59

37. \$76,704.11; \$50,814.62

39. 18

Exercises 8.2, page 499

1. 2; 3

3. $4x$; 4

5. $-\dfrac{4y}{x^3}, \dfrac{2}{x^2}$

7. $\dfrac{2y}{(u+v)^2}, -\dfrac{2u}{(u+v)^2}$

9. $3(2s-t)(s^2-st+t^2)^2$; $3(2t-s)(s^2-st+t^2)^2$

11. $\dfrac{4x}{3(x^2+y^2)^{1/3}}$; $\dfrac{4y}{3(x^2+y^2)^{1/3}}$

13. ye^{xy+1}; xe^{xy+1}

15. $\ln y + \dfrac{y}{x}, \dfrac{x}{y} + \ln x$

17. $e^u \ln v$; $\dfrac{e^u}{v}$

19. $yz + y^2 + 2xz$, $xz + 2xy + z^2$; $xy + 2yz + x^2$

21. ste^{rst}, rte^{rst}, rse^{rst}

23. $f_x(1, 2) = 8$; $f_y(1, 2) = 5$

25. $f_x(2, 1) = 1$; $f_y(2, 1) = 3$

27. $f_x(1, 2) = \frac{1}{2}$; $f_y(1, 2) = -\frac{1}{4}$

29. $f_x(1, 1) = e$; $f_y(1, 1) = e$

31. $f_x(1, 0, 2) = 0$; $f_y(1, 0, 2) = 8$; $f_z(1, 0, 2) = 0$

33. $f_{xx} = 2y, f_{xy} = 2x + 3y^2 = f_{yx}$; $f_{yy} = 6xy$

35. $f_{xx} = 2$; $f_{xy} = f_{yx} = -2$; $f_{yy} = 4$

37. $f_{xx} = \dfrac{y^2}{(x^2+y^2)^{3/2}}$; $f_{xy} = f_{yx} = -\dfrac{xy}{(x^2+y^2)^{3/2}}$;

  $f_{yy} = \dfrac{x^2}{(x^2+y^2)^{3/2}}$

39. $f_{xx} = \dfrac{1}{y^2}e^{-x/y}$, $f_{xy} = \dfrac{y-x}{y^3}e^{-x/y} = f_{yx}$, $f_{yy} =$

  $\dfrac{x}{y^3}\left(\dfrac{x}{y} - 2\right)e^{-x/y}$

41. a. $f_x = 7.5$; $f_y = 40$ **b.** Yes

43. a. $P_x = 10$; At $(0, 1)$, the price of land is changing at the rate of \$10 per sq ft per mile change to the left.

  b. $P_y = 0$; At $(0, 1)$, the price of land is constant per mile change upward.

45. Complementary commodities

47. \$30/unit change in finished desks; $-$\$25/unit change in unfinished desks. The weekly revenue increases by \$30/unit for each additional finished desk produced (beyond 300) when the level of production of unfinished desks remains fixed at 250; the revenue decreases by \$25/unit when each additional finished desk (beyond 250) is produced and the level of production of finished desks remains fixed at 300.

49. 0.039 l/degree; -0.015 l/mm of mercury; 0.0145; 0.00760. The volume increases by 0.039 l when the temperature increases by 1 degree (beyond 300°K) and the pressure is fixed at 800 mm of mercury. The volume decreases by 0.15 l when the pressure increases by 1 mm of mercury (beyond 800 mm) and the temperature is fixed at 300°K.

Exercises 8.3, page 510

1. $(0, 0)$; relative maximum value: $f(0, 0) = 1$

3. $(1, 2)$; saddle point: $f(1, 2) = 4$

5. $(8, -6)$; relative minimum value: $f(8, -6) = -41$

7. $(1, 2)$ and $(2, 2)$; saddle point: $f(1, 2) = -1$; relative minimum value: $f(2, 2), = -2$

9. $\left(-\frac{1}{3}, \frac{11}{3}\right)$ and $(1, 5)$; saddle point: $f\left(-\frac{1}{3}, \frac{11}{3}\right) = -\frac{319}{27}$; relative minimum value: $f(1, 5) = -13$

11. $(0, 0)$ and $(1, 1)$; saddle point: $f(0, 0) = -2$; relative minimum value: $f(1, 1) = -3$

13. $(2, 1)$; relative minimum value: $f(2, 1) = 6$

15. $(0, 0)$; saddle point: $f(0, 0) = -1$

17. $(0, 0)$; relative minimum value: $f(0, 0) = 1$

19. $(0, 0)$; relative minimum value: $f(0, 0) = 0$

21. 200 finished units and 100 unfinished units; \$10,500

23. Price of land (\$200/sq ft) is highest at $\left(\frac{1}{2}, 1\right)$.

25. $(0, 1)$ gives desired location

27. 18 inches $\times$ 18 inches $\times$ 36 inches

29. 6 inches $\times$ 4 inches $\times$ 2 inches

Exercises 8.4, page 520

1. a. $y = 2.3x + 1.5$
   b.

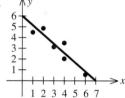

3. a. $y = -0.77x + 5.73$
   b.

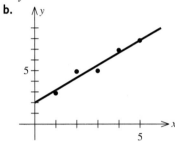

5. a. $y = 1.2x + 2$
   b.

7. a. $y = 0.34x - 0.9$
   b.

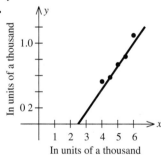

In units of a thousand
In units of a thousand

   c. 1276 applications
9. a. $y = -2.8x + 440$
   b.   **c.** 420

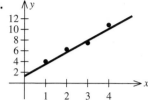

Years

11. a. $y = 6.23x + 167.6$ **b.** 541 acres
13. a. $y = 2.8x + 17.6$ **b.** $40,000,000
15. a. $y = 4.842x + 11.842$ **b.** 103.8 billion cans

17. a. $y = 13.321x + 72.571$ **b.** 192 million tons

Exercises 8.5, page 533

1. Min of $\frac{3}{4}$ at $\left(\frac{3}{4}, \frac{1}{4}\right)$ **3.** Max of $-\frac{7}{4}$ at $\left(2, \frac{7}{2}\right)$
5. Min of 4 at $\left(\sqrt{2}, \frac{\sqrt{2}}{2}\right)$ and $\left(-\sqrt{2}, -\frac{\sqrt{2}}{2}\right)$
7. Max of $-\frac{3}{4}$ at $\left(\frac{3}{2}, 1\right)$
9. Max of $2\sqrt{3}$ at $\left(\frac{\sqrt{3}}{3}, -\sqrt{6}\right)$ and $\left(\frac{\sqrt{3}}{3}, \sqrt{6}\right)$
11. Max of 8 at $(2\sqrt{2}, 2\sqrt{2})$ and $(-2\sqrt{2}, -2\sqrt{2})$;
   min of -8 at $(2\sqrt{2}, -2\sqrt{2})$ and $(-2\sqrt{2}, 2\sqrt{2})$
13. Max: $\frac{2\sqrt{3}}{9}$; min: $-\frac{2\sqrt{3}}{9}$
15. Min of $\frac{18}{7}$ at $\left(\frac{9}{7}, \frac{6}{7}, \frac{3}{7}\right)$
17. 140 finished and 60 unfinished units
19. $10\sqrt{2}$ ft $\times$ $40\sqrt{2}$ ft   **21.** 2 ft $\times$ 2 ft $\times$ 3 ft
23. $15,000 on newspaper and $45,000 on TV
25. 30 ft $\times$ 40 ft $\times$ 10 ft; $7200

Chapter 8 Review Exercises, page 536

1. $0, 0, \frac{1}{2}$; No   **3.** $2, -(e + 1), -(e + 1)$
5. The set of all ordered pairs (x, y) such that $y \neq -x$.
7. The set of all (x, y, z) such that $z \geq 0$ and $x \neq 1$,
   $y \neq 1$, and $z \neq 1$.
9.   **11.**

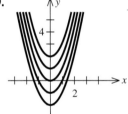

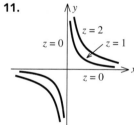

13. $f_x = \sqrt{y} + \dfrac{y}{2\sqrt{x}}$; $f_y = \dfrac{x}{2\sqrt{y}} + \sqrt{x}$
15. $f_x = \dfrac{3y}{(y + 2x)^2}$; $f_y = -\dfrac{3x}{(y + 2x)^2}$
17. $10y(2xy + 3y^2)^4$, $10(x + 3y)(2xy + 3y^2)^4$
19. $2x(1 + x^2 + y^2)e^{x^2+y^2}$; $2y(1 + x^2 + y^2)e^{x^2+y^2}$
21. $\dfrac{2x}{x^2 + y^2}$; $-\dfrac{2x^2}{y(x^2 + y^2)}$
23. $f_{xx} = 12x^2 + 4y^2$; $f_{xy} = 8xy = f_{yx}$; $f_{yy} = 4x^2 - 12y^2$
25. $g_{xx} = \dfrac{-2y^2}{(x + y^2)^3}$; $g_{xy} = \dfrac{2x(3y^2 - x)}{(x + y^2)^3} = g_{yx}$; $g_{yy} = \dfrac{2y(x - y^2)}{(x + y^2)^3}$
27. $h_{ss} = -\dfrac{1}{s^2}$; $h_{st} = h_{ts} = 0$; $h_{tt} = \dfrac{1}{t^2}$
29. $(2, 3)$; relative minimum value: $f(2, 3) = -13$

31. $(0, 0)$ and $\left(\frac{3}{2}, \frac{9}{4}\right)$; saddle point at $f(0, 0) = 0$;
relative minimum value: $f\left(\frac{3}{2}, \frac{9}{4}\right) = -\frac{27}{16}$

33. $(0, 0)$; relative minimum value: $f(0, 0) = 1$

35. $f\left(\frac{12}{11}, \frac{20}{11}\right) = -\frac{32}{11}$

37. Relative maximum value: $f(5, -5) = 26$;
relative minimum value: $f(-5, 5) = -24$

39. a. $R(x, y) = -0.02x^2 - 0.05y^2 - 0.2xy + 80x + 60y$

b. The set of all points (x, y) satisfying $0.02x + 0.1y \leq 80$, $0.1x + 0.05y \leq 60$, $x \geq 0$, $y \geq 0$

c. 15,300; When 100 16-speed blenders and 300 10-speed blenders are sold daily, the revenue is $15,300.

41. $11,000/month on advertising; employ 14 agents

43. 337.5 yd $\times$ 900 yd

INDEX